# GEOGRAPHIC INFORMATION
# SCIENCE & SYSTEMS

# GEOGRAPHIC INFORMATION
# SCIENCE & SYSTEMS

**FOURTH EDITION**

**PAUL A. LONGLEY**
UNIVERSITY COLLEGE LONDON, UK

**MICHAEL F. GOODCHILD**
UNIVERSITY OF CALIFORNIA, SANTA BARBARA, USA

**DAVID J. MAGUIRE**
UNIVERSITY OF GREENWICH, UK

**DAVID W. RHIND**
CITY UNIVERSITY LONDON, UK

# WILEY

| | |
|---|---|
| VP & DIRECTOR | Petra Recter |
| EXECUTIVE EDITOR | Ryan Flahive and Jessica Fiorillo |
| SPONSORING EDITOR | Marian Provenzano |
| MARKETING MANAGER | Christine Kushner |
| ASSOCIATE PRODUCTION MANAGER | Joyce Poh |
| COVER DESIGN | Kenji Ngieng |
| FRONT COVER (PHOTO) | Paul Longley |
| BACK COVER PHOTO | Jack Dangermond |

All photographs within this text that do not list a credit were either created by one of the authors or for which the author holds copyright.

This book was set in Avenir LT Std by Aptara®, Inc. and printed and bound by Courier Kendallville.

Founded in 1807, John Wiley & Sons, Inc. has been a valued source of knowledge and understanding for more than 200 years, helping people around the world meet their needs and fulfill their aspirations. Our company is built on a foundation of principles that include responsibility to the communities we serve and where we live and work. In 2008, we launched a Corporate Citizenship Initiative, a global effort to address the environmental, social, economic, and ethical challenges we face in our business. Among the issues we are addressing are carbon impact, paper specifications and procurement, ethical conduct within our business and among our vendors, and community and charitable support. For more information, please visit our website: www.wiley.com/go/citizenship.

Evaluation copies are provided to qualified academics and professionals for review purposes only, for use in their courses during the next academic year. These copies are licensed and may not be sold or transferred to a third party. Upon completion of the review period, please return the evaluation copy to Wiley. Return instructions and a free of charge return shipping label are available at: www.wiley.com/go/returnlabel. If you have chosen to adopt this textbook for use in your course, please accept this book as your complimentary desk copy. Outside of the United States, please contact your local sales representative.

*Library of Congress Cataloging in Publication Control Number: 2014040579*

ISBN: 978-1-118-67695-0 (Paperback)

Printed in the United States of America

10 9 8 7 6 5 4 3 2 1

# CONTENTS

## Introduction

## 1 Principles

## 5  Uncertainty  99

# 2 Techniques

## 6  GI System Software  128

## 7  Geographic Data Modeling  152

## Supplementary Materials Available Online

### SM SM Supplementary Materials

# FOREWORD

**Joe Lobley here, again.**

This is the fourth time the "Gang of Four" authors have asked me to write an introduction to their textbook. I was flattered, so again said yes. Now I know why they keep on doing this: I heard from a Wiley insider that market research shows that my stuff is the most read section of the book! And who can resist having his great thoughts read by 80,000 purchasers (so far) of the book? But next time I'll charge the Gang.

Looking at publisher's blurb for the fourth edition, the first thing my eagle eyes picked up was the "science and systems" goof right there on the front cover—rather than the other way round used previously. I know you won't pulp the print run just because of this, but y'know the new title might just be a blessing in disguise. If you Google "GIS," I've noticed that references to my general infantrymen colleagues keep popping up on the list, so perhaps the term doesn't denote the sunrise industry it once was. Today's bright young wannabes (and old tight-fisted cheapskates like me) are more likely to patch together free and open software than toe the corporate software line. At the end of the day, I buy the line the Gang have spun since I first started helping them write, namely that science is more exciting than this month's favorite software release.

Which brings me to my news. Those who have followed my "most read" contributions will know that life in the GI "system garden" has not always been smooth for me. I've tried all sorts of roles, worked in many countries as a consultant, started businesses, smooched with governments, and got marooned on a desert island for my pains. Despite all my entrepreneurial activities, I'm still not rich. In fact, I'm broke. I'm living in a battered caravan in an alcohol-free Islamic country. Because I worked for the U.S. military for a time, where I am had better be secret 'til I raise enough cash to move on.

So I've been rethinking what's gone wrong, despite my unrivalled experience and scientific skills. Partly it's the structure of our industry. I've noticed that almost all the job adverts are for relatively lowly paid technical roles, and there are not many highly paid employees that are data bashers. I want to be one of the top guys, not a technician—I'm too old to keep up with techie college graduates when the GIS world is changing so rapidly. If putting science before systems presents new market opportunities, count me in, guys.

But at the end of the day, science just isn't where the real action is. When Calvin Coolidge was President, he said that "the business of America is business." So I've retrained: I've used my GIS to acquire a three-month Masters in intellectual property law from a respected online learning provider—my life-experience credits put me on the fast track from the start, and they accepted my successful patent filing for the Lobley Precisional Adjustment to differential GPS instead of a dissertation. The only problem was the huge fee I had to pay an agent to get certified as having passed everything. There have been so many big legal cases of late between Apple, Samsung, Google, and the rest over infringement of patents that I must be able to make it big in the "law and GIS" domain. If I had done it a bit earlier I could have sued one of the street data providers on behalf of users of their error-prone mapping. All I would have needed is for the families of a few people drowned after driving into a river by following these maps to ask me to act for them. OK, timing is everything.

I can only see one problem with GIS and law. It comes, as you might guess, from government. In the United States, government—apart from the military—mostly and until recently hasn't seen data as an asset to be treasured, protected, and exploited (I could help them). Worse, this plague is spreading. Can you believe that 60 or so national governments—including some serious ones (but not yet China or India, and Russia changed its mind)—have signed up to something called the Open Government Partnership? The idea is to flagellate themselves by making public commitments to reform government, foster innovation, and make everything transparent. Making almost all government data free seems to be the way that they will enable armchair auditors to keep watch on their government and politicians. This madness could be a serious barrier to my wealth creation if everything everywhere is free. But hey, maybe I could become a super-auditor, identifying fraud through use of GIS to bring data together. It would need to include lots of personal data, but privacy is an outdated concept anyway. My ex-wife Lolita found that out when I tracked her philandering throughout Lincolnshire some years ago.

All this, of course, is about Big Data—another fashion. We GIS folk have been doing it for years but no one has listened to us. As is normal with new fashions, big consultancies have proclaimed they are experts in it and can change the world. I could try giving them the benefit of my experience. But my best hope is to work for the U.S. National Security Agency or another country's version of it. Those guys—as Snowden pointed out—are focused, with clear aims, limited accountability, and lots of money. My kind of folk in fact. The bad guys have to live somewhere so the good guys need GIS. . . .

**Joe Lobley**

# DEDICATION

We dedicate this fourth edition to Roger Tomlinson (1933–2014). Often called the "Father of GIS," Roger devoted most of his adult life to promoting the systems, technology, and science of geographic information (GI), as an integral part of the discipline of geography. In the 1960s he was the prime instigator behind the Canada Geographic Information System, a federal–provincial project to automate the measurement of Canada's land resource. In the 1970s he argued forcefully for a single, integrated technology for handling geographic information, completed a PhD at University College London, organized groundbreaking conferences through the aegis of the International Geographical Union, and founded a consulting practice to advise government agencies on the adoption of GI systems. His approach is ably detailed in his book, *Thinking about GIS: Geographic Information System Planning for Managers* (Esri Press), which is now in its fifth edition, and in the executive seminars he has led at the Esri International User Conference for many years.

Roger was an unflagging promoter of GI systems, which he saw as an essential part of humanity's interaction with its environment and the key to the solution of many of humanity's problems. He will be remembered for the force of his personality, his wit and charm, and his passionate support of the field, which he did more than perhaps anyone else to establish and support.

# PREFACE

It is an old but true adage that everything that happens, happens somewhere. Throughout the history of humankind, geography has played a central role in many types of decision-making, some of which have life or death, or at least major strategic, impacts. In the past 50 years decision-making has benefited enormously, and in very many ways, from access to geographic information (GI), the science that underpins it, and the systems technology that enables it.

The previous edition of this textbook was published in 2011. Since then our world has changed, in some respects dramatically. Many of our interactions with information now occur through mobile devices rather than desktops, laptops, or paper. Location- (i.e., geographic-) based services have been estimated to be worth between $150bn and $270bn annually. Open Data, Open Software, and Open Science have been developing rapidly. The emergence of Big Data—where our community has pioneered many developments—has been hailed by some as obviating many past constraints (such as ensuring that samples are representative of a known population). Virtually all data are now collected in digital form rather than on paper; it is claimed that more data are now collected every two years than in the whole of previous human history. Crowdsourcing has produced many new datasets and changes in the way we tackle some tasks—such as scanning satellite images of a huge area of the South Indian Ocean for wreckage from Malaysian Airlines MH370 flight, a project organized by DigitalGlobe using imagery from its Worldview-2 system. Many governments are at last disgorging the information they hold for general use. And social media data are providing the fuel for real-time analysis of the geotemporal activity patterns of hundreds of millions of citizens. Given all that, this edition attempts to identify, explain, and evaluate the key changes and portray a snapshot of the contemporary world of geographic information, GI science, and GI systems.

In times past we wrote about geographic information systems, or GIS. The world has moved on. Except where we are quoting from others, we no longer use the abbreviation GIS. GI systems continue to evolve rapidly in their functions, ease of use, and number and spread of their users. They continue to provide the tools to describe and analyze the physical or human environments, bringing together data and converting them into information and even evidence (see Section 1.2). But underpinning that use of dazzling new technologies is a rapidly developing GI science. Here we deal with principles, many of which have endured in changing guises ever since the first edition of this book appeared in 2001. Where they exist, we deal with laws akin to those in the physical sciences, but also address the statistical generalizations of the social and environmental sciences. The third driving force of our Gang of Four is geographic information itself: we need to know its many characteristics, including quality, if we are to accommodate the inevitable uncertainty that arises when we admix different data using a variety of algorithms.

## The New Vision

Reflecting this emerging GI ecosystem, we have made a subtle change of title in this, the fourth edition. The internal structure and content of the book reflects the change. After an introductory chapter, we develop a section on principles. This encompasses the nature of geographic data and information, representing geography, georeferencing, and uncertainty. We follow this with the "how"—a section on techniques, dealing with GI system software, data modeling, data collection, creating and maintaining geographic databases, and the Geoweb. The fourth section on analysis covers cartography, geovisualization, spatial data analysis, inferential spatial analysis, and spatial modeling. The fifth section covers human factors in relation to what we now term geographic information science and systems (GISS). It deals with information and decision-making, and with navigating the legal, ethical, and many other risks that GISS practitioners face. The concluding chapter—the Epilog—looks ahead. But it does this not by seeking to assess technological change, important as that is. Rather, it seeks to identify where we can use our GISS understanding, knowledge, skills, and tools to tackle major problems.

Throughout the book we emphasize the commonalities and the differences between groups of GI system users. Thus those in business, in governments at a variety of levels, in academia, and in not-for-profit organizations have overlapping concerns but some different drivers. This extends to differences between national and subnational cultures (and even between individuals), where our value systems and preferred modes of operating vary greatly. We have tried to give due credence to these similarities and differences.

Throughout the book we use examples and descriptions of luminaries whom we judge to have made a substantial contribution. We have tried throughout the text to provide detail because "the devil is in the detail" while also trying to highlight key points (such as through use of short tweet-like "factoids" that appear in bold), further reading, and a set of questions at the end of each chapter to test how much the student has gained from it and whether the student can develop new ideas or practice.

## Online Supplementary Materials

This fourth edition is available both in print and online. In addition to the full content of the print edition, the online Web site includes significant supplementary material:

- A detailed discussion of four examples of GI system application, chosen to illustrate both the breadth of applications of GI technology, and the importance of the scientific principles elaborated throughout the book.
- Powerpoint slides for each of the chapters of the book, designed to be used as the basis for a course of lectures on the book's contents.
- An Instructor's Manual, giving pointers to the most effective ways to use the book in courses.

## The Best of Times

In short, we are in the most exciting of times. Human ingenuity is transforming the way we can describe, analyze, and communicate what is occurring on the face of the Earth (and beyond). We have good enough science, information, and tools to make a real impact in improving societies, business performance, and much else—at all levels from the very local to the global. Central to all this is geographic variation and the awareness and skills to cope with it or even to reshape it. We authors are excited by what GISS practitioners have already achieved and by the prospects for the future. This book seeks to tell you why and convince you to join us.

## Acknowledgments

We take complete responsibility for all the material contained herein. But we have drawn on contributions made by many others from across the world, which have helped us create this edition. We especially thank Muhammad Adnan, Stuart Ashfield, Brian Baker, Lawrence Ball, Sir John Beddington, Ann Blake, James Borrell, Peter Benton, Budhu Bhaduri, Paul Boyle, Alisdair Calder, Folke Carlsson, Tao Cheng, James Cheshire, Keith Clarke, Steve Coast, David Cowen, Max Craglia, Jack Dangermond, Keith Dugmore, Sarah Elwood, Ryan Flahive, Chris Gale, Vishal Gaudhar, Paul Hancock, David Hand, Stuart Houghton, Liz Hughes, Indy Hurt, Pete Jones, Jens Kandt, Milan Konecny, Kira Kowalska, Guy Lansley, Vanessa Lawrence, Alistair Leak, Dan Lewis, Antonio Lima, Yu Liu, Ross Maciejewski, Sara McLafferty, Mirco Musolesi, Tomoki Nakaya, Ollie O'Brien, Giles Pavey, Joyce Poh, Marian Provenzano, Denise Pumain, Muttukrishnan "Raj" Rajarajan, Jonathan Rhind, Doug Richardson, James Russiello, Matt Sims, Alex Singleton, Karen Slaght, Carl Steinitz, Lynn Usery, Ruth Wembridge, Roy Wood, Michael Worboys, Bradley Williams, Qingling Wu, and Keiji Yano. We sincerely apologize to anyone else whose contribution we may have overlooked.

Many others have contributed to the content of previous editions; we reiterate our thanks to them.

*Paul Longley, University College London*
*Michael Goodchild, Emeritus Professor, University of California, Santa Barbara*
*David Maguire, University of Greenwich*
*David Rhind, Emeritus Professor, City University London*

# List of Acronyms and Abbreviations

AAG    Association of American Geographers
ABM    agent-based model
AEGIS    Advanced Emergency Geographic Information System
AHP    Analytical Hierarchy Process
AJAX    Asynchronous Javascript and XML
ALSM    airborne laser swath mapping
AM/FM    automated mapping/facilities management
AOL    America On-Line
API    application programming interface
AR    augmented reality
ARPANET    Advanced Research Projects Agency Network
ASCII    American Standard Code for Information Interchange
AVHRR    Advanced Very High Resolution Radiometer
AVIRIS    Airborne Visible Infrared Imaging Spectrometer
BLOB    binary large object
BS    Bachelor of Science
CA    cellular automaton
CAD    computer-assisted design
CAMS    Capacity Area Management System (Sears)
CARS    Computer-Aided Routing System (Sears)
CAS    Chinese Academy of Sciences
CASE    computer-aided software engineering
CCTV    closed-circuit television
CD    compact disk
CDR    carbon dioxide removal
CEN    Comité Européen de Normalisation
CERCO    Comité Européen de Responsables de la Cartographie Officielle
CERN    European Organization for Nuclear Research
CGIS    Canada Geographic Information System
CGS    Czech Geological Survey
CIA    Central Intelligence Agency (US)
CODATA    Committee on Data for Science and Technology (International Council for Science)
COGO    coordinate geometry
COM    component object model
COTS    commercial off-the-shelf
CPI    consumer price index
CPU    central processing unit
CSAIL    Laboratory for Computer Science and Artificial Intelligence (MIT)
CSDGM    Content Standards for Digital Geospatial Metadata
CSV    comma-separated values
DBA    database administrator
DBMS    database management system
DCL    Data Control Language (SQL)
DCM    digital cartographic model

DDL    Data Definition Language (SQL)
DEM    digital elevation model
DGPS    Differential GPS
DIG    Decentralized Information Group (MIT)
DIME    dual independent map encoding
DLG    digital line graph
DLM    digital landscape model
DMI    distance measuring instrument
DML    Data Manipulation Language (SQL)
DNA    deoxyribonucleic acid
DoD    Department of Defense (US)
dpi    dots per inch
DRASTIC    model named for its inputs: depth, recharge, aquifer, soils, topography, impact, conductivity
DRG    digital raster graphic
DVD    digital video disk
DWD    National Meteorological Service (Germany)
EC    European Commission
ECU    Experimental Cartography Unit
EDA    exploratory data analysis
EL.STAT    Hellenic Statistical Agency (Greece)
EOS    Earth Observing System
EOSDIS    Earth Observing System Data and Information System
EPA    Environmental Protection Agency (US)
EPSG    European Petroleum Study Group
ERDAS    Earth Resource Data Analysis System
ESDA    exploratory spatial data analysis
Esri    Environmental Systems Research Institute
Esri BIS    Esri Business Information Solutions
EU    European Union
ExCiteS    Extreme Citizen Science (University College London)
FEMA    Federal Emergency Management Agency (US)
FGDC    Federal Geographic Data Committee (US)
FOIA    freedom of information act
FOSS4G    Free and Open-Source Software for Geospatial
FSA    Forward Sortation Area
GA    genetic algorithm
GAO    Government Accountability Office (US)
GDP    gross domestic product
GDT    Geographic Data Technology Inc.
GEOINT    geospatial intelligence
GFIS    Geographic Facilities Information System (IBM)
GGIM    (Initiative on) Global Geospatial Information Management (UN)
GI    geographic information
GIF    Graphics Interchange Format
GIS    geographic information system
GISS    geographic information science and systems
GIST    Geographic Information Science and Technology group (ORNL)

GIS-T    geographic information systems for transportation
GITA    Geospatial Information and Technology Association
GLONASS    Global Orbiting Navigation Satellite System
GML    Geography Markup Language
GPS    Global Positioning System
GRASS    Geographic Resources Analysis Support System
GSDI    global spatial data infrastructure
GSN    Global Spatial Network
GUI    graphical user interface
GWR    geographically weighted regression
HIV-AIDS    Human Immunodeficiency Virus - Acquired Immune Deficiency Syndrome
HLS    hue, lightness, saturation
HTML    Hypertext Markup Language
HTTP    Hypertext Transfer Protocol
HUMINT    human intelligence
IARPA    Intelligence Advanced Research Projects Activity
IBRU    International Boundaries Research Unit
ICSU    International Council for Science
ICT    information and communication technology
ID    identifier
IDE    integrated developer environment
IDW    inverse-distance weighting
IGN    Institut Géographique National
IJDE    International Journal of Digital Earth
IM    Instant Messenger
INPE    Instituto Nacional de Pesquisas Espaciais (Brazil)
INSPIRE    Infrastructure for Spatial Information in the European Community (Europe)
IP    Internet Protocol
IPCC    Intergovernmental Panel on Climate Change
IPR    intellectual property rights
ISDE    International Society for Digital Earth
ISO    International Organization for Standardization
IT    information technology
ITS    intelligent transportation systems
ITT    invitation to tender
JPEG    Joint Photographic Experts Group
JPL    Jet Propulsion Laboratory
KML    Keyhole Markup Language
LAN    local-area network
LBS    location-based service
LDO    Local Delivery Office
LiDAR    light detection and ranging
LIESMARS    State Key Laboratory for Information Engineering in Surveying, Mapping, and Remote Sensing (China)
LMIS    Land Management Information System (South Korea)
MAT    (point of) minimum aggregate travel
MAUP    Modifiable Areal Unit Problem

MBR    minimum bounding rectangle
MCDM    multi-criteria decision making
MDGs    Millennium Development Goals
MER    minimum enclosing rectangle
MGCP    Multinational Geospatial Co-Production Program
MIDI    Musical Instrument Digital Interface
MIT    Massachusetts Institute of Technology
MOCT    Ministry of Construction and Transportation (South Korea)
MODIS    Moderate Resolution Imaging Spectroradiometer
MOOC    massive open online course
MP3    MPEG Audio Layer III
MPEG    Motion Picture Experts Group
MrSID    Multiresolution Seamless Image Database
MSC    Mapping Science Committee (US National Research Council)
NAD27    North American Datum of 1927
NAD83    North American Datum of 1983
NASA    National Aeronautics and Space Administration
NATO    North Atlantic Treaty Organization
NCGIA    National Center for Geographic Information and Analysis (US)
NGA    National Geospatial-Intelligence Agency (US)
NGO    non-governmental organization
NII    national information infrastructure
NIMA    National Imagery and Mapping Agency (US)
NIMBY    not in my backyard
NLS    National Land Survey
NMCA    national mapping and charting agency
NMO    national mapping organization
NMP    National Mapping Program
NOAA    National Oceanic and Atmospheric Administration (US)
NPWS    National Parks and Wildlife Service (Australia)
NSDI    National Spatial Data Infrastructure (US)
NSF    National Science Foundation (US)
OAC    Output Area Classification (UK Office of National Statistics)
OAS    Organization of American States
OCR    optical character recognition
OD    Open Data
ODBMS    object database management system
OGC    Open Geospatial Consortium
OGL    Open Government License (UK)
OGP    Open Government Partnership
OLM    object-level metadata
OLS    ordinary least-squares
OMB    Office of Management and Budget (US)
ORDBMS    object-relational database management system
ORNL    Oak Ridge National Laboratory
OS    Ordnance Survey (Great Britain, or Northern Ireland)
OSINT    open-source intelligence

OSM   Open Street Map
PAIGH   PanAmerican Institute of Geography and History
PAF   Postal Address File
PARC   Palo Alto Research Center (Xerox)
PB   petabyte
PC   personal computer
PCC   percent correctly classified
PCRaster   Personal Computer Raster (GIS)
PDA   personal digital assistant
PDF   Portable Document Format
PERT   Program Evaluation and Review Technique
PGIS   participatory geographic information systems
PLSS   Public Land Survey System
PPGIS   public-participation geographic information systems
PROTECT   Port Resilience for Operational/Tactical Enforcement to Combat Terrorism (US Coast Guard)
PSI   public-sector information
QA   quality assurance
QR   quick response (code)
RADI   Institute of Remote Sensing and Digital Earth (Chinese Academy of Sciences)
R&D   research and development
RDBMS   relational database management system
RDFa   Resource Description Framework in Attributes
REST   Representation State Transfer Protocol
RFI   request for information
RFID   radio frequency identification
RFP   request for proposals
RGB   red, green, blue
RGS   Royal Geographical Society (UK)
RMSE   root mean squared error
ROI   return on investment
RS   remote sensing
RSS   Rich Site Summary or Really Simple Syndication
SDE   Spatial Database Engine
SDI   spatial data infrastructure
SDSS   spatial decision support system
SDTS   Spatial Data Transfer Standard
SETI   search for extra-terrestrial intelligence
SIGINT   signals intelligence
SOA   service-oriented architecture
SOAP   Simple Object Access Protocol
SOHO   small office/home office
SPC   State Plane Coordinates
SPOT   Système Probatoire d'Observation de la Terre
SQL   Structured (or Standard) Query Language
SQL/MM   Structured (or Standard) Query Language/Multimedia
SRM   solar radiation management

SWMM   Storm Water Management Model
SWOT   strengths, weaknesses, opportunities, threats
TB   terabyte
TIFF   Tagged Image File Format
TIGER   Topologically Integrated Geographic Encoding and Referencing
TIN   triangulated irregular network
TOID   topographic identifier
TSP   traveling-salesperson problem
TV   television
UAM   Metropolitan Autonomous University (Mexico)
UAV   unmanned aerial vehicle
UCAS   University of the Chinese Academy of Sciences
UCGIS   University Consortium for Geographic Information Science (US)
UK   United Kingdom (of Great Britain and Northern Ireland)
UML   Unified Modeling Language
UN   United Nations
UNAM   National Autonomous University of Mexico
UNIGIS   University GIS Consortium
URI   uniform resource identifier
URL   uniform resource locator
US   United States (of America)
USA   United States of America
USGS   United States Geological Survey
USLE   Universal Soil Loss Equation
UTM   Universal Transverse Mercator projection
VACCINE   Visual Analytics for Command, Control, and Interoperability Environments (Purdue University)
VBA   Visual Basic for Applications
VfM   value for money
VGA   video graphics array
VGI   volunteered geographic information
VR   virtual reality
W3C   World Wide Web Consortium
WAN   wide-area network
WCS   Web Coverage Service
WFS   Web Feature Service
WGS84   World Geodetic System of 1984
WHO   World Health Organization
WIMP   windows, icons, menus, pointers
WMS   Web Map Service
WTO   World Trade Organization
WWF   World Wide Fund for Nature
WWW   World Wide Web
WYSIWYG   what you see is what you get
XML   Extensible Markup Language
XSEDE   Extreme Science and Engineering Discovery Environment

# Geographic Information: Science, Systems, and Society

**1**

This chapter sets the conceptual framework for and summarizes the content of the book by addressing several major questions:

- What exactly is geographic information (GI), and why is it important? What is special about it?
- What new technological developments are changing the world of GI?
- How do GI systems affect the lives of average citizens?
- What kinds of decisions make use of geographic information?
- What is a geographic information system (GI system), and how would you recognize one?
- What is geographic information science (GI science), and why is it important to GI systems?
- How do scientists and governments use GI systems, and why do they find them helpful?
- How do companies make money from GI systems?

## LEARNING OBJECTIVES

After studying this chapter you will:

- Know definitions of many of the terms used throughout the book.
- Be familiar with a brief history of GI science and GI systems.
- Recognize the sometimes invisible roles of GI systems in everyday life, business, and government.
- Understand the significance of GI science and how it relates to GI systems.
- Understand the many impacts that GI systems and its underpinning science are having on society and the need to study those impacts.

## 1.1 Introduction: What Are GI Science and Systems, and Why Do They Matter?

Almost everything that happens, happens somewhere. We humans confine our activities largely to the surface and near-surface of the Earth. We travel over it and through the lower levels of its atmosphere, and we go through tunnels dug just below the surface. We dig ditches and bury pipelines and cables, construct mines to get at mineral deposits, and drill wells to access oil and gas. We reside on the Earth and interact with others through work, leisure, and family pursuits. Keeping track of all this activity is important, and knowing where it occurs can be the most convenient basis for tracking. Knowing where something happens is of critical importance if we want to go there ourselves or send someone there, to find more information about the same place, or to inform people who live nearby. In addition, geography shapes the range of options that we have to address things that happen, and once they are made, decisions have geographic consequences. For example, deciding the route of a new high-speed railroad may be shaped by topographic and environmental considerations, and the chosen route will create geographic winners and losers in terms of access. Therefore geographic location is an important component of activities, policies, strategies, and plans.

**Almost everything that happens, happens somewhere. Knowing where something happens can be critically important.**

The focus of this book is on geographic information, that is, information that records *where* as well as *what* and perhaps also *when*. We use the abbreviation *GI* throughout the book. GI systems were originally conceived as something separate from the world they represent—a special kind of information system, often located on a user's desk, dedicated to performing special kinds of operations related to location. But today such information pervades the Internet, can be accessed by our smartphones and other personal devices, and is fundamental to the services provided by governments, corporations, and even individuals. Locations are routinely attached to health records, to Twitter feeds and photographs uploaded to Flickr, and to the movements of mobile phone users and vehicles. In a sense, then, the whole digital world has become one vast, interconnected GI system. This book builds on what users of this system already know—that use of GI services is integral to many of our interactions through the Internet. Later chapters will describe, for example, how storage and management of more and more data entail use of the Cloud, how Big Data and Open Data have become ubiquitous (but not necessarily useful), and how Web-based GI systems have become a fact of life.

Underlying these changes are certain fundamentals, however, and these have a way of persisting despite advances in technology. We describe them with the term *GI science*, which we define as the general knowledge and important discoveries that have made GI systems possible. GI science provides the structure for this book because as educators we believe that knowledge of principles and fundamentals—knowledge that will still be valid many years from now—is more important than knowledge of the technical details of today's versions of GI technology. We use the acronym *GISS*—geographic information science and systems—at various points in this book to acknowledge the interdependence between the underpinning science and the technology of problem solving.

At the outset, we also observe that GI science is also fundamentally concerned with solving applied problems in a world where business practices, or the realpolitik of government decision making, are important considerations. We also discuss the practices of science and social science that, although governed by clearly defined scientific principles, are imperfectly coupled in some fast-developing areas of citizen science.

## 1.1.1 The Importance of Location

Because location is so important, it is an issue in many of the problems society must solve. Some of these problems are so routine that we almost fail to notice them—the daily question of which route to take to and from work, for example. Others are quite extraordinary and require rapid, concerted, and coordinated responses by a wide range of individuals and organizations—such as responding to the major emergencies created by hurricanes or earthquakes (see Box 1.1). Virtually all aspects of human life involve location. Environmental and social scientists recognize the importance of recording location when collecting data; major information companies such as Google recognize the importance of providing mapping and driving directions and prioritizing searches based on the user's location; and citizens are increasingly familiar with services that map the current positions of their friends. Here are some examples of major decisions that have a strong geographic element and require GI:

- Health-care managers decide where to locate new clinics and hospitals.
- Online shopping companies decide the routes and schedules of their vehicles, often on a daily basis.
- Transportation authorities select routes for new highways and anticipate their impacts.
- Retailers assess the performance of their outlets and recommend how to expand or rationalize store networks.
- Forestry companies determine how best to manage forests, where to cut trees, where to locate roads, and where to plant new trees.
- National park authorities schedule recreational path creation, maintenance, and improvement (Figure 1.1).
- Governments decide how to allocate funds for building sea defenses.
- Travelers and tourists give and receive driving directions, select hotels in unfamiliar cities, and find their way around theme parks (Figure 1.2).
- Farmers employ new GI technology to make better decisions about the amounts of fertilizer and pesticides to apply to different parts of their fields.

If location and GI are important to the solution of so many problems, what distinguishes those problems from each other? Here are three bases for classifying problems. First, there is the question of *scale*, or level of geographic detail. The architectural design of a building involves GI, but only at a very detailed or local scale. The information needed to service the building is also local—the size and shape of the

Figure 1.1 Maintaining and improving footpaths in national parks is a geographic problem.

Figure 1.2 Navigating tourist destinations is a geographic problem.

parcel, the vertical and subterranean extent of the building, the slope of the land, and its accessibility using normal and emergency infrastructure. At the other end of the scale range, the global diffusion of epidemics and the propagation of tsunamis across the Pacific Ocean (Box 1.1) are phenomena at a much broader and coarser scale.

**Scale or level of geographic detail is an essential property of any project.**

Second, problems can be distinguished on the basis of *intent*, or *purpose*. Some problems are strictly practical in nature—they must often be solved as quickly as possible and at minimum cost to achieve such practical objectives as saving lives in an emergency, avoiding fines by regulators, or responding to civil disorder. Others are better characterized as driven by human curiosity. When GI is used to verify the theory of continental drift, to map distributions of glacial deposits, or to analyze the historic movements of people in anthropological or biosocial research (see Box 1.2 and Figure 1.5), there is no sense of an immediate problem that needs to be solved. Rather, the intent is to advance human understanding of the world, which we often recognize as the intent of science.

Although science and practical problem solving can be thought of as distinct human activities, it is

often argued that there is no longer any effective distinction between their methods. Many of the tools and methods used by a retail analyst seeking a site for a new store are essentially the same as those used by a scientist in a government agency to ensure the protection of an endangered species, or a transport planner trying to ameliorate peak-hour traffic congestion in a city. Each requires the most accurate measurement devices, employs terms whose meanings have been widely shared and agreed on, produces results that are replicable by others, and in general follows all the principles of science that have evolved over the past centuries. The knowledge-exchange activities carried out between research organizations and the government and business sectors can be used to apply many of the results of curiosity-driven science to the practical world of problem solving.

The use of GI systems in support of science, routine application, and knowledge exchange reinforces the idea that science and practical problem solving are no longer distinct in their methods, as we will discuss later. As a consequence, GI systems are used widely in all kinds of organizations, from academic institutions to government agencies, not-for-profit organizations, and corporations. The use of similar tools and methods across so much of science and problem solving is part of a shift from the pursuit of curiosity within traditional academic disciplines to solution-centered, interdisciplinary teamwork.

Nevertheless, in this book we distinguish between uses of GI systems that focus on applications such as inventory or resource management, or so-called normative uses, and uses that advance science, or so-called positive uses (a rather confusing meaning of that term, unfortunately, but the one commonly used by philosophers of science—its use implies that science confirms theories by finding positive evidence in support of them and rejects theories when negative

## The 2011 Tōhoku Earthquake and Tsunami

At 14.46 local time (05.56 GMT) on March 11, 2011, an undersea earthquake measuring 9.0 on the Richter scale occurred approximately 43 miles (70 kilometers) east of the Japanese coast of Tōhoku. This was the most powerful earthquake ever to have been scientifically documented in Japan, and the fifth most powerful earthquake in the world since modern record-keeping began in c. 1900. The earthquake moved Honshu (the main island of Japan) 2.4 m (8 ft) east and shifted the Earth on its axis by estimates of between 10 cm (4 in) and 25 cm (10 in). Of more immediate significance, the earthquake caused severe earth tremors on the main islands of Japan and triggered powerful tsunami waves that reached heights of up to 40.5 meters (133 ft) in Tōhoku Prefecture and traveled up to 10 km (6 mi) inland in Sendai.

Directly or indirectly, the earthquake led to at least 15,883 deaths and the partial or total collapse of over 380,000 buildings. It also caused extensive and severe structural damage in northeastern Japan (Figure 1.3B), including heavy damage to roads and railways, as well as fires in many areas and a dam collapse. In its immediate aftermath, 4.4 million households in northeastern Japan were left without electricity and 1.5 million without water. In the following days, the tsunami set in action events that led to cooling system failures, explosions, and major meltdowns at three reactors of the Fukushima Daiichi Nuclear Power Plant and the associated evacuation of hundreds of thousands of residents. The World Bank estimated the economic cost at US$235 billion, making it the costliest natural disaster in world history.

All of this happened to a very advanced economy in an earthquake-prone region, which was almost certainly the best prepared in the world for a natural disaster of this kind. GI systems had been used to assemble information on a full range of spatially distributed

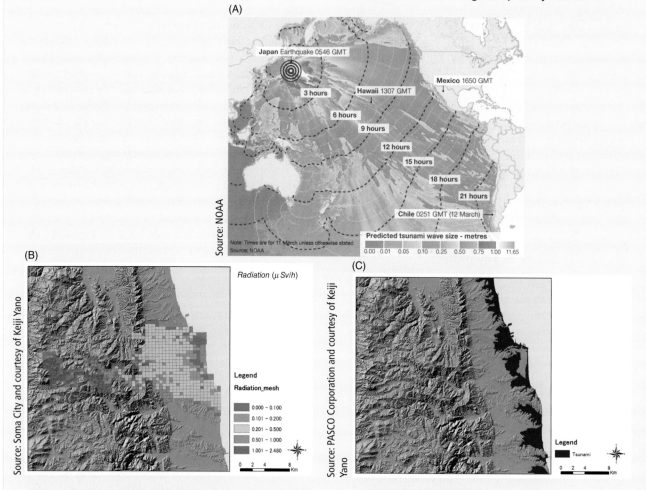

Figure 1.3 (A) The passage of the tsunami arising out of the Great East Japan (Tōhoku) earthquake of March 11, 2011. It had subsequent effects on Soma City in terms of (B) radiation (measured in μSv/h (micro Sievert per hour) and (C) tsunami inundation.

phenomena—including the human population, the built environment, and transportation infrastructure—in preparation for a major earthquake disaster and protection against many of its foreseeable consequences.

Yet the science of predicting the location, timing, and intensity of earthquakes has made little progress over the past century. A magnitude-9.0 earthquake is a very rare event and so did not fall within any disaster-management scenario prior to the event. For example, the Fukushima reactors had been built to withstand a magnitude-8.6 earthquake on the basis of historic occurrences plus a safety margin: but not an event of magnitude 9.0. However, even when major events are unforeseen, GI science and systems are integral to response and recovery in the short term (e.g., alerting populations to the imminent arrival of a tsunami, coordinating citizen reports of how localities have been affected, and organizing evacuation), the medium term (e.g., managing the disruption to industrial supply chains), and the long term (e.g., prioritizing repair and replacement of damaged transport infrastructure). All these actions take place in an organizational context. Early warning systems are very much an international effort. In terms of addressing effects after the event, the Tōhoku earthquake raised issues that were best addressed at the national level, whereas much of the implementation was best effected at local levels.

**The three Ps of disaster management are prevention, preparedness, and protection. GI science and systems are integral to each of them.**

evidence is found). Finding new locations for retailers, with its focus on design, is an example of a normative application of GI systems. But to predict how consumers will respond to new locations, it is necessary for retailers to analyze and model the actual patterns of behavior they exhibit. Therefore, the models they use will be grounded in observations of messy reality that have been tested in a positive manner.

Design is concerned with improving the world—with decisions that when implemented achieve certain desired objectives, such as constructing new housing subdivisions, developing conservation plans, or defining sales territories. In recent years the term *geodesign* has become a popular way of referring to design decisions at geographic scales, supported by GI systems. All of us would like to design improvements to the world, and GI systems are valuable tools for doing so. Although most work with GI systems is considerably more mundane, it is always good to bear its grander potential in mind. As we show in Section 14.4, geodesign combines two important functions of GI systems—the ability to capture new ideas through sketching (creating/editing new features) and the ability to evaluate them and assess their impacts. A user might sketch a design for a new development, for example, and ask the GI system to predict its impacts on transportation, groundwater, and air pollution.

**With a single collection of tools, GI systems are able to bridge the gap between curiosity-driven science and practical problem solving**

The third way in which problems can be distinguished is on the basis of their *time scale*, ranging in human terms from the dynastic (perhaps thousands of years; see Box 1.2) to the diurnal, but very much longer with respect to understanding geological or geomorphological change. At one end of the human time spectrum, some decisions are operational and are required for the smooth day-to-day functioning of an organization, such as how to control electricity inputs into grids that experience daily surges and troughs in usage. At slightly longer timescales, tactical decisions might include where to cut trees in next year's forest harvesting plan. Still other decisions are more infrequent and strategic in nature, such as those required to give an organization long-term direction, as when a retailer decides to expand or rationalize its store network (Figure 1.4). At the far end of the human time spectrum, Box 1.2 describes how the geographic

**Figure 1.4** Many store location principles are generic across different retail markets, as with Tesco's investment in Ostrava, Czech Republic.

distributions of family names, past and present, can be used to indicate how settled (or otherwise) is the population of different places, and even the geography of the DNA of long-settled residents consequent on population movements in early human history (see Box 1.4).

Although humans like to classify time frames into hours, days, years, centuries, and epochs, the real world is somewhat more complex than this, and these distinctions may blur—what is theoretically and statistically a 1000-year flood in a river system influences strategic and tactical considerations, but may arrive a year after the previous one! Other problems that interest geophysicists, geologists, or evolutionary biologists may occur on timescales that are much longer than a human lifetime, but are still geographic in nature, such as predictions about the future physical environment of Japan or about the animal populations of Africa. GI databases are often *transactional* (see Section 9.9.1), meaning that they are constantly being updated as new information arrives, unlike paper maps, which stay the same once printed.

Applications are discussed to illustrate particular principles, techniques, analytic methods, and management practices (such as risk minimization) as these arise throughout the book.

## 1.1.2 Spatial Is Special

The adjective *geographic* refers to the Earth's surface and near surface, at scales from the architectural to the global. This defines the subject matter of this book, but other terms have similar meaning. *Spatial* refers to any space, not only the space of the Earth's surface; this term is used frequently in the book, almost always with the same meaning as *geographic*. But many of the methods used in GI systems are also applicable to other nongeographic spaces, including the surfaces of other planets, the space of the cosmos, and the space of the human body that is captured by medical images. Techniques that are integral to GI systems have even been applied to the analysis of genome sequences on DNA. So the discussion of analysis

---

### Applications Box 1.2

#### Researching Family Histories and Geo-Genealogy

As individuals, many of us are interested in *where*, in general terms, we came from at different points in recorded human history—for example, whether we are of Irish, Spanish, or Italian descent. More specific locational information can provide clues about the work and other lifestyle characteristics of our ancestors. Some of the best clues to our ancestry may come from our surnames (family names) because many surnames indicate geographic origins to greater or lesser degrees of precision (such clues are less important in some Eastern societies, where family histories are generally much better documented). Research at University College London uses GI systems to analyze historic and present-day lists of names to investigate the changing local and regional geographies of surnames across the world. Figure 1.5 illustrates how the bearers of four selected Anglo-Saxon names in Great Britain (the ancestors of the authors of this book) have mostly stayed put in those parts of the island where the names first came into common parlance at some point between the 12th and 14th centuries—although some have evidently migrated to urban centers.

It also turns out that the mix of names with similar geographic origins in any given area can provide a good indication of regional identity. Figure 1.6, derived from the PhD thesis of Jens Kandt, presents a regionalization of Great Britain on the basis of the present-day

residences of bearers of different surnames. (This is essentially a geography of rural Britain. Note that the major urban areas have been excluded because they are characterized by mixes of names arising from urban–rural, interregional, and international migration over the last 200 or so years).

All of this is most obviously evident for Great Britain and many of the countries of Europe, where populations have remained settled close to the locations at which their names were first coined. But there is also evidence to suggest that the spatial patterning of names in former colonies, such as North America, Australia, and New Zealand, is far from random. Figure 1.7 illustrates this for the surname Singleton, which can be used to build evidence about the migration patterns of bearers of this name from their documented origins in northwest England.

Fundamentally, this is curiosity-driven research, driven by the desire among amateur genealogists to discover their roots. But the same techniques can be used to represent the nature and depth of affiliation that people feel toward the places in which they live. Moreover, the work of Sir Walter Bodmer and colleagues (Box 1.4) is highlighting probable links between surnames and genetics, rendering this curiosity-driven research relevant to the development of drug and lifestyle interventions.

▶

---

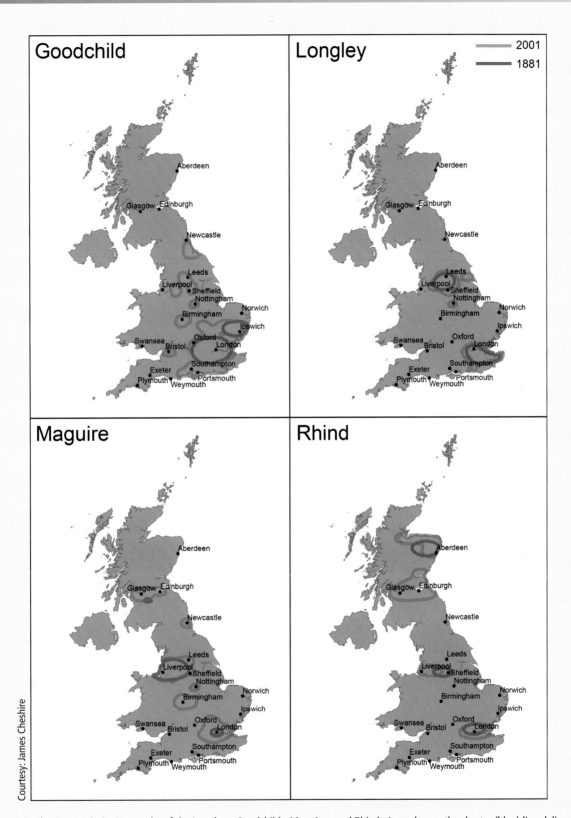

**Figure 1.5** The Great Britain Geography of the Longleys, Goodchilds, Maguires, and Rhinds. In each case the shorter (blue) line delineates the smallest possible area within which 95% of name bearers reside, based on 1881 Census of Population figures, and the outer (red) line encloses the smallest area that accommodates the same proportion of adult name bearers according to a recent address register.

**Figure 1.6** A regionalization based on the coincidence of distinctive patterns of surnames, showing the southern part of Great Britain. Major urban areas do not fit into this regional pattern because their residents are drawn from a wide range of national and international origins.

Courtesy: Jens Kandt

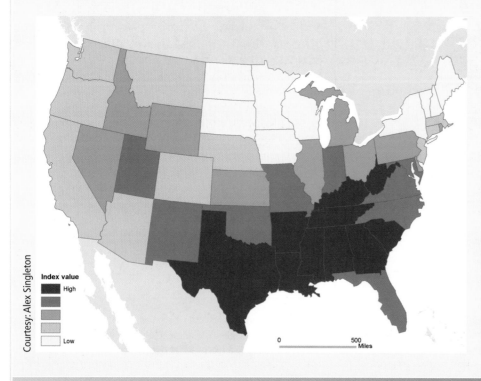

Courtesy: Alex Singleton

Index value

High

Low

**Figure 1.7** The Singleton family name derives from a place in northwest England, and understandably the greatest concentration of this name today still occurs in this region. But why should the name be disproportionately concentrated in the south and west of the United States? Geographical analysis of the global pattern of family names can help us to hypothesize about the historic migrations of families, communities, and cultural groups.

### Some Technical Reasons Why Geographic Information Is Special and Why GI Science and Systems Have Developed

- It is multidimensional, because *two* coordinates must be specified to define a location, whether they be *x* and *y* or latitude and longitude; and a third coordinate is needed when elevation is important.

- It is voluminous because a geographic database can easily reach a terabyte in size (see Table 1.2).

- It may be collected by citizens, governments, or other organizations, and it may prove useful to pool information from these diverse sources.

- It may be represented at different levels of spatial resolution, for example, by using a representation equivalent to a 1:1 million-scale map or a 1:24,000-scale one (see Section 3.7).

- It may be represented in different ways inside a computer (see Chapter 3), and how this is done can strongly influence the ease of analysis and the end results.

- It must often be projected onto a flat surface, for reasons identified in Section 4.8.

- It requires many special methods for its analysis (see Chapters 13 and 14).

- It may be transformed to present different views of the world, for example, to aid interpretation.

---

in this book is of *spatial* analysis (see Chapters 13 and 14), not geographic analysis, to emphasize this versatility.

Another term that has been growing in usage in recent years is *geospatial*—implying a subset of spatial applied specifically to the Earth's surface and near surface. In this book we have tended to avoid *geospatial*, preferring *geographic*, and we use *spatial* where we need to emphasize generality.

> **Although there are subtle distinctions between the terms *geographic(al), spatial,* and *geospatial,* for many practical purposes they can be used interchangeably.**

People who encounter GI for the first time are sometimes driven to ask why geography is so important; why, they ask, is spatial special? After all, there is plenty of information around about geriatrics, for example, and in principle one could create a geriatric information system. So why has GI spawned an entire industry, if geriatric information has not done so to anything like the same extent? Why are there unlikely to be courses in universities specifically in geriatric information science and systems? Part of the answer should be clear already: almost all human activities and decisions involve a location component, and the location component is important. Another reason will become apparent in Chapter 2, where we will see that working with GI involves complex and difficult choices that are also largely unique. Other, more technical reasons will

become clear in later chapters and are briefly summarized in Box 1.3.

## 1.2 Data, Information, Evidence, Knowledge, and Wisdom

Information systems help us to manage *what we know*, by making it easy to organize and store, access and retrieve, manipulate and synthesize, and apply to the solution of problems. We use a variety of terms to describe what we know, including the five that head this section and that are shown in Table 1.1. There are no universally agreed-on definitions of these terms. Nevertheless it is worth trying to come to grips with their various meanings because the differences between them can often be significant, and what follows draws on many sources and thus provides the basis for the use of these terms throughout the book. Data clearly refers to the most mundane kind of information and wisdom to the most substantive. *Data* consist of numbers, text, or symbols, which are in some sense neutral and almost context-free. Raw geographic facts, such as sensor measurements of temperature at a specific time and location, are examples of data. When data are transmitted, they are treated as a stream of bits; a crucial requirement is to preserve the integrity of the data set. The internal meaning of the data is irrelevant in such considerations. Data (the noun is the plural of datum) are assembled together in a

**Table 1.1** A ranking of the support infrastructure for decision making.

| Decision-making support infrastructure | Ease of sharing with everyone | GIS example |
|---|---|---|
| **Wisdom** ↑ | *Impossible* | Policies developed and accepted by stakeholders |
| **Knowledge** ↑ | *Difficult, especially tacit knowledge* | Personal knowledge about places and issues |
| **Evidence** ↑ | *Often not easy* | Results of GIS analysis of many data sets or scenarios |
| **Information** ↑ | *Easy* | Contents of a database assembled from raw facts |
| **Data** | *Easy* | Raw geographic facts |

database (see Chapter 9), and the volumes of data that are required for some typical applications are shown in Table 1.2.

The term *information* can be used either narrowly or broadly (and we use both in this book). In a narrow sense, information can be treated as devoid of meaning and therefore as essentially synonymous with data as defined in the previous paragraph. Others define information as *anything* that can be digitized, that is, represented in digital form (see Chapter 3), but also argue that information is differentiated from data by implying some degree of selection, organization, and preparation for particular purposes—information is data serving some *purpose* or data that have been given some degree of interpretation. Information is often costly to produce, but once digitized, it is cheap to reproduce and distribute. Geographic data sets, for example, may be very expensive to collect and assemble, but very cheap to copy and disseminate. One other characteristic of information is that it is easy to add value to it through processing and through merger with other information. GI systems are very useful for

doing the latter because of the tools they provide for combining information from different sources.

**GI systems do a better job of sharing data and information than knowledge, which is more difficult to detach from the knower.**

*Knowledge* does not arise simply from having access to large amounts of information. It can be considered as information to which value has been added by interpretation based on a particular context, experience, and purpose. Put simply, the information available in a book or on the Internet or on a map becomes knowledge only when it has been read and understood, as when an experienced hiker chooses not to set off into unfamiliar terrain having read about it and taken stock of the weather forecast. How the information is interpreted and used will be different for different readers depending on their previous experience, expertise, and needs. It is important to distinguish two types of knowledge: *codified* and *tacit*. Knowledge is codifiable if it can be written down and transferred relatively easily to others. Tacit

**Table 1.2** Potential GI database volumes in bytes for some typical applications (volumes estimated to the nearest order of magnitude). Strictly, bytes are counted in powers of 2—1 kilobyte is 1024 bytes, not 1000.

| | | |
|---|---|---|
| 1 megabyte | 1 000 000 ($2^{20}$) | Single data set in a small project database |
| 1 gigabyte | 1 000 000 000 ($2^{30}$) | Entire street network of a large city or small country |
| 1 terabyte | 1 000 000 000 000 ($2^{40}$) | Elevation of entire Earth surface recorded at 30 m intervals |
| 1 petabyte | 1 000 000 000 000 000 ($2^{50}$) | Satellite image of entire Earth surface at 1 m resolution |
| 1 exabyte | 1 000 000 000 000 000 000 ($2^{60}$) | A possible 3-D representation of the entire Earth at 10 m resolution |
| 1 zettabyte | 1 000 000 000 000 000 000 000 ($2^{70}$) | One-fifth of the capacity (in 2013) of U.S. National Security Agency Utah Data Center |

knowledge is often slow to acquire and much more difficult to transfer. Examples include the knowledge built up during an apprenticeship, understanding of how a particular market works, or familiarity with using a particular technology or language. This difference in transferability means that codified and tacit knowledge need to be managed and rewarded quite differently. Because of its nature, tacit knowledge is often a source of competitive advantage.

Some have argued that knowledge and information are fundamentally different in at least three important respects:

- Knowledge entails a knower. Information exists independently, but knowledge is intimately related to people.

- Knowledge is harder to detach from the knower than information; shipping, receiving, transferring it between people, or quantifying it are all much more difficult than for information.

- Knowledge requires much more assimilation—we digest it rather than hold it. We may hold conflicting information, but we rarely hold conflicting knowledge.

*Evidence* is considered a halfway house between information and knowledge. It seems best to regard it as a multiplicity of information from different sources, related to specific problems, and with a consistency that has been validated. Major attempts have been made in medicine to extract evidence from a welter of sometimes contradictory sets of information, drawn from different geographic settings, in what is known as meta-analysis, or the comparative analysis of the results of many previous studies.

*Wisdom* is even more elusive to define than the other terms. Normally, it is used in the context of decisions made or advice given, which is disinterested, based on all the evidence and knowledge available. It is given with some understanding of the likely consequences of various actions and assessment of which is or are most beneficial. Almost invariably, knowledge is highly individualized rather than being easy to create and share within a group. Wisdom is in a sense the top level of a hierarchy of decision-making infrastructure.

## 1.3 GI Science and Systems

GI systems are computer-based tools for collecting, storing, processing, analyzing, and visualizing geographic information. They are tools that improve the efficiency and effectiveness of handling information about objects and events located in geographic space. They can be used to carry out many useful tasks, including storing vast amounts of GI in data-

bases, conducting analytical operations in a fraction of the time they would take to do by hand, and automating the process of making useful maps. GI systems also process information, but there are limits to the kinds of procedures and practices that can be automated when turning data into useful information.

The question of whether and how such selectivity and preparation for purpose actually adds value, or whether the results add insight to interpretation in geographic applications, falls into the realm of GI science. This rapidly developing field is concerned with the concepts, principles, and methods that are put into practice using the tools and techniques of GI systems. It provides sound principles for the sample designs used to create data and the ways in which data can be turned into information that is representative of a study area. GI science also provides a framework within which new evidence, knowledge, and ultimately wisdom about the Earth can be created, in ways that are efficient, effective, and safe to use.

Like all sciences, an essential requirement of GI science is a method for discovering new knowledge. The GI scientific method must support:

- Transparency of assumptions and methods so that other GI scientists can determine how previous knowledge has been discovered and how they might themselves add to the existing body of knowledge

- Best attempts to attain objectivity through a detached and independent perspective that avoids or accommodates bias (unintended or otherwise)

- The ability of any other qualified scientist to reproduce the results of an analysis

- Methods of validation using the results of the analysis (internal validation) or other information sources (external validation)

- Generalization from partial representations that are developed for analytical purposes to the wider objective reality that they purport to represent

How, then, are problems solved using a scientific method, and are geographic problems solved in ways different from other kinds of problems? We humans have accumulated a vast storehouse of knowledge about the world, including information both on how it *looks*—that is, its *forms*—and how it *works*—that is, its dynamic *processes*. Some of those processes are natural and built into the design of the planet, such as the processes of tectonic movement that lead to earthquakes and the processes of atmospheric circulation that lead to hurricanes. Others are human in origin, reflecting the increasing influence that we have on ecosystems,

**Figure 1.8** Social processes, such as carbon dioxide emissions, modify the Earth's environment independent of location.

through the burning of fossil fuels, the felling of forests, and the cultivation of crops (Figure 1.8). Still others are imposed by us, in the form of laws, regulations, and practices: for example, zoning regulations affect the ways in which specific parcels of land can be used.

> **Knowledge about how the world works is more valuable than knowledge about how it looks. This is because knowledge about how it works can be used to predict.**

These two types of information differ markedly in their degree of generality. Form varies geographically, and the Earth's surface looks dramatically different in different places; compare the settled landscape of northern England with the deserts of the U.S. Southwest (Figure 1.9). But processes can be very general. The ways in which the burning of fossil fuels affects the atmosphere are essentially the same in China as in Europe, although the two land-

scapes look very different. Science has always valued such general knowledge over knowledge of the specific, and hence has valued process knowledge over knowledge of form. Geographers in particular have witnessed a long-standing debate, lasting centuries, between the competing needs of *idiographic* geography, which focuses on the description of form and emphasizes the unique characteristics of places, and *nomothetic* geography, which seeks to discover general processes. Both are essential, of course, because knowledge of general process is useful in solving specific problems only if it can be combined effectively with knowledge of form. For example, we can only assess the risk of roadside landslip if we know both how slope stability is generally affected by such factors as shallow subsurface characteristics and porosity and where slopes at risk are located (Figure 1.10).

One of the most important merits of a GI system as a problem-solving tool lies in its ability to combine the general with the specific, as in this example. A GI system designed to solve this problem would contain knowledge of local slopes, in the form of computerized maps, and the programs executed by the GI system would reflect general knowledge of how slopes affect the probability of mass movement under extreme weather conditions. The *software* of a GI system captures and implements general knowledge, whereas the *database* of a GI system represents specific information. In that sense, a GI system resolves the long-standing debate between nomothetic and idiographic camps by accommodating both.

> **GI systems solve the ancient problem of combining general scientific knowledge with specific information and give practical value to both.**

**Figure 1.9** The form of the Earth's surface shows enormous variability, for example, between (A) the deserts of the southwest United States and (B) the settled landscape of Northern England.

(A)

(B)

© BernardAllum/iStockphoto

**Figure 1.10** Predicting landslides requires general knowledge of processes and specific knowledge of the area—both can be brought together in a GI system.

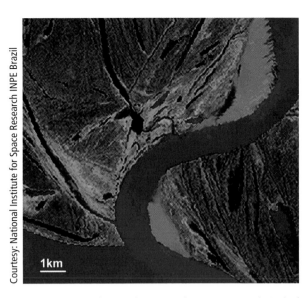

Courtesy: National Institute for Space Research INPE Brazil

1km

**Figure 1.11** A classified Landsat image (at 30-meter resolution) of part of the Amazon region of Brazil.

This perspective is consistent with our understanding of *places* in the world as sites at which unique relations develop among people and the locations that they occupy and the accumulated effects of these relations over time. GI systems provide ways of generalizing about and between places, albeit in ways that acknowledge differences between them. Place-based methods in GI systems make it possible to think of geography as repetitive (where in the world is like this place?) while at the same time remaining sensitive to the unique context of unique places.

General knowledge about unique places comes in many forms. Classification is perhaps the simplest and most rudimentary and is widely used in problem solving. In many parts of the United States and other countries, efforts have been made to limit the development of wetlands in the interest of preserving them as natural habitats and avoiding excessive impact on water resources. To support these efforts, resources have been invested in mapping wetlands, largely from aerial photography and satellite imagery. These maps simply classify land, using established rules that define what is and what is not a wetland (Figure 1.11).

More sophisticated forms of knowledge include *rule sets*—for example, rules that determine what use can be made of wetlands, or what areas in a forest can be legally logged. The U.S. Forest Service has rules to define wilderness and to impose associated regulations regarding the use of wilderness, including prohibition on logging and road construction. Such rules can be captured in the data model of a GI database (see Chapter 7).

Much of the knowledge gathered by the activities of scientists suggests the term *law*. The work of

Sir Isaac Newton established the Laws of Motion, according to which all matter behaves in ways that can be perfectly predicted. From Newton's Laws we are able to predict the motions of the planets almost perfectly, although Einstein later showed that certain observed deviations from the predictions of the Laws could be explained with his Theory of Relativity. Laws of this level of predictive quality are few and far between in the geographic world of the Earth's surface. The real world is the only laboratory that is available for understanding the effects of many factors on unique places in the social and environmental sciences, and considerable uncertainty is generated when we are unable to control for all conditions. These problems are compounded in the social realm, where the role of human agency makes it almost inevitable that any attempt to develop rigid laws will be frustrated by isolated exceptions. Thus, whereas market researchers use spatial interaction models, in conjunction with GI systems, to predict how many people will shop at each shopping center in a city, substantial errors will occur in the predictions—because people are in significant part autonomous agents. Nevertheless, the results are of great value in developing location strategies for retailing. The Universal Soil Loss Equation, used by soil scientists in conjunction with GI systems to predict soil erosion, is similar in its rather low predictive power, but again the results are sufficiently accurate to be very useful in the right circumstances. "Good" usually means "good enough for this specific application" in GI systems applications.

Solving problems involves several distinct components and stages. First, there must be an *objective*, or a goal that the problem solver wishes to achieve. Often this is a desire to maximize or minimize—find the solution of least cost, shortest distance, least time, greatest profit or make the most accurate prediction possible. These objectives are all expressed in *tangible* form; that is, they can be measured on some well-defined scale. Others are said to be *intangible* and involve objectives that are much harder, if not impossible, to measure. They include maximizing *quality of life* and *satisfaction* and minimizing *environmental impact*. Sometimes the only way to work with such intangible objectives is to involve human subjects, through surveys or focus groups, by asking them to express a preference among alternatives. A large body of knowledge has been acquired about such human-subjects research, and much of it has been employed in connection with the design of GI systems. For discussion of the use of such mixed objectives see Section 15.4. This topic is taken up again in Chapter 16 in the context of estimating the return on investment of GI systems.

Often a problem will have *multiple objectives*, each of which is measured in a different way. For example, a company providing a mobile snack service to construction sites will want to maximize the number of sites that can be visited during a daily operating schedule and will also want to maximize the expected returns by visiting the most lucrative sites. An agency charged with locating a corridor for a new power transmission line may decide to minimize cost, while at the same time seeking to minimize environmental impact. Such problems employ methods known as *multicriteria decision making* (MCDM).

**Many geographic problems involve multiple goals and objectives, which often cannot be expressed in commensurate terms.**

## 1.4 The Technology of Problem Solving

Today it is a truism to reflect that geographic information is everywhere and that we access and divulge it from many different sources and in many different contexts. A system is usually thought of as a *bounded* set of components, and in a world in which geographic information is transmitted and shared across physical, public/private, political, and institutional *networks*, it hardly seems to make sense to think in terms of bounded systems at all. However, although geographic information may be pervasive and ubiquitous, the notion of a networked system remains useful in understanding the compo-

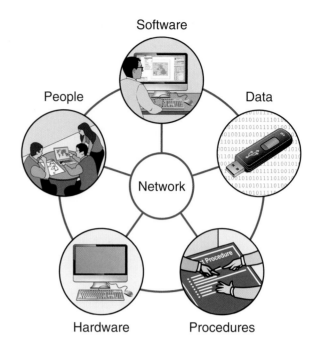

Figure 1.12 The six component parts of a GI system.

nents of the tools that in turn underpin GI science (Figure 1.12).

Today, almost all GI software products are designed as components of a network. *Cloud computing* (see Chapter 10) is a colloquial expression that is widely used in business to describe the supply of hosted services to industry and commerce, using computer infrastructure that is located remotely. Networks of large numbers of computers in different locations may be used for collection, storage, and analysis of data in real time. Cloud computing makes it possible to gain convenient, on-demand network access to a shared pool of computer hardware, software, data storage, and other services. Many of these components were previously colocated prior to the innovations of fast wide-area networks; powerful, inexpensive server computers; and high-performance virtualization of computer hardware.

In terms of hardware, the user's device is the *client*, connected through the network to a *server*, or a *server farm* in the Cloud, that is designed to handle many other user clients simultaneously. The client may be *thick*, if it performs a large part of the work locally, or *thin* if it does little more than link the user to the server (as with a mobile phone application, for example). In Cloud computing applications, most or all of the computation is performed remotely.

Uptake and use of the Internet to link computers has been remarkably quick, diffusion being considerably faster than almost all comparable innovations (for

**Table 1.3** World Internet usage and penetration statistics as of June 30 2012. (Source: www.internetworldstats.com)

| World Region | 2012 Population | Internet Users (Dec. 2000) | Internet Users (June 2012) | Penetration (% Population) | Growth 2000–12 | Users % of all Table |
|---|---|---|---|---|---|---|
| Africa | 1,073,380,925 | 4,514,400 | 167,335,676 | 15.6% | 3,607% | 7.0 |
| Asia | 3,922,066,987 | 114,304,000 | 1,076,681,059 | 27.5% | 842% | 44.8 |
| Europe (inc. EU) | 820,918,446 | 105,096,093 | 518,512,109 | 63.2% | 393% | 21.5 |
| Middle East | 223,608,203 | 3,284,800 | 90,000,455 | 40.2% | 2,640% | 3.7 |
| North America (excl. Canada) | 348,280,154 | 108,096,800 | 273,785,413 | 78.6% | 153% | 11.4 |
| Latin America/ Caribbean | 593,688,638 | 18,068,919 | 254,915,745 | 42.9% | 1,311% | 10.6 |
| Oceania/Australia | 35,903,569 | 7,620,480 | 24,287,919 | 67.6% | 219% | 1.0 |
| WORLD TOTAL | 7,017,846,922 | 360,985,492 | 2,405,518,376 | 34.3% | 566% | 100.0 |

example, the radio, the telephone, and the television). RealTimeStatistics.org estimated that in 2013 some 2.4 billion of the world's 7 billion population were Internet users, although stark variations in Internet availability and usage remain—see Table 1.3.

Many of the early Internet applications of GI systems remain in use, in updated form, today. They range from using GI systems on the Internet to disseminate information on the location of businesses (e.g., www.yell.com), to consolidated lists of available goods and services, to direct revenue generation through subscription services, to helping members of the public to participate in important local, regional, and national debates. The Internet has become very popular as a vehicle for delivering business GI system applications for several reasons. It provides an established, widely used platform and accepted standards for interacting with information of many types. It also offers a cost-effective way of linking distributed users (for example, telecommuters and office workers, customers and suppliers, students and teachers). From the early days onward, the interactive and exploratory nature of navigating linked information became a great hit with users.

Internet-enabled devices became portable in the early 2000s (see Section 10.3) with the wide diffusion of location-aware smartphones and other handheld devices and the availability of wireless networks in public places such as airports and railway stations. The subsequent innovation of 3G and 4G mobile broadband now routinely allows portable and in-vehicle devices to deliver *location-based services* (see Section 10.3.2) to users on the move. Users receive real-time geographic services such as mapping, routing, traffic congestion, and geographic yellow pages. These services are usually funded directly or indirectly through advertising, with Google perhaps the most obvious exponent of understanding the importance of location in delivering targeted advertising.

We now turn to consider the other components of a GI system. First, the user's *hardware* is the device that the user interacts with directly in carrying out GI system operations, by typing, pointing, clicking, or speaking, and that returns information by displaying it on the device's screen or generating meaningful sounds. Traditionally, this device sat on an office desktop, but today's user has more options and much more freedom because GI system functions can also be delivered through smartphones, notebooks, and in-vehicle devices.

The second component is the software programs that represent the world by running locally in the user's machine or remotely in the Cloud. Increasing numbers of users manipulate geographic information using executable *open-source software* code that is often freely available for download across the Web. Users can execute this code and also modify it if they wish. Other *open software* is also available for use as linked executable files, although the computer code that was used to generate it is not made available by its authors and so cannot be modified by other users. Both of these types of software may be made available by their authors in the interests of solving particular problems, or they may be made available as part of larger linked software libraries, such as the R project for statistical computing and graphics (www.r-project.org/). Some open software libraries have a focus on geographic problem solving and as such are described as GI systems—with the Quantum GIS Project (www.qgis.org/) providing perhaps the best contemporary example. The international "Geo for All" initiative (www.geoforall.org/) seeks to combine the potential of e-learning tools and open-source software to strengthen education in GI science, with particular emphasis on fast-changing needs in low-income countries.

Still other software is sold as closed commercial packages by established GI-system vendors, such as Autodesk Inc. (San Rafael, California; www.autodesk.com), Esri, Inc. (Redlands, California; www.esri.com), Intergraph Corp. (Huntsville, Alabama; www.intergraph.com/), or MapInfo Corp. (Troy, New York; www.mapinfo.com). Each vendor offers a range of products, designed for different levels of sophistication, different volumes of data, and different application niches. Idrisi (Clark University, Worcester, Massachusetts, www.clarklabs.org) is an example of a GI system produced and marketed by an academic institution rather than by a commercial vendor (for further information on GI system sources see Chapter 6).

Michael de Smith, along with two of the authors of this book, has produced an online guide (www.spatialanalysisonline.com) and book that is intended to raise awareness of the range of commercial and open software options that are available and the quality of the results that may be produced.

**It is not always easy to compare the software solutions suggested by Internet searches.**

The third component of a GI system is the data, which provide the foundations for digital representation of selected aspects of some specific area of the Earth's surface or near surface. A database might be built for one major project, such as the location of a new high-voltage power transmission corridor, or it might be continuously maintained, fed by the daily transactions that occur in a major utility company (e.g., installation of new underground pipes, creation of new customer accounts, and daily service-crew activities). As Open Data (see Section 17.4) become more freely available for download, the data that are downloaded for particular projects are frequently obtained from different sources, and thus the constituents of a GI database may have originally been assembled or collected for widely varying purposes and to widely varying standards. We discuss some of the implications of this in Chapters 5 and 17. The size of a project database may be as small as a few megabytes (a few million bytes, easily stored on a DVD) or as large as many terabytes (see Table 1.2).

*Big Data* is a term that has come to describe individual or linked data sets that are too large and complex to process using standard data-processing software or database-management tools on standard computer servers (see Box 17.2). Geographic databases are often big, not least of all because they include large numbers of location coordinates and sometimes many raster images. This poses significant challenges of data capture, storage, maintenance, sharing, visualization, and analysis. Scientists regularly encounter limitations in their abilities to manage and analyze Big Data in fields such as genomics (see Box 1.4) and meteorology. These discipline-specific problems are becoming more pervasive as more and more data are gathered using ubiquitous information-sensing mobile devices, remote sensing, radio-frequency identification (RFID) tagging, and wireless sensor networks. The world's technological per capita capacity to store information has roughly doubled every 40 months since the 1980s, and as of 2012, an average of 2.5 petabytes of data were created every day. This poses important management challenges for organizations that need to decide who should own Big Data initiatives that straddle their operation. The management and analysis of Big Data are closely associated with developments in Cloud computing.

Major GI applications also require management. An organization must establish procedures, lines of reporting, control points, and other mechanisms for ensuring that its GI activities meet its needs, stay within budgets, maintain high quality, avoid breaking the law, and generally meet the needs of the organization. These issues are explored in Chapters 16, 17, and 18.

Finally, a GI system is useless without the people who design, program, and maintain it, supply it with data, and interpret its results. The people of a GI system will have various skills, depending on the roles they perform. Almost all will have the basic knowledge needed to work with geographic data—knowledge of such topics as data sources, scale and accuracy, and software products—and will also have a network of acquaintances in the GI community. Most important, they will have a capacity for critical spatial thinking, allowing them to filter the message of spatial data through the medium of a GI system.

## 1.5 The Disciplinary Setting of GI Science and Systems (GISS)

At this point we review the emergence of GI science, the ways in which it is used, and its relationship with other disciplines. We discuss the significance of its underpinning technologies to business as well as some of the issues arising from its use in government. It should already be apparent that computer science is important because GI systems are computer applications, and its perspective is addressed in Section 1.5.4. Similarly geography, the science and

## Sir Walter Bodmer, Human Geneticist

Sir Walter Bodmer (Figure 1.13A) is a German-born British human geneticist. He studied mathematics and statistics at Cambridge University, doing his PhD under the renowned statistician and geneticist Sir Ronald A. Fisher before joining Nobel-Prize-winning microbiologist Joshua Lederberg's laboratory in the Genetics Department of Stanford University in 1961. He was an early pioneer of the use of computing to study population genetics, and after a period as a faculty member at Stanford, he left to become the first Professor of Genetics at Oxford University in 1970.

Population genetics is the study of the changing distribution of gene variants (or alleles) under the influence of four important evolutionary processes: natural selection, genetic drift, mutation, and gene flow. Walter was one of the first to suggest the idea of identifying the physical and functional characteristics of the 20,000–25,000 genes of the human genome. This idea was subsequently pursued in the Human Genome Project, which in important respects remains the ultimate investigative analysis using Big Data.

Geography is central to study of the human genome, for the very good reason that most families remain settled in one part of the world for many generations. In 2005 Walter was appointed to lead a major (£2.3 million/$3.8 million) project to examine geographic variations in the genetic makeup of the people of the British Isles. The aim of this project is to measure the genetic profiles of long-established families who can trace their recent ancestry to particular locations and to relate this to historical and archaeological evidence of invasion and settlement. The DNA samples of thousands of volunteers have been analyzed in ways that reveal the biological traces of successive waves of colonizers of Britain—such as the original ancient British settlers, the Anglo-Saxons, and the Vikings. The resulting genetic map (Figure 1.13B) shows, for example, that the Viking invasion of Britain was predominately by Danish Vikings, whereas the Orkney Islands were settled by Norwegian Vikings.

Walter was a pioneering advocate of public engagement with science and technology and remains very active in this field. He was elected a Fellow of the Royal Society in 1974 and was awarded the Society's Royal Medal in 2013 for seminal contributions to population genetics, gene mapping, and our understanding of familial genetic disease.

Check out the People of the British Isles project at www.peopleofthebritishisles.org.

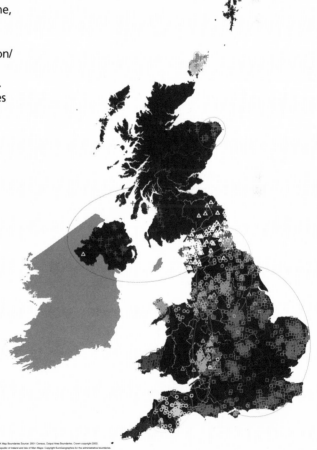

**Figure 1.13** (A) Walter Bodmer, human geneticist; and (B) a genetic map of the United Kingdom, with different colors denoting the genetic groups of long-settled residents.

study of phenomena distributed over the surface and near-surface of the Earth, also provides much of the disciplinary context of GI science and is addressed in Section 1.5.5. Finally, we consider some of the ways in which the use of GI is embedded in society.

## 1.5.1 The Historical Perspective

Although the coining of the term "GI science" can be traced to Goodchild's paper published in 1992, greater controversy surrounds the emergence of GI systems because parallel developments occurred in North America, Europe, and Australia (at least). Until recently, it was convenient to think of geographic information handling as confined to a freestanding, self-contained, computer-based system, like many other pieces of equipment. Indeed, prior to the innovation of the Internet, or the intranets of large organizations, the system was a physically isolated system of computer hardware, software, and data, such as a desktop computer, with no connections to the rest of the world. It was the extraction of simple geographic measures that largely drove the development of the first GIS to be described as such, the Canada Geographic Information System or CGIS, in the mid-1960s. The Canada Land Inventory was a massive effort by the federal and provincial governments to identify the nation's land resources and their existing and potential uses. The most useful results of such an inventory are measures of area, yet area was (and still is) notoriously difficult to measure accurately from a paper map (see Section 14.1.1). CGIS was planned and developed as a measuring tool, a producer of tabular information, rather than as a mapping tool.

**The first GI system was the Canada Geographic Information System, designed in the mid-1960s as a computerized map-measuring system.**

A second burst of innovation occurred in the late 1960s in the U.S. Bureau of the Census, in planning the tools needed to conduct the 1970 Census of Population. The DIME (Dual Independent Map Encoding) program created digital records of all U.S. streets to support automatic referencing and aggregation of census records. The similarity of this technology to that of CGIS was recognized immediately and led to a major program at Harvard University's Laboratory for Computer Graphics and Spatial Analysis to develop a general-purpose GIS that could handle the needs of both applications—a project that led eventually to the ODYSSEY GIS software of the late 1970s.

**Early GI system developers recognized that the same basic needs were present in many different application areas, from resource management to the census.**

In a largely separate development during the latter half of the 1960s, cartographers and mapping agencies had begun to ask whether computers might be adapted to their needs and possibly to reducing the costs and shortening the time of map creation. The UK Experimental Cartography Unit (ECU) pioneered high-quality computer mapping in 1968; it published the world's first computer-made map in a regular series in 1973 with the British Geological Survey (Figure 1.14). National mapping agencies, such as Britain's Ordnance Survey, France's Institut Géographique National, and the U.S. Geological Survey and Defense Mapping Agency (now the National Geospatial-Intelligence Agency) began to investigate the use of computers to support the editing and updating of maps, to avoid the expensive and slow process of hand correction and redrafting. The first automated cartography developments occurred in the 1960s, and by the late 1970s most major cartographic agencies were already computerized to some degree. But the limits of the technology of the time ensured that it was not until 1995 that the first country (Britain) achieved complete and detailed digital map coverage in a database.

Remote sensing also played a part in the development of GI systems, as a source of technology as well as a source of data. The first military satellites of the 1950s were developed and deployed in great secrecy to gather intelligence, and although the early spy satellites used conventional film cameras to record images, digital remote sensing began to replace them in the 1960s. By the early 1970s civilian remote-sensing systems such as Landsat were beginning to provide vast new data resources on the appearance of the planet's surface from space and to exploit the technologies of image classification and pattern recognition that had been developed earlier for military applications. The military was also responsible for the development in the 1950s of the world's first uniform system of measuring location, driven by the need for accurate targeting of intercontinental ballistic missiles, and this development led directly to the methods of positional control in use today (see Section 4.7). Military needs were also responsible for the initial development of the Global Positioning System (GPS; see Section 4.9 and Box 17.7).

**Many technical developments in GI systems originated in the Cold War.**

GI systems really began to take off in the early 1980s, when the price of computing hardware had fallen to a level that could sustain a significant software industry and cost-effective applications. Among the first customers were forestry companies and natural-resource agencies, driven by the need to keep track of vast timber resources and to regulate their use effectively.

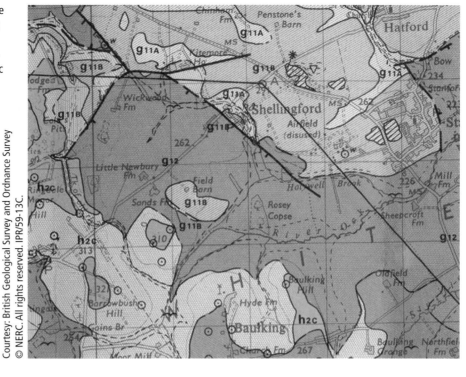

**Figure 1.14** Section of the 1:63,360 scale geological map of Abingdon, UK—the first known example of a map produced by automated means and published in a standard map series to established cartographic standards.

At the time a modest computing system—far less powerful than today's personal computer—could be obtained for about $250,000 (about $750,000 in 2015 prices) and the associated software for about $100,000 ($300,000 today). Even at these prices, the benefits of consistent management using GI systems and the decisions that could be made with these new tools substantially exceeded the costs. The market for GI software continued to grow, computers continued to fall in price and increase in power, and the GI software industry has been growing ever since.

> **The modern history of GI systems dates from the early 1980s, when the price of sufficiently powerful computers fell below a critical threshold.**

As indicated earlier, the history of GI systems is a complex story, much more complex than can be described in this brief history, but Table 1.4 summarizes developments from this early history to more recent commercial developments and the wide advent of Open Data and open-source software.

## 1.5.2 The Business Perspective

GI systems provide the underpinning technology for GI science, and many people play many roles in their development. Important activities range from software development to championing the importance of the spatial dimension in the activities of an organization through spatial data analysis. The business activities of many established companies are based on the collection of geographically referenced data or activities that build on this, such as value-added reselling of data, consulting, app development, training, system integration, software design, and so forth. Many new opportunities (particularly for start-up companies) are arising from use of open-source software or Open Data (see Section 1.5.3), where the underpinning costs of data or software production are met through volunteer activity or are underwritten by research institutions such as universities.

Such activity has mushroomed with the ever-wider use of the Internet to disseminate software and data, as well as the innovation and use of search engines, blogs, and social networking sites to spread news about what is available. This section looks at the diverse roles that people play in the business of GI systems, and is organized by the major areas of human activity associated with it.

### 1.5.2.1 The Software Industry

Many of the roots of the development of GI systems can be traced to the commercial development of off-the-shelf software packages that began in the 1980s. Today's commercial solutions are manifest in a wide range of forms. Thus, for example, the major software vendor Esri, Inc. (Redlands, California) today sells a family of GI system products under the ArcGIS brand name to service the disparate needs of its diverse user base. At its core are three niche desktop systems: Basic, for viewing spatial data, creating layered maps, and performing rudimentary spatial

**Table 1.4** Major events that shaped the development of GI systems.

### The Era of Innovation

| | |
|---|---|
| 1963 | Canada Geographic Information System is developed by Roger Tomlinson and colleagues for Canadian Land Inventory. This project pioneers much technology and introduces the term *GIS*. |
| 1964 | The Harvard Laboratory for Computer Graphics and Spatial Analysis is established under the direction of Howard Fisher. SYMAP, the first raster system for automated cartography, is created by Harvard researchers in 1966. |
| 1967 | The U.S. Bureau of Census develops DIME-GBF (Dual Independent Map Encoding—Geographic Base File), a data structure and street-address database for the 1970 U.S. Census. |
| 1969 | ESRI (Environmental Systems Research Institute) Inc. formed by Jack Dangermond, previously at the Harvard Lab, and his wife Laura. |
| 1969 | M&S Computing (subsequently renamed Intergraph Corp.) formed by Jim Meadlock and four others, who worked on guidance systems for Saturn rockets. |
| 1969 | Publication of *Design with Nature* by Ian McHarg; introduces many of the basic concepts of geographic analytics, including the map overlay process (see Section 13.2.4). |
| 1972 | Landsat 1 launched—the first of many civilian remote-sensing satellites. |
| 1973 | First digitizing of maps by a national mapping agency in a production system (Ordnance Survey, Great Britain). |

### The Era of Commercialization

| | |
|---|---|
| 1981 | ESRI ArcInfo launched—the first major commercial GI system based on the vector data structure and a relational database. |
| 1985 | The Global Positioning System gradually becomes a major source of data for navigation, surveying, and mapping. |
| 1986 | MapInfo Corp.'s software develops into first major desktop GI system. It defines a new standard for GI systems, complementing earlier software. |
| 1986 | The British Broadcasting Corporation (BBC) launches the Domesday machine to mark the 900th anniversary of the original Domesday Survey. Based on a microcomputer system with some GI system functionality, it held data and information provided by a million volunteers. The 21,000 GI files comprised maps, millions of words of text, and photographs of all of Britain, all cross-referenced and accessible by location and theme. |
| 1988 | TIGER (Topologically Integrated Geographic Encoding and Referencing), a follow-up from DIME, is described by the U.S. Census Bureau. Low-cost TIGER data stimulate rapid growth in U.S. business GI systems. |
| 1992 | The 1.7 GB Digital Chart of the World, sponsored by the U.S. Defense Mapping Agency (now NGA), is the first integrated 1:1 million-scale database offering global coverage. |
| 1994 | Executive Order 12906, signed by President Clinton, leads to creation of U.S. National Spatial Data Infrastructure (NSDI), clearinghouses, and the Federal Geographic Data Committee (FGDC). |
| 1994 | OpenGIS (subsequently Open Geospatial) Consortium of vendors and users established to improve interoperability between software. |
| 1995 | Complete conversion of the 240,000 topographic maps in the Great Britain national map coverage at 1:1,250, 1:2,500 and 1:10,000 scales into digital form. |
| 1996 | Innovation of Internet GI system products by Autodesk, ESRI, Intergraph, and MapInfo. |
| 1996 | MapQuest Internet mapping service launched. |
| 1999 | New generation of commercial satellites launched with submeter resolution capability (e.g., IKONOS and Quickbird). |

### The Era of Openness and Pervasive Use

| | |
|---|---|
| 2000 | The United States ceases the deliberate degradation (or "selective availability") of U.S. Global Positioning System (GPS) signals for national security purposes to encourage commercial and civilian applications using GPS technology. |
| 2003 | U.S. Federal e-government initiative provides "One-Stop" access to geospatial data and information (now part of geo.data.gov/geoportal/). |

**Table 1.4** (*continued*)

| | The Era of Openness and Pervasive Use |
|---|---|
| 2004 | OpenStreetMap founded by Steve Coast to create citizen-enabled mapping for the UK. The OpenStreetMap Foundation is subsequently established to encourage the creation, development, and distribution of free geospatial data (notably through "mapping parties") to provide geospatial data for anybody to use and share. |
| 2004 | Biggest GI system user in the world, National Imagery and Mapping Agency (NIMA), renamed National Geospatial-Intelligence Agency to signify emphasis on geo-intelligence. |
| 2005 | Launch of Google Earth service, the first major virtual 3-D globe with 150 million downloads in first 12 months. |
| 2006 | Launch of Amazon Web Services (AWS) as a "cloud," a distributed utility computing service. |
| 2007 | Purchase by Nokia mobile phone company of NAVTEQ street data provider for $8.1 billion. |
| 2007 | Launch of first touch-screen iPhone, including GPS-enabled mobile mapping with touch-screen functions. |
| 2008 | TeleAtlas street data provider purchased by TomTom for $2.9 billion. |
| 2009 | Cost of RFID tags falls to the point that the "Internet of Things" becomes a widespread reality. |
| 2009 | Quantum GIS (QGIS; initially developed by Gary Sherman) launched. Developed through the Open Source Geospatial Foundation (OSGeo) and making extensive use of open-source plugins, this user-friendly multiple-operating-system software is available under public license. |
| 2010 | ESRI Inc. launches www.arcgis.com (trialed in 2009 as www.arcgisonline.com), a Cloud-based GI platform offering a range of GI services. |
| 2010 | Advent of the UK Open Government Licence, allowing public bodies to publish a range of previously copyright material for public use. These are compatible with Creative Commons licences and are contributing to the wider dissemination of geographically referenced Open Data by governments around the world (see Section 17.4). |
| 2011 | Google launches indoor maps (floor plans) to enhance Google Maps service. |

analysis; Standard, which offers more advanced tools for manipulating spatial databases; and Advanced, which has further additional facilities for data manipulation, editing, and analysis. Other vendors specialize in certain niche markets, such as the utility industry, or military and intelligence applications. The software industry employs several thousand programmers, software designers, systems analysts, application specialists, and sales staff, with backgrounds that include computer science, geography, and many other disciplines. Given the all-pervasive nature of GI systems and their linkage to other forms of software, it no longer makes much sense to guesstimate the global value of the GI system industry, but where such estimates are attempted, it seems clear that the industry continues to grow rapidly—for example, estimates of GI system sales in China, where GeoStar (www.geostar.com.cn/Eng/) is widely used, suggest that the value of the industry has increased 50-fold over the last decade. GI systems are a dynamic and evolving field, and their future continues to offer many exciting developments, some of which we will discuss in the final chapter.

**Today a single GI system vendor offers many different products for distinct applications.**

### 1.5.2.2 The Data Industry

The acquisition, creation, maintenance, dissemination, and sale of GI data also account for a huge volume of economic activity. Traditionally, a large proportion of GI data have been produced centrally by national mapping agencies, such as the U.S. Geological Survey (USGS) or Great Britain's Ordnance Survey. In response to U.S. federal government policy, USGS supplies Open Data, defined as being priced at no more than the cost of reproduction of data. Elsewhere, some or all of available national mapping agency data are charged for to recoup the costs of their collection, although this is changing under various government Open Data initiatives. In Great Britain, for example, many Ordnance Survey mapping series have been made available free of charge since 2010; other, commercially valuable, data sets are not made available, however, because the government is unwilling to bear the projected $45 million cost of losing future revenue streams for highly detailed mapping products used by corporations such as utility companies. The innovation of free-to-view mapping services such as Google Maps (maps.google.com) and Microsoft Bing Maps and Virtual Earth, with their business model based on advertising revenues, is having profound implications for the provision of

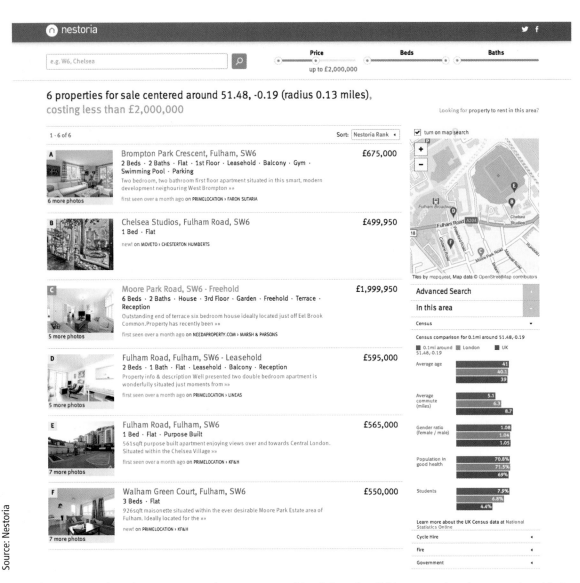

**Figure 1.15** The Nestoria Website (www.nestoria.com) presents a consolidated view of available property for sale or rent, alongside Open Data about neighborhoods. It is funded by the realtors who supply the linked listings and who "pay by click" for users searching for property.

map data (see Chapter 18). Specialist mapping sites, such as those used in the real-estate industry, present a consolidated map view of all available properties for sale or rent, along with Open Data pertaining to neighborhood quality (Figure 1.15). Volunteer-driven, open-source approaches to online cartography such as OpenStreetMap (www.openstreetmap.org) are also revolutionizing online cartography with their novel approach to map production and have stimulated the growth of commercial firms such as CloudMade (cloudmade.com).

**Open Data are revolutionizing the business model for GI applications.**

Geographically referenced attribute data relating to customer transactions are collected by many organizations in pursuit of their day-to-day operational activities. Subject to legal data protection requirements, such data may be reused in pursuit of the tactical or strategic objectives of the organization, and many retailers, energy suppliers, and financial-service providers employ teams of analysts to devise better ways of servicing customers and increasing market share. Still other companies (including cell phone companies) fulfill the role of *value-added data resellers* of commercial and public-sector data. Important applications for such services are opening up, including improving public-service delivery. The GeoWeb (see Chapter 10) is creating fertile environments in which a very wide range of public- and private-sector data sources can be combined, analyzed, and displayed.

Private companies are now also licensed to collect fine-resolution data using satellites and to sell them to customers—as, for example, with GeoEye's (geofuse.geoeye.com) IKONOS satellite data (see Table 1.4). Other companies collect similar data from aircraft. Still other companies specialize in the production of good-quality data on street networks, a basic requirement of many delivery companies. TomTom (www.tomtom.com) is an example of this industry, employing over 1800 staff in producing, maintaining, and marketing good-quality street-network data worldwide.

### 1.5.2.3 GI Services

As developments in the information economy gather still further momentum, many organizations are becoming focused on delivering integrated business solutions rather than raw or value-added data. The Cloud makes possible easy user access to data from sites that may be remote from locations where more specialized analysis and interpretation functions are performed. In these circumstances, it is no longer incumbent on an organization to manage either its own data or those that it buys from value-added resellers. For example, Esri Inc. offers a geographic data management service, in which data are managed and maintained for a range of clients that are at liberty to analyze them in quite separate locations. This may in time lead to greater vertical integration of the software and data industry— for example, Esri Inc. has developed a business information solutions division and acquired its own geodemographic system (called Tapestry) to service a range of business needs. This merging of the software and data industries comes together in the business of providing *GI systems as a service*, in which the Cloud is used as a remote platform on which software and data reside and GI system operations are performed.

## 1.5.3 The Government Perspective

Most government administration and decision-making activities have spatial implications, and governments—including the military and security services—remain by some margin the biggest users of commercial GI systems. Governments play an important role in many of the issues discussed throughout this book—including the development of data standards and interoperability, the creation of Web portals to disseminate GI (e.g., catalog.data.gov/dataset and data.gov.uk/), and the development of apps that automate key workflows such as zoning, property tax collection, or planning. Governments (such as those of Brazil and Spain) are also involved in the creation of open GI software.

One of the most profound changes concerning government and GI in recent years is in the ways in which government information is shared with citizens and external organizations in many parts of the world (see Chapter 17). Public-sector information (PSI) is defined by the wide range of information that government bodies collect, produce, reproduce, and disseminate across various areas of activity to fulfill their public-task functions. In terms of the nomenclature of Table 1.1, PSI is most conveniently thought of as information that has not been subjected to processing or any other manipulation beyond that necessary for its first use by government. Given the multiplicity of government functions, information that has been processed for one function (such as estimating local levels of homelessness or overcrowding) may be further refined and reused in the process of devising different measures for other functions (such as the composite measures of barriers to housing and public services used in various deprivation, or hardship, indices).

It is usual to think of PSI as having high fixed costs of assembly or collection and much smaller variable costs of copying and dissemination to users across the Web or through other media. It is also increasingly the case that the effort of rendering PSI into machine-readable or linked form can be absorbed into the fixed costs of information provision (see the star ratings schema discussed in Section 17.4.1). The small variable costs of disseminating some PSI have usually been absorbed in the United States, but this practice has been by no means universal, or even common, in other parts of the world. It is only in recent years that many governments worldwide have committed to greater availability of Open Data—which can be defined as data that can be used, reused, and redistributed freely by anyone, subject only at most to the requirement to attribute and share alike. Open Data are not necessarily free at the point of delivery, but any charge is usually no more than the cost of reproduction.

> **Open Data can be used, reused, and redistributed free of charge, subject only to Creative Commons or Open-Government licensing**

The most obvious motivation for government supply of Open Data has been to improve transparency and accountability of government decision making, as well as to improve economic efficiency and to stimulate growth. It is often left to the private sector to spot opportunities for leveraging value and to lobby for new Open Data sources to be made available for the common good or commercial gain.

> **Open Data improve the transparency, accountability, and efficiency of decision making**

The greater availability of PSI nevertheless raises a number of important issues for users of GI:

- Data are usually aggregated and cross tabulated to anonymize unit records, often using quite conservative procedures that mask important small-area variations.

- Government organizations are wary of the potential risk of *deanonymization*, that is, the combination of auxiliary data to link data to the individuals they characterize. GI systems provide the software environment in which such combination might take place

- Governments have traditionally emphasized data quality over timeliness, but are reluctant to extend quality-assurance procedures to rapidly delivered feeds of Open Data. There are likely to be issues with the provenance of Open Data as a consequence.

- Some of the core reference data produced by government provide the authoritative or definitive frameworks that are necessary to use other information—as, for example, with ZIP (post-) codes. Yet in many jurisdictions these are not open because they have considerable commercial value. There is considerable ongoing debate on whether government Open Data sources should build together into national (or international) information frameworks.

## 1.5.4 Computer-Science and Information-Science Perspectives

Computer scientists have been central to the development of GI technology, particularly in software development. Although sometimes mistakenly taken to be a branch of computer science, information science is a related but broad multidisciplinary area with a practical focus on the collection, classification, manipulation, storage, retrieval, analysis, movement, and dissemination of information. This brings with it concerns with database concepts, devising efficient algorithms for representing and accessing information, improving user interfaces, and developing computer architectures that are appropriate to different organizational settings. Like GI science and systems, these perspectives share concerns with the interactions between people, organizations, and computerized information systems.

All of this raises the question of whether GI is a fundamental part of computer science, or whether it merely presents a class of applications. The concepts underpinning spatiotemporal databases and computational geometry are core areas of computer science and information science that are directly relevant to GI technology, as are specific concerns with efficient computer processing, indexing schemes, database design, and computational geometry. Yet in each

case, the treatment of such issues is more focused on the technical aspects of GI technology than the context to which it is applied.

Context is important from a geographic perspective, defined by its concern with the human and environmental properties of the Earth's surface and near surface (Section 1.1), at scales from the architectural to the global. These concerns bring special focus to issues such as the nature of geographic data, their representation in GI systems, the ways in which they are georeferenced, and the uncertainties that arise when working with real-world places. They are not the priority of computer-science and information-science perspectives when these emphasize technology at the expense of context. Thus although GI science can be viewed as a branch of information science, the special nature of GI ensures that many of the principles of GI science have only a tenuous relationship to the broader principles of information science, and the same can be said about computer science.

In recent years much attention has been devoted to what is often termed *data science*, the issues arising from society's increasing dependence on data. It has been argued that science is increasingly data driven (sometimes called the *Fourth Paradigm* for research) and is based on the vast quantities of data that are now becoming available from ground-based and space-based sensors of various kinds, from social media, and from a host of other sources. From this perspective, some application domains produce such vast quantities of data that new and specialized techniques for storing, accessing, processing, and visualizing data are necessary to handle these enormous data problems. It has even been argued that such research neither needs nor produces theory, but instead mines data for patterns that may be useful in solving humanity's problems. We hear about the *exaflood*, the flood of information in quantities of exabytes (Table 1.2) and talk of science "drinking from a fire-hose." Data science studies the principles and techniques involved in managing these vast quantities of information, which include acquiring, sharing, documenting, managing, and archiving them.

Insofar as a substantial proportion of these data include information about location, it is clear that the issues of GI science have significant overlap with those of data science. Yet we maintain that the special characteristics of GI demand that many of these issues be treated separately and that education in the generic principles of data science is not necessarily sufficient qualification. Moreover, the kinds of infrastructure concerns that are often raised to prominence in data science, ranging from systematic documentation of data (metadata) to archiving and sharing, have a long history of development in GI science and may already be more advanced within this specialized domain than in generic data science (see also Chapters 17 and 19).

## 1.5.5. The Geography Perspective

Geographers have long agonized over the content, coherence, and relevance of their discipline, and set against this background, it is perhaps surprising that they have nonetheless made time to investigate and innovate in the use of new research methods in their discipline. Throughout the period documented in Table 1.4, GI science and systems have developed and maintained a special relationship to the academic discipline of geography and other disciplines that deal with the Earth's surface, including geodesy, landscape architecture, planning, and surveying. Yet any special relationship is multifaceted, nuanced, and sometimes tense. This is even more so, given that the roots to much of geography lie in mapping for warfare, that the military remain heavy users of GI systems, and that commercial software applications are rarely open to full academic scrutiny in deference to business priorities.

The idiographic and nomothetic concerns of geography (Section 1.3) have recently been the focus of thinking about the representation of *place* and *place effects*. Place is a social construct of space by humans and is key to the way that they understand their surroundings. Thus, although space exists independent of the existence of people, it is human interactions and experiences that crystallize space into place through shared perceptions and recognition. Figure 1.16 shows one delineation of the place known as "Paris" (France) through mapping the extent of attribution of this label to photographs uploaded to Flickr, along with the more limited extents of other more localized places identified by users.

Geography also tells us that widely scattered places (for example, in Washington, California, Florida, or Vermont) share important social or physical similarities that are manifest in different ways and to differing degrees. GI systems provide a way of representing these similarities, using standardized quantitative measures and, hence, the repetition of place effects across space. Geodemographic classifications provide one prominent example of the assignment of dispersed locations to place-relevant typologies, independent of their locational proximity to one another (see Figure 1.17).

In both of these examples, absolute location is not the principal focus, but it is the relative locations of places and the connectivity and relations between them that are important.

Recent years have seen the popularization of the term *neogeography* to describe developments in Web mapping technology and spatial data infrastructures that have greatly enhanced our abilities to assemble, share, and interact with geographic information online. Allied to this is the increased crowd sourcing by online communities of *volunteered geographic information* (VGI), discussed in Section 1.5.6 below. Neogeography is founded on the two-way, many-to-many interactions between users and Web sites that have emerged under Web 2.0, as embodied in projects such as Wikimapia (www.wikimapia.org) and OpenStreetMap

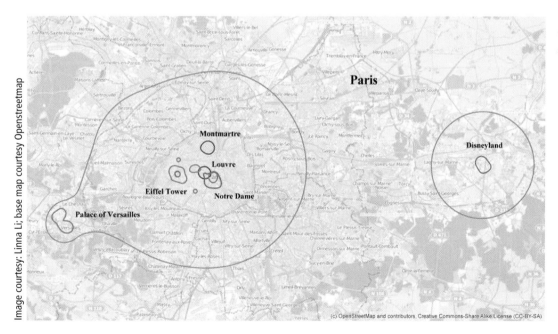

**Figure 1.16** Delineation of the extent of Paris, France, and locations associated with it, based on the density of geotagged Flickr photos for each place name. Contours of the place surfaces are depicted by choosing a threshold value of point density visually (e.g., the threshold value for the contour of Louvre is 500 points per square kilometer).

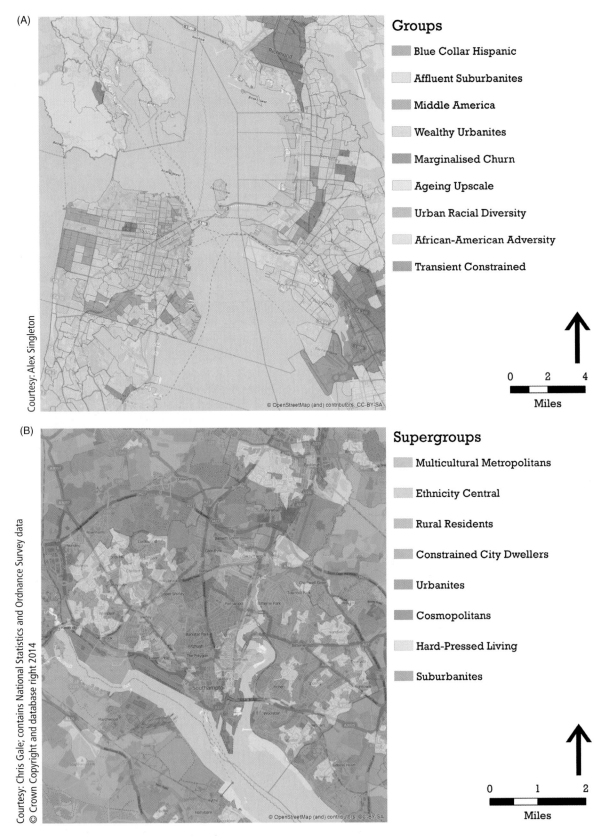

**(A)**

**Groups**

Blue Collar Hispanic

Affluent Suburbanites

Middle America

Wealthy Urbanites

Marginalised Churn

Ageing Upscale

Urban Racial Diversity

African-American Adversity

Transient Constrained

Courtesy: Alex Singleton

© OpenStreetMap (and) contributors CC-BY-SA

0   2   4
Miles

**(B)**

**Supergroups**

Multicultural Metropolitans

Ethnicity Central

Rural Residents

Constrained City Dwellers

Urbanites

Cosmopolitans

Hard-Pressed Living

Suburbanites

Courtesy: Chris Gale; contains National Statistics and Ordnance Survey data
© Crown Copyright and database right 2014

© OpenStreetMap (and) contributors CC-BY-SA

0   1   2
Miles

**Figure 1.17** Geodemographic neighborhood classifications make it possible to quantify how similar one place is to another, subject to the availability of consistent data across a jurisdiction. (A) A classification based on 2010 U.S. Census data, showing part of San Francisco; and (B) the UK 2011 Output Area Classification, showing part of Southampton.

Source: OpenStreetMap

**Figure 1.18** A crowd-sourced street map of part of São Paulo, Brazil.

(www.openstreetmap.org). Today, Wikimapia contains user-generated entries for more places than are available in any official list of place-names, whereas OpenStreetMap is well on the way to creating a free-to-use global map database through assimilation of digitized satellite photographs with GPS tracks supplied by volunteers (see Figure 1.18). This has converted many new users to the benefits of creating, sharing, and using geographic information, often through ad hoc collectives and interest groups. The creation, maintenance, and distribution of databases is no less than a "wikification of GI systems." Neogeography brings GI systems and some uses of spatial data infrastructures to the masses, while also presenting new challenges to citizen privacy and confidentiality. The empowerment of many nonexpert GI system users also brings with it the new challenges of ensuring that tools are used efficiently, effectively, and safely (see Chapter 17) and reemphasizes that technology can never offer more than a partial solution to the effective deployment of GI systems.

A core motivation for GI science is to better understand why GI systems applications are successful in practice. In this way, it should be possible to develop still better research practices for the future. Yet the applied nature of these endeavors and their core concerns with data, technology, and procedures inevitably distances any application from the investigator's direct experience. For some, this redefines the subject in ways that are inherently unacceptable in their reductionism; and no engagement with ever-more immersive technologies, richer data, or more sensitive analytical method can accommodate what are seen as failures inherent in empiricist methodology. In recent years protagonists of "critical GIS" have claimed

some success in illustrating how better technology, data, and methods make it possible to go beyond simplistic depiction of raw data. Yet for others, the deployment of computer systems inherently implies greater conviction than is warranted about the attributes and behaviors of the subjects of research. If this is the case, the "critical" prefix can never be more than a sop that implies, but does not deliver, acceptable levels of humility when engaging with the subjects of research.

**Some geographers remain suspicious of the use of GI systems in geography.**

Still broader issues arise when research is sponsored, directly or indirectly, by vested interests of commerce or the military. The latter was thrown into sharp focus in 2013 in Joel Wainwright's *Geopiracy*—a critique of recent U.S. military funding of academic geography to fund expeditions to map the human terrain of other countries. The Bowman expeditions—named after Isaiah Bowman, an early Twentieth Century American geographer identified with both empiricism and empire building—attracted criticism from Mexican authorities for alleged nondisclosure of U.S. military funding. To some, this was an uncomfortable reminder of the role of mapping as a tool of warfare (see Box 11.1) and the historic role of military motivations in the development of geography. We discuss this further in Chapter 18.

## 1.5.6 The Societal Perspective

The history of the development of GI systems has been one in which a technology of problem solving has evolved from specialized (and expensive) standalone configurations of hardware, software, and data to a background technology that is very much part of the information-technology mainstream. Today, when most of us casually use computers to pose "Where?" questions, we are rarely if ever cognizant that we are "doing GIS." We rarely contemplate the costs that an organization, somewhere, incurs in providing the GI services that we consume—because the costs of using the geographically enabled search of online mapping systems, for example, are absorbed through the advertising and data-sharing revenues of corporations such as Google, Apple, or Microsoft, or the costs of real-time public-transit information systems are borne by the operator out of ticket revenues. Internet behemoths foist use of cookies on us as users that betray our locations and many characteristics of our online identities, and we may resent the sophistication of the resulting advertising that is targeted on us. It seems that opting out of default options of disclosure requires greater diligence and doggedness than in the pre-Internet era, and we discuss privacy and legal concerns in Chapter 17. But most of us accept this intrusiveness, most of the time,

## The Top-Level Categories of the Geographic Information Science and Technology Body of Knowledge

Following are the 73 top-level categories of knowledge in the first edition (2006), and pointers (in brackets) to related chapters in this book where appropriate.

### Knowledge Area: Analytical Methods (AM)

Unit AM1 Academic and analytical origins (13)

Unit AM2 Query operations and query languages (13 and 14)

Unit AM3 Geometric measures (13 and 14)

Unit AM4 Basic analytical operations (13 and 14)

Unit AM5 Basic analytical methods (13 and 14)

Unit AM6 Analysis of surfaces (14)

Unit AM7 Spatial statistics (13)

Unit AM8 Geostatistics (13)

Unit AM9 Spatial regression and econometrics (13 and 14)

Unit AM10 Data mining (13 and 14)

Unit AM11 Network analysis (14)

Unit AM12 Optimization and location-allocation modeling (14)

### Knowledge Area: Conceptual Foundations (CF)

Unit CF1 Philosophical foundations (1)

Unit CF2 Cognitive and social foundations (1)

Unit CF3 Domains of geographic information (1)

Unit CF4 Elements of geographic information (2 and 3)

Unit CF5 Relationships (2 and 3)

Unit CF6 Imperfections in geographic information (5)

### Knowledge Area: Cartography and Visualization (CV)

Unit CV1 History and trends (11 and 12)

Unit CV2 Data considerations (11 and 12)

Unit CV3 Principles of map design (11)

Unit CV4 Graphic representation techniques (11 and 12)

Unit CV5 Map production (11)

Unit CV6 Map use and evaluation (11)

### Knowledge Area: Design Aspects (DA)

Unit DA1 The scope of GI system design (6)

Unit DA2 Project definition (16)

Unit DA3 Resource planning (16)

Unit DA4 Database design (9)

Unit DA5 Analysis design (13 and 14)

Unit DA6 Application design (1)

Unit DA7 System implementation (16)

### Knowledge Area: Data Modeling (DM)

Unit DM1 Basic storage and retrieval structures (7)

Unit DM2 Database management systems (6)

Unit DM3 Tessellation data models (7)

Unit DM4 Vector and object data models (7)

Unit DM5 Modeling 3-D, temporal, and uncertain phenomena (7)

### Knowledge Area: Data Manipulation (DN)

Unit DN1 Representation transformation (7 and 8)

Unit DN2 Generalization and aggregation (11 and 12)

Unit DN3 Transaction management of geospatial data (8)

### Knowledge Area: Geocomputation (GC)

Unit GC1 Emergence of geocomputation (15)

Unit GC2 Computational aspects and neurocomputing (15)

Unit GC3 Cellular Automata (CA) models (15)

Unit GC4 Heuristics (14)

Unit GC5 Genetic algorithms (GA) (15)

Unit GC6 Agent-based models (15)

Unit GC7 Simulation modeling (15)

Unit GC8 Uncertainty (5)

Unit GC9 Fuzzy sets (5)

### Knowledge Area: Geospatial Data (GD)

Unit GD1 Earth geometry (4)

Unit GD2 Land partitioning systems (4)

▶

either in ignorance or in implicit recognition of the convenience of the search engine and other services that such companies provide. Disclosure of online identity (of which location is an integral part) helps us to deal with the tyranny of small decisions in everyday life, even if GI service providers then try to use this to shape aspects of our identities as consumers in the interests of their advertising clients.

Choosing whether or not to use a particular search engine is an individual decision, but other GI applications are fundamentally underpinned by collective attitudes to creation and use. VGI is the collective harnessing of tools to create, assemble, and disseminate geographic data provided voluntarily by individuals—such as street locations (e.g., www.openstreetmap.org), geotagged photographs (e.g., www.flickr.com), or locationally referenced restaurant reviews. The motivations for supply of such information are multifaceted and may ebb and flow over time—recent years have seen much talk of "Wiki fatigue"—but all such projects are dependent on funding models for creating and maintaining GI systems—for example through individual donations or corporate sponsorship.

The funding imperative holds in all other societal deployments of GI. Governments pay the cost of ensuring the availability of Open Data and public-sector information (PSI). The costs of creating open software are often borne by research-grant-awarding organizations or funding by universities using student-fee income (justified as enhancing the public-facing profile of the organization). We discuss these different funding models in greater detail in Chapter 18.

The diversity of motivations underpinning the creation, maintenance, and dissemination of GI and the software to analyze it leads to the following societal concerns (see also the contribution of Sarah Elwood, discussed in Box 10.4):

● The links between knowledge and power. The ways in which GI systems represent the Earth's surface, and particularly human society, have been observed to privilege certain people, phenomena, and perspectives, at the expense of others. Minority views, and the views of individuals, can be submerged in this process, as can information that differs from the official or consensus view. GI systems often force knowledge into forms that are more likely to reflect the view of the majority, or the official view of government, and as a result *marginalize* the opinions of minorities or the less powerful. Countering this outlook is the view that greater access to PSI (see Section 17.4), especially over the Web, has enlivened debate and has gone some considerable way toward leveling the playing field in terms of data access. Moreover, the near-ubiquitous availability of mapping services and virtual Earths with limited GI system functionality (e.g., maps.google.com, apple.com/ios/maps/, and www.microsoft.com/virtualearth) brings rudimentary mapping and analysis capabilities to almost anyone with an Internet connection.

- In principle it is possible to use GI systems for any purpose, and so in practice they may sometimes be used for purposes that may be ethically questionable or may invade individual privacy—such as surveillance and the gathering of military and industrial intelligence. The technology may appear neutral, but it is always used in a social context. Although most would agree that the consequences of misuse of GI systems are not cataclysmic, there are nevertheless some similarities with the debates over the atomic bomb in the 1940s and 1950s: the scientists who develop and promote the use of GI systems surely bear some responsibility for how they are eventually used to the net benefit of humankind. The idea that a tool can be inherently neutral, and its developers therefore immune from any ethical debates, is now strongly questioned.

- The very success of GI systems is a cause of concern. There are qualms about a field that appears to be led by technology and the marketplace, rather than by human need. There are fears that GI systems have become *too* successful in modeling socioeconomic distributions and that as a consequence GI systems have become a tool of the "surveillance society." And there are fears that many users of services such as social media do not understand the uses to which their data may be put and the ways that they may be linked to other sources. Geography provides a powerful means of linking rich data sources pertaining to individual behavior.

- There remains an underrepresentation of applications of GI systems in *critical* research that focuses on the connections between human agency and particular social structures and contexts. Some detractors of GI systems have also suggested that such connections are not amenable to digital representation in whole or in part. Applications of "critical GIS" (see Section 1.5.5) have been developed that investigate, for example, the compatibility between GI systems and feminist epistemologies and politics; yet for some, the "critical" prefix does not address fundamental concerns with empiricism and the use of technologies to analyze the subjects of critical research.

- Some view the association of GI systems with the scientific and technical project as fundamentally flawed. More narrowly, there is a view that GI systems are inextricably bound to the philosophy and assumptions of the approach to science known

as *logical positivism* (see also the reference to "positive" in Section 1.1.1). As such, the argument goes, GI systems can never be more than a positivist tool and a normative instrument and cannot enrich other more critical perspectives in social science and even core areas of science itself. This is a criticism not just of GI systems, but also of the application of the scientific method to the analysis of social systems.

## 1.6 GI Science and Spatial Thinking

GI systems are useful tools, helping everyone from scientists to citizens to solve geographic problems. But like many other kinds of tools, such as computers themselves, their use raises questions that are sometimes frustrating and sometimes profound. For example, how does a GI system user know that the results obtained are accurate? What principles might help a GI system user to design better maps? How can location-based services be used to help users to navigate and understand human and natural environments? Some of these questions concern GI system design, and others are about the nature of GI and the methods used to analyze it. Taken together, we can think of them as questions that arise from the use of GI systems—that are stimulated by exposure to GI systems in their many guises. Many of them are addressed in detail in this book, and the book's title emphasizes the enduring importance of the science of problem solving.

Science is a systematic enterprise that builds and organizes knowledge in the form of testable explanations and predictions about the world (and beyond). It also refers to a body of knowledge itself, of the type that can be rationally explained and reliably applied. The U.S. University Consortium for Geographic Information Science (www.ucgis.org) is an organization of roughly 70 research universities that engages in research agenda setting, lobbying for research funding, and related activities. In 2006 it developed and published *Geographic Information Science and Technology Body of Knowledge* (BoK) in collaboration with the Association of American Geographers (a revised second edition is under development). The *BoK* is intended to be a comprehensive summary of knowledge in the field, to be used to evaluate the content of courses and programs, to set standards for job descriptions, and many other uses.

Box 1.5 lists the top-level contents of the 2006 first edition, which is currently available from www.aag.org/galleries/publications-files/GIST_Body_of_Knowledge.pdf. Each of the top-level topics is elaborated in detail in the document.

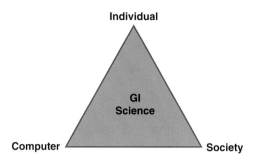

**Figure 1.19** The remit of GI science, according to Project Varenius (www.ncgia.ucsb.edu/varenius/varenius.html).

One disarmingly simple way of viewing this taxonomy is in the framework of the Varenius project (www.ncgia.ucsb.edu/varenius/varenius.html; Figure 1.19). Here, GI science is viewed as anchored by three concepts: the individual, the computer, and society. These form the vertices of a triangle, and GI science lies at its core. The various terms that are used to describe GI-science activity can be used to populate this triangle. Thus research about the individual is dominated by cognitive science, with its concern for understanding of spatial concepts, learning and reasoning about geographic data, and interaction with the computer. Central to such research is the desire to engineer more readily intelligible user interfaces, in the interest of improved human–computer interaction. Allied to this is the desire to improve geovisualization and render intelligible the results of spatiotemporal analysis. Research about the computer is dominated by issues of representation, the adaptation of new technologies, computation, and visualization. And finally, research about society addresses issues of impact and societal context.

## 1.7 GI Systems and Science in Society

GI science is often about using the software environment of GI systems to redefine, reshape, and resolve problems that are even older than the first GI systems. The need for methods of spatial analysis, for example, dates from the first maps, and many methods were developed long before the first GI system appeared on the scene in the mid-1960s. Another way to look at GI science is to see it as the body of knowledge that GI systems implement and exploit. Map projections (Chapter 4), for example, are part of GI science and are used and transformed through use of GI systems. A further area of great importance to GI systems is cognitive science, particularly the scientific

understanding of how people think about their geographic surroundings. If GI systems are to be easy to use, they must fit with human ideas about such topics as driving directions or how to construct useful and understandable maps.

**Many roots to GI systems can be traced to the spatial analysis tradition in the discipline of geography.**

Yet if there are enduring qualities to the body of knowledge that constitutes GI science, so there is also profound change in the setting in which it is applied. Astonishing improvements in our technical capacity to create and manipulate digital information have profoundly changed the ways in which GI systems are understood and used, both in business and in government. But perhaps the most profound changes of all have taken place in society at large. Just as a GI "system" is no longer a readily identifiable and clearly bounded "thing," so the notion of a "GI system user" has become a similarly anomalous conception—for the simple reason that we are now all consumers of GI and applications that use it. Other changes characterize the environment in which GI science is applied, given our vastly improved abilities to identify and monitor the movements of objects and (consenting) individuals, the rise of citizen science, and the availability of vast arrays of data that make it possible to devise rich yet comparable depictions of unique places.

We wonder where this will all end. Resolving the "Where?" question places unending demands on society's evolving information infrastructures and, in turn, begs many more questions concerning the sourcing, assembly, maintenance, and deployment of GI. How might the quality and reliability of volunteered geographic information be ascertained? How can users best understand the standards by which the information that they rely on has been created? How might user interfaces be improved to render GI intelligible to all relevant users? What are the obligations of governments to supply Open Data that are timely, comprehensive, and safe to use? What issues arise when commercial spatial data infrastructures (such as Google Earth) are reused for scientific or policy purposes? And what rights does the citizen have to privacy, or to be forgotten?

These are some of the key questions of GI science that we address in this book. They have profound implications for the ways in which we interact with one another and the ways in which we interact with places. For billions of casual users, GI systems have become an integral part of the information technology mainstream, yet this begs important questions about the roles, functioning, and status of pervasive GI technologies in society.

Although we continue to refer to "GI systems," an emphasis throughout much of this book is on this new societal vision of the emerging globally connected "GI system." As the software and procedures that are used to manipulate and analyze data continue to evolve, GISS provides focus to the way that better system design advances science and vice versa. Much of this book is about the more enduring principles of GI science, which are applied through society's use of geographic information, using the standards and procedures that are integral to GI system design.

The title of the first three editions of this book, which appeared between 2001 and 2011, was *Geographic Information Systems and Science*. Our revision of the title, as well as the content, of this fourth edition is designed to emphasize the enduring importance of scientific principles against a background of improvements in GI system design. Over this period, GI systems have evolved from isolated, monolithic software to society's resource of choice when the question "Where?" arises in any of its many guises. We believe strongly that effective users of GI systems require some awareness of *all* aspects of geographic information, from the basic principles and techniques to concepts of management and familiarity with applications. We hope this book provides that kind of awareness.

## Questions for Further Study

1. Examine the geographic data available for the area within 80 km (50 miles) of either where you live or where you study. Use it to produce a short (2500-word) illustrated profile of either the socio-economic or the physical environment. (See, for example, www.data.gov/ or data.gov.uk/data)

2. What are the distinguishing characteristics of the scientific method? Discuss the relevance of each to the history of GI systems.

3. Has society-wide use of GI systems transformed the research agenda for GI science? Give reasons for your answer.

4. Locate a subset of the issues identified in Box 1.5 in two triangular "GI science" diagrams like that shown in Figure 1.17—one for themes that predominantly relate to developments in science and one for themes that predominantly relate to developments in technology. Further details on each of the themes is available in the online *BoK*. If time is available, give short written reasons for your assignments. Compare the distribution of issues within each of your triangles to assess the relative importance of the individual, the computer, and society in the development of GI science and of GI technologies.

## Further Reading

De Smith, M., Goodchild, M. F., and Longley, P. A. 2009. *Geospatial Analysis: A Comprehensive Guide to Principles, Techniques and Software Tools* (3rd ed.). Leicester: Troubador and www.spatialanalysisonline.com.

DiBiase, D., deMers, M., Johnson, A., Kemp, K., Taylor Luck, A., Plewe, B. and Wentz, E. (eds.). 2006. *Geographic Information Science & Technology Body of Knowledge*. Washington, DC: Association of American Geographers. www.aag.org/galleries/publications-files/GIST_Body_of_Knowledge.pdf

Foresman, T. W. (ed.). 1998. *The History of Geographic Information Systems: Perspectives from the Pioneers.* Upper Saddle River, NJ: Prentice Hall.

Goodchild, M. F. 1992. Geographical information science. *International Journal of Geographical Information Systems* 6: 31–45.

Wainwright, J. 2013. *Geopiracy: Oaxaca, Militant Empiricism, and Geographical Thought*. New York: Palgrave Macmillan.

# The Nature of
# Geographic Data

**2**

This chapter elaborates on the *spatial is special* theme by examining the nature of geographic data. It sets out the distinguishing characteristics of geographic data and suggests a range of guiding principles for working with them. Many geographic data are correctly thought of as sample observations, selected from the larger universe of possible observations that could be made. This chapter describes the main principles that govern scientific sampling and the principles that are invoked to infer information about the gaps between samples. It also discusses the relevance of these principles when using crowd-sourced data and other Big Data sources. When devising spatial sample designs, it is important to be aware of the nature of spatial variation, and here we learn how this is formalized and measured as spatial autocorrelation. Another key property of geographic information is the level of detail that is apparent at particular scales of analysis. The concept of fractals provides a solid theoretical foundation for understanding scale when building geographic representations.

## LEARNING OBJECTIVES

After studying this chapter you will understand:

- How Tobler's First Law of Geography is formalized through the concept of spatial autocorrelation.
- The relationship between scale and the level of geographic detail in a representation.
- The principles of building representations around geographic samples.
- How the properties of smoothness and continuous variation can be used to characterize geographic data.
- How fractals can be used to measure and simulate apparently irregular geographic phenomena.

## 2.1 Introduction

In Chapter 1 we identified the central motivation for geographic information (GI) science as the development of representations, not only of how the world *looks*, but also how it *works*. In this chapter we develop a fuller understanding of the ways in which we think about the *nature of spatial variation*. We do this by asserting three principles:

1. Proximity effects are key to representing and understanding spatial variation and to joining incomplete representations of unique places.

2. Issues of geographic scale and level of detail are key to building appropriate representations of the world.

3. Different measures of the world *covary*, and understanding the nature of covariation can help us to predict.

Implicit in all this is one further principle that we will develop in Chapters 3 and 5: because we rarely if ever have complete information about the world, we must selectively sample from it and relate what we know to what we are less sure about. GI science is about representing spatial and temporal phenomena in the observable world, and because the observable world is complicated, this task is difficult, error prone, and in the terminology of Chapter 5, *uncertain*. The observable world provides an intriguing laboratory in which to examine phenomena, but is one in which it can be impossible to control for variation in all characteristics—be they relevant to landscape evolution, consumer behavior, urban growth, or whatever. As a consequence, in the terminology of Section 1.3, generalized *laws* governing spatial distributions and temporal dynamics are therefore most unlikely to work perfectly.

We choose to describe the three points listed above as principles rather than laws because, like most applications of GI science, this chapter is grounded in empirical generalization about the real world. A more elevated discussion of the way that these principles build into fundamental laws of GI science has been published by Goodchild.

## 2.2 The Fundamental Problem

It is instructive to reflect for a moment on how we might describe our lives to date to a friend or colleague. A human lifetime is infinitesimally small compared with the geographic extent and history of the world, but as we move to finer spatial and temporal scales, it nevertheless remains very intricate in detail. Viewed in aggregate, human behavior in space appears structured when we aggregate the outcomes of day-to-day (often repetitive) decisions about where to go, what to do, or how much time to spend doing it. Over the longer term, structure also arises out of (one-off) decisions about where to live, how to achieve career objectives, and how to balance work, leisure, and family pursuits. But just as giving clear narrative accounts of our lives requires clear recall of the key decisions that have very much made us what we are, so the GI scientist must be competent in discarding (or not feeling troubled to measure) the inessentials while retaining the salient characteristics of the observable world.

GI scientists distinguish between *controlled* and *uncontrolled* variation over time. Controlled variation in our lives oscillates around a steady state (daily, weekly) pattern, whereas uncontrolled variation (career changes, residential moves) does not. The same might be said, respectively, of seasonal change in climate and the phenomenon of global warming. When relating our own daily regimens and life histories, or indeed any short- or long-term *time series* of events, we are usually mindful of the contexts in which our decisions (to go to work, to change jobs, to marry) are made; "the past is the key to the present" aptly summarizes the effect of temporal context on our actions. The day-to-day operational context to our activities is very much determined by where we live and work. The longer-term strategic context may well be provided by where we were born, grew up, or went to college.

**Our behavior in geographic space often reflects past patterns of behavior.**

The relationship between consecutive events in *time* can be formalized in the concept of *temporal autocorrelation*. The analysis of time series data is in some senses straightforward because the direction of causality is only one way: past events are sequentially related to the present and to the future. This chapter (and book) is principally concerned with spatial, rather than temporal, autocorrelation. Spatial autocorrelation shares some similarities with its temporal counterpart. Yet time moves in one direction only (forward), making temporal autocorrelation one dimensional, whereas spatial events can potentially have consequences anywhere in two-dimensional or even three-dimensional space.

**Explanation in time need only look to the past, but explanation in space must look in all directions simultaneously.**

Just as the expectation is that temporal autocorrelation will be strongest between events that happen at about the same time as one another, so we expect spatial autocorrelation to be stronger for occurrences that are located close to one another. This guiding principle is often elevated to the status of a First Law of Geography, attributed to Waldo Tobler that *all places are similar, but nearby places are more similar than distant places.*

**Tobler's First Law of Geography: Everything is related to everything else, but near things are more related than distant things.**

Assessment of spatial autocorrelation can be informed by knowledge of the degree and nature of *spatial heterogeneity*—the tendency of geographic places and regions to be different from each other. Everyone would recognize the extreme difference of landscapes between such regions as the Antarctic, the Nile Delta, the Sahara Desert, or the Amazon Basin, and many would recognize the more subtle differences between the Central Valley of California, the Northern Plain of China, and the valley of the Ganges in India. As with change over time, variation in space may be controlled or uncontrolled: the spatial variation in some processes simply oscillates about an average (controlled variation), whereas other processes vary ever more the longer they are observed (uncontrolled variation). For example, controlled variation characterizes the operational environment of GI applications in utility management, or the tactical environment of retail promotions, whereas longer-term processes such as global warming or deforestation may lead to uncontrolled variation.

As a general rule, spatial data exhibit an increasing range of values, hence increased heterogeneity, with increased distance. In this chapter we focus on the ways in which phenomena vary across space and on the general nature of geographic variation. Later, in Section 13.2.1, we will return to the techniques for measuring spatial heterogeneity. This requires us to move beyond thinking of GI as abstracted only from the continuous spatial distributions implied by Tobler's Law and from sequences of events over continuous time. Some events, such as the daily rhythm of the journey to work, are clearly incremental extensions of past practice, whereas others, such as residential relocation, constitute sudden breaks with the past. Similarly, landscapes of gently undulating terrain are best thought of as smooth and continuous, whereas others (such as the landscapes developed about fault systems or mountain ranges) are best conceived as discretely bounded, jagged, and irregular. Smoothness and irregularity turn out to be among the most important distinguishing characteristics of geographic data.

**Some geographic phenomena vary smoothly across space, whereas others can exhibit extreme irregularity, in violation of Tobler's Law.**

It is highly likely that a representation of the real world that is suitable for predicting future change will need to incorporate information on how two or more factors *covary*. For example, planners seeking to justify improvements to a city's public transit system might wish to point out how house prices increase with proximity to existing rail stops. It is highly likely that patterns of spatial autocorrelation in one variable will, to a greater or lesser extent, be mirrored in another. However, although this is helpful in building representations of the real world, we will see in Section 14.5.1 that the property of spatial autocorrelation can frustrate our attempts to build inferential statistical models of the covariation of geographic phenomena.

**Spatial autocorrelation helps us to build representations but frustrates our efforts to predict.**

The nature of geographic variation, the scale at which uncontrolled variation occurs, and the way in which different geographic phenomena covary all help us to understand the form and functioning of the real world. These principles are of practical importance and guide us toward answering questions such as: What is an appropriate scale or level of detail at which I should measure the relevant and observable characteristics of the world? How do I design my spatial sample? How do I generalize from my sample measurements? And what formal methods and techniques can I use to relate key spatial events and outcomes to one another?

Each of these questions is a facet of the fundamental problem of the analysis of GI, that is, of selecting what to measure and record, using the scales of measurement described in Box 2.1. The Tobler Law

## Technical Box (2.1)

### Types of Attributes

The simplest type of attribute, termed *nominal*, is one that serves only to identify or distinguish one entity from another. Place-names are a good example, as are names of houses, or the numbers on a driver's license—each serves only to identify the particular instance of a class of entities and to distinguish it from other members of the same class. Nominal attributes include numbers, letters, and even colors. Even though a nominal attribute can be numeric, it makes no sense to apply arithmetic operations to it: adding two nominal attributes, such as two drivers' license numbers, creates nonsense.

Attributes are *ordinal* if their values have a natural order. For example, Canada rates its agricultural land by classes of soil quality, with Class 1 being the best, Class 2 not so good, and so on. Adding or taking the ratios of such numbers makes little sense because 2 is not twice as much of anything as 1, but at least ordinal attributes have inherent order. Averaging makes no sense either, but the *median*, or the value such that half of the attributes are higher ranked and half are lower ranked, is an effective substitute for the average for ordinal data as it gives a useful central value.

Attributes are *interval* if the differences between values make sense. The scale of Celsius temperature is interval because it makes sense to say that 30 and 20 are as different as 20 and 10.

Attributes are *ratio* if the ratios between values make sense. Weight is ratio because it makes sense to say that a person of 100 kg is twice as heavy as a person of 50 kg, but Celsius temperature is only interval because 20 is not twice as hot as 10 (and this argument applies to all scales that are based on arbitrary zero points, including longitude).

It is sometimes necessary to deal with GI that fall into categories beyond these four. For example, data can be directional or *cyclic*, including flow direction on a map, or compass direction, or longitude, or month of the year. The special problem here is that the number following 359 degrees is 0. Averaging two directions such as 359 and 1 yields 180, so the average of two directions close to north can appear to be south. Because the cyclic case sometimes occurs in GI, and few designers of GI software have made special arrangements for them, it is important to be alert to the problems that may arise.

amounts to a succinct definition of spatial autocorrelation. A prior understanding of the nature of the spatial autocorrelation that characterizes a GI application helps us *deduce* how best to collect and assemble data for a representation and also how best to develop inferences between events and occurrences. The concept of geographic *scale* or level of detail will be fundamental to observed measures of the likely strength and nature of autocorrelation in any given application. Together, the scale and spatial structure of a particular application suggest ways in which we should *sample* geographic reality and should *weight* sample observations to build our representation. We will return to the key concepts of scale, sampling, and weighting throughout much of this book and further discuss how representations are conceived and built from samples in Chapter 3. If data are collected for one purpose and reused for another, secondary,

---

## Technical Box (2.2)

### Types of Spatial Objects

Geographic objects are classified according to their *topological dimension*, which provides a measure of the way they fill space. For present purposes we assume that these dimensions are restricted to *integer* (whole number) values, though later (Section 2.8) we relax this constraint and consider geographic objects of noninteger (fractional, or *fractal*) dimension.

Geometric objects can be used to represent occurrences of recognizable features in space (sometimes described as *natural* objects), or they may be used to summarize spatial distributions of recognizable features (creating what are sometimes described as *artificial* objects).

A *point* has neither length nor breadth nor depth, and hence it is said to be of dimension 0. Points may be used to indicate spatial occurrences or events and their spatial patterning. *Point pattern analysis* is used to identify whether occurrences or events are interrelated— as in analyzing the incidence of crime or in identifying whether patterns of disease infection might be related to environmental or social factors (see Section 13.3.3). The *centroid* of an area object is an artificial point reference, which is located to provide a summary measure of the location of the object (see Section 14.2.1).

*Lines* have length, but not breadth or depth and hence are of dimension 1. They are used to represent linear entities such as roads, pipelines, and cables, which frequently build together into networks. They can also be used to measure distances between spatial objects, as in the measurement of intercentroid distance. To reduce the burden of data capture and storage, lines are often held in GI systems in *generalized* form (see Section 3.8).

*Area* objects have the two dimensions of length and breadth, but not depth. They may be used to represent natural objects, such as agricultural fields, but are also commonly used to represent artificial aggregations, such as census tracts (see below). Areas may bound linear features and enclose points, and GI systems can be used to identify whether a given area encloses a given point (Section 13.2.3).

*Volume* objects have length, breadth, and depth and hence are of dimension 3. They are used to represent natural objects such as river basins or artificial phenomena such as the population potential of shopping centers or the density of resident populations (Section 13.3.5).

*Time* is often considered to be the fourth dimension of spatial objects, although GI science remains poorly adapted to the modeling of temporal change.

Lower-dimension objects can be derived from those of higher dimension but not vice versa.

Certain phenomena may be represented in a GI database as either natural or artificial spatial objects— a human individual may be represented as a point or as part of a population represented as living within a census tract, for example. As will be discussed in Chapter 3, the chosen way of representing objects in space defines not only the apparent nature of geographic variation, but also the ways in which geographic variation may be analyzed. Some objects, such as agricultural fields or digita zero l terrain models, are represented in their natural, observable state. Others are transformed from one spatial object class to another, which is what happens if population data are aggregated from individual points to census tract areas for reasons of confidentiality, for example. Some high-order representations are created by interpolation between lower-order objects, as in the creation of digital elevation models (DEMs) from spot height data (see Section 13.3.6).

The classification of spatial phenomena into object types is dependent fundamentally on scale. For example, on a less-detailed map of the world, New York is represented as a zero-dimensional point. On a more detailed map such as a road atlas, it will be represented as a two-dimensional area. Yet if we visit the city, it is very much experienced as a three-dimensional entity, and virtual reality systems seek to represent it as such (see Section 10.3.1).

purpose, it is important that the scale, sampling, and weighting of the data are appropriate to the secondary application.

# 2.3 Spatial Autocorrelation and Scale

Objects existing in space are described by locational (spatial) descriptors and are conventionally classified using the taxonomy shown in Box 2.2. Spatial autocorrelation measures attempt to deal simultaneously with similarities in the location of spatial objects (Box 2.2) and their attributes (Box 2.1). If features that are similar in location are also similar in attributes, then the pattern as a whole is said to exhibit *positive spatial autocorrelation*. Conversely, *negative spatial autocorrelation* is said to exist when features that are close together in space tend to be more dissimilar in attributes than features that are further apart (in opposition to Tobler's Law). Zero autocorrelation occurs when attributes are independent of location.

Figure 2.1 presents some simple field representations of a geographic variable in 64 cells that can each take one of two values, coded blue and white. Each of the five illustrations contains the same set of attributes, 32 white cells and 32 blue cells, yet the spatial arrangements are very different. Figure 2.1A presents the familiar chess board and illustrates extreme negative spatial autocorrelation between neighboring cells. Figure 2.1E presents the opposite extreme of positive autocorrelation, when blue and white cells cluster together in homogeneous regions. The other illustrations show arrangements that exhibit intermediate levels of autocorrelation. Figure 2.1C corresponds to spatial independence, or no autocorrelation; Figure 2.1B shows a relatively dispersed arrangement; and Figure 2.1D shows a relatively clustered one.

**Spatial autocorrelation is determined by similarities in both position and attributes.**

The patterns shown in Figure 2.1 are examples of a particular case of spatial autocorrelation. In terms of the measurement scales described in Box 2.1, the attribute data are *nominal* (blue and white simply

---

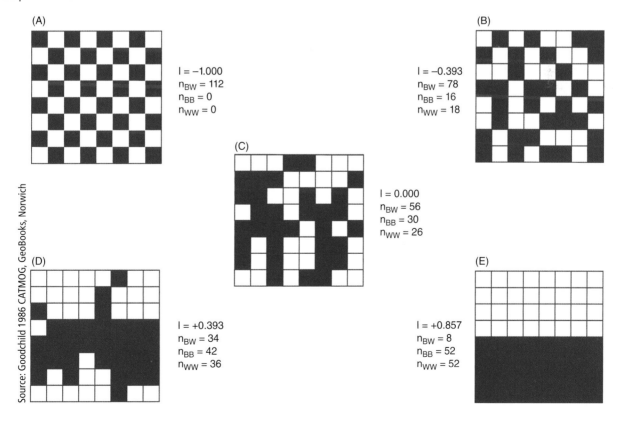

**Figure 2.1** Field arrangements of blue and white cells exhibiting: (A) extreme negative spatial autocorrelation, (B) a dispersed arrangement, (C) spatial independence, (D) spatial clustering, and (E) extreme positive spatial autocorrelation. The values of the I statistic are calculated using the equation in Section 2.6.

(A)

$I = -1.000$
$n_{BW} = 112$
$n_{BB} = 0$
$n_{WW} = 0$

(B)

$I = -0.393$
$n_{BW} = 78$
$n_{BB} = 16$
$n_{WW} = 18$

(C)

$I = 0.000$
$n_{BW} = 56$
$n_{BB} = 30$
$n_{WW} = 26$

(D)

$I = +0.393$
$n_{BW} = 34$
$n_{BB} = 42$
$n_{WW} = 36$

(E)

$I = +0.857$
$n_{BW} = 8$
$n_{BB} = 52$
$n_{WW} = 52$

Source: Goodchild 1986 CATMOG, GeoBooks, Norwich

identify two different possibilities, with no implied order and no possibility of difference or ratio). The figure gives no clue to the true dimensions of the area being represented. In other cases, similarities in attribute values may be more precisely measured on higher-order measurement scales, enabling continuous measures of spatial variation (see Section 5.3.2.2 and Box 5.4 for a discussion of precision).

The way in which we define what we mean by *neighboring* in investigating spatial arrangements may be more or less sophisticated. In considering the various arrangements shown in Figure 2.1, we have only considered the relationship between the attributes of a cell and those of its four *immediate* neighbors. But we could include a cell's four diagonal neighbors in the comparison (see Figure 2.2), and more generally there is no reason why we should not interpret Tobler's Law in terms of a gradual, incremental attenuating effect of distance as we traverse successive cells.

We began this chapter by considering a time series analysis of events that are highly, even perfectly, repetitive in the short term. Activity patterns often exhibit strong positive temporal autocorrelation (where you were at this time last week or this time yesterday is likely to affect where you are now), but only if measures are made at the same time every day—that is, at the temporal scale of the daily interval. If, say, sample measurements were taken every 17 hours, measures of the temporal autocorrelation of your activity patterns would likely be much lower. Similarly, if the measures of the blue/white property were made at intervals that did not coincide with the dimensions

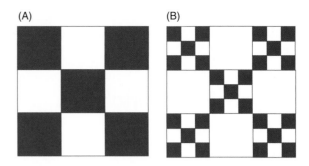

**Figure 2.2** A Sierpinski carpet at two levels of resolution: (A) coarse scale, and (B) finer scale. In general, measures of spatial and temporal autocorrelation are scale dependent (see Box 2.7; this point is explored in connection with the chessboard example in Section 13.2.5).

of the squares of the chess boards in Figure 2.1, then the spatial autocorrelation measures would be different. Thus the issue of *sampling interval* is of direct importance in the measurement of spatial autocorrelation because spatial events and occurrences may or may not accommodate spatial structure.

Scale (see Box 2.3) is often integral to the trade-off between the level of spatial resolution and the degree of attribute detail that can be stored in a given application—as in the trade-off between spatial and spectral resolution in remote sensing (see Section 8.2.1). The scale at which data from different sources are usually made available is discussed in Chapter 8.

A particular instance of the importance of scale is illustrated using a mosaic of squares in Figure 2.2. Figure 2.2A is a coarse-scale representation of attributes in nine squares and a pattern of negative spatial autocorrelation. However, the pattern is

---

## Technical Box (2.3)

### The Many Meanings of Scale

The concept of *scale* is fundamental to GI science, but unfortunately the word has acquired too many meanings in the course of time. Because they are to some extent contradictory, it is best to use other terms that have a clearer meaning where appropriate.

*Scale is in the details.* Many scientists use scale in the sense of spatial resolution, or the level of spatial detail in data. Data are fine scaled (or are at a fine level of granularity) if they include records of small objects and coarse-scaled (coarse-grained) if they do not.

*Scale is about extent.* Scientists also use scale to talk about the geographic extent or scope of a project: a large-scale project covers a large area, and a small-scale project covers a small area. Scale can also refer to other

aspects of the project's scope, including the cost or the number of people involved.

*The scale of a map.* Geographic data are often obtained from maps and often displayed in map form. Cartographers use the term *scale* to refer to a map's *representative fraction* (the ratio of distance on the map to distance on the ground; see Section 3.7). Unfortunately, this leads to confusion (and often bemusement) over the meaning of *large* and *small* with respect to scale. To a cartographer a large scale corresponds to a large representative fraction, in other words to plenty of geographic detail. This is exactly the opposite of what an average scientist understands by a large-scale study. In this book we have tried to avoid this problem by using the terms *coarse* and *fine* instead.

© Paulo Ferreira/iStockphoto

**Figure 2.3** Individual rocks may resemble the forms of larger structures, such as rock outcrops or eroded coastlines.

self-replicating at finer scales, and in Figure 2.2B, a finer-scale representation reveals that the smallest blue cells replicate the pattern of the whole area in a recursive manner. The pattern of spatial autocorrelation at the coarser scale is replicated at the finer scale, and the overall pattern is said to exhibit the property of *self-similarity*. Self-similar structure is characteristic of natural as well as social systems. For example, a rock may resemble the physical form of the mountain or coastline from which it was broken (Figure 2.3), small coastal features may resemble larger bays and inlets in structure and form, and neighborhoods may be of similar population size and each offer similar ranges of retail facilities across a metropolitan area. Self-similarity is a core concept of fractals, a topic introduced in Section 2.8.

## 2.4 Spatial Sampling

The quest to generalize about the myriad complexity of the real world requires us to abstract, or sample, events and occurrences from the universe of eligible elements of interest, which is known as the *sample*

*frame.* A spatial sampling frame might be defined by the rectangle formed by four pairs of coordinates, or by the combined extent of a set of natural or artificial objects (see Box 2.2). We can think of spatial sampling as the process of selecting points from within this bounding rectangle or mosaic of objects. The process of sampling requires us to select some points or objects and to discard the others. The procedure of selecting some elements rather than others from a sample frame can very much determine the quality of the representation that is built using them.

Scientific sampling requires that each element in the sample frame have a known and prespecified chance of selection. In some important senses, we can think of any geographic representation as a kind of sample, in that the elements of reality that are retained are abstracted from the observable world in accordance with some overall design. This is the case in remote sensing, for example (see Sections 3.6.1 and 8.2.1), in which each pixel value takes a spatially averaged reflectance value calculated at the spatial resolution characteristic of the sensor. In many situations, we will need to consciously select some observations, and not others, to create a generalizable abstraction. This is because, as a general rule, the resources available to any given project do not stretch to measuring every single one of the elements (soil profiles, migrating animals, shoppers) that we know to make up our population of interest. And even if resources were available, science tells us that this would be wasteful because procedures of *statistical inference* allow us to infer from samples to the populations from which they were drawn. We will return to the process of statistical inference in Section 14.5. Here, we will confine ourselves to the question, how do we ensure a good sample?

**Geographic data are only as good as the sampling scheme used to create them.**

Classical statistics often emphasizes the importance of randomness in sound sample design. The purest form, simple random sampling, is well known: each element in the sample frame is assigned a unique number, and a prespecified number of elements is selected using a random number generator. In the case of a spatial sample from continuous space, $x, y$ coordinate pairs might be randomly sampled within the range of $x$ and $y$ values (see Section 4.8 for information on coordinate systems). Because each randomly selected element has a known and prespecified probability of selection, it is possible to make robust and defensible generalizations about the population from which the sample was drawn. A spatially random sample is shown in Figure 2.4A.

Random sampling is integral to probability theory, and this enables us to use the distribution of values in our sample to tell us something about the likely distribution of values in the parent population from which the sample was drawn. However, sheer bad luck can mean that randomly drawn elements are disproportionately concentrated among some parts of the population at the expense of others, particularly when the size of our sample is small relative to the population. For example, a survey of household incomes might happen to select households with unusually low incomes. Systematic spatial sampling aims to circumvent this problem and ensure greater evenness of coverage across the sample frame. This is achieved by identifying a regular sampling interval $k$ (equal to the reciprocal of the sampling fraction $N/n$, where $n$ is the required sample size and $N$ is the size of the population) and proceeding to select every $k$th element. In spatial terms, the sampling interval of spatially systematic samples maps into a regularly spaced grid, as shown in Figure 2.4B.

---

Figure 2.4 Spatial sample designs: (A) simple random sampling, (B) stratified sampling, (C) stratified random sampling, (D) stratified sampling with random variation in grid spacing, (E) clustered sampling, (F) transect sampling, and (G) contour sampling.

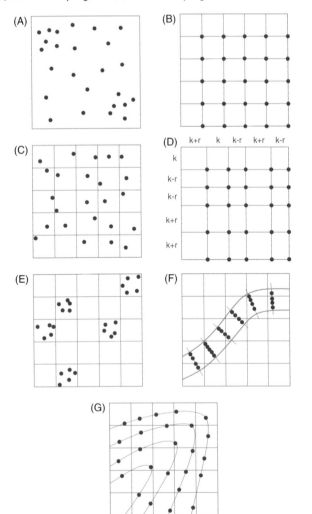

The advantage this offers over simple random sampling may be two-edged, however, if the sampling interval and the spatial structure of the study area coincide, that is, if the sample frame exhibits *periodicity*. A sample survey of urban land use along streets originally surveyed under the U.S. Public Land Survey System (PLSS; Section 4.6) would be ill-advised to take a sampling interval of one mile, for example, for this was the interval at which blocks within townships were originally laid out, and urban structure is still likely to be repetitive about this original design. In such instances, there may be a consequent failure to detect the true extent of heterogeneity of population attributes. For example, it is extremely unlikely that the attributes of street intersection locations would be representative of land uses elsewhere in a block structure. A number of hybrid sample designs have been devised to get around the vulnerability of spatially systematic sample designs to periodicity and the danger that simple random sampling may generate freak samples. These include stratified random sampling to ensure evenness of coverage (Figure 2.4C) and periodic random changes in the grid width of a spatially systematic sample (Figure 2.5D), perhaps subject to minimum spacing intervals.

In certain circumstances, it may be more efficient to restrict measurement to a specified range of sites—because of the prohibitive costs of transport over large areas, for example. Clustered sample designs, such as that shown in Figure 2.4E, may be used to generalize about attributes if the cluster presents a microcosm of surrounding conditions. In fact, this provides a legitimate use of a comprehensive study of one area to say something about conditions beyond it—as long as the study area is known to be representative of the broader study region. For example, political opinion polls are often taken in shopping centers, where shoppers can be deemed broadly representative of the population at large. However, instances where they provide a comprehensive detailed picture of spatial structure are likely to be the exception rather than the rule, and in practice increased sample sizes are often used to mitigate this fact.

Use of either simple random or spatially systematic sampling presumes that each observation is of equal importance, and hence of equal weight, in building a representation. As such, these sample designs are suitable for circumstances in which spatial structure is weak or nonexistent, or where (as in circumstances fully described by Tobler's Law) the attenuating effect of distance is constant in all directions. They are also suitable in circumstances where spatial structure is unknown. Yet in most practical applications, spatial structure is (to some extent at least) known, even if it cannot be wholly explained by Tobler's Law. These circumstances make it both more efficient and necessary to devise application-specific sample designs. This makes for improved

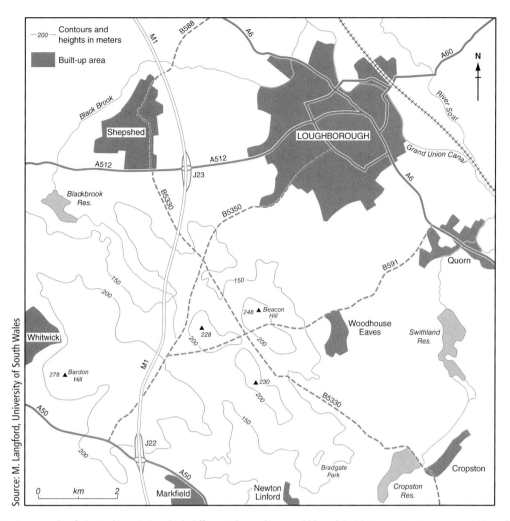

Source: M. Langford, University of South Wales

**Figure 2.5** An example of physical terrain in which differential sampling would be advisable to construct a representation of elevation.

quality of representation, with minimum resource costs of collecting data. Relevant sample designs include sampling along a transect, such as a soil profile (Figure 2.4F), or along a contour line (Figure 2.4G).

Consider the area of Leicestershire, UK, illustrated in Figure 2.5. It depicts a landscape in which the hilly relief of an upland area falls away sharply toward a river's floodplain. In identifying the sample spot heights that we might measure and hold in a GI database to create a representation of this area, we would be well advised to sample a disproportionate number of observations in the upland part of the study area where the local variability of heights is greatest. In a socioeconomic context, consider the task of ascertaining the total repair cost of bringing all housing in a city up to a specified standard. (Such applications are common, for example, in forming bids for federal or central government funding.) If spatial data are available relating to the time period during which different neighborhoods were developed, these would provide a useful guide to effective use of sampling resources. Newer houses are all likely to be in more or less the

same condition, whereas the repair costs of the older houses are likely to be much more heterogeneous and dependent on the different levels of attention that the occupants have lavished on them. As a general rule, older neighborhoods warrant a higher sampling interval than the newer ones, but other considerations may also be accommodated into the sampling design as well—such as construction type (duplex versus apartment, etc.) and local geology (as an indicator of risk of subsidence).

In any application where the events or phenomena that we are studying are spatially heterogeneous, we will require a large sample to capture the full variability of attribute values at all possible locations. Other parts of the study area may be much more homogeneous in attributes, and a sparser sampling interval may thus be more appropriate. Both simple random and systematic sample designs (and their variants) may be adapted to allow a differential sampling interval over a given study area (see Section 5.3.2.1 for more on this issue with respect to sampling vegetation cover). Thus it may be sensible to partition the sample frame into subareas, based on our knowledge of spatial structure—specifically our knowledge of the likely variability of the attributes that we are measuring. Other application-specific special circumstances include the following:

- Whether source data are ubiquitous or must be specially collected
- The resources available for any survey undertaking
- The accessibility of all parts of the study area to field observation (still difficult even in the era of ubiquitous availability of Global Positioning System receivers; see Section 4.9)

**Stratified sampling designs accommodate the unequal abundance of different phenomena on the Earth's surface.**

It is very important to be aware that this discussion of sampling is appropriate to problems where there is a large hypothetical population of evenly distributed locations (elements, in the terminology of sampling theory, or atoms of information in the terminology of Section 3.4) and that each has a known and prespecified probability of selection. Random selection of elements plays a part in each of the sample designs illustrated in Figure 2.4, albeit that the probability of selecting an element may be greater for clearly defined subpopulations that lie along a contour line or across a soil transect, for example. In circumstances where spatial structure is either weak or is explicitly incorporated through clear definition of subpopulations, standard statistical theory provides a robust framework for inferring the attributes of the population from those of the sample.

But reality is somewhat messier. In most practical applications, the population of elements (animals, glacial features, voters) may not be large, and its distribution across space may be far from random and independent. In these circumstances, conventional wisdom suggests a number of rules of thumb to compensate for the likely increase in error in estimating the true population value—as in clustered sampling, where slightly more than doubling the sample size is usually taken to accommodate the effects of spatial autocorrelation within a spatial cluster. However, the existence of spatial autocorrelation may fundamentally undermine the inferential framework and invalidate the process of generalizing from samples to populations. We examine this issue in more detail in our discussion of inference and hypothesis testing in Section 14.5.1.

## 2.5 Sampling and VGI

Thus far, our discussion has assumed that we have the luxury of collecting our own data for our own particular purpose. The reality of analysis in a data-rich world that is increasingly dominated by Big and Open Data (Section 17.4) is that more and more of the data that we use are collected by other parties, often for very different purposes. These parties may variously be nonexpert, self-selecting, or self-interested. The almost inevitable consequence is bias in the nature of the data that are available, and the source and effects of such bias may be far from obvious. Even if properly detected, sampling theory tells us that seeking to accommodate the effects of bias by reweighting under- and overrepresented observations is a very dangerous practice. For example, if few successful interviews of resident opinions are collected from addresses in gated communities, for reasons of access, then it is not acceptable to assign high weight to the few successful responses. In such cases the metadata of the data set are crucially important in establishing their provenance for the particular investigation that we may wish to undertake (see Section 10.2).

Still more insidious problems can characterize *volunteered geographic information* (VGI; see Section 1.5.6, and Box 3.5). At its best, it may be possible to invoke automated procedures that are robust and open to scrutiny to ensure that the data provided are unbiased in terms of their content, coverage, and collection procedures. But in the absence of knowledge about the proficiency and knowledge of every volunteer, the devil may be in the details. The wider "Wikification of GI systems" entails few safeguards about the reliability of ways in which the properties of places are represented in social media or community Web services.

**The content, coverage, and collection procedures of VGI are often ill-defined**

In the broad sense, VGI can be defined to include, for example, data that we "volunteer" to supermarket chains, telephone companies, Internet search engines, or financial services providers about our purchasing behavior and intentions to receive discounts or other incentives. Such data differ from "pure" VGI in that they are not supplied for purely altruistic purposes, but if the broader market niche of the businesses that collect such data is known, it may be possible to establish generalizations from such samples to the broader populations from which they are drawn.

Here are some questions to ask about the generalizability of VGI, some of which have been discussed in relation to the emergence of "citizen science" by researchers such as Muki Haklay and Dan Sui (Box 2.4):

1. Are all places equally accessible to volunteers, or is access to some areas difficult (e.g., physically remote areas) or constrained (e.g., gated communities)?

2. Are volunteers more likely to provide data on places or properties that interest them than those that do not? For example, social class bias among

OpenStreetMap volunteers appears to have led to a lack of coverage of low-status, peripheral housing developments in the early stages of the endeavor.

3. Where volunteers supply information about themselves, is this representative of other nonvolunteers? Retail-store loyalty card data, for example, may not be representative of those who do not patronize the store chain, or those who choose not to participate in loyalty programs.

4. Are volunteers at liberty to collect locationally sensitive data? Some governments are resistant to free use of social media or remain sensitive to record taking near military installations.

5. Should volunteers supply data that are open to malevolent use? This choice may appear straightforward in some circumstances (e.g., not revealing the locations of rare bird nesting sites to possible egg collectors) but in other instances is less clear-cut (e.g., mapping the geography of mobility-

impaired individuals will help in developing a neighborhood fire evacuation strategy, but the information may also be of interest to criminals).

6. Is it socially acceptable to make available *any* observations of uniquely identifiable individuals—especially if they might be undertaking behavior that might be construed as socially unacceptable)?

7. Is the information date-stamped, as an indicator of its provenance and current reliability?

## 2.6 Distance Decay

In selectively abstracting, or sampling, reality, judgment is required to fill in the gaps between the observations that make up a representation. This requires us to specify the likely attenuating effect of distance between the sample observations based on our understanding of the nature of geographic

## Dan Sui, Geographer and GI Scientist

Daniel Sui (隋殿志 or 老隋 in Chinese; Figure 2.6) is Distinguished Professor of Social and Behavioral Sciences at The Ohio State University, where he has also served as Chair of Geography, as a Director of the Institute of Population Research, and as Director of the Center for Urban and Regional Analysis. Dan undertook his undergraduate and master's degrees at Peking University and completed his PhD at the University of Georgia in 1993.

Dan's early work focused on new approaches to using GI in urban, environment, and health-related studies. In the early 2000s, he also worked on integrating GI systems with the mainstream of academic media studies to probe the challenging social, legal, and ethical issues related the emerging geographies of today's information society. Most recently, he has conducted

research into the use of volunteered geographic information (VGI) and social media for crowdsourcing geographic knowledge. His 2012 coedited book provides a comprehensive primer on the phenomenon of VGI, seeing it as part of a profound transformation in the ways in which geographic data, information, and knowledge are produced and circulated (see Section 1.2). VGI contributes to the Big Data deluge (Section 1.4), and the way that it is handled is thus an important priority for data science (Section 1.5.4). Dan's work also makes clear that the uncertainties inherent in the use of VGI for geographic knowledge production need to be systematically evaluated if its contribution to science is to be both convincing and enduring.

The supply and use of VGI is part of a much wider trend in the practice of science. Dan says: "After 50 years of development, both the social and technological environments for GI have changed dramatically. What happens in the next five years will be critical for the next phase of their development. A post–GI systems age needs to fully embrace the value of our emerging open culture—including Open Data, open software, open hardware, open standards, open research collaboration, open publication, open funding, and open education/learning. This offers us the best opportunities for responding to the challenges of Big Data, for better understanding our changing planet, and for reaping the benefits of VGI through citizen science. New technology-driven, application-led, science-inspired, and education-focused opportunities will propel GI systems to a new level of excellence, but we must also be mindful of the academic, legal, political, and environmental barriers for the development of open GI systems."

Courtesy: Dan Sui

**Figure 2.6** Dan Sui, geographer and GI scientist.

**Figure 2.7** We require different ways of interpolating between points, as well as different sample designs, for representing mountains and forested hillsides.

data (Figure 2.7). That is to say, we need to make an informed judgment about an appropriate *interpolation* function and how to *weight* adjacent observations. A literal interpretation of Tobler's Law implies a continuous, smooth, attenuating effect of distance on the attribute values of adjacent or contiguous spatial objects, or incremental variation in attribute values as we traverse a surface. The polluting effect of a chemical spillage decreases in a predictable fashion with distance from the point source, aircraft noise decreases on a linear trend with distance from the flight path, and the number of visits from localities (suitably normalized by population) to a national park decreases at a regular rate as we traverse the counties that adjoin it. This section focuses on principles and introduces some of the functions that are used to describe attenuation effects over distance, or the nature of geographic variation. Section 13.3.6 discusses ways in which the principles of distance decay are embodied in techniques of spatial interpolation.

The precise nature of the function used to represent the effects of distance is likely to vary between applications, and Figure 2.8 illustrates several hypothetical types. In mathematical terms, if $i$ is a point for which we have a recorded measure of an attribute and $j$ is a point with no recorded measurement, we use $b$ as a parameter that determines the rate at which the weight $w_{ij}$ assigned to point $j$ declines with distance from $i$. A small $b$ value produces a slower decrease than a large one. In most applications, the choice of distance attenuation function is the outcome of past experience, the fit of a particular application data set, and convention. Figure 2.8A presents the simple case of linear distance decay, given by the expression:

$$w_{ij} = a - bd_{ij}$$

for $d_{ij} < a/b$. This function might, for example, reflect the noise levels experienced across a transect perpendicular to an aircraft flight path. Figure 2.8B presents a negative power distance decay function, given by the expression:

$$w_{ij} = d_{ij}^{-b}$$

which some researchers have used to describe the decline in the density of resident population with distance from historic central business district (CBD) areas. Figure 2.8C illustrates a negative exponential statistical fit, given by the expression:

$$w_{ij} = e^{-bd_{ij}}$$

conventionally used in human geography to represent the decrease in retail store patronage with distance from it ($e$ here denotes an exponential term, approximately equal to 2.71828, sometimes written *exp*).

Each of the attenuation functions illustrated in Figure 2.8 is idealized in that the effects of distance are presumed to be regular, continuous, and *isotropic* (uniform in every direction). This notion of smooth and continuous variation underpins many of the representational traditions in cartography, as in the creation of *isopleth* (or isoline) maps (see Box 2.5). It is

**Figure 2.8** The attenuating effect of distance: (A) linear distance decay, $w_{ij} = a - bd_{ij}$; (B) negative power distance decay, $w_{ij} = d_{ij}^{-b}$; and (C) negative exponential distance decay, $w_{ij} = e^{-bd_{ij}}$.

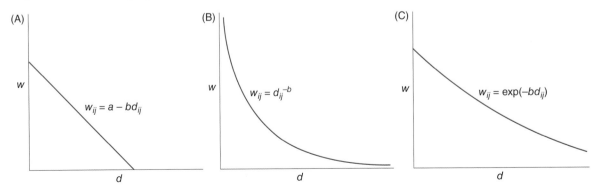

also consistent with our experience of high-school math, which is redolent of a world in which variation is continuous and best represented by interpolating smooth curves between everything. Yet even casual observation of the real world tells us that geographic variation is often far from smooth and continuous. The Earth's surface and geology, for example, are discontinuous at cliffs and fault lines, whereas the socioeconomic patterning of city neighborhoods can be similarly characterized by abrupt changes. Some illustrative issues pertaining to the catchment of a doctor's surgery general practice are apparent from Figure 2.9. Naïve GI analysis might assume that a map of 10-, 20-, and 30-minute travel times to see the doctor depicts a series of equidistant concentric circles. On this basis we might assume an isotropic linear distance decay function (Figure 2.8A), although the GI representation of the accessibility of the general practice is affected by

- Mode of transport—the 10-minute travel time buffer (Section 13.3.2) in effect measures walk time, whereas bus, car, and rail modes increase the extent of the 20- and 30-minute buffers.

- The quality and capacity of road and rail infrastructure, congestion issues at different times of day, and access constraints (e.g., to railway stations or the effects of road traffic calming in residential areas and one-way streets).

- Availability of public transport—by road or rail.

- Informal rights of way and the navigability of public open space (note that the public parks in Figure 2.9 appear to impede accessibility rather than enhance it).

- Physical barriers to movement, such as rivers.

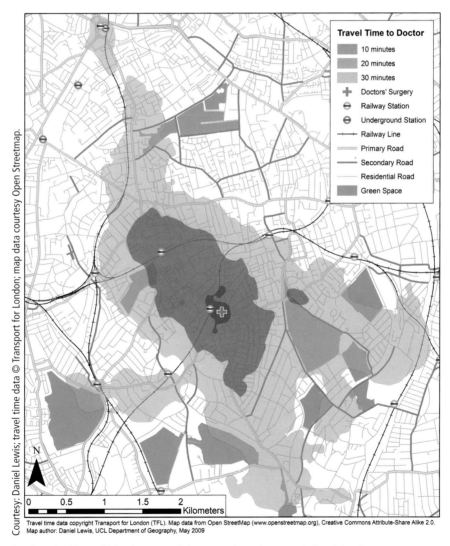

Courtesy: Daniel Lewis; travel time data © Transport for London; map data courtesy Open Streetmap.

Travel time data copyright Transport for London (TFL). Map data from Open StreetMap (www.openstreetmap.org), Creative Commons Attribute-Share Alike 2.0. Map author: Daniel Lewis, UCL Department of Geography, May 2009

**Figure 2.9** Map showing 10-, 20-, and 30-minute travel times to a doctor's surgery in South London.

## Isopleth and Choropleth Maps

*Isopleth* maps are used to visualize phenomena that vary continuously over space. An *isoline* connects points with equal attribute values, such as contour lines (equal height above sea level), *isohyets* (points of equal precipitation), *isochrones* (points of equal travel time), or *isodapanes* (points of equal transport cost). Figure 2.10 illustrates the procedures that are used to create a surface about a set of point measurements (Figure 2.10A), such as might be collected from rain gauges across a study region (see Section 13.3.6 for more technical detail on the process of spatial interpolation). A parsimonious number of user-defined values is identified to define the contour intervals (Figure 2.10B). A standard GI system operation is to interpolate a contour between point observations of greater and lesser value (Figure 2.10C) using standard procedures of distance decay, and the other contours are then interpolated using the same procedure (Figure 2.10D). Hue or shading can be added to improve user interpretability (Figure 2.10E).

**Choropleth** maps are constructed from values describing the properties of nonoverlapping areas, such as counties or census tracts. Each area is colored, shaded, or cross-hatched to symbolize the value of a specific variable, as in the 2011 UK Census data shown in Figure 2.11. Two types of variables can be used, termed *spatially extensive* and *spatially intensive*. Spatially extensive variables are those whose values are true only of entire areas, such as total population or total number of children under 5 years of age. Spatially intensive variables are those that could potentially be true of every part of an area, if the area were homogeneous; examples include densities, rates, or proportions. Conceptually, in terms of the field-object distinction, which we will introduce in the next chapter, a spatially intensive variable is a field, averaged over each area, whereas a spatially extensive variable is a field of density whose values are summed or integrated to obtain each area's value.

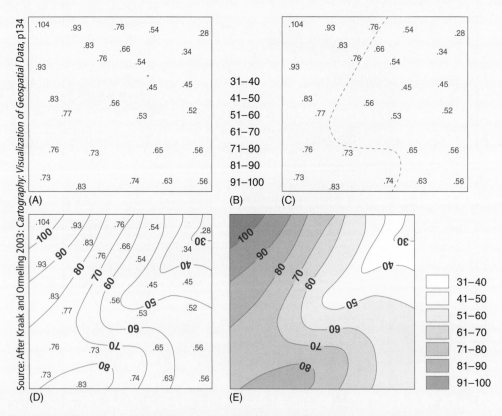

Source: After Kraak and Ormeling 2003: *Cartography: Visualization of Geospatial Data*, p134

**Figure 2.10** The creation of isopleth maps: (A) point attribute values, (B) user-defined classes, (C) interpolation of class boundary between points, (D) addition and labeling of other class boundaries, and (E) use of hue to enhance perception of trends.

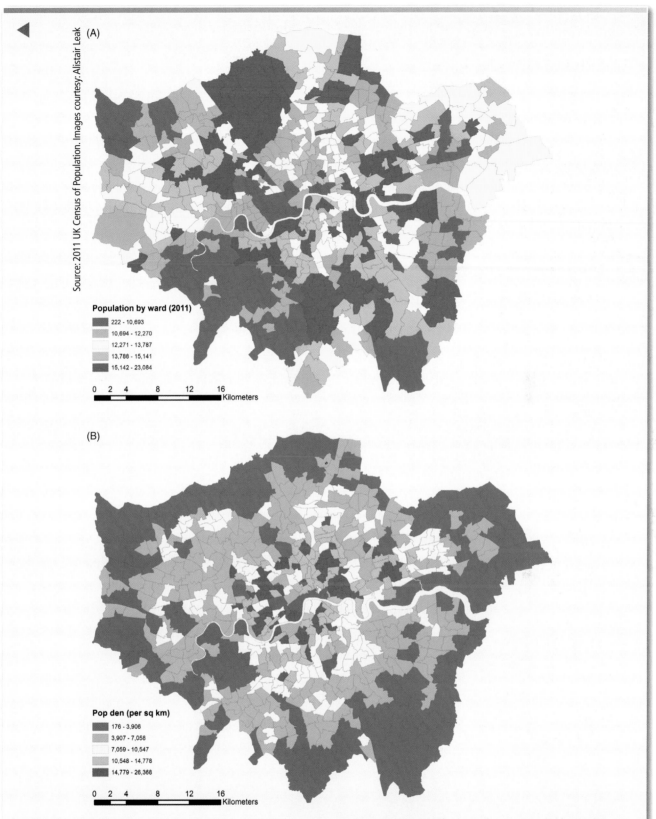

Source: 2011 UK Census of Population. Images courtesy: Alistair Leak

(A)

**Population by ward (2011)**
- 222 - 10,693
- 10,694 - 12,270
- 12,271 - 13,787
- 13,788 - 15,141
- 15,142 - 23,084

0  2  4    8    12    16
Kilometers

(B)

**Pop den (per sq km)**
- 176 - 3,906
- 3,907 - 7,058
- 7,059 - 10,547
- 10,548 - 14,778
- 14,779 - 26,368

0  2  4    8    12    16
Kilometers

**Figure 2.11** Choropleth maps of (A) a spatially extensive variable, total population, and (B) a related but spatially intensive variable, population density. Many cartographers would argue that (A) is misleading and that spatially extensive variables should always be converted to spatially intensive form (as densities, ratios, or proportions) before being displayed as choropleth maps.

In practice, the number of patients who use the general practice is also likely to depend on patient characteristics, as measured by

- Health care needs, their match with medical practice specialisms, and other socioeconomic characteristics
- The availability of medical services from other local providers.
- A demand constraint, which requires that the probabilities of any individual attending each of the available practices sum to 1 (unless people opt out of the health-care system)

## 2.7 Measuring Distance Effects as Spatial Autocorrelation

Understanding spatial structure is key to building understanding of other real-world structures because it helps us to *deduce* a good sampling strategy and to use an appropriate means of interpolating between sampled points that is fit for purpose. Knowledge of the actual or likely nature of spatial autocorrelation is thus important deductive understanding. However, in many applications we do not understand enough about geographic variability, distance effects, and spatial structure to make reliable deductions. A further branch of spatial

analysis thus emphasizes the *measurement* of spatial autocorrelation as an end in itself. This amounts to a more *inductive* approach to developing an understanding of the nature of a geographic data set.

**Induction reasons from data to build up understanding, whereas deduction begins with theory and principle as a basis for looking at data.**

In Section 2.3 we saw that spatial autocorrelation measures the extent to which similarities in position match similarities in attributes. Methods of measuring spatial autocorrelation depend on the types of objects used as the basis of a representation, and as we saw in Section 2.2, the scale of attribute measurement is important too. Interpretation depends on how the objects relate to our conceptualization of the phenomena they represent.

Figure 2.12 shows examples of each of the four object types described in Box 2.2, with associated attributes, chosen to represent situations in which a scientist might wish to measure spatial autocorrelation. The point data in Figure 2.12A comprise data on wellbores over an area of 30 km$^2$ and together provide information on the depth of an aquifer beneath the surface (the blue shading identifies those within a given threshold). We would expect values to exhibit strong spatial autocorrelation, with departures from this indicative of changes in bedrock structure or form. The line data in Figure 2.12B present numbers of accidents for links of road over a lengthy survey

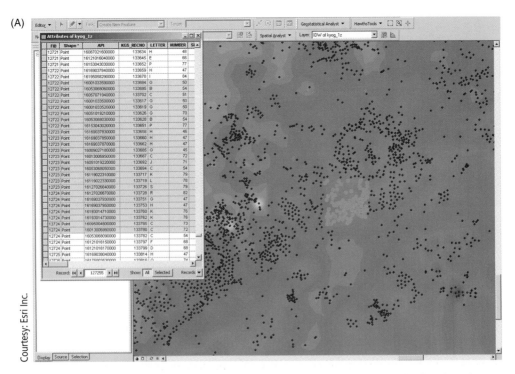

Courtesy: Esri Inc.

**Figure 2.12** Situations in which a scientist might want to measure spatial autocorrelation: (A) point data (wells with attributes stored in a spreadsheet; linear extent of image 0.6 km); (*continued*)

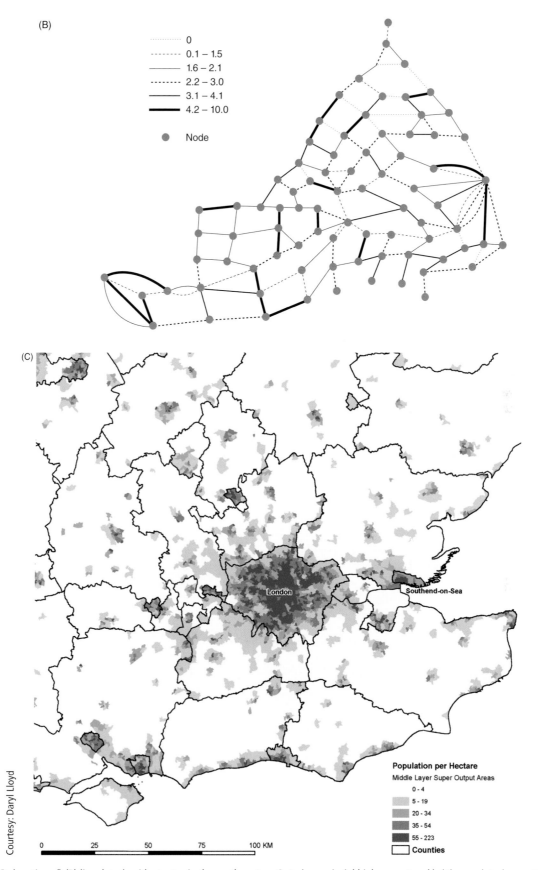

**Figure 2.12** (*continued*) (B) line data (accident rates in the southwestern Ontario provincial highway network); (C) area data (percentage of population that are old age pensioners in southeast England);

(D)

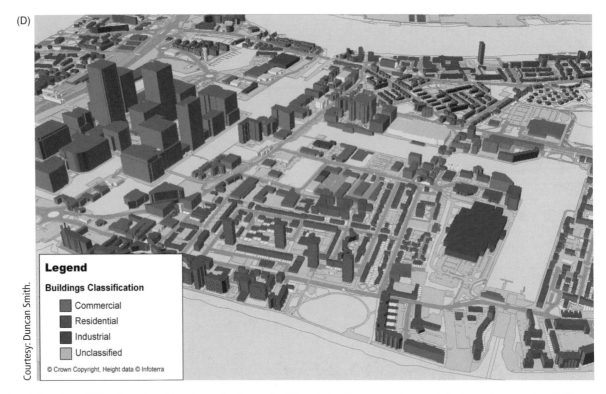

Legend

**Buildings Classification**

- ■ Commercial
- ■ Residential
- ■ Industrial
- ■ Unclassified

© Crown Copyright. Height data © Infoterra

Courtesy: Duncan Smith.

**Figure 2.12** (*continued*) (D) volume data (elevation and volume of buildings in London: east to the top of the image; image extent 1.6 km north-south, 1.5 km east-west).

period in the southwestern Ontario (Canada) provincial highway network. Low spatial autocorrelation in these statistics implies that local causative factors (such as badly laid out junctions) account for most accidents, whereas strong spatial autocorrelation would imply a more regional scale of variation, implying a link between accident rates and lifestyles, climate, or population density. The area data in Figure 2.12C illustrate the settlement structure of southeast England and invite analysis of the effectiveness of land use zoning (green belt) policies, for example. The volume data in Figure 2.12D allow some measure

of the spatial autocorrelation of high-rise structures to be made as part of a study of the relationship between the built form and economic function of London, UK. The way that spatial autocorrelation statistic might actually be calculated for the data used to construct Figure 2.12C is described in Box 2.6.

Spatial autocorrelation measures tell us about the interrelatedness of phenomena across space, one attribute at a time. Its measurement is key to formalizing and understanding many geographic problems, and it is central to locational analysis in geography. Another important facet to the nature of geographic

## Technical Box (2.6)

### Measuring Similarity between Neighbors

In the simple example shown in Figure 2.13, we compare neighboring values of spatial attributes by defining a weights matrix **W** in which each element $w_{ij}$ measures the locational similarity of $i$ and $j$ ($i$ identifies the row and $j$ the column of the matrix). We use a simple measure of contiguity, coding $w_{ij} = 1$ if regions $i$ and $j$ are contiguous and $w_{ij} = 0$ otherwise. $w_{ii}$ is set equal to 0 for all $i$. This is shown in Table 2.1.

The weights matrix provides a simple way of representing similarities between location and attribute values in a region of contiguous areal objects. Autocorrelation is identified by the presence of neighboring cells or zones that take the same (binary) attribute value. More sophisticated measures of $w_{ij}$ might include a decreasing function (such as one of those shown in Figure 2.7) of the straight-line distance between points at the centers of zones, or the lengths

▶

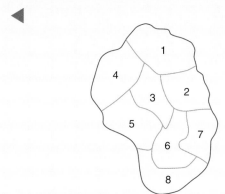

**Figure 2.13** A simple mosaic of zones.

**Table 2.1** The weights matrix **W** derived from the zoning system shown in Figure 2.12

|   | 1 | 2 | 3 | 4 | 5 | 6 | 7 | 8 |
|---|---|---|---|---|---|---|---|---|
| 1 | 0 | 1 | 1 | 1 | 0 | 0 | 0 | 0 |
| 2 | 1 | 0 | 1 | 0 | 0 | 1 | 1 | 0 |
| 3 | 1 | 1 | 0 | 1 | 1 | 1 | 0 | 0 |
| 4 | 1 | 0 | 1 | 0 | 1 | 0 | 0 | 0 |
| 5 | 0 | 0 | 1 | 1 | 0 | 1 | 0 | 1 |
| 6 | 0 | 1 | 1 | 0 | 1 | 0 | 1 | 1 |
| 7 | 0 | 1 | 0 | 0 | 0 | 1 | 0 | 1 |
| 8 | 0 | 0 | 0 | 0 | 1 | 1 | 1 | 0 |

of common boundaries (although in this case weight would be an increasing function of length, rather than the decreasing function of distance implied by $b$ in Figure 2.8). A range of different spatial metrics may also be used, such as existence of linkage by air, or a decreasing function of travel time by air, road, or rail, or the strength of linkages between individuals or firms on some (nonspatial) network.

The weights matrix makes it possible to develop measures of spatial autocorrelation using attributes measured on the nominal, ordinal, interval, or ratio scales (see Box 2.1) and the dimensioned classes of spatial objects shown in Box 2.2. Any measure of spatial autocorrelation seeks to compare a set of locational similarities $w_{ij}$ (contained in a weights matrix) with a corresponding set of attribute similarities $c_{ij}$, combining them into a single index in the form of a cross product:

$$\sum_i \sum_j c_{ij} w_{ij}$$

This expression is the total obtained by multiplying every cell in the **W** matrix with its corresponding entry in the **C** matrix and summing.

There are different ways of measuring the attribute similarities, $c_{ij}$, depending on whether they are measured on the nominal, ordinal, interval, or ratio scale (see Box 2.1). For nominal data, the usual approach is to set $c_{ij}$ to 1 if $i$ and $j$ take the same attribute value, and zero otherwise. For ordinal data, similarity is usually

based on comparing the ranks of $i$ and $j$. For interval and ratio data, the attribute of interest is denoted $z_i$, and the product $(z_i - \bar{z})(z_j - \bar{z})$ is calculated, where $\bar{z}$ denotes the average of the $z$'s.

One of the most widely used spatial autocorrelation statistics for the case of area objects and interval scale attributes is the Moran Index. This is positive when nearby areas tend to be similar in attributes, negative when they tend to be more dissimilar than one might expect, and approximately zero when attribute values are arranged randomly and independently in space. It is given by the expression:

$$I = \frac{n \sum_i \sum_j w_{ij}(z_i - \bar{z})(z_j - \bar{z})}{\sum_i \sum_j w_{ij} \sum_i (z_i - \bar{z})^2}$$

where $n$ is the number of areal objects in the set. This brief exposition is provided at this point to emphasize the way in which spatial autocorrelation measures are able to accommodate attributes scaled as nominal, ordinal, interval, and ratio data and to illustrate that there is flexibility in the nature of contiguity (or adjacency) relations that may be specified. Further techniques for measuring spatial autocorrelation are reviewed in connection with spatial interpolation in Section 13.3.6.

data is the tendency for relationships to exist between different phenomena at the same location. The interrelatedness of the various properties of a location that together constitute place (Section 1.3) is an important aspect of the nature of geographic data and is key to understanding how the world works (Section 1.3). But it is also a property that defies conventional statistical analysis because most such methods assume zero spatial autocorrelation of sampled observations—in direct contradiction to Tobler's Law.

## 2.8 Taming Geographic Monsters

Thus far in our discussion of the nature of geographic data, we have assumed that spatial variation is smooth and continuous, apart from when we encounter abrupt truncations and discrete shifts at boundaries. However, much spatial variation does not appear to possess these properties, but rather is jagged and apparently irregular. The processes that give rise to the form of a mountain range produce features that are spatially autocorrelated (for example, the highest peaks tend to be clustered), yet it would be wholly inappropriate to represent a mountainscape using smooth interpolation between peaks or between valley troughs.

Jagged irregularity turns out to be a property that is also often observed across a range of scales, and detailed irregularity may resemble coarse irregularity in shape, structure, and form. We commented on this in Section 2.3 when we suggested that a rock broken off a mountain may, for reasons of lithology, represent the mountain in form; this property has been termed *self-similarity*. Urban geographers also recognize that cities and city systems are also self-similar in organization across a range of scales, and Batty and Longley have discussed the ways in which this echoes many of the earlier ideas of Walter Christaller's Central Place Theory. It is unlikely that idealized smooth curves and conventional mathematical functions will provide useful representations for self-similar, irregular spatial structures: at what scale, if any, does it become meaningful to approximate the San Andreas Fault system by a continuous curve? Urban geographers, for example, have long sought to represent the apparent decline in population density with distance from historic central business districts (CBDs) as a continuous curve (Figure 2.8B), yet the three-dimensional profiles of cities (Figure 2.12D) are clearly characterized by urban canyons between irregularly spaced high-rise buildings. Each of these phenomena is characterized by spatial trends (the largest faults, the largest mountains, or the largest skyscrapers each tend to be close to one another), but they are not contiguous and smoothly joined, and the kinds of surface functions shown in Figure 2.8 present inappropriate generalizations of their structure.

For many years, such features were considered geometrical monsters that defied intuition. More recently, however, a more general geometry of the irregular, termed *fractal geometry* by Benoît Mandelbrot, has come to provide a more appropriate and general means of summarizing the structure and character of spatial objects. Fractals can be thought of as geometric objects that are, literally, between Euclidean dimensions, as described in Box 2.7.

---

## Technical Box (2.7)

### The Strange Story of the Lengths of Geographic Objects

How long is the coastline of Maine (Figure 2.14)? (Benoît Mandelbrot, a French mathematician of Polish origin, posed a similar question in 1967 with regard to the coastline of Great Britain.) Consider the stretch of coastline shown in Figure 2.14A. With dividers set to measure 100-km intervals, we would take approximately 4.4 swings and record a length of 340 km (Figure 2.14B).

If we then halved the divider span to measure 50-km swings, we would take approximately 7.1 swings, and the measured length would increase to 355 km (Figure 2.14C). If we halved the divider span once again to measure 25-km swings, we would take approximately 16.6 swings, and the measured length would increase still further to 415 km (Figure 2.14D).

And so on until the divider span was so small that it picked up all of the detail on this particular representation of the coastline. But that would not be the end of the story.

If we were to resort instead to field measurement, using a tape measure or the distance measuring instruments (DMIs) used by highway departments, the length would increase still further, as we picked up wiggles in the coast that even the most detailed maps do not seek to represent. If we were to use dividers, or even microscopic measuring devices, to measure every last grain of sand or earth particle, our recorded length measurement would stretch toward infinity, without apparent limit.

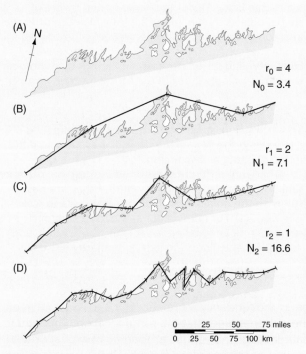

(A) N

$r_0 = 4$
$N_0 = 3.4$

(B)

$r_1 = 2$
$N_1 = 7.1$

(C)

$r_2 = 1$
$N_2 = 16.6$

(D)

| 0 | 25 | 50 | 75 miles |
| 0 | 25 | 50 | 75 | 100 km |

**Figure 2.14** The coastline of Maine, at three levels of recursion: (A) the base curve of the coastline, (B) approximation using 100-km steps, (C) 50-km step approximation, and (D) 25-km step approximation.

**In a self-similar object, each part has the same structure as the whole.**

Ideas from fractal geometry are important, and for many phenomena a measurement of fractal dimension is as important as measures of spatial autocorrelation, or of medians and modes in standard statistics. An important application of fractal concepts is discussed in Section 14.3.1, and we return again to the issue of length estimation in Section 13.3.1. Ascertaining the fractal dimension of an object involves identifying the scaling relation between its length or extent and the yardstick (or level of detail) that is used to measure it. Regression analysis, which can be thought of as putting a best fit line through a scatter of points, provides one (of many) means of establishing this relationship. If we return to the Maine coastline example in Figure 2.14, we might obtain scale-dependent coast-length estimates ($L$) of 14.6 ($4 \times 3.4$), 14.1 ($2 \times 7.1$), and 16.6 ($1 \times 16.6$) units for the step lengths ($r$) used in Figures 2.14B, 2.14C, and 2.14D, respectively. (It is arbitrary whether the steps are measured in kilometers or miles.)

If we then plot the natural log of $L$ (on the $y$-axis) against the natural log of $r$ for these and other pairs of values, we will build up the scatterplot shown in Figure 2.15. If the points lie more or less on a straight line and we fit a trend (regression) line through it, the value of the slope ($b$) parameter is equal to $(1 - D)$, where

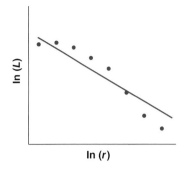

**Figure 2.15** The relationship between recorded length ($L$) and step length ($r$).

$D$ is the fractal dimension of the line. This method for analyzing the nature of geographic lines was originally developed by Lewis Fry Richardson (Box 2.8).

## 2.9 Induction and Deduction and How It All Comes Together

The continual message of this chapter is that spatial is special—that geographic data have a unique nature. Tobler's Law presents an elementary general rule about spatial structure and provides a starting point for the measurement and simulation of spatially autocorrelated structures. This in turn assists us in devising

---

**Biographical Box (2.8)**

### Lewis Fry Richardson

Lewis Fry Richardson (1881–1953; Figure 2.16) was one of the founding fathers of the ideas of scaling and fractals. He was brought up a Quaker and, after earning a degree at Cambridge University, went to work for the Meteorological Office, but his pacifist beliefs forced him to leave in 1920 when the Meteorological Office was militarized under the Air Ministry. His early work on how atmospheric turbulence is related at different scales established his scientific reputation. Later he became interested in the causes of war and human conflict, and in order to pursue one of his investigations, he found that he needed a rigorous way of defining the length of a boundary between two states. Unfortunately, published lengths tended to vary dramatically, a specific instance being the difference between the lengths of the Spanish–Portuguese border as stated by Spain and by Portugal. Richardson developed a method of walking a pair of dividers along a mapped line and analyzed the relationship between the length estimate and the setting of the dividers, finding remarkable predictability. In the

Source: Wikipedia

**Figure 2.16** Lewis Fry Richardson.

1960s Benoît Mandelbrot's concept of fractals finally provided the theoretical framework needed to understand this result.

appropriate spatial sampling schemes and creating improved representations, which tell us still more about the real world and how we might represent it.

Spatial data provide the foundations for operational and strategic applications of GI, and these foundations must be developed creatively, yet rigorously, if they are to support the spatial analysis super-structure that we wish to erect on them. This entails much more than technical competence with software. An understanding of the nature of geographic data allows us to use induction (reasoning from observations) and deduction (reasoning from principles and theory) alongside each other to develop effective spatial representations that are safe to use.

## Questions for Further Study

1. Many jurisdictions tout the number of miles of shoreline in their community—for example, Ottawa County, Ohio, claims 107 miles of Lake Erie shoreline. What does this mean, and how could you make it more meaningful?

2. With reference to Figure 2.11, list the design considerations that should be incorporated into GI software to measure accessibility of (a) a neighborhood medical center to wheelchair-bound pedestrians, (b) a grocery store to high-income customers, and (c) all residential buildings in a small town from a single fire service station.

3. The apparatus of inference was developed by statisticians because they wanted to be able to reason from the results of experiments involving small samples to make conclusions about the results of much larger, hypothetical experiments—for example, in using samples to test the effects of drugs. Summarize the problems inherent in using this apparatus for geographic data, in your own words.

4. "Can geometry deliver what the Greek root of its name [geo-] seemed to promise—truthful measurement, not only of cultivated fields along the Nile River but also of untamed Earth?" Discuss this challenge posed by Mandelbrot, offered in his 2012 autobiography, *The Fractalist: Memoir of a Scientific Maverick* (New York: Pantheon Books, p. xii).

## Further Reading

Batty, M. and Longley, P. A. 1994. *Fractal Cities: A Geometry of Form and Function*. London: Academic Press. Available for download (free of charge) from www.fractalcities.org.

De Smith, M., Goodchild, M. F., and Longley, P. A. 2009. *Geospatial Analysis: A Comprehensive Guide to Principles, Techniques and Software Tools*. Chapter 5: Data Exploration and Spatial Statistics. Leicester: Troubador. See also www.spatialanalysisonline.com.

Goodchild, M. F. 2004. The validity and usefulness of laws in geographic information science and geography. *Annals of the Association of American Geographers* 94(2): 300–304.

Haklay, M. 2010. *Interacting with Geospatial Technologies*. Chichester: Wiley.

Harvey, F. J., Tobler, W., Chrisman, N. R., Frank, A. U., Sui, D., and Goodchild, M. F. 2012. *Are There Fundamental Principles in Geographic Information Science?* www.createspace.com.

Mandelbrot, B. B. 1984. *The Fractal Geometry of Nature*. San Francisco: Freeman.

# Representing Geography

## LEARNING OBJECTIVES

This chapter introduces the concept of representation, or the construction of a digital model of some aspect of the Earth's surface. Representations have many uses, allowing us to learn, think, and reason about places and times that are outside our immediate experience. This is the basis of scientific research, planning, and many forms of day-to-day problem solving.

The geographic world is extremely complex, revealing more detail the closer one looks, seemingly ad infinitum. In order to build a representation of any part of it, it is necessary to make choices about what to represent, at what level of detail, and over what time period. The large number of possible choices creates many opportunities for designers of geographic information (GI) systems.

Generalization methods are used to remove detail that is unnecessary for an application, in order to reduce data volume and speed up operations.

After studying this chapter you will understand:

- What representation is and why it is important.
- The concepts of fields and objects and their fundamental significance.
- What raster and vector representations entail, and how these data structures affect many GI principles, techniques, and applications.
- The similarities and differences between online map services and paper maps.
- Why map generalization methods are important, and how they are based on the concept of representational scale.
- The art and science of representing real-world phenomena in GI databases.

## 3.1 Introduction

We live on the surface of the Earth and spend most of our lives in a relatively small fraction of that space. Of the approximately 500 million sq. km of surface, only one-third is land, and only a fraction of that is occupied by the cities and towns in which most of us live. The rest of the Earth, including the parts we never visit, the atmosphere, and the solid ground under our feet, remains unknown to us except through the information that is communicated to us through books, newspapers, television, the Web, or the spoken word. We live lives that are almost infinitesimal in comparison with the 4.5 billion years of Earth history, or the over 10 billion years since the universe began, and we know about the Earth as it was before we were born only through the evidence compiled by geologists, archaeologists, historians, and other specialists.

Similarly, we know nothing about the world that is to come, where we have only predictions to guide us.

Because we can observe so little of the Earth directly, we rely on a host of methods for learning about its other parts, deciding where to go as tourists or shoppers, choosing where to live, running the operations of corporations, agencies, and governments, and many other activities. Almost all human activities at some time require knowledge (see Section 1.2) about parts of the Earth that are outside our direct experience because they occur either elsewhere in space or elsewhere in time.

Sometimes this knowledge is used as a *substitute* for directly sensed information, creating a *virtual* reality. Increasingly, it is used to *augment* what we can see, touch, hear, feel, and smell through the use of mobile information systems that can be carried around (see Section 12.4.3). Our knowledge of the

Earth is not created entirely freely, but must fit with the mental concepts we began to develop as young children—concepts such as containment (Paris is *in* France) or proximity (Dallas and Fort Worth are *close*). In digital representations, we formalize these concepts through *data models* (Chapter 7), the structures and rules that are programmed into a GI system to accommodate data. These concepts and data models together constitute our *ontologies*, the frameworks that we use for acquiring knowledge of the world.

**Almost all human activities require knowledge about the Earth—past, present, or future.**

One such ontology, a way to structure knowledge of movement through time, is a three-dimensional diagram, in which the two horizontal axes denote a location on the Earth's surface, and the vertical axis denotes time. Figure 3.1 presents the *time space aquarium* that largely contains the activity space of three children living in Cheshunt, UK. The icons on the three trajectories identify the travel modes (walking, cycling, or automobile) that they use as they move through space (the base of the aquarium is rendered using OpenStreetMap: see Box 3.5) and time (the vertical dimension) on a single weekend day. The spatial and temporal granularity of the diagram is set in such a way that even quite small events are recorded, but the granularity could be reduced, for example, by taking only a single GPS signal on

the hour or by recording only those changes in location that exceed a predetermined distance from the preceding point. When we view this diagram on the page of a book, much of the fine detail of the three activity patterns is lost: it is not possible to discern, for example, the precise trajectory of a child playing football on a field, or whether the cyclist dismounts for short sections of an uphill journey, or how long a car waits at traffic lights before continuing its journey. Indeed, some of these details cannot be recovered even if the GPS is set to record changes in location every 5 minutes. Closer perspectives could display more information, but such detail rarely adds much that is useful, and a vast storehouse would be required to capture the precise trajectories of many children throughout even a single day.

The real trajectories of the individuals shown in Figure 3.1 are complex, yet the figure is only a representation of them—a model on a piece of paper or computer screen, generated by a computer from a database. We use the terms *representation* and *model* because they imply a simplified relationship between the contents of the figure and the database and the real-world trajectories of the individuals.

Such representations or models serve many useful purposes and occur in many different forms. For example, representations occur

- in the human mind, when our senses capture information about our surroundings, such as

---

**Figure 3.1** Schematic representation of weekend activities of three children in Cheshunt, UK. The horizontal dimensions represent geographic space (rendered using OpenStreetMap), and the vertical dimension represents time of day. Each person's track plots as a three-dimensional line, beginning at the base in the morning and ending at the top in the evening.

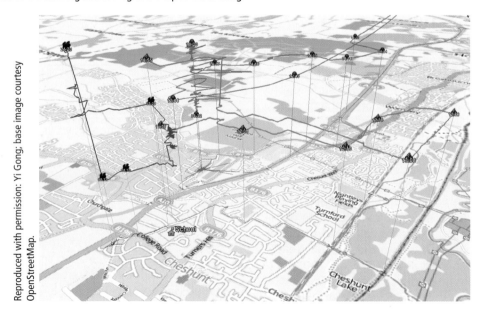

the images captured by the eye, or the sounds captured by the ear, and memory preserves such representations for future use;

- in photographs, which are two-dimensional models of the light emitted or reflected by objects in the world into the lens of a camera;

- in spoken descriptions and written text, in which people describe some aspect of the world in language, in the form of travel accounts or diaries; or

- in the numbers that result when aspects of the world are measured, using such devices as thermometers, rulers, or speedometers.

By building representations, we humans can assemble far more knowledge about our planet than we ever could as individuals. We can build representations that serve such purposes as planning, resource management and conservation, travel, or the day-to-day operations of a parcel delivery service.

**Representations help us assemble far more knowledge about the Earth than is possible on our own.**

Representations are reinforced by the rules and laws that we humans have learned to apply to the unobserved world around us. When we encounter a fallen log in a forest, we are willing to assert that it once stood upright and once grew from a small shoot, even though no one may actually have observed or reported either of these stages. We predict the future occurrence of eclipses based on the laws we have discovered about the motions of the Solar System.

# 3.2 Digital Representation

This book is about one particular form of representation that has become very important in our society—representation in digital form. Today, almost all communication between people through such media as the telephone, Web pages, microblogs, music, television, newspapers and magazines, or e-mail is at some time in its life in digital form. Information technology based on digital representation is moving into all aspects of our lives, from science to commerce to daily existence. The smartphone is the most pervasive digital information-processing device, and it is estimated that as of June 2013 there were 2.1 billion mobile Internet users (29.5% of the global population). Computers are also the mainstay of most office work, and digital technology has pervaded many devices that we use every day, from the microwave oven to the automobile.

One important characteristic of digital technology is that the representation itself is rarely if ever seen by the user because only a few technical experts ever see the individual elements of a digital representation. What we see instead are *views* designed to present the contents of the representation in a form that is meaningful to us. The term *digital* derives from *digits*, or the fingers, and our system of counting based on the 10 digits of the human hand. But although the counting system has 10 symbols (0 through 9), the representation system in digital computers uses only two (0 and 1). In a sense, then, the term *digital* is a misnomer for a system that represents all information using some combination of the two symbols 0 and 1, and the more exact term *binary* is more appropriate. In this book we follow the convention of using *digital* to refer to electronic technology based on binary representations.

**Computers represent phenomena as binary digits. Every item of useful information about the Earth's surface is ultimately reduced by a GI database to some combination of 0s and 1s.**

Over the years many standards have been developed for converting information into digital form. Box 3.1 shows the standards that are commonly used to store GI, whether they consist of whole or decimal numbers or text. There are many competing coding standards for images and photographs (GIF, JPEG, TIFF, etc.) and for movies (e.g., MPEG) and sound (e.g., MIDI, MP3). Much of this book is about the coding systems used to represent geographic data, especially Chapter 7, and as you might guess, this turns out to be quite complicated.

Digital technology is successful for many reasons, not the least of which is that all kinds of information share a common basic format (0s and 1s) and can be handled in ways that are largely independent of their actual meaning (see Box 3.1). The Internet, for example, operates on the basis of packets of information, consisting of strings of 0s and 1s, which are sent through the network based on the information contained in the packet's header. The network needs to know only what the header means and how to read the instructions it contains regarding the packet's destination. The rest of the contents are no more than a collection of bits, representing anything from an e-mail message to a short burst of music or highly secret information on its way from one military installation to another and are almost never examined or interpreted during transmission. This allows one digital communications network to serve every need, from electronic commerce to social networking sites, and it allows manufacturers to build processing and storage technology for vast numbers of users who have very different applications in mind. Compare this to earlier ways of communicating, which required printing presses and delivery trucks for one application (newspapers) and networks of copper wires for another (telephone).

## The Binary Counting System

The binary counting system uses only two symbols, 0 and 1, to represent numerical information. A group of eight binary digits is known as a *byte*, and volume of storage is normally measured in bytes rather than bits (see Table 1.1). There are only two options for a single digit, but there are four possible combinations for two digits (00, 01, 10, and 11), eight possible combinations for three digits (000, 001, 010, 011, 100, 101, 110, 111), and 256 combinations for a full byte. Digits in the binary system (known as binary digits, or *bits*) behave like digits in the decimal system but use powers of two. The rightmost digit denotes units, the next digit to the left denotes twos, the next to the left denotes fours, and so on. For example, the binary number 11001 denotes one unit, no twos, no fours, one eight, and one sixteen and is equivalent to 25 in the normal (decimal) counting system. We call this the *integer* digital representation of 25 because it represents 25 as a whole number and is readily amenable to arithmetic operations. Whole numbers are commonly stored in GI databases using either *short* (2-byte or 16-bit) or *long* (4-byte or 32-bit) options. Short integers can range from –32767 to +32767, and long integers from –2147483647 to +2147483647.

The 8-bit ASCII (American Standard Code for Information Interchange) system assigns codes to each symbol of text, including letters, numbers, and common symbols. The number 2 is assigned ASCII code 50 (00110010 in binary), and the number 5 is 53 (00110101), so if 25 were coded as two characters using 8-bit ASCII its digital representation would be 16 bits long (0011001000110101). The characters 2=2 would be coded as 50, 61, 50 (001100100011110100110010). ASCII is used for coding text, which consists of mixtures of letters, numbers, and punctuation symbols.

Numbers with decimal places are coded using *real* or *floating-point* representations. A number such as 123.456 (three decimal places and six significant digits) is first transformed by powers of 10 so that the decimal point is in a standard position, such as the beginning (e.g., $0.123456 \times 10^3$). The fractional part (0.123456) and the power of 10 (3) are then stored in separate sections of a block of either 4 bytes (32 bits, *single* precision) or 8 bytes (64 bits, *double* precision). This gives enough precision to store roughly 7 significant digits in single precision, or 14 in double precision.

Integer, ASCII, and real conventions are adequate for most data, but in some cases it is desirable to associate images or sounds with places in GI databases, rather than text or numbers. To allow for this, GI system designers have included a BLOB option (standing for binary large object), which simply allocates a sufficient number of bits to store the image or sound, without specifying what those bits might mean.

---

Digital representations of geography hold enormous advantages over previous types—paper maps, written reports from explorers, or spoken accounts. We can use the same cheap digital devices—the components of smartphones, PCs, the Internet, or mass storage devices—to handle every type of information, independent of its meaning. Digital data are easy to copy, they can be transmitted at close to the speed of light, they can be stored at high density in very small spaces, and they are less subject to the physical deterioration that affects paper and other physical media.

Perhaps more important, data in digital form are easy to transform, process, and analyze. GI systems allow us to do things with digital representations that we were never able to do with paper maps: to measure accurately and quickly, to overlay and combine, and to change scale, zoom, and pan without respect to map sheet boundaries. The vast array of possibilities for processing that digital representation opens up is reviewed in Chapters 13 through 15 and is also covered in the applications that appear throughout this book.

**Digital representation has many uses because of its simplicity and low cost.**

## 3.3 Representation of What and for Whom?

Thus far we have seen how humans are able to build representations of the world around them, but we have not yet discussed why representations are useful, and why humans have become so ingenious at creating and sharing them. The emphasis here and throughout the book is on one type of representation, termed *geographic*, and defined as a representation of some part of the Earth's surface or near-surface, at scales ranging from the architectural to the global.

**Geographic representation is concerned with the Earth's surface or near-surface at scales from the architectural to the global.**

Geographic representations are among the most ancient, having their roots in the needs of very early societies. The tasks of hunting and gathering can be much more efficient if hunters are able to communicate the details of their successes to other members of their group—the locations of edible roots or game, for example. Maps must have originated in the sketches early people made in the dirt of campgrounds or on cave walls, long before language became sufficiently sophisticated to convey equivalent information through speech. We know that the peoples of the Pacific built representations of the locations of islands, winds, and currents from simple materials to guide each other, and that social insects such as bees use very simple forms of representation to communicate the locations of food resources.

Hand-drawn maps and speech are effective media for communication between members of a small group, but much wider communication became possible with the invention of the printing press in the Fifteenth Century. Now large numbers of copies of a representation could be made and distributed, and for the first time it became possible to imagine that something could be known by every human being—that knowledge could be the common property of humanity. Only one major restriction affected what could be distributed using this new mechanism: the representation had to be flat. If one were willing to accept that constraint, however, paper proved to be enormously effective; it was cheap, light and thus easily transported, and durable. Only fire and water proved to be disastrous for paper, and human history is replete with instances of the loss of vital information through fire or flood, from the burning of the Alexandria Library in the Seventh Century that destroyed much of the accumulated knowledge of classical times to the major conflagrations of London in 1666, San Francisco in 1906, or Tokyo in 1945, and the flooding of the Arno that devastated Florence in 1966.

One of the most important periods for geographic representation began in the early Fifteenth Century in Portugal and had parallels in China. Henry the Navigator (Box 3.2) is often credited with originating the Age of Discovery, the period of European history that led to the accumulation of large amounts of information about other parts of the world through sea voyages and land explorations. Maps became the medium for sharing information about new discoveries and for administering vast colonial empires, and their value was quickly recognized. Although detailed representations now exist of all parts of the world, including Antarctica, in a sense the spirit of the Age of Discovery continues in the explorations of the oceans, caves, and outer space, and in the constant process of remapping that is needed to keep up with frequent changes in the human and natural worlds.

It was the creation, dissemination, and sharing of accurate representations that distinguished the Age of Discovery from all previous periods in human history (and it would be unfair to ignore its distinctive negative consequences, notably the spread of European diseases and the growth of the slave trade). Information about other parts of the world was assembled in the form of maps and journals, reproduced in large numbers using the recently invented printing press, and distributed on paper. Even the modest costs associated with buying copies were eventually addressed through the development of free public lending libraries in the Nineteenth Century, which gave access to virtually everyone. Today, we benefit from what is now a long-standing tradition of free and open access to much of humanity's accumulated store of knowledge about the geographic world, in the form of paper-based representations, through the institution of libraries and the copyright doctrine that gives people rights to material for personal use (see Chapter 17 for a discussion of laws affecting ownership and access). The Internet is the present-day delivery channel that provides distributed access to geographic information through online virtual Earths (specifically those of Google, Apple, and Microsoft) and other GI services. In several countries, the Open Data (Section 17.4) initiatives of recent years have done much to provide individuals and organizations with the digital materials to produce their own representations, and wide dissemination of open software and GPS-enabled technologies has empowered communities to create mapping representations for general or specific community purposes (see Box 3.5).

**In the Age of Discovery, maps became extremely valuable representations of the state of geographic knowledge.**

Not surprisingly, representation also lies at the heart of our ability to solve problems using the digital tools that are available in GI systems. Any application of GI requires clear attention to questions of *what* should be represented and *how*. There is a multitude of possible ways of representing the geographic world in digital form, none of which is perfect and none of which is ideal for all applications.

**The key GI representation issues are what to represent and how to represent it.**

One of the most important criteria for the usefulness of a representation is its *accuracy*. Because the geographic world is seemingly of infinite complexity, choices are always to be made in building

## Prince Henry the Navigator and Admiral Zheng He

Prince Henry of Portugal (Figure 3.2), who died in 1460, was known as Henry the Navigator because of his keen interest in exploration. In 1433 Prince Henry sent a ship from Portugal to explore the west coast of Africa in an attempt to find a sea route to the Spice Islands. This ship was the first from Europe to travel south of Cape Bojador (latitude 26 degrees 20 minutes North). To make this and other voyages Prince Henry assembled a team of mapmakers, sea captains, geographers, ship builders, and many other skilled craftsmen. Prince Henry showed the way for Vasco de Gama and other famous Fifteenth-Century explorers. His management skills could be applied in much the same way in today's GI projects.

Admiral Zheng was born in 1371 in what is now China's Yunnan Province. In a series of seven expeditions between 1405 and 1433, he explored the coasts of Thailand, Indonesia, India, Arabia, and East Africa, using massive fleets of up to 200 ships (Figure 3.3), mapping, trading, and settling Chinese along the way. His last two voyages were the most extensive, and speculation persists about the areas that he might have discovered (see the book *1421: The Year the Chinese Discovered America* by Gavin Menzies). Unfortunately, the Ming Emperor destroyed the records of these expeditions.

**Figure 3.2** Prince Henry the Navigator, originator of the Age of Discovery in the Fifteenth Century and promoter of a systematic approach to the acquisition, compilation, and dissemination of geographic knowledge.

(A)

(B)

**Figure 3.3** (A) Admiral Zheng and (B) two of his ships.

any representation—what to include and what to leave out. When U.S. President Thomas Jefferson dispatched Meriwether Lewis to explore and report on the nature of the lands from the upper Missouri to the Pacific, he said Lewis possessed "a fidelity to the truth so scrupulous that whatever he should report would be as certain as if seen by ourselves." But Jefferson clearly did not expect Lewis to report everything he saw in complete detail: Lewis exercised a large amount of judgment about what to report and what to omit. (The related question of the accuracy of what is reported is taken up at length in Chapter 5.)

One more vital interest drives our need for representations of the geographic world, as well as the need for representations in many other human activities. When a pilot must train to fly a new type of aircraft, it is much cheaper and less risky for him or her to work with a flight simulator than with the real aircraft. Flight simulators can represent a much wider range of conditions than a pilot will normally experience in flying. Similarly, when decisions have to be made about the geographic world, it is effective to experiment first on models or representations, exploring different scenarios. Of course, this works only if the representation behaves as the real aircraft or world does, and a great deal of knowledge must be acquired about the world before an accurate representation can be built that permits such simulations. But the use of representations for training, exploring future scenarios, and for re-creating the past is now common in many fields, including surgery, chemistry, and engineering, and with GI technologies it is becoming increasingly common in dealing with the geographic world.

**Many plans for the real world can be tried out first on models or representations.**

## 3.4 The Fundamental Problem

Geographic data are built up from atomic elements or from facts about the geographic world. At its most primitive, an atom of geographic data (strictly, a datum) links a place, often a time, and some descriptive property. The first of these, place, is discussed at greater length in Chapter 4, and there are also many ways of specifying the second, time. We often use the term *attribute* to refer to the last of these three descriptive properties. For example, consider the statement "The temperature at local noon on December 2, 2014, at latitude 34 degrees 45 minutes north, longitude 120 degrees 0 minutes west, was 19 degrees Celsius." It ties location and time to the property or attribute of atmospheric temperature.

**Geographic data link place, time, and attributes.**

Other facts can be broken down into their primitive atoms. For example, the statement "Mount Everest is 8848 m high" can be derived from two atomic geographic facts, one giving the location of Mount Everest in latitude and longitude, and the other giving the elevation at that latitude and longitude. Note, however, that the statement would not be a geographic fact to a community that had no way of knowing where Mount Everest is located.

Many aspects of the Earth's surface are comparatively static and slow to change. Height above sea level changes slowly because of erosion and the movements of the Earth's crust, but these processes operate on scales of hundreds or thousands of years, and for most applications except geophysics we can

safely omit time from the representation of elevation. In contrast, atmospheric temperature changes daily, and dramatic changes sometimes occur in minutes with the passage of a cold front or thunderstorm. Thus time is distinctly important, though such climatic variables as mean annual temperature can be represented as static over periods of a decade or two.

There is a vast range of attributes in geographic information. We have already seen that some attributes vary slowly and some rapidly. Some attributes are physical or environmental in nature, whereas others are social or economic. Some simply *identify* a place or an entity, distinguishing it from all other places or entities; examples include street addresses, social security numbers, or the parcel numbers used for recording land ownership. Other attributes measure something at a location and perhaps at a time (e.g., atmospheric temperature or elevation), whereas others classify into categories (e.g., the class of land use, differentiating between agriculture, industry, or residential land). The standard terms for the different types of attributes were discussed in Box 2.1.

But this idea of recording atoms of geographic information, combining location, time, and attribute, misses a fundamental problem, which is that the world is in effect infinitely complex, and the number of atoms required for a complete representation is similarly infinite. The closer we look at the world, the more detail it reveals—and it seems that this process extends ad infinitum. The shoreline of Maine appears complex on a map, but even more complex when examined in greater detail, and as more detail is revealed the shoreline appears to get longer and longer, and more and more convoluted (see Figure 2.14).

To characterize the world completely, we would have to specify the location of every person, every blade of grass, and every grain of sand—in fact, every subatomic particle, which is clearly an impossible task, because the Heisenberg Uncertainty Principle places limits on the ability to measure precise positions of subatomic particles. Thus in practice any representation must be partial—it must limit the level of detail provided, or ignore change through time, or ignore certain attributes, or simplify in some other way.

**The world is infinitely complex, but computer systems are finite. Representations must somehow limit the amount of detail captured.**

One very common way of limiting detail is to throw away or ignore information that applies only to small areas—in other words not look too closely. The image you see on a computer screen is composed of a million or so picture elements or *pixels*, and if the whole Earth were displayed at once, each pixel would cover an area roughly 10 km on a side, or about 100 sq km.

Courtesy: NOAA: Liam Gumley, MODIS Atmosphere Group, University of Wisconsin-Madison

**Figure 3.4** An image from NASA's Terra satellite showing a large plume of smoke streaming southward from the remnants of the burning World Trade Towers in downtown Manhattan on September 11, 2001. The image was acquired using the Moderate-Resolution Imaging Spectroradiometer (MODIS), which has a spatial resolution of about 250 m. The red pixels in this scene show the location of vegetation. Light blue-white pixels show where there are concrete surfaces. The slightly darker blue pixels streaming southward toward the New Jersey coast are the smoke from the destroyed World Trade Center towers. The large black areas to the east are the waters of the Atlantic. The Hudson River runs northward just west of Manhattan Island.

At this level of detail the island of Manhattan occupies roughly 10 pixels, and virtually everything on it is a blur. We would say that such an image has a *spatial resolution* of about 10 km and know that anything much less than 10 km across is virtually invisible. Figure 3.4 shows Manhattan at a spatial resolution of 250 m, detailed enough to pick out the shape of the island and Central Park.

It is easy to see how this helps with the problem of too much information. The Earth's surface covers about 500 million sq km, so if this level of detail (a spatial resolution of 10 km) is sufficient for an application, a property of the surface such as elevation can be described with only 5 million pieces of information, instead of the 500 million it would take to describe elevation with a resolution of 1 km, and the 500 trillion (500,000,000,000,000) it would take to describe elevation with 1-m resolution.

Another strategy for limiting detail is to observe that many properties remain constant over large areas. For example, in describing the elevation of the Earth's surface we could take advantage of the fact that roughly two-thirds of the surface is covered by water, with its surface at sea level. Of the 5 million pieces of information needed to describe elevation at 10-km resolution, approximately 3.4 million will be recorded

as zero, a colossal waste. If we could find an efficient way of identifying the area covered by water, then we would need only 1.6 million real pieces of information.

Humans have found many ingenious ways of describing the Earth's surface efficiently because the problem we are addressing is as old as representation itself and as important for paper-based representations as it is for binary representations in computers. But this ingenuity is itself the source of a substantial problem for GI: there are many ways of representing the Earth's surface, and users of GI thus face difficult and at times confusing choices. This chapter discusses some of those choices, and the issues are pursued further in subsequent chapters on uncertainty (Chapter 5) and data modeling (Chapter 7). Representation remains a major concern of GI science, and researchers are constantly looking for ways to extend GI representations to accommodate new types of information.

## 3.5 Discrete Objects and Continuous Fields

### 3.5.1 Discrete Objects

The level of detail as a fundamental choice in representation has already been mentioned. Another, perhaps even more fundamental, choice is between two conceptual schemes. There is good evidence that we as humans like to simplify the world around us by naming things and by seeing individual things as instances of broader categories. We prefer a world of black and white, of good guys and bad guys, to the real world of shades of gray.

> **The two fundamental ways of representing geography are discrete objects and continuous fields.**

This preference is reflected in one way of viewing the geographic world, known as the *discrete object* view. In this view, the world is empty, except where it is occupied by objects with well-defined boundaries that are instances of generally recognized categories. Just as the desktop is littered with books, pencils, or computers, the geographic world is littered with cars, houses, lampposts, and other discrete objects. In a similar vein the landscape of Minnesota is littered with lakes and that of Scotland with mountains. One characteristic of the discrete object view is that objects can be counted, so license plates issued by the State of Minnesota carry the legend "10,000 lakes," and climbers know that there are exactly 284 mountains in Scotland over 3,000 ft (the so-called Munros, from Sir Hugh Munro, who originally listed 277 of them in 1891—the count was expanded to 284 in 1997).

**The discrete object view represents the geographic world as objects with well-defined boundaries in otherwise empty space.**

Biological organisms fit this model well, allowing us to count the number of residents in an area of a city or to describe the behavior of individual bears. Manufactured objects also fit the model, and we have little difficulty counting the number of cars produced in a year or the number of airplanes owned by an airline. But other phenomena are messier. It is not at all clear what constitutes a mountain, for example, or exactly how a mountain differs from a hill, or when a mountain with two peaks should be counted as two mountains.

Geographic objects are identified by their dimensionality. We saw in Box 2.2 that GI objects are conceived as filling zero (points), one (lines), two (areas), and three (volumes) dimensions, and that fractal geometry allowed us to think of objects as filling intermediate amounts of space between these integer dimensions (Section 2.8). Of course, in reality, all objects that are perceptible to humans are three dimensional, and their representation in fewer dimensions can be at best an approximation. But the ability of GI systems to handle truly three-dimensional objects as volumes with associated surfaces remains limited. GI databases increasingly allow for a third (vertical) coordinate to be specified for all point locations. Buildings are sometimes represented by assigning height as an attribute, though if this option is used it is impossible to distinguish flat roofs from any other kind. Various strategies have been used for representing overpasses and underpasses in transportation networks because this information is vital for navigation but not normally present in strictly two-dimensional network representations. One common strategy is to represent turning options at every intersection, so that an overpass appears in the database as an intersection with no turns (Figure 3.5).

The discrete object view leads to a powerful way of representing geographic information about

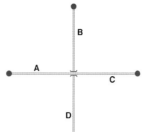

| From link | To link | Turn? |
|-----------|---------|-------|
| A | B | No |
| A | C | Yes |
| A | D | No |
| B | C | No |
| B | D | Yes |
| B | A | No |
| C | D | No |
| C | A | Yes |
| C | B | No |
| D | A | No |
| D | B | Yes |
| D | C | No |

**Figure 3.5** The problems of representing a three-dimensional world using a two-dimensional technology. The intersection of links A, B, C, and D is an overpass, so no turns are possible between such pairs as A and B.

objects. Think of a class of objects of the same dimensionality—for example, all the brown bears (Figure 3.6) in the Kenai Peninsula of Alaska. We would naturally think of these objects as points. We might want to know the sex of each bear and its date of birth, if our interests were in monitoring the bear population. We might also have a collar on each bear that transmitted the bear's location at regular intervals. All this information could be expressed in a table, such as the one shown in Table 3.1, with each

**Figure 3.6** Bears are easily conceived as discrete objects, maintaining their identities as objects through time and surrounded by empty space.

**Table 3.1** Example of representation of geographic information as a table: the locations and attributes of each of four brown bears in the Kenai Peninsula of Alaska. Locations have been obtained from radio collars. Only one location is shown for each bear, at noon on July 31, 2014 (imaginary data).

| Bear ID | Sex | Estimated year of birth | Date of collar installation | Location, noon on 31 July 2014 |
|---------|-----|-------------------------|-----------------------------|--------------------------------|
| 001 | M | 2008 | 02242009 | −150.6432, 60.0567 |
| 002 | F | 2006 | 03312009 | −149.9979, 59.9665 |
| 003 | F | 2013 | 04212009 | −150.4639, 60.1245 |
| 004 | F | 2010 | 04212009 | −150.4692, 60.1152 |

row corresponding to a different discrete object and each column to an attribute of the object. To reinforce a point made earlier, this is a very efficient way of capturing raw geographic information on brown bears.

But it is not perfect as a representation for all geographic phenomena. Imagine visiting the Earth from another planet and asking the humans what they chose as a representation for the infinitely complex and beautiful environment around them. The visitor would hardly be impressed to learn that they chose tables, especially when the phenomena represented were natural phenomena such as rivers, landscapes, or oceans. Nothing on the natural Earth looks remotely like a table. It is not at all clear how the properties of a river or the properties of an ocean should be represented as a table. So although the discrete object view works well for some kinds of phenomena, it misses the mark badly for others.

## 3.5.2 Continuous Fields

Although we might think of terrain as composed of discrete mountain peaks, valleys, ridges, slopes, and the like and might list them in tables and count them, there are unresolvable problems of definition for all these objects. Instead, it is much more useful to think of terrain as a continuous surface in which elevation can be defined rigorously at every point (see Box 3.3). Such continuous surfaces form the basis of the other common view of geographic phenomena, known as the *continuous field* view (not to be confused with other meanings of the word field). In this view the geographic world can be described by a number of *variables*, each measurable at any point on the Earth's surface and changing in value across the surface.

> **The continuous field view represents the real world as a finite number of variables, each one defined at every possible position.**

Objects are distinguished by their dimensions and naturally fall into categories of points, lines, or areas. Continuous fields, on the other hand, can be distinguished by what varies and how smoothly. A continuous field of elevation, for example, varies much more smoothly in a landscape that has been worn down by glaciation or flattened by blowing sand than one recently created by cooling lava. Cliffs are places in continuous fields where elevation changes suddenly rather than smoothly. Population density is a kind of continuous field, defined everywhere as the number of people per unit area, though the definition breaks down if the field is examined so closely that the individual people become visible. Continuous fields can also be created from classifications of land into categories of land use or soil type. Such fields change suddenly at the boundaries between different classes. Other types of fields can be defined by continuous variation along lines rather than across space. Traffic density, for example, can be defined everywhere on a road network, and flow volume can be defined everywhere on a river. Figure 3.7 shows some examples of field-like phenomena.

Continuous fields can be distinguished by *what* is being measured at each point. As described in Box 2.1, the variable may be nominal, ordinal, interval, ratio, or cyclic. A *vector* field assigns two variables, magnitude and direction, at every point in space and is used to represent flow phenomena such as winds or currents; fields of only one variable are termed *scalar* fields.

Here is a simple example illustrating the difference between the discrete object and field conceptualizations. Suppose you were hired for the summer to count the number of lakes in Minnesota and were promised that your answer would appear on every license plate issued by the state. The task sounds simple, and you were happy to get the job. But on the first day you started to run into difficulty (Figure 3.8). What about small ponds—do they count as lakes? What about wide stretches of rivers? What about swamps that dry up in the summer? And is a lake with a narrow section connecting two wider parts one lake or two? Your biggest dilemma concerns the scale of

---

### Technical Box (3.3)

#### Dimensions

Areas are two-dimensional objects, and volumes are three dimensional, but GI users sometimes talk about 2.5-D. Almost without exception the elevation of the Earth's surface has a single value at any location (exceptions include overhanging cliffs or caves). So elevation is conveniently thought of as a continuous field, a variable with a value everywhere in two dimensions, and a full 3-D representation is only necessary in areas with an abundance of overhanging cliffs or caves, if these are important features. The idea of dealing with a three-dimensional phenomenon by treating it as a single-valued function of two horizontal variables gives rise to the term 2.5-D. Figure 3.7B shows an example, in this case an elevation surface.

(A)

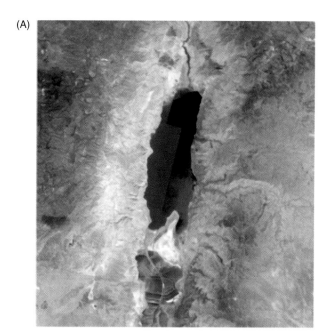

(B)

Courtesy: NASA/JPL–Caltech

**Figure 3.7** Examples of field-like phenomena. (A) Image of part of the Dead Sea in the Middle East. The lightness of the image at any point measures the amount of radiation captured by the satellite's imaging system. (B) A simulated image derived from the Shuttle Radar Topography Mission. The image shows the Carrizo Plain area of Southern California, with a simulated sky and with land cover obtained from other satellite sources.

mapping because the number of lakes shown on a map clearly depends on the map's level of detail; a more detailed map almost certainly will show more lakes.

Your task clearly reflects a discrete object view of the phenomenon. The action of counting implies that lakes are discrete, two-dimensional objects littering an otherwise empty geographic landscape. In a continuous field view, however, all points are either lake or nonlake. Moreover, we could refine the scale a little to take account of marginal cases; for example, we might define the scale shown in Table 3.2, which

**Figure 3.8** Lakes (this one is in Helsinki, Finland) are difficult to conceptualize as discrete objects because it is often difficult to tell where a lake begins and ends, or to distinguish a wide river from a lake.

has five degrees of lakeness. The complexity of the view would depend on how closely we looked, of course, and so the scale of mapping would still be important. But all the problems of defining a lake as a discrete object would disappear (though there would still be problems in defining the levels of the scale). Instead of counting, our strategy would be to lay a grid over the map and assign each grid cell a score on the lakeness scale. The size of the grid cell would determine how accurately the result approximated the value we could theoretically obtain by visiting every one of the infinite number of points in the state. At the end, we would tabulate the resulting scores, counting the number of cells having each value of lakeness or averaging the lakeness score. We could even design a new and scientifically more reasonable license plate— "Minnesota, 12% lake" or "Minnesota, average lakeness 2.02." The vagaries of defining such objects in Finland (Figure 3.8), another lake-strewn territory, form

**Table 3.2** A scale of lakeness suitable for defining lakes as a continuous field.

| Lakeness | Definition |
| --- | --- |
| 1 | Location is always dry under all circumstances. |
| 2 | Location is sometimes flooded in spring. |
| 3 | Location supports marshy vegetation. |
| 4 | Water is always present to a depth of less than 1 m. |
| 5 | Water is always present to a depth of more than 1 m. |

the backdrop to some of the work of cartographer and GI scientist Kirsi Virrantaus (see Box 3.4).

The difference between objects and fields is also well illustrated by photographs. Paper images produced from old-fashioned photographic film are created by variation in the chemical state of the material in the photographic film. In early photography, minute particles of silver were released from molecules of silver nitrate when the unstable molecules were exposed to light, thus darkening the image in proportion to the amount of incident light. We think of the image as a field of continuous variation in color or darkness. But when we look at the image, the eye and brain begin to infer the presence of discrete objects, such as people, rivers, fields, cars, or houses, as they interpret the content of the image.

A continuous field view still potentially contains an infinite amount of information if it defines the value of the variable at every point because there are an infinite number of points in any defined geographic area. Discrete objects can also require an infinite amount of information for full description. As we saw in Section 2.8, a coastline contains an infinite amount of information if it is mapped in infinite detail. Thus continuous fields and discrete objects are no more than conceptualizations, or ways in which we think about geographic phenomena; they are not designed to deal with the limitations of computers.

Two methods are used to reduce geographic phenomena to forms that can be coded in computer databases, and we call these methods raster and vector. In principle, both can be used to code both fields and discrete objects, but in practice a strong association exists between raster and fields, and between vector and discrete objects.

## 3.6 Rasters and Vectors

Continuous fields and discrete objects define two conceptual views of geographic phenomena, but they do not solve the problem of digital representation.

**Raster and vector are two methods of representing geographic data in digital computers.**

# 3.6.1 Raster Data

In a raster representation space is divided into an array of rectangular (usually square) cells (Figure 3.10). All geographic variation is then expressed by assigning properties or attributes to these cells. The cells are sometimes called pixels (short for *picture elements*).

> **Raster representations divide the world into arrays of cells and assign attributes to the cells.**

One of the most common forms of raster data comes from remote-sensing satellites, which capture information in this form and send it to the ground to be distributed and analyzed. Data from the Landsat Thematic Mapper, for example, which are commonly used in GI applications, come in cells that are 30 m a side on the ground, or approximately 0.1 ha (hectare) in area. Similar data can be obtained from sensors mounted on aircraft. Imagery varies according to the spatial resolution (expressed as the length of a cell side as measured on the ground) and also according to the timetable of image capture by the sensor. Some satellites are in *geostationary* orbit over a fixed point on the Earth and capture images constantly. Others pass over a fixed point at regular intervals (e.g., every 12 days). Finally, sensors vary according to the part or parts of the spectrum that they sense. The visible parts of the spectrum are most important for remote sensing, but some invisible parts of the spectrum are particularly useful in detecting heat and the phenomena that produce heat, such as volcanic activities. Many sensors capture images in several areas of the spectrum, or *bands*, simultaneously because the relative amounts of radiation in different parts of the spectrum are often useful indicators of certain phenomena, such as green leaves, or water, on the Earth's surface. The AVIRIS (Airborne Visible InfraRed Imaging Spectrometer) captures no fewer than 224 different parts of the spectrum and is being used to detect particular minerals in the soil, among other applications. Remote sensing is a complex topic, and further details are available in Chapter 8.

Square cells fit together nicely on a flat table or a sheet of paper, but they will not fit together neatly on the curved surface of the Earth. So just as representations on paper require that the Earth be flattened, or projected, so too do rasters. (Because of the distortions associated with flattening, the cells in a raster can never be perfectly equal in shape or area on the Earth's surface.) Projections, or ways of flattening the Earth, are described in Section 4.8. Many of the terms that describe rasters suggest the laying of a tile floor on a flat surface—we talk of raster cells *tiling* an area, and a raster is said to be an instance of a *tessellation*, derived from the word for a mosaic. The mirrored ball hanging above a dance floor recalls the impossibility of covering a spherical object like the Earth perfectly with flat, square pieces.

When information is represented in raster form, all detail about variation within cells is lost, and instead the cell is given a single value. Suppose we wanted to represent the map of the counties of Texas as a raster. Each cell would be given a single value to identify a county, and we would have to decide on the rule to apply when a cell falls in more than one county. Often the rule is that the county with the *largest share* of the cell's area gets the cell. Sometimes the rule is based on the *central point* of the cell, and the county at that point is assigned to the whole cell. Figure 3.11 shows these two rules in operation. The largest-share rule is almost always preferred, but the central-point rule is sometimes used in the interests of faster computing and is often used in creating raster datasets of elevation.

**Figure 3.11** Effect of a raster representation using (A) the largest-share rule and (B) the central-point rule.

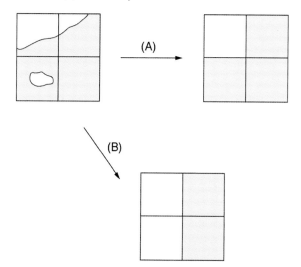

**Figure 3.10** Raster representation. Each color represents a different value of a nominal-scale variable denoting land cover class.

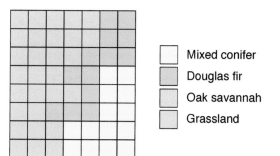

Mixed conifer
Douglas fir
Oak savannah
Grassland

## 3.6.2 Vector Data

In a vector representation, all lines are captured as points connected by precisely straight lines. (Some GI systems allow points to be connected by curves rather than straight lines, but in most cases curves have to be approximated by increasing the density of points.) An area is captured as a series of points or *vertices* connected by straight lines as shown in Figure 3.12. The straight edges between vertices explain why areas in vector representation are often called *polygons*, and in GI-speak the terms *polygon* and *area* are often used interchangeably. Lines are captured in the same way, and the term *polyline* has been coined to describe a curved line represented by a series of straight segments connecting vertices.

To capture an area object in vector form, we need only specify the locations of the points that form the vertices of a polygon. This seems simple and is also much more efficient than a raster representation, which would require us to list all the cells that form the area. These ideas are captured succinctly in the comment "Raster is vaster, but vector is correcter." To create a precise approximation to an area in raster, it would be necessary to resort to using very small cells, and the number of cells would rise proportionately (with every

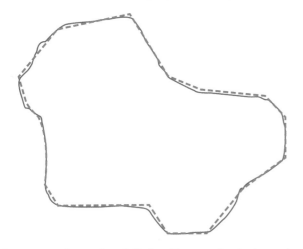

**Figure 3.12** An area (purple line) and its approximation by a polygon (dashed blue line).

halving of the width and height of each cell resulting in a quadrupling of the number of cells). But things are not quite as simple as they seem. The apparent precision of vector is often not real because many geographic phenomena simply cannot be located with high accuracy (see Section 5.3). So although raster data may look less attractive, they may be more honest to the inherent quality of the data. Also, various methods exist for compressing raster data that can greatly reduce the capacity needed to store a given dataset (see Chapter 7). So the choice between raster and vector is often complex, as summarized in Table 3.3.

## 3.6.3 Representing Continuous Fields

Although discrete objects lend themselves naturally to representation as points, lines, or areas using vector methods, it is less obvious how the continuous variation of a field can be expressed in a digital representation. GI systems commonly implement six alternatives (Figure 3.13):

A. Capturing the value of the variable at each of a grid of regularly spaced sample points (for example, elevations at 30-m spacing in a DEM).

B. Capturing the value of the field variable at each of a set of irregularly spaced sample points (for example, variation in surface temperature captured at weather stations).

C. Capturing a single value of the variable for a regularly shaped cell (for example, values of reflected radiation in a remotely sensed scene).

D. Capturing a single value of the variable over an irregularly shaped area (for example, vegetation cover class or the name of a parcel's owner).

E. Capturing the linear variation of the field variable over an irregularly shaped triangle (for example, elevation captured in a triangulated irregular network or TIN, see Section 7.2.3.4).

F. Capturing the isolines of a surface as digitized lines (for example, digitized contour lines representing surface elevation).

**Table 3.3** Relative advantages of raster and vector representation.

| Issue | Raster | Vector |
|---|---|---|
| Volume of data | Depends on cell size | Depends on density of vertices |
| Sources of data | Remote sensing, imagery | Social and environmental data |
| Applications | Resources, environmental | Social, economic, administrative |
| Software | Raster GI systems, image processing | Vector GI systems, automated cartography |
| Resolution | Fixed | Variable |

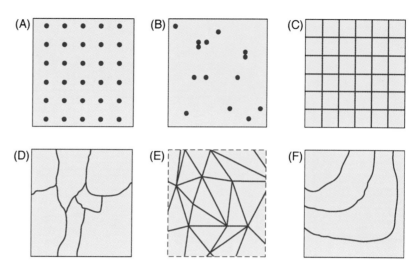

**Figure 3.13** The six approximate representations of a field used in GI systems. (A) Regularly spaced sample points. (B) Irregularly spaced sample points. (C) Rectangular cells. (D) Irregularly shaped polygons. (E) Irregular network of triangles, with linear variation over each triangle (the Triangulated Irregular Network or TIN model; the bounding box is shown dashed in this case because the unshown portions of complete triangles extend outside it). (F) Polylines representing contours (see the discussion of isopleth maps in Box 2.3).

(A) (B) (C) (D) (E) (F)

Each of these methods succeeds in compressing the potentially infinite amount of data in a continuous field to a finite amount, using one of the six options, two of which (A and C) are raster and four (B, D, E, and F) are vector. Of the vector methods one (B) uses points, two (D and E) use polygons, and one (F) uses lines to express the continuous spatial variation of the field in terms of a finite set of vector objects. But unlike the discrete object conceptualization, the objects used to represent a field are not real, but simply artifacts of the representation of something that is actually conceived as spatially continuous. The triangles of a TIN representation (E), for example, exist only in the digital representation and cannot be found on the ground, and neither can the lines of a contour representation (F).

Representation also affects the conception of spatial autocorrelation (Section 2.3). If a phenomenon of interest is conceived as a field, then spatial autocorrelation measures the smoothness of the field using data from the sample points, lines, or areas that represent the field. If the phenomenon is conceived as a set of discrete objects, then spatial autocorrelation measures how the attribute values are distributed among the objects, distinguishing between arrangements that are clustered, random, and locally contrasting.

## 3.7 The Paper Map

The paper map has long been a powerful and an effective means of communicating geographic information. In contrast to digital data, which use coding schemes such as ASCII, it is an instance of an *analog* representation, or a physical model in which the real world is scaled; in the case of the paper map, part of the world is scaled to fit the size of the paper. A key property of a paper map is its *scale* or *representative fraction*, defined as the ratio of distance on the map to distance on the Earth's surface. For example,

a map with a scale of 1:24,000 reduces everything on the Earth to one-24,000th of its real size. This is a bit misleading because the Earth's surface is curved and a paper map is flat, so scale cannot be exactly constant.

**A paper map is a source of data for geographic databases, an analog product from a GI system, and an effective communication tool.**

Maps have been so important, particularly prior to the development of digital technology, that many of the ideas associated with GI are actually inherited directly from paper maps. For example, scale is often cited as a property of a digital database, even though the definition of scale makes no sense for digital data—ratio of distance *in the computer* to distance on the ground; how can there be distances in a computer? What is meant is a little more complicated: when a scale is quoted for a digital database, it is usually the scale of the map that formed the source of the data. So if a database is said to be at a scale of 1:24,000, one can safely assume that it was created from a paper map at that scale and includes representations of the features that are found on maps at that scale. The many meanings of scale were discussed in Box 2.2, and we discuss the importance of scale to the concept of uncertainty in Chapter 5.

There is a close relationship between the contents of a map and the raster and vector representations discussed in the previous section. The U.S. Geological Survey, for example, distributes two digital versions of its topographic maps, one in raster form and one in vector form, and both attempt to capture the contents of the map as closely as possible. In the raster form, or *digital raster graphic* (DRG), the map is scanned at a very high density, using very small pixels, so that the raster looks very much like the original (Figure 3.14). The coding of each pixel simply records

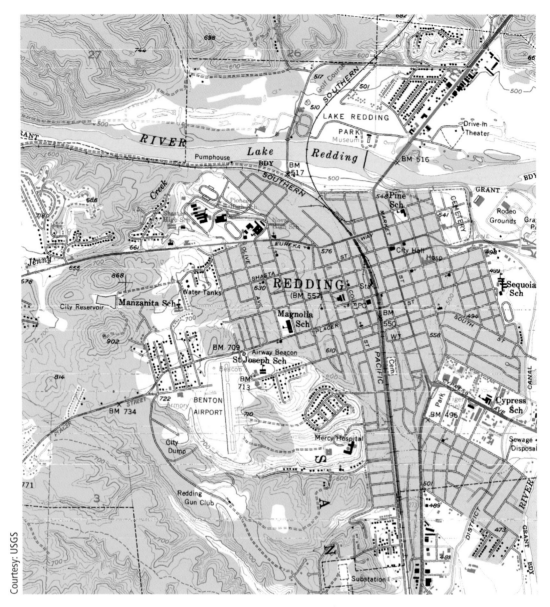

**Figure 3.14** Part of a digital raster graphic, a scan of a U.S. Geological Survey 1:24 000 topographic map.

the color of the map picked up by the scanner, and the dataset includes all the textual information surrounding the actual map.

In the vector form, or *digital line graph* (DLG), every geographic feature shown on the map is represented as a point, polyline, or polygon. The symbols used to represent point features on the map, such as the symbol for a windmill, are replaced in the digital data by points with associated attributes and must be regenerated when the data are displayed. Contours, which are shown on the map as lines of definite width, are replaced by polylines of no width and given attributes that record their elevations. In both cases, and especially in the vector case, there is a significant difference between the analog representation of the map and its digital equivalent.

So it is quite misleading to think of the contents of a digital representation as a map and to think of a GI database as a container of digital maps. Digital representations can include information that would be very difficult to show on maps. For example, they can represent the curved surface of the Earth, without the need for the distortions associated with flattening. They can represent changes, whereas maps must be static because it is very difficult to change their contents once they have been printed or drawn. Digital databases can represent all three spatial dimensions, including the vertical, whereas maps must always show two-dimensional views. So although the paper map is a useful metaphor for the contents of a geographic database, we must be careful not to let it limit our thinking about what is possible in the way of representation. This issue is pursued at greater length in Chapter 7, and map production is discussed in detail in Chapter 11.

# 3.8 Generalization

In Section 3.4 we saw how thinking about geographic information as a collection of atomic links—between a place, a time (not always, because many geographic facts are stated as if they were permanently true), and a property—led to an immediate problem because the potential number of such atomic facts is infinite. If seen in enough detail, the Earth's surface is unimaginably complex, and its effective description impossible. So instead, humans have devised numerous ways of simplifying their view of the world. Rather than making statements about each and every point, we describe entire areas, attributing uniform characteristics to them, even when areas are not strictly uniform; we identify features on the ground and describe their characteristics, again assuming them to be uniform; or we limit our descriptions to what exists at a finite number of sample points, hoping that these samples will be adequately representative of the whole (Section 2.4).

**A geographic database cannot contain a perfect description; instead, its contents must be carefully selected to fit within the limited capacity of computer storage devices.**

## 3.8.1 Generalization about Places

From this perspective some degree of generalization is almost inevitable in all geographic data. But cartographers often take a somewhat different approach, for which this observation is not necessarily true. Suppose we are tasked to prepare a map at a specific scale, say 1:25,000, using the standards laid down by a national mapping agency, such as the Institut Géographique National (IGN) of France. Every scale used by IGN has its associated rules of representation. For example, at a scale of 1:25,000 the rules specify that individual buildings will be shown only in specific circumstances, and similar rules apply to the 1:24,000 series of the U.S. Geological Survey (see Figure 3.14). These rules are known by various names, including *terrain nominal* in the case of IGN, which translates roughly but not very helpfully to "nominal ground," and is perhaps better translated as "specification." From this perspective a map that represents the world by following the rules of a specification precisely can be perfectly accurate *with respect to the specification*, even though it is not a perfect representation of the full detail on the ground.

**A map's specification defines how real features on the ground are selected for inclusion on the map.**

Consider the representation of vegetation cover using the rules of a specification. For example, the rules might state that at a scale of 1:100,000, a vegetation cover map should not show areas of vegetation that cover less than 1 ha (hectare). But small areas of vegetation almost certainly exist, so deleting them inevitably results in information loss. But under the principle discussed earlier, a map that adheres to this rule must be accurate, *even though it differs substantively from the truth as observed on the ground.*

The level of detail of a GI dataset is one of its most important properties, as it determines both the degree to which the dataset approximates the real world and the dataset's complexity. In the interests of compressing data, it is often necessary to remove detail, fitting them into a storage device of limited capacity, processing them faster, or creating less confusing visualizations that emphasize general trends. Consequently, many methods have been devised for generalization, and several of the more important are discussed in this section.

McMaster and Shea identify the following types of generalization rules:

- *Simplification*, for example, by weeding out points in the outline of a polygon to create a simpler shape
- *Smoothing*, or the replacement of sharp and complex forms by smoother ones
- *Collapse*, or the replacement of an area object by a combination of point and line objects
- *Aggregation*, or the replacement of a large number of distinct symbolized objects by a smaller number of new symbols
- *Amalgamation*, or the replacement of several area objects by a single area object
- *Merging*, or the replacement of several line objects by a smaller number of line objects
- *Refinement*, or the replacement of a complex pattern of objects by a selection that preserves the pattern's general form
- *Exaggeration*, or the relative enlargement of an object to preserve its characteristics when these would be lost if the object were shown to scale
- *Enhancement*, through the alteration of the physical sizes and shapes of symbols
- *Displacement*, or the moving of objects from their true positions to preserve their visibility and distinctiveness

The differences between these types of rules are much easier to understand visually, and Figure 3.15 is based upon McMaster's and Shea's

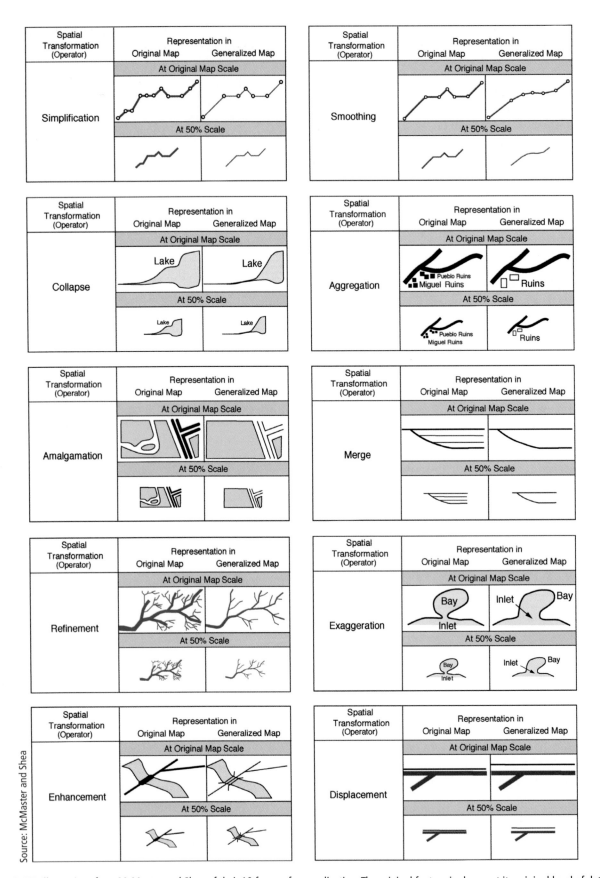

**Figure 3.15** Illustrations from McMaster and Shea of their 10 forms of generalization. The original feature is shown at its original level of detail, and below it at 50% coarser scale. Each generalization technique resolves a specific problem of display at coarser scale and results in the acceptable version shown in the lower right.

illustrative drawings. In addition, the drawings describe two forms of generalization of attributes, as distinct from geometric forms of generalization. *Classification* generalization reclassifies the attributes of objects into a smaller number of classes, whereas *symbolization* generalization changes the assignment of symbols to objects. For example, it might replace an elaborate symbol including the words "Mixed Forest" with a color identifying that class.

One of the most common forms of generalization of GI is the process known as weeding, or the simplification of the representation of a line represented as a polyline. The process is an instance of McMaster and Shea's simplification. Standard methods exist in GI systems for doing this, and the most common by far is the method known as the Douglas–Poiker algorithm after its inventors, David Douglas and Tom Poiker. The operation of the Douglas–Poiker weeding algorithm upon features such as that shown in Figure 3.16 is shown in Figure 3.17.

**Weeding is the process of simplifying a line or an area by reducing the number of points in its representation.**

Figure 3.16 The Douglas–Poiker algorithm is designed to simplify complex objects like this shoreline by reducing the number of points in its polyline representation.

iStockphoto

Note that the algorithm relies entirely on the assumption that the line is represented as a polyline—in other words, as a series of straight-line segments. GI systems increasingly support other representations,

Figure 3.17 The Douglas–Poiker line simplification algorithm in action. The original polyline has 15 points. In (A) Points 1 and 15 are connected (red), and the furthest distance of any point from this connection is identified (blue). This distance to Point 4 exceeds the user-defined tolerance. In (B) Points 1 and 4 are connected (green). Points 2 and 3 are within the tolerance of this line. Points 4 and 15 are connected, and the process is repeated. In the final step 7 points remain (identified with green disks), including 1 and 15. No points are beyond the user-defined tolerance distance from the line.

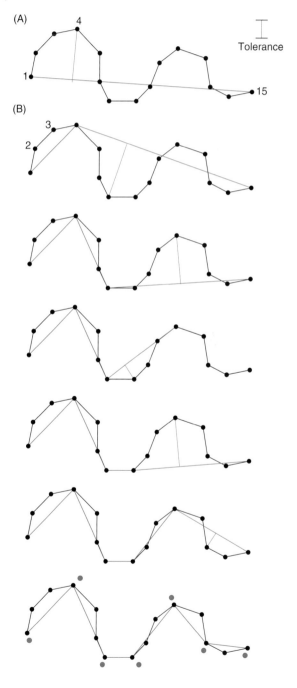

including arcs of circles, arcs of ellipses, and Bézier curves, but there is little consensus to date on appropriate methods for weeding or generalizing them, or on methods of analysis that can be applied to them. The general procedures of generalization of map labels and features such as coastlines in Web mapping systems are explored in Box 3.5.

## 3.8.2 Generalization about Properties

We saw in Sections 1.3 and 1.4 how the pace of scientific research had accelerated over the past two decades, in no small part because the Internet now provides a globally linked network of data warehouses and clearinghouses. In theory, the provenance of each of these various sources is known and is documented using metadata ("data about data": see Section 10.2). Many GI representations bring together multiple properties of places as composite *indicators* of conditions at particular locations, in order to fulfill particular needs. Thus, for example, summary measures of social "deprivation" (often described as "hardship" in the United States) combine representations of conditions with respect to heath, employment, and housing, among other social phenomena. In a similar vein, multiple criteria might be combined in order to define "wilderness" in an environmental application (see Section 1.3). Problems arise if some

---

## Applications Box (3.5)

### Online Map Generalization with OpenStreetMap, Mapnik, and Leaflet

OpenStreetMap is a digital map of the world (www. openstreetmap.org), created using crowd-sourced volunteered geographic information (VGI: Section 2.5). Just as with commercial (advertising-based) products like Google (maps.google.com), Apple (www.apple.com/ios/maps/), and Bing (www.bing.com/maps/), when a user zooms through successive levels of detail, so different features are revealed.

In OpenStreetMap, this effect is achieved using Mapnik (mapnik.org), a free software tool for developing mapping applications. The user views a rasterized version of the vector data that were used to create the digital base map through a JavaScript-powered map interface called Leaflet (leafletjs.com/). The setting of Mapnik's scale-dependent cartographic rules enables the display of some features earlier and others later in the map style. Rasterized versions of the original vector map data are used because they load much quicker when users pan and zoom across the map base.

One exception to this rule is the representation of coastlines, which are derived from two different simplified coastline datasets rather than crowd-sourced data—one for coarsely granular levels with simplified boundaries, and the second for fine levels of granularity with more complex, detailed coastlines (see the discussion of scale in Box 2.3). Figure 3.18 illustrates several different zoom levels of OpenStreetMap for Times Square, New York—the noncommercial nature of this Open Data (Section 17.4) product means that the map image includes a mix of points of interest uploaded by volunteers and map content is not determined by commercial sponsorship.

Similar methods for serving rasterized versions of vector data were used in early versions of services provided by the commercial online map services. How-ever, the ever-increasing use of handheld GPS-enabled devices such as smartphones has brought four issues to the fore. First, the output of software like Mapnik is a series of rasterized maps at fixed zoom levels; this means that map labels (such as points of interest or street names) are fixed in position and orientation, so when the map is rotated (for example, using the compass facility of a smartphone), the labels may appear upside down. Second, map labels may be obscured by overlay information such as the direction path rendered on top of the map. Third, the jumps between discrete prerendered representations that are each at fixed scale are not conducive to the pinching gestures that users have become accustomed to using on recent-generation devices. Fourth, although raster data can be served very efficiently and rapidly by the map server, slow mobile data networks may lack the bandwidth to serve up successive tiles fast enough for user requirements. (Users can avoid this fourth problem by loading a map of an area they will visit and then navigating around using GPS even without a mobile connection. Cached maps are much used in tourist maps products, e.g., offmaps.com.)

More recent versions of commercial online mapping services use vector maps, in which all coastlines, roads, labels, and other data are represented as mathematical lines rather than as fixed rasterized graphic images. (The rendering of the map image is still done by the server, so the client smartphone or other device still receives raster images.) This means that when a mobile user rotates a map, the text labels dynamically reorient in order to remain legible, and the text size of labels scales smoothly when a user zooms in or out. This is possible because the labels are rendered live as dynamic text, rather than as a graphic image that must be "repainted" for every zoom level. ▶

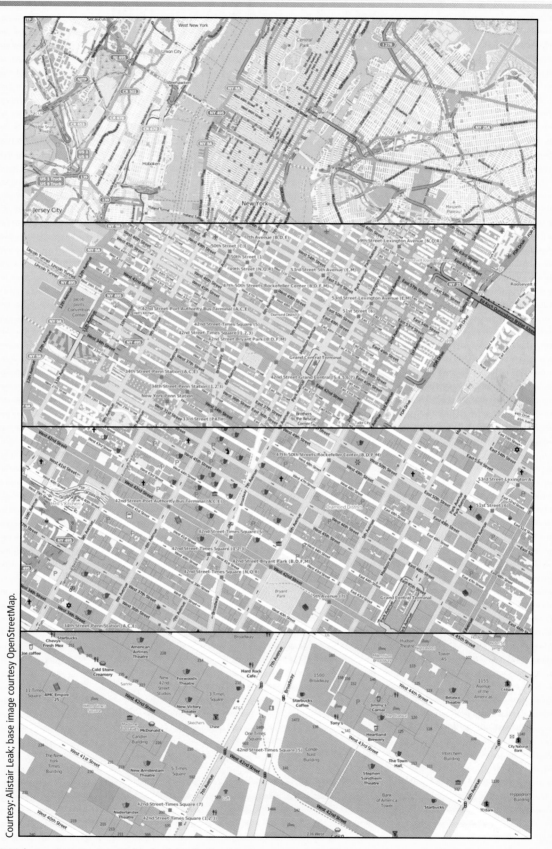

**Figure 3.18** Web mapping at successive levels of detail using OpenStreetMap, centered on Times Square, New York.

of these constituent representations of places have been collected at scales, or using areal units, that are less appropriate to the purpose than others. Thus a standard soil map may not be appropriate to inform a gardener whether a land parcel is suitable for growing sunflowers, yet a terrain model of aspect and elevation may well be suitable for this purpose. The validity of the sample design in drawing generalizations from representations, introduced in Section 2.4, is relevant here, and we return to the procedures and problems for generalizing about the properties of places in Section 5.3.2. These questions compound the fundamental problem that representations are necessarily incomplete (Section 3.4), and a representation may not be fit for its purpose if it is based upon inappropriate indicators of real-world conditions.

## 3.9 Conclusion

Representation, or more broadly *ontology*, is a fundamental issue in GI because it underlies all our efforts to express useful information about the surface of the Earth in a digital computer. The fact that there are so many ways of doing this makes GI systems at once complex and interesting, a point that will become much clearer on reading the technical chapter on data modeling (Chapter 7). But the broader issues of representation, including the distinction between field and object conceptualizations, underlie not only that chapter but many other issues as well, including uncertainty (Chapter 5) and analysis and modeling (Chapters 13 through 15).

## Questions for Further Study

1. What fraction of the Earth's surface have you experienced in your lifetime? Make diagrams like that shown in Figure 3.1, at appropriate levels of detail, to show (a) where you have lived in your lifetime, and (b) how you spent last weekend. How would you describe what is missing from each of these diagrams?

2. Table 3.3 summarized some of the relative merits of raster and vector representations. Expand on these, and use your enlarged table to assess the suitability of the two representational forms for a range of GI applications.

3. The early explorers had limited ways of communicating what they saw, but many were very effective at it. Examine the published diaries, notebooks, or dispatches of one or two early explorers, and look at the methods they used to communicate with others. What words did they use to describe unfamiliar landscapes, and how did they mix words with sketches?

4. With reference to Google Maps (maps.google.com) or another mapping site, identify the extent of your own neighborhood. Mark on it the discrete objects you are familiar with in the area. What features are hard to think of as discrete objects? For example, how will you divide the various roadways in the neighborhood into discrete objects—where do they begin and end?

## Further Reading

Haklay, M. 2010. How good is volunteered geographical information? A comparative study of OpenStreetMap and Ordnance Survey datasets. *Environment and Planning B*, 37: 682–703.

Kraak, M. J. and Ormeling, F. J. 2011. *Cartography: Visualization of Geospatial Data*, 3rd ed. Chapter 9. London and New York: Pearson Education.

McMaster, R. B. and Shea, K. S. 1992. *Generalization in Digital Cartography*. Washington, DC: Association of American Geographers.

Sui, D., Elwood, S., and Goodchild, M. F. 2013. *Crowdsourcing Geographic Knowledge: Volunteered Geographic Information in Theory and Practice*. Berlin: Springer.

# 4 Georeferencing

Geographic location is the element that distinguishes geographic information (GI) from all other types of information, so methods for specifying location on the Earth's surface are essential to the creation of useful GI. Many such techniques have been developed over the centuries, but in recent years it has become possible to convert from one to another with ease using GI systems and Web-based services. This chapter provides a basic guide to georeferencing for GISS students— what you need to know about georeferencing to succeed. The first section lays out the principles of georeferencing, including the requirements that any effective system must satisfy. Subsequent sections discuss commonly used systems, starting with the ones closest to everyday human experience, including place-names and street addresses, and moving to the more accurate scientific methods that form the basis of geodesy and surveying, and the most recent methods developed for the Web. The final sections deal with issues that arise over conversions between georeferencing systems, with the Global Positioning System (GPS), with georeferencing of computers and cell phones, and with the concept of a gazetteer.

## LEARNING OBJECTIVES

After studying this chapter you will:

- Know the requirements for an effective system of georeferencing.

- Be familiar with the problems associated with place-names, street addresses, and other systems used every day by humans to define locations that are important in their daily lives.

- Know how the Earth is measured and modeled for the purposes of positioning.

- Know the basic principles of map projections and the details of some commonly used projections.

- Know about conversion between different systems of georeferencing.

- Understand the principles behind GPS and learn some of its applications.

## 4.1 Introduction

Chapter 3 introduced the idea of an atomic element of GI: an atom made up of location, time (optionally), and attribute. To make a GI system work there must be techniques for assigning values to all three aspects in ways that are commonly understood by people who wish to communicate. Almost all the world agrees on a common calendar and time system, so there are only minor problems associated with communicating that element of the atom when it is needed (although

different time zones, different names of the months in different languages, the annual switch to summer or Daylight Saving Time, and systems such as the classical Japanese convention of dating by the year of the emperor's reign or the Islamic tradition of dating from the prophet's arrival in Medina all sometimes manage to confuse us).

Specification of time is optional in GI systems, but location is not, so this chapter focuses on techniques for specifying location and the problems and issues that arise. Locations are the basis for many of the benefits of GI systems: the ability to map, to tie different kinds of information together because they refer to the same place, to measure distances and areas, or to make decisions.

### Time is an optional element in geographic information, but location is essential.

Several terms are commonly used to describe the act of assigning locations to atoms of information. We use the verbs *georeference*, *geolocate*, and *geocode* and say that facts have been *georeferenced* or *geocoded*. We talk about *tagging* records with geographic locations, or about *locating* them. The term *georeference* will be used throughout this chapter.

The primary requirements of a georeference are (1) that it be *unique*, so that there is only one location associated with a given georeference, and therefore no confusion about the location that is referenced; and (2) that its meaning be *shared* among all of the people who wish to work with the information. For example, the georeference 3334 NE Blakeley St, Seattle, Washington, USA, points to a single house—there is no other house anywhere on Earth with that address—and its meaning is shared sufficiently widely to allow mail to be delivered to the address from virtually anywhere on the planet. The address may not be meaningful to everyone living in China, but it will be meaningful to a sufficient number of people within China's postal service, so that a letter mailed from China to that address will likely be delivered successfully, even if all of the address is written in the Roman alphabet and Arabic numerals. Uniqueness and shared meaning are sufficient also to allow people to link different kinds of information based on common location: for example, home address could be used to link information from databases that store that information as part of each record.

To be as useful as possible, a georeference must be *persistent through time* because it would be very confusing if georeferences changed frequently, and very expensive to update all the records that depend on them. This can be problematic when a georeferencing system serves more than one purpose or is used by more than one agency with different priorities. For example, a municipality may change its boundaries by incorporating more land, creating problems for mapping agencies and for researchers who wish to study the municipality through time. Street names sometimes change, and postal agencies sometimes revise postal codes. Changes even occur in the names of cities (Saigon to Ho Chi Minh City) or in their conventional transcriptions into the Roman alphabet (Peking to Beijing).

### To be most useful, georeferences should stay constant through time.

Although some georeferences are based on simple names, others are based on various kinds of *measurements* and are called *metric* georeferences. They include latitude and longitude and various kinds of coordinate systems, many of which are discussed in more detail later in this chapter and are essential to the making of maps, the display of mapped information, and any kind of numerical analysis. One enormous advantage of such coordinate systems is that they provide the potential for unlimited accuracy. Provided we have sufficiently accurate measuring devices and use enough decimal places, such systems will allow us to locate information to any level of accuracy. Another advantage is that from measurements of two or more locations, it is possible to compute distances, a very important requirement of georeferencing in GI systems.

### Metric georeferences are much more useful because they allow maps to be made and distances to be calculated.

Every georeference has an associated uncertainty. In some cases this is a form of spatial resolution, equal to the size of the area that is assigned to that georeference. For example, knowing that an address is somewhere in Alaska clearly has much greater uncertainty than knowing that an address is somewhere in Rhode Island. When georeferences are measured, for example, by using GPS to georeference a tweet, the uncertainty is the result of errors of measurement, which we might express as the area around the measured location in which the true location might exist. For example, a tweet georeferenced using GPS with a measurement error of 100 m might be said to have an uncertainty equal to the area of a circle of radius 100 m, that is, 3.14 hectares. A mailing address could be said to have an uncertainty equal to the size of the mailbox, or perhaps to the area of the parcel of land or structure assigned that address. Many other systems of georeferencing have similarly wide-ranging uncertainties.

In each of the examples in the previous paragraph, uncertainty was expressed in area measure. Often, however, we think of uncertainty of position in linear measure. But it is easy to connect the two by taking the square root of area. Thus the positional uncertainty inherent in the georeference "in Rhode Island" might be expressed either as the area of Rhode Island (3140 sq km) or as its square root (56 km).

Many systems of georeferencing are unique only within an area or *domain* of the Earth's surface. For example, many populated places in the United States have the name Springfield (62 according to the official online Geographic Names Information System geonames.usgs.gov; similarly, there are in the United Kingdom nine populated places called Whitchurch in the Geonames database geonames.org that integrates official and volunteered sources). However,

**Figure 4.1** Place-names are not necessarily unique at the global level. This map shows the locations of 40 places named Santa Barbara in the Geonames database (geonames.org). Additional information would be needed (e.g., limiting the search to California) to locate a specific Santa Barbara.

Basemap courtesy: Google. Imagery copyright 2013 by NASA and TerraMetrics, map data copyright 2013 by MapLink and Tele Atlas

the advent of postal systems in many countries led to the renaming of many duplicates, so that in the UK city names are unique within counties. Today there is no danger that there are two Springfields in Massachusetts; a driver can therefore confidently ask for directions to "Springfield, Massachusetts" in the knowledge that there is no danger of being sent to the wrong Springfield. But people living in London, Ontario, Canada, are well aware of the dangers of talking about "London" without specifying the appropriate domain. Even in Toronto, Ontario, a reference to "London" may be misinterpreted as a reference to the older (UK) London on a different continent rather than to the one 200 km away in the same province (Figure 4.1). Street name is unique in the United States within municipal domains, but not within larger domains such as county or state. There are 120 places on the Earth's surface with the same Universal Transverse Mercator coordinates (see Section 4.8.2), and a zone number and hemisphere must be added to make a reference unique in the global domain.

This section has reviewed some of the general properties of georeferencing systems, and Table 4.1 shows some commonly used systems. The following sections discuss the specific properties of the systems that are most important in GI practice.

**Table 4.1** Some commonly used systems of georeferencing.

| System | Domain of uniqueness | Metric? | Example | Positional uncertainty |
|---|---|---|---|---|
| Place-name or POI | varies | no | London, Ontario, Canada | varies by type of feature |
| Postal address | global | no, but ordered along streets in some countries | 3334 NE Blakeley St, Seattle, WA, USA | size of one mailbox |
| Postal code | country | no | 98104 (U.S. ZIP code) | area occupied by a defined number of mailboxes |
| Telephone calling area | country | no | 804 | varies |
| Cadastral system | local authority | no | Parcel 01442944, City of Springfield, MA | area occupied by a single parcel of land |
| Public Land Survey System | Western Canada and United States only, unique to Principal Meridian | yes | Sec 4, Township 4N, Range 6E | defined by level of subdivision |
| Latitude/ longitude | global | yes | 119 degrees 44 minutes West, 34 degrees 40 minutes North | infinitely fine (2.8 sq km in this example) |
| Universal Transverse Mercator | zones six degrees of longitude wide, and N or S hemisphere, but not polar latitudes | yes | 463146E, 4346732N | infinitely fine (1 sq m in this example) |
| State Plane Coordinates | U.S. only, unique to state and to zone within state | yes | 4408634E, 7421076N | infinitely fine (1 sq ft in this example) |

## Lynn Usery

E. Lynn Usery (Figure 4.2) is a Research Physical Scientist and Director of the Center of Excellence for Geospatial Information Science (CEGIS) with the U.S. Geological Survey (USGS). Lynn's work has spanned traditional field-surveying, photogrammetric, and cartographic processes for topographic mapping in the 1970s to modern all-digital processes of GI systems and Internet mapping today. He was involved in the automation of topographic mapping for the USGS and then explored feature-based GI concepts at the University of Wisconsin–Madison and the University of Georgia. Returning to the USGS and topographic mapping in 1999, Lynn began examining problems of map projections and coordinate transformations, particularly for large raster datasets. Treating the raster cells as areas rather than points, which was the approach used in commercial transformation software at that time, a USGS research team under Lynn's direction provided solutions to projection problems of raster data, including pixel loss and gain in categorical datasets and repeated areas, the so-called wrap-around problem (Figure 4.3). The large data volumes of fine resolution and global geospatial data, and the complex calculations of map-projection transformations, are well suited to parallel and grid computing approaches. With its grid cell structure, raster data are ideally partitioned by rows, columns, or blocks for multiple processors. Lynn and his USGS research team paired

Courtesy: E. Lynn Usery

**Figure 4.2** E. Lynn Usery, cartographer and GI scientist for the U.S. Geological Survey.

with the CyberGIS community to develop parallel solutions to map projection transformations. Results of that effort include a map projections package, pRasterBlaster, that operates in parallel and can be used with the XSEDE supercomputer network of the National Science Foundation. In an era of multiple petabytes of data for USGS mapping operations, CyberGIS provides a practical solution. In addition to the projections research, Lynn is also engaged in researching semantics for geographic information, particularly for terrain features, such as hills, mountains, and valleys.

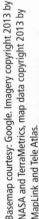

Basemap courtesy: Google. Imagery copyright 2013 by NASA and TerraMetrics, map data copyright 2013 by MapLink and Tele Atlas.

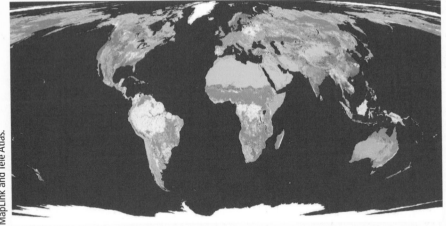

**Figure 4.3** The wrap-around problem of map projection transformations resulting from improper framing of the output transformation space. Note the repetition of Alaska and Siberia on both east and west edges of this Hammer projection.

## 4.2 Place-Names and Points of Interest

Giving names to places is the simplest form of georeferencing and was most likely the first one developed by early hunter–gatherer societies. Any distinctive feature on the landscape, for example, a particularly old tree or a restaurant, can serve as a point of reference for two people who wish to share information, such as the existence of good game in the tree's vicinity or a proposed rendezvous. Human landscapes rapidly became littered with names, as people sought distinguishing labels to use in describing aspects of their surroundings and as other people adopted them. Today, of course, we have a complex system of naming oceans, continents, cities, mountains, rivers, and other prominent features, plus a rich collection of named points of interest. Each country maintains a system of authorized naming, often through national or state committees assigned with the task of standardizing geographic names. Nevertheless, multiple names are often attached to the same feature, for example, when cultures try to preserve the names given to features by the original or local inhabitants (Mount Everest to many, but Chomolungma to many Tibetans), or when city names are different in different languages (Florence in English, Firenze in Italian). Even the question of what merits naming is culturally related and depends on how different cultures make use of the landscapes they occupy. Collections of official names are termed *gazetteers*, but in recent years much larger databases of points of interest have been assembled to support online services for mapping and navigation, and the term *point-of-interest* (POI) database is now more appropriate. In what follows the term *place-name* will be used to refer both to officially recognized features and also to points of interest.

**Many commonly used place-names and named points of interest have meanings that vary between people and with the context in which they are used.**

But place-names are sometimes of limited use as georeferences. First, they often imply very substantial uncertainty. "Asia" covers over 43 million sq km, so the information that something is located "in Asia" is not very helpful in pinning down its location. One approach to this issue is to use a point to represent the area. This can lead to unintended consequences, as, for example, when Google Maps is asked to find a route from "Colorado" to "Wyoming." The area of Colorado is represented by a central point west of Denver and Wyoming by a central point west of Casper—and a detailed route of 634 km is recommended by the service. But of course a single step is all that might be needed to move between these neighboring states if one were located at the border.

Second, only certain place-names are officially authorized by national or subnational agencies. Many more are recognized only locally, so their use is limited to communication between people in the local community. Place-names may even be lost through time: Although there are many contenders, we do not know with certainty where the "Camelot" described in the English legends of King Arthur was located, if indeed it ever existed.

**The meaning of certain place-names can become lost through time.**

The growing Web phenomenon of volunteered geographic information (VGI) is changing this situation, however, and creating an alternative to the traditional top-down system of naming. Individuals are now able to assign names to features quite independent of officialdom by using sites such as Wikimapia (www.wikimapia.org; see Figure 4.4). At time of writing, individuals worldwide had added descriptions of over 20,000,000 features, from the largest cities to the smallest buildings. Descriptions can be in any language and a wide range of scripts and can include photographs and links to other sources of information.

Wikimapia's mantra is "Let's describe the whole world," and it is one example of a growing phenomenon that draws individual citizens into the process of creating geographic knowledge. In essence, it echoes an earlier era before the establishment of national mapping agencies and authoritative committees, when place-names were created by explorers and local citizens. Many names fell by the wayside, but some were adopted and shown on maps. The process by which America was named, in 1507, was essentially of this nature: an act by an individual cartographer, Martin Waldseemüller, in St-Dié-des-Vosges in France. Waldseemüller and his colleague, Vautrin Lud, had recently read letters from the Florentine explorer Amerigo Vespucci claiming credit for recognizing that the lands discovered to the west of Europe formed a previously unknown continent. Accordingly, he invented a name by feminizing Vespucci's first name, believing that all continents should have feminine names (Figure 4.5). Copies of the map were distributed across Europe, and the name stuck.

**Figure 4.4** Wikimapia coverage of the area of Lhasa, Tibet. Each of the hundreds of polygons in white represents one entry, where a volunteer has provided a description and possible images and links to other information, all of which can be exposed by clicking on the polygon. The identification of the highlighted feature as "underground missile site" is clearly open to question, given the voluntary nature of the Wikimapia project.

Basemap courtesy: Google. Imagery copyright 2013 by TerraMetrics.

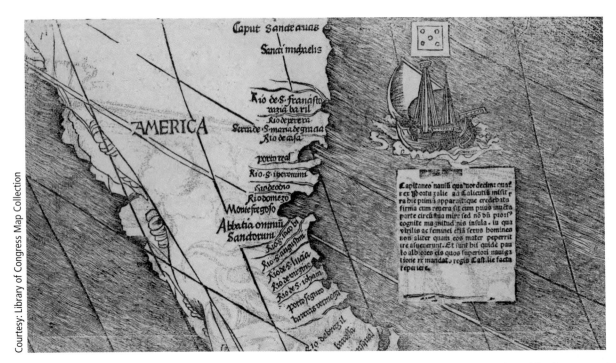

**Figure 4.5** Detail of the Waldseemüller map of 1507, which for the first time showed the name the cartographer selected for the new continent.

## 4.3 Postal Addresses and Postal Codes

Postal addresses were introduced after the development of mail delivery in the Nineteenth Century. They rely on several assumptions:

- Every dwelling and office is a potential destination for mail.

- Dwellings and offices are arrayed along paths, roads, or streets and are numbered sequentially.

- Paths, roads, and streets have names that are unique within local areas.

- Local areas have names that are unique within larger regions.

- Regions have names that are unique within countries.

If the assumptions are true, then mail address provides a unique identification for every dwelling and office on Earth.

Today, postal addresses are an almost universal means of locating many kinds of human activity: delivery of mail, place of residence, or place of business. They fail, of course, in locating anything that is not a potential destination for mail, including almost all kinds of natural features (Mount Everest does not have a postal address and neither does Manzana Creek in Los Padres National Forest in California).

They are not as useful when dwellings are not numbered consecutively along streets, as happens in some cultures (notably in Japan, where street numbering can reflect date of construction, not sequence along the street; it is temporal, rather than spatial) and in large building complexes like condominiums. Many applications of GI systems rely on the ability to locate activities by postal address and to convert addresses to some more universal system of georeferencing, such as latitude and longitude, for mapping and analysis.

**Postal addresses work well to georeference dwellings and offices, but not natural features.**

Postal codes were introduced in many countries in the late Twentieth Century in order to simplify the sorting of mail. In the UK system, for example, the first characters of the code identify an Outward Code, and mail is initially sorted so that all mail directed to a single Local Delivery Office (LDO) is together. Incoming mail is accumulated in a local sorting station and sorted a second time by the last characters of the code so that it can be delivered by an LDO. Figure 4.6 shows a map of the Outward Codes for Southend-on-Sea (SS). The full six or seven characters of the postal code (e.g., SS5 4PJ) are unique to roughly 13 houses, a single large business, or a single building. Different countries are more or less insistent on completeness of postal codes; in the UK, for

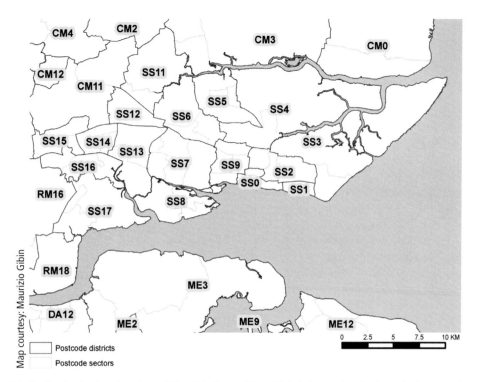

**Figure 4.6** Outward Codes for the Southend-on-Sea, UK, Local Delivery Offices (LDOs). Outward codes form the first two, three, or four characters of the UK postal code and are separated from the three characters that identify the Inward Codes.

example, sorters will deliver mail with only an Outward Code because LDOs are required to know their local communities.

Postal codes have proven very useful for many purposes besides the sorting and delivery of mail. Although the area covered by a Canadian FSA (Forward Sortation Area), a U.S. ZIP code, or a UK postcode varies and can be changed whenever the postal authorities want, it is sufficiently constant to be useful for mapping purposes, and many businesses routinely make maps of their customers by counting the numbers present in each postal code area, as well as by dividing by total population to get a picture of market penetration. Figure 4.7 shows an example of summarizing data by postal code. In Figure 4.7A it has been necessary to suppress the locations of postcodes where there are fewer than five resident patients; it is not necessary to hide any data in the choropleth map shown in Figure 4.7B because privacy is not violated at this coarser level of granularity. Most people know the postal code of

**Figure 4.7** The use of Outward Code boundaries as a convenient basis for summarizing data. In this instance (A) points identifying the residential unit (Inward) postcode have been used to plot the locations of a doctor's patients, and (B) the Outward Code areas have been shaded according to the density of patients per square kilometer.

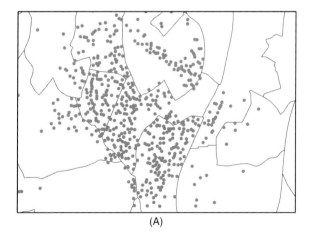

(A)

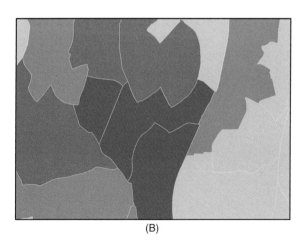

(B)

their home, whereas almost no one knows his or her latitude and longitude, and in some instances postal codes have developed popular images (the ZIP code for Beverly Hills, California, 90210, became the title of a successful television series).

## 4.4 IP Addresses

Every device (computer, printer, etc.) connected to the Internet has a unique IP (Internet Protocol) address, such as 128.111.106.183, the address used by the computer of one of the authors when in his office. IP addresses are allocated to organizations, and the street address included in the registration record can provide an approximate location. The IP address of the user's computer is provided whenever the computer is used to access a Web site, allowing the operators of major sites to determine the user's location. Positional uncertainty will vary, however, because all the IP addresses allocated to a university campus may be georeferenced to a single point on the campus, and all the addresses allocated to an Internet Service Provider that operates a broadband service may be georeferenced to a point that is many kilometers from some customers.

The ability to determine even approximate locations for computers has led to some powerful applications. It allows search engines to order the results of search by proximity to the user's location, and it also permits sites such as Wikimapia to open centered on the user's location. Many Web services offer conversion between IP address and geographic coordinates. For example, www.networldmap.com resolves the address 128.111.106.183 to 34.4119 north, 119.7280 west, approximately 10 km from the location of the office at the University of California, Santa Barbara, where the IP address is used. Wikimapia does rather better, opening centered on a location only 1 km away.

## 4.5 Linear Referencing Systems

A linear referencing system identifies location on a network by measuring distance from a defined point of reference along a defined path in the network. Figure 4.8 shows an example, an accident whose location is reported as being a measured distance from a street intersection, along a named street. Linear referencing is closely related to street address, but uses an explicit measurement of distance rather than the much less reliable surrogate of street address number. Linear referencing is widely used in applications that depend on a linear network. This includes highways (e.g., Mile 1240 of the Alaska Highway), railroads

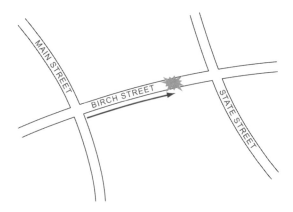

**Figure 4.8** Linear referencing—an incident's position is determined by measuring its distance (87 m) along one road (Birch Street) from a well-defined point (its intersection with Main Street).

(e.g., 25.9 miles from Paddington Station in London on the main line to Bristol, England), electrical transmission lines, pipelines, and canals. Highway agencies use linear references to define the locations of bridges, signs, potholes, and accidents and to record pavement condition.

**Linear referencing systems are widely used in managing transportation infrastructure and in dealing with emergencies.**

Linear referencing provides a sufficient basis for georeferencing for some applications. GI systems have many applications in transportation that are known collectively as GIS-T, and in the developing field of intelligent transportation systems, or ITS. But for other applications it is important to be able to convert between linear references and other forms, such as latitude and longitude. For example, the OnStar system that is installed in many Cadillacs sold in the United States is designed to radio the position of a vehicle automatically as soon as it is involved in an accident. When the airbags deploy, a GPS receiver determines position, which is then relayed to a central dispatch office. Emergency response centers often use street addresses and linear referencing to define the locations of accidents, so the latitude and longitude received from the vehicle must be converted before an emergency team can be sent to the accident.

Linear referencing systems are often difficult to implement in practice in ways that are robust in all situations. In an urban area with frequent intersections, it is relatively easy to measure distance from the nearest one (e.g., on Birch St 87 m east of the intersection with Main Street). But in rural areas an incident may be a long way from the nearest intersection. Even in urban areas it is not uncommon for two streets to intersect more than once (e.g., Birch may have two intersections with Columbia Crescent), or for a street to intersect with itself (Figure 4.9)

**Figure 4.9** The intersection of a street with itself (Calgary, Alberta, Canada), causing problems for linear referencing.

**Figure 4.10** Google Earth simulation of the view looking east along Baseline Road in Ontario, California—the road that follows the original survey baseline for Southern California laid out by Colonel Henry Washington in 1842. The monument marking the intersection between the baseline and the principal meridian is atop Mount San Bernardino, which appears on the horizon.

# 4.6 Cadasters and the U.S. Public Land Survey System

The *cadaster* is defined as the map of land ownership in an area, maintained for the purposes of taxing land or of creating a public record of ownership. The process of *subdivision* creates new parcels by legally subdividing existing ones.

Parcels of land in a cadaster are often uniquely identified, by number or by code and are also reasonably persistent through time, thereby satisfying the requirements of a georeferencing system. Indeed, it has often been argued that the cadaster could form the universal basis of mapping (a *multipurpose cadaster*), with all other geographic information tied to it. But very few people know the identification code of their home parcel; thus use of the cadaster as a georeferencing system is limited largely to local officials, with one major exception.

The U.S. Public Land Survey System (PLSS) evolved out of the need to survey and distribute the vast land resources of the Western United States, starting in the early Nineteenth Century, and expanded to become the dominant system of cadaster for all the United States west of Ohio, and all of Western Canada. Its essential simplicity and regularity make it useful for many purposes and understandable by the general public. Its geometric regularity also allows it to satisfy the requirement of a metric system of georeferencing because each georeference is defined by measured distances.

**The Public Land Survey System defines land ownership over much of western North America and is a useful system of georeferencing.**

To implement the PLSS in an area, a surveyor first laid out an accurate north–south line or *principal meridian*. An east–west *baseline* was then laid out perpendicular to this (Figure 4.10). Rows were laid out 6 miles apart and parallel to the baseline, to become the *townships* of the system. Then blocks or *ranges* were laid out in 6 mile by 6 mile squares on either side of the principal meridian (see Figure 4.11). Each square is referenced by township number, range number, whether it is to the east or to the west, and the name of the principal meridian. Thirty-six *sections* 1 mile by 1 mile were laid out inside each township and numbered using a standard system (note how the numbers reverse in every other row). Each section was divided into four quarter-sections of 1/4 square mile, or 160 acres, the size of the nominal family farm or homestead in the original conception of the PLSS. The process can be continued by subdividing into four to obtain any level of spatial resolution.

The PLSS would be a wonderful system if the Earth were flat and if survey measurements were always exact; but unfortunately neither of these assumptions is true. To account for the Earth's curvature, the squares are not perfectly 6 miles by 6 miles, and the rows must be offset frequently; errors in the original surveying complicate matters still further, particularly in rugged landscapes. Figure 4.11 shows the offsetting exaggerated for a small area. Nevertheless, the PLSS remains an efficient system and one with which many people

(A)

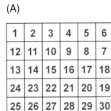

(B)

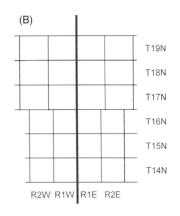

**Figure 4.11** Portion of the Township and Range system (Public Lands Survey System) widely used in the Western United States as the basis of land ownership (B). Townships are laid out in 6-mile squares on either side of an accurately surveyed Principal Meridian. The offset shown between T16 N and T17 N is needed to accommodate the Earth's curvature (shown much exaggerated). The square mile sections within each township are numbered as shown in (A).

in the Western United States and Western Canada are familiar. It is often used to specify location, particularly in managing natural resources in the oil and gas industry and in mining, and in agriculture. Services have been built to convert PLSS locations automatically to and from latitude and longitude (Section 4.7).

# 4.7 Measuring the Earth: Latitude and Longitude

The most powerful systems of georeferencing are those that provide the potential for very accurate measurement of position, that allow distance to be computed between pairs of locations, and that support other forms of spatial analysis (see Chapters 13 and 14). The system of latitude and longitude is in many ways the most comprehensive and is often called the *geographic* system of coordinates. It is based on the Earth's rotation about its center of mass.

To define latitude and longitude, we first identify the *axis* of the Earth's rotation. The Earth's center of mass lies on the axis, and the plane through the center of mass perpendicular to the axis defines the *equator*. Slices through the Earth parallel to the axis, and perpendicular to the plane of the equator, define lines of constant longitude (Figure 4.12), rather like the segments of an orange. In 1884 a conference attended by delegates from 25 nations agreed that zero longitude, the *prime meridian*, should be defined by a line marked on the ground at the Royal Observatory

in Greenwich, England; the angle between this slice and any other slice defines the latter's measure of longitude. Each of the 360 degrees of longitude is divided into 60 minutes and each minute into 60 seconds. But it is more conventional to refer to longitude by degrees East or West, so longitude ranges from 180 degrees West to 180 degrees East of the prime meridian. Finally, because computers are designed to handle numbers ranging from very large and negative to very large and positive, we normally store longitude in computers as if West was negative and East was positive, and we store parts of degrees using decimals rather than minutes and seconds. A line of constant longitude is termed a *meridian*.

Longitude can be defined in this way for any rotating solid, no matter what its shape, because the axis of rotation and the center of mass are always defined. But the definition of latitude requires that we know something about the Earth's shape. The *geoid* is defined as a surface of equal gravity formed by the oceans at rest, and by an imaginary extension of this surface under the continents; it has a complex shape that is only approximately spherical (a radius of 6378 km is a reasonable approximation if the Earth is assumed to be spherical). A much better mathematical approximation or *figure of the Earth* is the *ellipsoid of rotation*, the figure formed by taking a mathematical ellipse and rotating it about its shorter axis (Figure 4.13). The term *spheroid* is also commonly used.

The difference between the ellipsoid and the sphere is measured by its *flattening*, or the reduction in the minor axis relative to the major axis. Flattening is defined as:

$$f = (a - b)/a$$

**Figure 4.12** Definition of longitude. The Earth is seen here from above the North Pole, looking along the axis, with the equator forming the outer circle. The location of Greenwich defines the Prime Meridian. The longitude of the point at the center of the red cross is determined by drawing a plane through it and the axis and measuring the angle between this plane and the Prime Meridian.

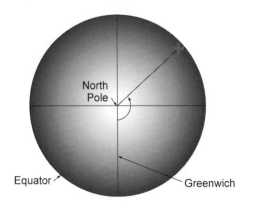

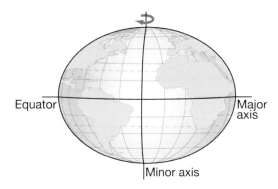

Figure 4.13 Definition of the ellipsoid, formed by rotating an ellipse about its minor axis (corresponding to the axis of the Earth's rotation).

where *a* and *b* are the lengths of the major and minor axes, respectively (we usually refer to the *semi*-axes, or half the length of the axes, because these are comparable to radii). The actual flattening is about 1 part in 300.

> **The Earth is slightly flattened, such that the distance between the poles is about 1 part in 300 less than the diameter at the equator.**

Much effort was expended over the past 200 years in finding ellipsoids that best approximated the shape of the Earth in particular countries, so that national mapping agencies could measure position and produce accurate maps. Early ellipsoids varied significantly in their basic parameters and were generally not centered on the Earth's center of mass. But the development of intercontinental ballistic missiles

in the 1950s and the need to target them accurately, as well as new data available from satellites, drove the push to a single international standard. Without a single standard, the maps produced by different countries using different ellipsoids could never be made to fit together along their edges, and artificial steps and offsets were often necessary in moving from one country to another (navigation systems in aircraft would have to be corrected, for example).

The ellipsoid known as WGS84 (the World Geodetic System of 1984) is now widely accepted (though minor adjustments continue to be made, they are largely insignificant in magnitude for GIS applications), and North American mapping has been brought into conformity with it through the adoption of the virtually identical North American Datum of 1983 (NAD83). It specifies a semimajor axis (distance from the center to the equator) of 6378137 m and a flattening of 1 part in 298.257, and its line of zero longitude passes about 100 m to the east of the Greenwich Observatory. But many other ellipsoids remain in use in certain parts of the world, and much older data still adhere to earlier standards, such as the North American Datum of 1927 (NAD27). For locations in the continental United States the difference between two points with identical latitude and longitude, but determined according to the NAD27 and NAD83 datums, can be as much as 100 m. Thus GIS users sometimes need to convert between datums, and functions to do that are commonly available.

## Applications Box 4.2

### Newton, Descartes, and the Shape of the Earth

Isaac Newton's understanding of centrifugal forces led him to conclude that a planet such as the Earth rotating about its axis should bulge at the equator, where the diameter should therefore be greater than the distance between the poles. The French mathematician René Descartes, on the other hand, argued that the reverse should be true—a greater distance between the poles than the diameter at the equator. Given the definition illustrated in Figure 4.14, lines of latitude should grow further apart as one moves away from the equator if Newton is right, and closer together if Descartes is right. In 1734 the French Academy of Sciences dispatched expeditions to Finland and Peru to make accurate determinations of the distance between lines of latitude. The results proved conclusively that Newton was right.

One of the members of the expedition to Peru, Jean Godin, married a Peruvian and promised to take her home to France via a difficult route over the Andes and down the Amazon, at a time when the Spanish

authorities in Peru were at odds with the Portuguese authorities in Brazil. He left first, and the 20-year story of Isabel's epic and ultimately successful adventure to rejoin him is admirably told in Robert Whitaker's *The Mapmaker's Wife*.

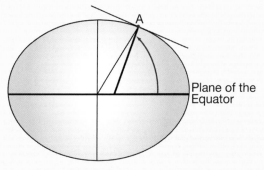

Figure 4.14 Definition of the latitude of Point A, as the angle between the equator and a line drawn perpendicular to the ellipsoid.

We can now define latitude. Figure 4.14 shows a line drawn through a point of interest perpendicular to the ellipsoid at that location. The angle made by this line with the plane of the equator is defined as the point's latitude and varies from 90 South to 90 North. Again, south latitudes are usually stored as negative numbers and north latitudes as positive. Latitude is often symbolized by the Greek letter phi ($\phi$) and longitude by the Greek letter lambda ($\lambda$), so the respective ranges can be expressed in mathematical shorthand as: $-180 \leq \lambda \leq 180$; $-90 \leq \phi \leq 90$. A line of constant latitude is termed a *parallel*.

It is important to have a sense of what latitude and longitude mean in terms of distances on the surface. Ignoring the flattening, two points on the same north–south line of longitude and separated by one degree of latitude are 1/360 of the circumference of the Earth apart, or about 111 km apart. One minute of latitude corresponds to 1.86 km and also defines one nautical mile, a unit of distance that is still commonly used in navigation. One second of latitude corresponds to about 30 m. But things are more complicated in the east–west direction, and these figures only apply to east–west distances along the equator, where lines of longitude are furthest apart. Away from the equator the length of a line of latitude gets shorter and shorter, until it vanishes altogether at the poles. The degree of shortening is approximately equal to the cosine of latitude, or $\cos \phi$, which is 0.866 at 30 degrees North or South, 0.707 at 45 degrees, and 0.500 at 60 degrees. So degrees of longitude are only 55 km apart along the northern boundary of the Canadian province of Alberta (exactly 60 degrees North).

In GI systems, latitude and longitude are often expressed as decimals of degrees, rather than degrees, minutes, and seconds. It is helpful to know that the 5th decimal place of degrees of latitude is about 1 m on the Earth's surface. In GIS it is very uncommon to know positions to greater accuracy, so any additional decimal places that may be displayed or recorded are probably beyond the limits of accuracy and therefore meaningless. Unfortunately it is all too common for online services to display latitudes and longitudes with precisions that are so far beyond the accuracy of measurement systems as to be almost laughable.

**Lines of latitude and longitude are equally far apart only at the equator; toward the poles lines of longitude converge.**

Given latitude and longitude, it is possible to determine distance between any pair of points, not just pairs along lines of longitude or latitude. It is easiest to pretend for a moment that the Earth is spherical because the flattening of the ellipsoid makes the equations more complex. On a spherical Earth the shortest path between two points is a *great*

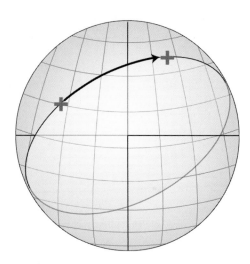

**Figure 4.15** The shortest distance between two points on the sphere is an arc of a *great circle*, defined by slicing the sphere through the two points and the center (all lines of longitude, and the equator, are great circles on a spherical Earth). The circle formed by a slice that does not pass through the center is a *small circle* (all lines of latitude except the equator are small circles).

*circle*, or the arc formed if the Earth is sliced through the two points and through its center (Figure 4.15; an off-center slice creates a *small circle*). The length of this arc on a spherical Earth of radius R is given by:

$$R \arccos [\sin \phi_1 \sin \phi_2 + \cos \phi_1 \cos \phi_2 \cos(\lambda_1 - \lambda_2)]$$

where the subscripts denote the two points. For example, the distance from a point on the equator at longitude 90 East (in the Indian Ocean between Sri Lanka and the Indonesian island of Sumatra) and the North Pole is found by evaluating the equation for $\phi_1 = 0$, $\lambda_1 = 90$, $\phi_2 = 90$, $\lambda_2 = 90$. It is best to work in radians (1 radian is 57.30 degrees, and 90 degrees is $\pi/2$ radians). The equation evaluates to $R \arccos 0$, or $\pi R /2$, or one-quarter of the circumference of the Earth. Using a radius of 6378 km, this comes to 10,018 km, or close to 10,000 km (not surprising because the French originally defined the meter in the late Eighteenth Century as one ten-millionth of the distance from the equator to the pole).

# 4.8 Projections and Coordinates

Latitude and longitude define location on the Earth's surface in terms of angles with respect to well-defined references: the prime meridian, the center of mass, and the axis of rotation. As such, they constitute the most comprehensive system of georeferencing and support a range of forms of analysis, including the calculation of distance between points, on the curved surface of the Earth. But many technologies for working with geographic data are inherently flat, including paper and printing, which evolved over many centuries long before the advent of digital geographic data and GI systems. For various reasons, therefore, much work in GI science deals with a flattened or *projected* Earth, despite the price we pay in the distortions that are an inevitable consequence of flattening. Specifically, the Earth is often flattened because:

- Paper is flat, and paper is still used as a medium for inputting data to GI systems by scanning or digitizing (see Section 9.3) and for outputting data in map or image form.

- Rasters (Section 7.2.2) are inherently flat because it is impossible to cover a curved surface with equal squares without gaps or overlaps.

- Photographic film is flat, and film cameras are still used widely to take images of the Earth from aircraft to use as GI.

- When the Earth is seen from space, the part in the center of the image has the most detail, and detail drops off rapidly, the back of the Earth being invisible. In order to see the whole Earth at once with approximately equal detail, it must be distorted in some way, and it is most convenient to make it flat.

The Cartesian coordinate system (Figure 4.16) assigns two coordinates to every point on a flat surface by measuring distances from an origin parallel to two axes drawn at right angles. We often talk of the two axes as *x* and *y*, and of the associated coordinates as the *x*- and *y*-coordinate, respectively. Because it is common to align the *y*-axis with North in geographic applications, the coordinates of a projection on a flat sheet are often termed *easting* and *northing*.

**Figure 4.16** A Cartesian coordinate system, defining the location of the blue cross in terms of two measured distances from the origin, parallel to the two axes.

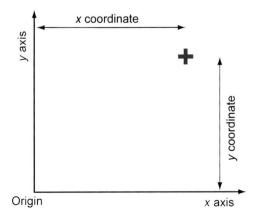

**Although projections are not absolutely required, there are several good reasons for using them to flatten the Earth.**

One way to think of a map projection, therefore, is that it transforms a position on the Earth's surface identified by latitude and longitude ($\phi$, $\lambda$) into a position in Cartesian coordinates ($x$, $y$). Every recognized map projection, of which there are many, can be represented as a pair of mathematical functions:

$$x = f(\phi, \lambda)$$
$$y = g(\phi, \lambda)$$

For example, the famous Mercator projection uses the functions:

$$x = \lambda$$
$$y = \ln \tan[\phi/2 + \pi/4]$$

where ln is the natural log function. The inverse transformations that map Cartesian coordinates back to latitude and longitude are also expressible as mathematical functions: in the Mercator case they are

$$\lambda = x$$
$$\phi = 2 \arctan e^y - \pi/2$$

where $e$ denotes the constant 2.71828. Many of these functions have been implemented in GI systems, allowing users to work with virtually any recognized projection and datum and to convert easily between them.

**Two datasets can differ in both the projection and the datum, so it is important to know both for every dataset.**

Projections necessarily distort the Earth, so it is impossible in principle for the scale (distance on the map compared to distance on the Earth; for a discussion of scale see Box 2.3) of any flat map to be perfectly uniform or for the pixel size of any raster to be perfectly constant. Although this point is unimportant for maps of small areas, for maps at global scale the result can be very misleading. In Figure 4.1, for example, which was generated from a Web service that uses Google Maps, there is a scale bar shown in the bottom left corner, indicating the distances on the map that correspond to 2000 km and 1000 mi. But the projection used in the map is Web Mercator (see Section 4.8.3), with a scale that varies continuously. At high latitudes the Mercator projection is acknowledged to produce a very distorted impression of area, Greenland and Russia appearing much larger in relation to countries in lower latitudes than they really are. The 2000-km bar should be positioned at the equator and should be roughly twice as long at the top and bottom of the map (approximately 60 North and 60 South respectively; at latitude $\phi$ the scale factor is $1/\cos\phi$) as it is in the middle.

### Gerard Mercator

Gerard Mercator was born Gerard Kremer on March 4, 1412, in Gangelt, close to the Dutch border in what is now Germany. (Kremer is German and Dutch for merchant, mercator in Latin; Gerard changed his name at age eighteen). Like many others before and since, he was intrigued by the problem of portraying the curved surface of the Earth on a flat sheet of paper. His solution, known today as the Mercator Projection, was originally designed to aid navigation; a course set on a constant compass bearing will be straight on the projection. It also displays the conformal property, meaning that scale is locally constant in all directions and making it ideally suited to many Web applications. It achieves these properties at some cost, however, and the Mercator Projection is today often derided. The poles are at infinity, and scales at high latitudes are badly distorted, making countries such as Greenland and Iceland appear much larger in relation to low-latitude countries.

Mercator was imprisoned by the Inquisition and survived plague, war, and famine, living to the remarkable age (for his time) of 82,

But although projections must distort the Earth, they can preserve certain properties. Two such properties are particularly important, although any projection can achieve at most one of them, not both:

- The *conformal* property, which ensures that the shapes of small features on the Earth's surface are preserved on the projection: in other words, that the scales of the projection in the *x*- and *y*-directions are always equal

- The *equal area* property, which ensures that areas measured on the map are always in the same proportion to areas measured on the Earth's surface

The conformal property of the Mercator projection is useful for navigation because a straight line drawn on the map has a constant bearing (the technical term for such a line is a *rhumb line* or *loxodrome*). The equal area property is useful for various kinds of analysis involving areas, such as the computation of the area of someone's property.

Besides their distortion properties, another common way to classify map projections is by analogy to a physical model of how positions on the map's flat surface are related to positions on the curved Earth. There are three major classes (Figure 4.17), but note that they do not cover all known projections:

- *Cylindrical* projections, which are analogous to wrapping a cylinder of paper around the Earth, projecting the Earth's features onto it, and then unwrapping the cylinder

- *Azimuthal* or *planar* projections, which are analogous to touching the Earth with a sheet of flat paper

- *Conic* projections, which are analogous to wrapping a sheet of paper around the Earth in a cone

In each case, the projection's *aspect* defines the specific relationship, for example, whether the paper

**Figure 4.17** The basis for three types of map projections—cylindrical, planar, and conic. In each case a sheet of paper is wrapped around the Earth, and positions of objects on the Earth's surface are projected onto the paper. The cylindrical projection is shown in the tangent case, with the paper touching the surface, but the planar and conic projections are shown in the secant case, where the paper cuts into the surface.

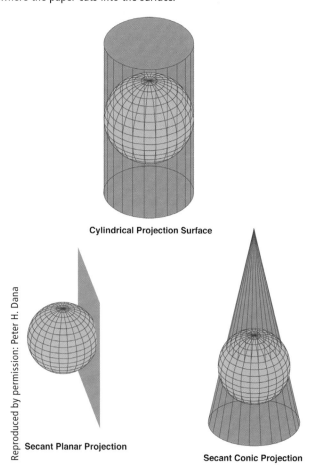

**Cylindrical Projection Surface**

Reproduced by permission: Peter H. Dana

**Secant Planar Projection**

**Secant Conic Projection**

is wrapped around the equator or touches at a pole. Where the paper coincides with the surface the scale of the projection is 1, and where the paper is some distance outside the surface the projected feature will be larger than it is on the Earth. *Secant* projections attempt to minimize distortion by imagining the paper cutting through the surface, so that scale can be both greater and less than 1 (Figure 4.17; projections for which the paper touches the Earth and in which scale is always 1 or greater are called *tangent*).

All three types can have either conformal or equal area properties, but of course not both. Figure 4.18 presents examples of several common projections and shows how the lines of latitude and longitude map onto the projection, in a (distorted) grid known as a *graticule*.

The next sections describe several particularly important projections in detail, together with the coordinate systems that they produce. Each is important to GI systems, and users are likely to come across them frequently. The map projection (and datum) used to make a dataset is sometimes not known to the user of the dataset, so it is helpful to know enough about map projections and coordinate systems to make intelligent

guesses when trying to combine such a dataset with other data. Several excellent books on map projections are listed in Further Reading.

## 4.8.1 The Plate Carrée or Cylindrical Equidistant Projection

The simplest of all projections maps longitude as *x* and latitude as *y*; for that reason this map is also known informally as the *unprojected* projection. The result is a heavily distorted image of the Earth, with the poles smeared along the entire top and bottom edges of the map and a very strangely shaped Antarctica. Nevertheless, it is the view that we most often see when images are created of the entire Earth from satellite data (for example, in illustrations of sea-surface temperature that show the El Niño or La Niña effects). The projection is not conformal (small shapes are distorted) and not equal area, though it does maintain the correct distance between every point and the equator. It is normally used only for the whole Earth, and maps of parts of the Earth, such as the United States or Canada, look distinctly odd in this projection. Figure 4.19

**Figure 4.18** Examples of some common map projections. The Mercator projection is a tangent cylindrical type, shown here in its familiar equatorial aspect (cylinder wrapped around the equator). The Lambert Conformal Conic projection is a secant conic type. In this instance, the cone onto which the surface was projected intersected the Earth along two lines of latitude: 20 North and 60 North.

**Figure 4.19** (A) The so-called unprojected or Plate Carrée projection, a tangent cylindrical projection formed by using longitude as *x* and latitude as *y*. (B) A comparison of three familiar projections of the United States. The Lambert Conformal Conic is the one most often encountered when the United States is projected alone and is the only one of the three to curve the parallels of latitude, including the northern border on the 49th Parallel.

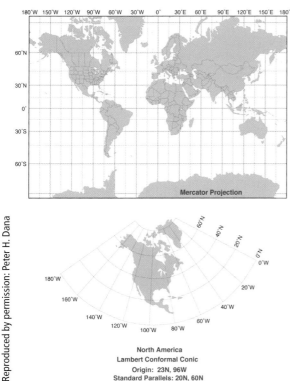

North America
Lambert Conformal Conic
Origin: 23N, 96W
Standard Parallels: 20N, 60N

(A)

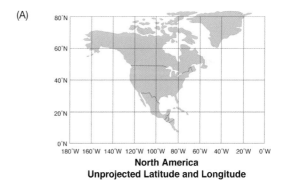

North America
Unprojected Latitude and Longitude

(B) Three Map Projections Centered at 39 N and 96 W

shows the projection applied to the world and also presents a comparison of three familiar projections of the United States: the Plate Carrée, Mercator, and Lambert Conformal Conic.

**When longitude is assigned to *x* and latitude to *y*, a very odd-looking Earth results.**

Serious problems can occur when doing analysis using this projection. Moreover, because most methods of analysis in GI systems are designed to work with Cartesian coordinates rather than latitude and longitude, the same problems can arise in analysis when a dataset uses latitude and longitude, or so-called geographic coordinates. For example, a command to generate a circle of radius one unit in this projection will create a figure that is two degrees of latitude across in the north–south direction and two degrees of longitude across in the east–west direction. On the Earth's surface this figure is not a circle at all, and at high latitudes it is a very squashed ellipse.

**It is wise to be careful when using a GI system to analyze data in latitude and longitude rather than in projected coordinates because serious distortions of distance, area, and other properties may result.**

## 4.8.2 The Universal Transverse Mercator (UTM) Projection

The UTM system is often found in military applications and in datasets with global or national coverage. It is based on the Mercator projection, but in the *transverse* rather than equatorial aspect, meaning that the projection is analogous to wrapping a cylinder around the poles rather than around the equator. There are 60 zones in the system, and each zone corresponds to a half cylinder wrapped along a particular line of longitude, each zone being 6 degrees wide. Thus Zone 1 applies to longitudes from 180 W to 174 W, with the half cylinder wrapped along 177 W; Zone 10 applies to longitudes from 126 W to 120 W centered on 123 W, and so on (Figure 4.20).

The UTM system is secant, with lines of scale 1 located some distance out on both sides of the Central Meridian. Because the projection is conformal, small features appear with the correct shape, and scale at each point is the same in all directions. Scale is 0.9996 at the Central Meridian and at most 1.0004 at the edges of the zone, so the maximum distortion of distances using this projection is about 4/100 of 1%. Both parallels and meridians are curved on the projection, with the exception of the zone's Central Meridian and the equator. Figure 4.21 shows the major features of one zone.

**Figure 4.20** The system of zones of the Universal Transverse Mercator system. The zones are identified at the top. Each zone is six degrees of longitude in width.

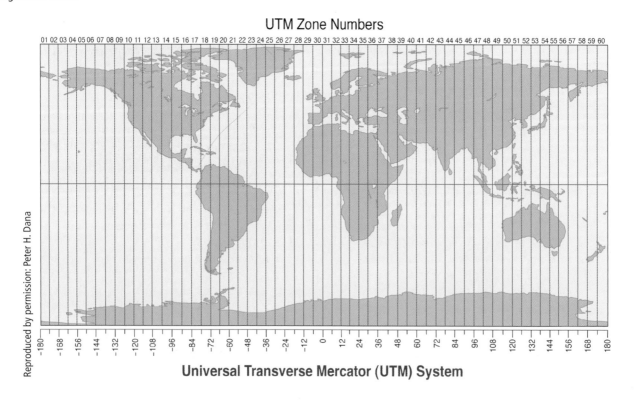

UTM Zone Numbers

Universal Transverse Mercator (UTM) System

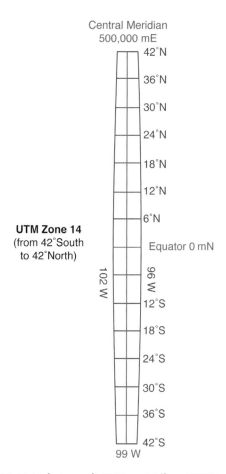

Central Meridian
500,000 mE

UTM Zone 14
(from 42°South
to 42°North)

**Figure 4.21** Major features of UTM Zone 14 (from 102 W to 96 W). The Central Meridian is at 99 W. Scale factors vary from 0.9996 at the Central Meridian to 1.0004 at the zone boundaries. See text for details of the coordinate system. (Reproduced by permission: Peter H. Dana)

The coordinates of a UTM zone are defined in meters and are set up such that the Central Meridian's easting is always 500,000 m (a *false* easting), so easting varies from near zero to near 1 million m. In the Northern Hemisphere the equator is the origin of northing, so a point at northing 5 million m is approximately 5000 km from the equator. In the Southern Hemisphere the equator is given a false northing of 10 million m, and all other northings are less than this.

**UTM coordinates are in meters, making it easy to make accurate calculations of short distances between points.**

Because there are effectively 60 different projections in the UTM system, maps will not fit together across a zone boundary. Zones become so much of a problem at high latitudes that the UTM system is normally replaced with azimuthal projections centered on each pole (known as the UPS or Universal Polar Stereographic system) above 80 degrees latitude. The problem is especially critical for cities that cross zone boundaries, such as Calgary, Alberta, Canada (which

crosses the boundary at 114 W between Zone 11 and Zone 12). In such situations one zone can be extended to cover the entire city, but this results in distortions that are larger than normal. Another option is to define a special zone, with its own Central Meridian selected to pass directly through the city's center. Italy is split between Zones 32 and 33, and many Italian maps carry both sets of eastings and northings.

UTM coordinates are easy to recognize because they commonly consist of a six-digit integer followed by a seven-digit integer (and decimal places if precision is better than a meter) and sometimes include zone numbers and hemisphere codes. They are an excellent basis for analysis because distances can be calculated from them for points within the same zone with no more than 0.04% error. But they are complicated enough that their use is effectively limited to professionals except in applications where they can be hidden from the user. UTM grids are marked on many topographic maps, and many countries project their topographic maps using UTM. It is therefore easy to obtain UTM coordinates from maps for input to digital datasets, either by hand or automatically using scanning or digitizing (Section 8.3).

### 4.8.3 Web Mercator

The advent of Web mapping services has created a need for a special projection that satisfies several specific requirements. First, it should be accurate enough to support simple kinds of analysis, such as calculation of distances when routing vehicles. Second, it should be fast to compute because users of Web services have little tolerance for response delays (latencies) of more than a fraction of a second. Third, it should be conformal so that local scale is the same in all directions. Finally, Web mapping services are most often used in areas of high population density, so high latitudes can be safely ignored.

Over the past decade the solution to this problem has emerged as what is sometimes called the Web Mercator projection and at other times the Google projection. The projection is also identified with the European Petroleum Study Group (EPSG), a standards organization, and has been assigned a number of EPSG identifiers over the years. Like any version of the Mercator projection it shows high latitudes greatly expanded, but preserves local shapes. It uses the equations for the ellipsoid, rather than the simpler equations for a spherical Earth, but its algorithms allow it to operate at speeds that satisfy user requirements. Figure 4.1 uses the Web Mercator projection, as do many other illustrations in this book derived from Web mapping services, and the misleading nature of its scale bar has already been discussed.

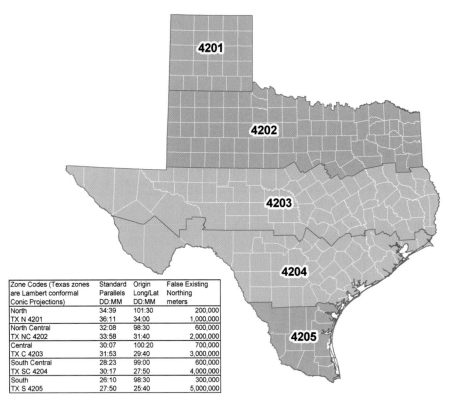

| Zone Codes (Texas zones are Lambert conformal Conic Projections) | Standard Parallels DD:MM | Origin Long/Lat DD:MM | False Existing Northing meters |
|---|---|---|---|
| North | 34:39 | 101:30 | 200,000 |
| TX N 4201 | 36:11 | 34:00 | 1,000,000 |
| North Central | 32:08 | 98:30 | 600,000 |
| TX NC 4202 | 33:58 | 31:40 | 2,000,000 |
| Central | 30:07 | 100:20 | 700,000 |
| TX C 4203 | 31:53 | 29:40 | 3,000,000 |
| South Central | 28:23 | 99:00 | 600,000 |
| TX SC 4204 | 30:17 | 27:50 | 4,000,000 |
| South | 26:10 | 98:30 | 300,000 |
| TX S 4205 | 27:50 | 25:40 | 5,000,000 |

**Figure 4.22** The five State Plane Coordinate zones of Texas. Note that the zone boundaries are defined by counties, rather than parallels, for administrative simplicity.

## 4.8.4 State Plane Coordinates and Other Local Systems

Although the distortions of the UTM system are small, they are nevertheless too great for some purposes, particularly in accurate surveying. Zone boundaries also are a problem in many applications because they follow arbitrary lines of longitude rather than boundaries between jurisdictions. In the 1930s each U.S. state agreed to adopt its own projection and coordinate system, generally known as State Plane Coordinates (SPC), in order to support these high-accuracy applications. Projections were chosen to minimize distortion over the area of the state, so choices were often based on the state's shape. Some large states decided that distortions were still too great, and so they designed their SPCs with internal zones (for example, Texas has five zones based on the Lambert Conformal Conic projection [Figure 4.22], whereas Hawaii has five zones based on the Transverse Mercator projection). Many GI systems have details of SPCs already stored, so it is easy to transform between them and UTM, or latitude and longitude. The system was revised in 1983 to accommodate the shift to the new North American Datum (NAD83).

**All U.S. states have adopted their own specialized coordinate systems for applications such as surveying that require very high accuracy.**

Many other countries have adopted coordinate systems of their own. For example, the UK uses a single projection and coordinate system known as the National Grid that is based on the Oblique Mercator projection and is marked on all topographic maps. Canada uses a uniform coordinate system based on the Lambert Conformal Conic projection, which has properties that are useful at mid- to high latitudes, for applications where the multiple zones of the UTM system would be problematic.

## 4.9 Measuring Latitude, Longitude, and Elevation: GPS

The Global Positioning System and its analogs (GLONASS in Russia, the Galileo system in Europe, and Beidou in China) have revolutionized the measurement of position, for the first time making it possible for people to know almost exactly where they are anywhere on the surface of the Earth. The GPS consists of a system of 24 satellites (plus some spares), each orbiting the Earth every 12 hours on distinct orbits at a height of 20,200 km and transmitting radio pulses at precisely timed intervals. To determine position, a receiver must make exact calculations from the signals, the known positions of the satellites, and the velocity of light. Positioning in three dimensions (latitude, longitude, and elevation) requires that at least four satellites are above the horizon, and accuracy depends on the number of such satellites and their positions (if elevation is not required, then only three satellites need be above the horizon). Several different versions of GPS exist, with distinct accuracies.

A simple GPS receiver, such as one might buy in an electronics store for $100, acquire in an iPhone or Android, or install as an optional addition to a laptop or vehicle, has an accuracy within 10 m. This accuracy will degrade in cities with tall buildings, or under trees, and GPS signals will be lost entirely under bridges or indoors. Differential GPS (DGPS) combines GPS signals from satellites with correction signals received via radio or telephone from base stations. Networks of such stations now exist, at precisely known locations, constantly broadcasting corrections; corrections are computed by comparing each known location to its apparent location determined from GPS. With DGPS correction, accuracies improve to 1 m or better. Even better accuracies are possible by using various sophisticated techniques or by remaining fixed and averaging measured locations over several hours.

GPS is very useful for recording ground control points when building GI databases; for locating objects that move (for example, combine harvesters, tanks, cars, and shipping containers); and for direct capture of the locations of many types of fixed objects, such as utility assets, buildings, geological deposits, and sample points. Other applications of GPS are discussed in Chapter 10.

Some care is needed in using GPS to measure elevation. First, accuracies are typically poorer, and a position determined to 10 m in the horizontal may be no better than plus or minus 50 m in the vertical. Second, a variety of reference elevations or vertical datums are in common use in different parts of the world and by different agencies; for example, in the United States the topographic and hydrographic definitions of the vertical datum are significantly different.

## 4.10 Converting Georeferences

GI systems are particularly powerful tools for converting between projections and coordinate systems because these transformations can be expressed as numerical operations. In fact, this ability was one of the most attractive features of early systems for handling digital geographic data and drove many early applications. But other conversions, for example, between place-names and geographic coordinates, are much more problematic. Yet they are essential operations. Almost everyone knows their mailing address and can identify travel destinations by name, but few are able to specify these locations in coordinates or to interact with GI systems on that basis. GPS technology is attractive precisely because it allows its user to determine his or her latitude and longitude, or UTM coordinates, directly at the touch of a button.

Methods of converting between georeferences are important for:

- Converting lists of customer addresses to coordinates for mapping or analysis (the task known as *geocoding*; see Box 4.3)

- Combining datasets that use different systems of georeferencing

---

### Technical Box (4.4)

#### Geocoding: Conversion of Street Addresses to Coordinates

Geocoding is the name commonly given to the process of converting street addresses to latitude and longitude, or some similarly universal coordinate system. It is widely used, as it allows any database containing addresses, such as a company mailing list or a set of medical records, to be input to a GI system and mapped. Geocoding requires a database containing records representing the geometry of street segments between consecutive intersections, and the address ranges on each side of each segment (a street centerline database; see Section 7.2.3.3). Addresses are geocoded by finding the appropriate street segment record and estimating a location based on linear interpolation within the address range. For example, 950 West Broadway in Columbia, Missouri, lies on the side of the segment whose address range runs from 900 to 998, or 50/98 = 51.02% of the distance from the start of the segment to the end. The segment starts at 92.3503 West longitude, 38.9519 North latitude and ends at 92.3527 West, 38.9522 North. Simple arithmetic gives the address location as 92.3515 West, 38.9521 North. Four decimal places suggests an accuracy of about 10 m, but the estimate also depends on the accuracy of the assumption that addresses are uniformly spaced, as well as on the accuracy of the street centerline database.

- Converting to projections that have desirable properties for analysis, for example, no distortion of area
- Searching the Internet or other distributed data resources for data about specific locations
- Positioning map displays by recentering them on places of interest that are known by name (these last two are sometimes called *locator* services)

Place-name databases (including gazetteers and POI databases) make it easy to convert named features to coordinates, though in many cases, and as already illustrated, the result may be no more than a central point representing an extensive line or area. In some cases conventions have emerged for such representative points—a common convention for rivers is to use the location of the mouth, for example. Such databases are vitally important to Web services that search for features based on user input because users will almost always refer to a feature by its name. Many Web services have become very sophisticated at helping the user to identify features, using suggestion and auto-completion, and accommodating spelling errors and alternative syntax. It is also possible to employ sophisticated software to detect place-names in text and to convert them to georeferences (e.g., www.metacarta.com).

## 4.11 Geotagging and Mashups

Online gazetteers, geocoding sites, and other services for converting georeferences have made it extremely easy to determine the geographic locations associated with many types of online data. It is now easy, for example, to take a mailing list containing street addresses, geocode them, and create maps. Services such as Google Earth, Google Maps, and their Yahoo! and Microsoft cousins provide the mapping capabilities and are readily invoked through their application programming interfaces (APIs). The term *mashup*, which derives originally from the popular-music industry, has been adopted as a way of describing the joining of two or more online services to create something that neither was able to do on its own. One of the best known mashups is www.housingmaps.com, which combines real-estate (property) listings from Craigslist (www.craigslist.org) with cartographic data from Google Maps (maps.google.com). A similar service for the UK is provided by www.nestoria.co.uk, and the www.londonprofiler.org site allows georeferenced property listings to be viewed against a range of thematic data. Today hundreds of thousands of such mashups have been created, many of them dealing with georeferenced data.

In many online services, such as Wikipedia (www.wikipedia.org), georeferences are embedded in text in the form of *geotags*, codes representing latitude and longitude in compressed form. Geotags make it easy to create mashups with mapping services.

## 4.12 Georegistration

Converting between georeferences is often needed during the process of assembling a database, when two datasets use different and perhaps unknown coordinate systems. For example, two datasets were needed in a study of fire alarms in London, Ontario: the locations of each of several thousand alarms and the boundaries of the tracts defined by the Canadian census. Linking the two datasets would allow the number of alarms to be determined in each tract and to be compared to statistics about the tract's housing and residents. The map of tract boundaries used UTM coordinates, but the map of alarms had been created from a street map using a digitizer (Section 8.3.2.1).

The standard approach in situations like this is to find a number of points that can serve as *registration* points or *tics*, using them to find equations that will convert coordinates. In this case it was convenient to use 10 major street intersections scattered over the city because these could be readily located both in the tract boundaries and on the map of alarms and to use UTM as the common system. Table 4.2 shows the coordinates of these intersections on the alarm map ($x$ and $y$) and in UTM. Note that the UTM coordinates in meters are rounded to the nearest hundred, an appropriate decision given the limited positional accuracy of street intersections.

To convert the alarm map to UTM, we look for the simplest possible equations and often adopt an *affine transformation* of the form:

$$\text{UTM north} = a + bx + cy$$
$$\text{UTM east} = d + ex + fy$$

where $a$ through $f$ are values to be determined from the tic points. Standard methods exist in GI systems for doing this, yielding the following values:

$$\text{UTM north} = 4744134 + 401.0\ x + 1222.2\ y$$
$$\text{UTM east} = 473668 + 1207.9\ x - 464.6\ y$$

Finally, these two equations are used to convert all the fire alarm data points to UTM so that the analysis can proceed.

Figure 4.23 shows another example of georegistration. The highlighted area was the focus of a proposed project on the campus of the University of California, Santa Barbara, which was intended to restore the area to its natural wetland state. A number of wells were drilled to monitor groundwater and

**Table 4.2** Registration points, with coordinates in both systems.

| Intersection | x | y | UTM north | UTM east |
|---|---|---|---|---|
| Oxford and Sanatorium | 2.90 | 9.80 | 4757500 | 472700 |
| Wonderland and Southdale | 4.86 | 5.96 | 4753800 | 476500 |
| Wharncliffe and Stanley | 7.32 | 8.56 | 4758200 | 478600 |
| Oxford and Wharncliffe | 7.58 | 9.67 | 4759800 | 478400 |
| Wellington and Southdale | 8.24 | 4.90 | 4754300 | 481500 |
| Highbury and Hamilton | 11.32 | 6.67 | 4758200 | 484100 |
| Trafalgar and Clarke | 13.17 | 7.56 | 4759700 | 486200 |
| Adelaide and Dundas | 9.49 | 8.59 | 4759400 | 481100 |
| Highbury and Fanshawe | 11.50 | 12.68 | 4765400 | 481500 |
| Richmond and Huron | 8.28 | 10.73 | 4761500 | 478800 |

accurately located using GPS. It was then necessary to merge these data with the campus database of detailed fine-resolution imagery and ground elevations, which used a different coordinate system. Registration points were established on features, such as the corners of buildings and streetlights, which could be easily located in both datasets. Note the difficulty in this case of registering the lower part of the highlighted area, where there are no readily identifiable features to use for registration.

The figure shows the result, displayed in Google Earth. Note the obvious displacement between the highlighted area and the Google Earth imagery, but the apparent agreement between the highlighted

**Figure 4.23** Registration of a fine-resolution image (highlighted) of part of the campus of the University of California, Santa Barbara, shown as a Google Earth mashup.

© 2009 Google, Image U.S. Geological Survey

area and the Google Earth roads. It seems clear that the roads and the highlighted area are registered much more accurately to the Earth than the Google Earth imagery, which is misplaced by more than 10 m.

## 4.13 Summary

This chapter has looked in detail at the complex ways in which humans refer to specific locations on the planet and how they measure locations. Any form of geographic information must involve some kind of georeference, and so it is important to understand the common methods, together with their advantages and disadvantages. Many of the benefits of GI systems rely on accurate georeferencing—the ability to link different items of information together through common geographic location; the ability to measure distances and areas on the Earth's surface, and to perform more complex forms of analysis; and the ability to communicate geographic information in forms that others can understand.

Georeferencing was introduced in early societies to deal with the need to describe locations. As humanity has progressed, we have found it more and more necessary to describe locations accurately and over wider and wider domains, so that today our methods of georeferencing are able to locate phenomena unambiguously and accurately anywhere on the Earth's surface. Today, with modern methods of measurement, it is possible to direct another person to a point on the other side of the Earth to an accuracy of a few centimeters. This level of accuracy and referencing is regularly achieved in such areas as geophysics and civil engineering.

But georeferences can never be perfectly accurate, and it is always important to know something about uncertainty. Questions of measurement accuracy are discussed at length in Chapter 5, and Section 5.2.3 deals with techniques for representation of phenomena that are inherently fuzzy, such that it is impossible to say with certainty whether a given point is inside or outside the georeference.

### Questions for Further Study

1. Visit your local map library, and determine: (1) the projections and datums used by selected maps; and (2) the coordinates of your house in several common georeferencing systems.

2. Summarize the arguments for and against a single global figure of the Earth, such as WGS84.

3. Access several online mapping services such as Google Maps, Google Earth, or Yahoo! Maps. What projection does each one use, how does it display map scale, and how does map scale vary over the map?

4. Chapter 14 discusses various forms of measurement in GI systems. Review each of those methods and the issues involved in performing analysis on databases that use different map projections. Identify the map projections that would be best for measurement of (1) area, (2) length, and (3) shape.

### Further Reading

Bugayevskiy, L. M. and Snyder, J. P. 1994. *Map Projections: A Reference Manual*. London: Taylor and Francis.

Kennedy, M. 2010. *The Global Positioning System and ArcGIS*. 3rd ed. Boca Raton, FL: CRC Press.

Maling, D. H. 1992. *Coordinate Systems and Map Projections*. 2nd ed. Oxford: Pergamon.

Snyder, J. P. 1997. *Flattening the Earth: Two Thousand Years of Map Projections*. Chicago: University of Chicago Press.

Sobel, D. 1994. *Longitude: The True Story of a Lone Genius Who Solved the Greatest Scientific Problem of His Time*. New York: Walker.

Steede-Terry, K. 2000. *Integrating GIS and the Global Positioning System*. Redlands, CA: Esri Press.

Whitaker, R. W. 2004. *The Mapmaker's Wife: A True Tale of Love, Murder, and Survival in the Amazon*. New York: Basic Books.

# 5 Uncertainty

Uncertainty in geographic representation arises because, of necessity, almost all representations of the world are incomplete. As a result, data in a geographic information (GI) system can be subject to measurement error, out of date, excessively generalized, or just plain wrong. This chapter identifies many of the sources of geographic uncertainty and the ways in which they operate in GI representations. Uncertainty arises from the way that GI users conceive of the world, how they measure and represent it, and how they analyze their representations of it. We investigate a number of conceptual issues in the creation and management of uncertainty, before reviewing the ways in which it may be measured using statistical and other methods. The propagation of uncertainty through geographical analysis is then considered. Uncertainty is an inevitable characteristic of GI usage, and one that users must learn to live with. In these circumstances, it becomes clear that all decisions based on GI systems are also subject to uncertainty.

## LEARNING OBJECTIVES

After studying this chapter you will:

- Understand the concept of uncertainty and the ways in which it arises from imperfect representation of geographic phenomena.

- Be aware of the uncertainties introduced in the three stages (conception, measurement and representation, and analysis) of database creation and use.

- Understand the concepts of vagueness and ambiguity and the uncertainties arising from the definition of key GI attributes.

- Understand how and why scale of geographic measurement and analysis can both create and propagate uncertainty.

## 5.1 Introduction

We can think of representations of the real world in GI databases as reconciling science with practice (how closely we can conform to the most appropriate scientific procedures: Section 2.4), concepts with applications (the normative versus the positive: Section 1.1.1), and analytical methods with the social context in which they are applied (see Section 1.7). Yet, almost always, such reconciliation is imperfect because, necessarily, representations of the world are incomplete (Section 3.4). In this chapter we will use *uncertainty* as an umbrella term to describe the problems that arise out of these imperfections. Occasionally, representations may approach perfect accuracy and precision (terms that we will define in Section 5.3.2.2)—as might be the case, for example, in the detailed site layout layer of a utility management system in which strenuous efforts are made to reconcile fine-scale multiple measurements of built environments. Yet perfect, or nearly perfect, representations of reality are the exception rather than the norm.

More often, the inherent complexity and detail of our world makes it virtually impossible to capture every single facet, at every possible scale, in a digital representation. (Neither is this usually desirable; see the discussion of spatial sampling in Section 2.4.) Furthermore, different individuals see the world in different ways, and in practice no single view is likely to be seen universally as the best or to enjoy uncontested status. In this chapter we discuss how the processes and procedures of abstraction create differences between the contents of our (geographic and attribute) databases and the observable world that we purport them to represent. Such differences are almost inevitable, and understanding them can help us to manage uncertainty and to live with it.

**It is impossible to make a perfect representation of the world, so uncertainty about it is inevitable.**

Various terms are used to describe differences between the real world and how it appears in a GI database, depending on the context. The concept of error in statistics arises in part from omission of some relevant aspects of a phenomenon—as in the failure to fully specify all the predictor variables in a multiple regression model, for example. Similar problems arise when one or more variables are omitted from the calculation of a composite indicator—as, for example, in omitting road accessibility in an index of land value or omitting employment status from a measure of social deprivation (see Sections 3.8.2 and 15.2.1 for discussions of indicators). The established scientific notion of measurement *error* focuses on differences between observers or between measuring instruments. This raises issues of *accuracy*, which can be defined as the difference between reality and *our* representation of reality. Although such differences are often principally addressed in formal mathematical terms, the use of the word *our* acknowledges the varying perspectives that different observers may take upon a complex, multiscale, and inherently uncertain world.

Yet even this established framework is too simple for understanding quality or the defining standards of geographic data. The terms *ambiguity* and *vagueness* (defined in Section 5.2.2) identify further considerations that need to be taken into account in assessing the *quality* of a GI representation. Many geographic representations depend on inherently vague definitions and concepts. Quality is an important topic in GI systems, and many attempts have been made to identify its basic dimensions. The U.S. Federal Geographic Data Committee's (FGDC's) various standards list five components of quality: attribute accuracy, positional accuracy, logical consistency, completeness,

and lineage. Definitions and other details on each of these and several more can be found on the FGDC's Web pages (www.fgdc.gov). Error, inaccuracy, ambiguity, and vagueness all contribute to the notion of uncertainty in the broadest sense, and uncertainty may thus be defined as a measure of the user's understanding of the difference between the contents of a dataset and the observable phenomena the data are believed to represent. This definition implies that phenomena are real, but includes the possibility that we are unable to describe them exactly. In GI systems, the term *uncertainty* has come to be used as the catchall term to describe situations in which the digital representation is simply incomplete and as a measure of the general quality of the representation.

**Uncertainty accounts for the difference between the contents of a dataset and the phenomena that the data are supposed to represent.**

The views outlined in the previous paragraph are themselves controversial and provide a rich ground for endless philosophical discussions. Some would argue that uncertainty can be inherent in phenomena themselves rather than just in their description. Others would argue for distinctions between *vagueness, uncertainty, fuzziness, imprecision, inaccuracy,* and many other terms that most people use as if they were essentially synonymous. Geographer Peter Fisher has provided a useful and wide-ranging discussion of these terms. We take the catchall view here and leave these arguments to further study.

In this chapter, we will discuss some of the principal sources of uncertainty and some of the ways in which uncertainty degrades the quality of a spatial representation. The way in which we conceive of a geographic phenomenon very much prescribes the way in which we are likely to set about representing (or measuring) it. Representation, in turn, heavily conditions the ways in which it may be analyzed within a GI system. This chain of events sequence, in which *conception* prescribes *representation*, which in turn prescribes *analysis*, is a succinct way of summarizing much of the content of this chapter and is summarized in Figure 5.1. In this diagram, U1, U2, and U3 each denote *filters* that can be thought of as selectively distorting or transforming the real world when it is stored and analyzed in a GI system. A later chapter (Section 12.2.1) introduces a fourth filter that mediates interpretation of analysis and the ways in which feedback may be accommodated through improvements in representation.

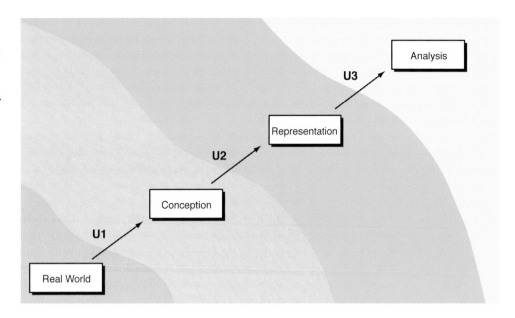

**Figure 5.1** A conceptual view of uncertainty. The three filters, U1, U2, and U3, distort the way in which the complexity of the real world is conceived, represented, and analyzed in a cumulative way.

## 5.2 U1: Uncertainty in the Conception of Geographic Phenomena

In Chapter 3 we defined an atom of geographic information as linking a descriptive property or *attribute*, a *place*, and a *time* (Section 3.4). We have acknowledged that what is left out of a representation may be important, but have nonetheless assumed that what is included in a representation is founded on clear conceptions of places and attributes. In fact, it often turns out that this is not the case, and the working definitions that are used to represent places or attributes may or may not be fit for purpose.

### 5.2.1 Conceptions of Place: Units of Analysis

The first component of an atom of geographic information is a *place*. The Tobler Law (Section 2.2), spatial autocorrelation (Section 2.7), and fractal geometry (Section 2.8) all testify to the special nature of spatial relationships between attributes, but this begs the important question of how these relationships may be best represented. In practice, this requires the creation of areal units of analysis with boundaries between them. Philosopher Barry Smith has argued that the basic typology of spatial boundaries involves distinguishing between *bona fide* (or physical) boundaries and *fiat* boundaries that are induced through human demarcation. Many geographic boundaries fall into the latter category and are said to demarcate areas that are *fiat objects*. Only rarely can geographic units of analysis be described as innate or *natural* to

the purpose of geographic enquiry. What is the natural unit of measurement for a soil profile? What is the spatial extent of a *pocket* of high unemployment, or a *cluster* of cancer cases? How might we delimit the polluting effect of a coal-fired power station?

The questions become still more difficult in bivariate (two-variable) and multivariate (more than two variable) studies. At what scale is it appropriate to investigate any relationship between background radiation and the incidence of leukemia? Or to assess any relationship between labor-force qualifications and unemployment rates? Figure 5.2 shows some "hotspots" (local smoothed aggregations, designed to maintain the confidentiality of individual patient records) of the incidence of diabetes in the London Borough of Southwark. The raw data do indeed suggest local concentrations of the problem, but do not immediately suggest any areal basis to attempt interventions such as promoting healthier diets.

**In many cases there are no natural units for geographic analysis.**

The discrete object view of geographic phenomena relies far more on the idea of natural units of analysis than the field view. As such, this problem is more likely to be manifest in vector GI applications, such as those identified in Table 3.3. Things we manipulate, such as pencils, books, or screwdrivers, are obvious natural units. Biological organisms are almost always natural units of analysis, as are groupings such as households or families—though even here there are certainly difficult cases, such as the massive networks of fungal strands that are often claimed to be the largest living organisms on Earth, or extended families of human individuals. Most of the difficult cases fall into one of two categories—they are either instances of

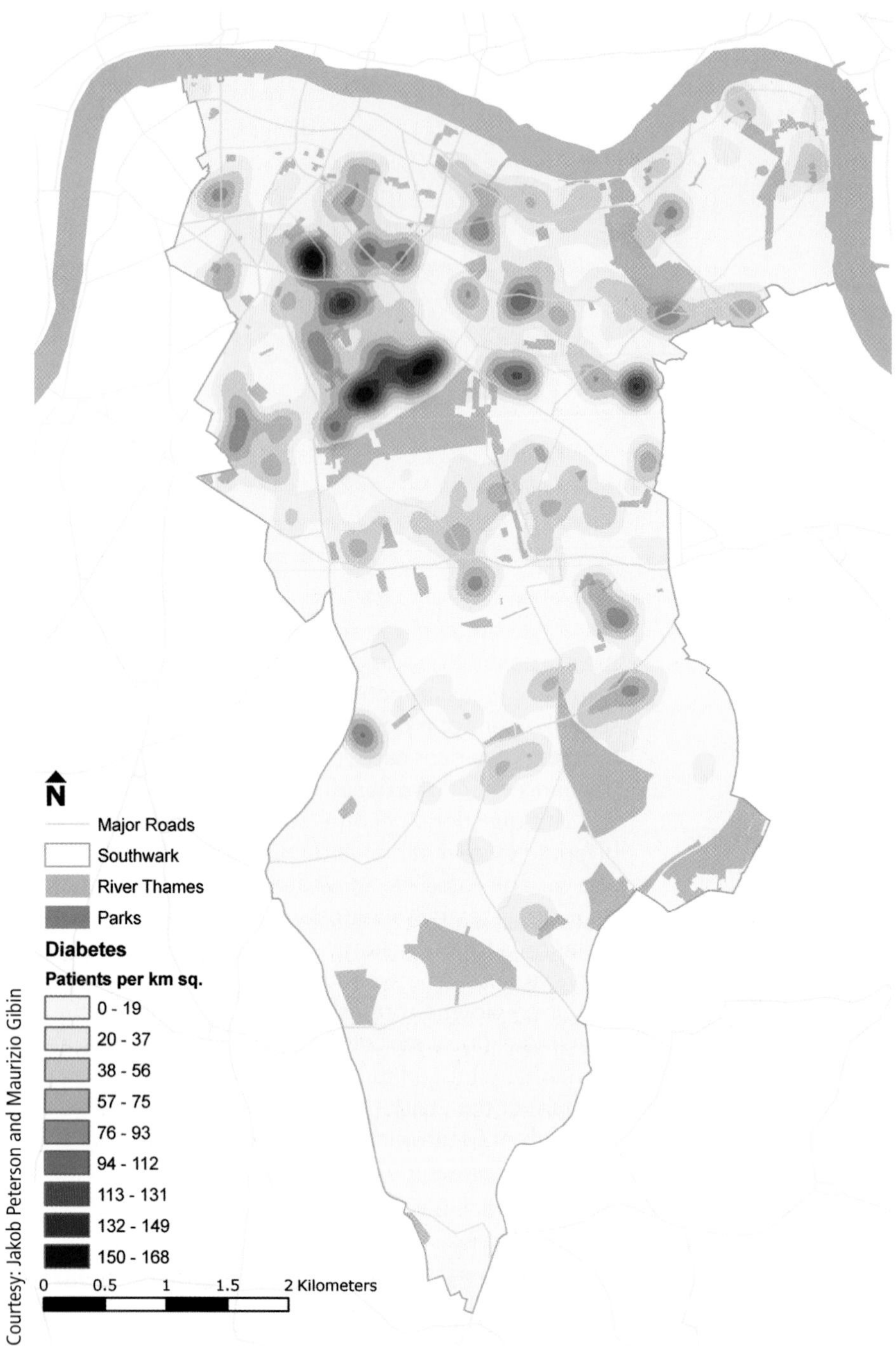

Courtesy: Jakob Peterson and Maurizio Gibin

**Figure 5.2** A map of local concentrations of diabetes in the London Borough of Southwark.

Legend:

Major Roads
Southwark
River Thames
Parks

**Diabetes**

**Patients per km sq.**

0 - 19
20 - 37
38 - 56
57 - 75
76 - 93
94 - 112
113 - 131
132 - 149
150 - 168

0    0.5    1    1.5    2 Kilometers

fields, where variation can be thought of as inherently continuous in space, or they are instances of poorly defined aggregations of discrete objects. In both of these cases it is up to the investigator to make the decisions about units of analysis, making the identification of the objects of analysis inherently subjective.

The absence of objective or uncontested definitions of place has not prevented geographers from attempting to classify them. The long-established regional geography tradition is fundamentally concerned with the quest to delineate zones that are internally homogeneous (with respect to climate, economic development, or agricultural land use, for example), set within a zonal scheme that maximizes between-zone heterogeneity—such as the map shown in Figure 5.3. Regional geography is fundamentally about delineating *uniform* zones, and many employ multivariate statistical techniques such as

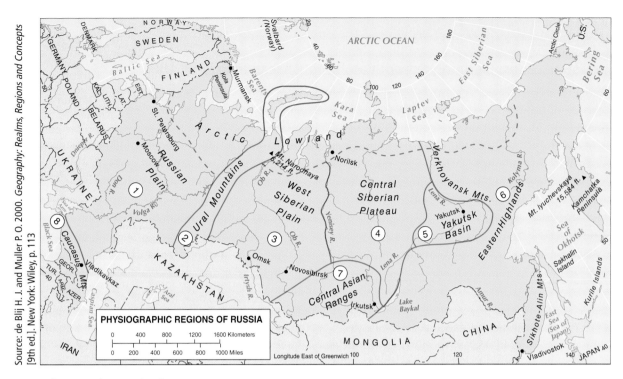

Source: de Blij H. J. and Muller P. O. 2000. Geography: Realms, Regions and Concepts [9th ed.]. New York: Wiley, p. 113

**Figure 5.3** The regional geography of Russia.

cluster analysis to supplement intuition—or sometimes to post-rationalize it.

> **Identification of homogeneous zones and spheres of influence lies at the heart of traditional regional geography as well as contemporary data analysis.**

Other geographers have tried to develop *functional* zonal schemes in which zone boundaries delineate the breakpoints between the spheres of influence of adjacent facilities or features—as in the definition of travel-to-work areas, for example, or the definition of a river catchment. Zones may be defined such that there is maximal interaction within zones and minimal interaction between zones. Any functional zoning system is likely to prove contentious, if spending public funds results in different outcomes for individuals in different locations. Nowhere is this more apparent than in the contentious domain of defining and implementing community school catchment areas (Box 5.1).

## 5.2.2 Conceptions of Attributes: Vagueness and Ambiguity

### 5.2.2.1 Vagueness
The frequent absence of objective geographic individual units means that, in practice, the labels

that we assign to zones and the ways in which we draw zone boundaries are often only vague best guesses. What absolute or relative incidence of oak trees in a forested zone qualifies it for the label *oak woodland* (Figure 5.5)? Or, in a developing country context in which aerial photography rather than ground enumeration is used to estimate population size, what rate of incidence of domestic properties indicates a zone of *dense* population? In each of these instances, it is expedient to transform point-like events (individual trees or individual properties) into area objects, and pragmatic decisions must then be taken in order to create a working definition of a spatial distribution. These decisions rarely have any absolute validity, and they raise two important questions:

- Is the defining boundary of a zone crisp and well defined?
- Is the assignment of a particular label to a given zone robust and defensible?

> **Uncertainty can exist both in the positions of the boundaries of a zone and in its attributes.**

The idiographic tradition in geography (see Section 1.3) has a long-held preoccupation with defining

## Functional Zones: Defining School Catchment Areas in Bristol, UK

In September 2007, the City of Bristol opened an attractive public-funded school at Redland Green, a wealthy area in which the inadequacies of provision had previously resulted in many parents seeking alternatives in the private sector. The location of the new school was itself the outcome of intense local lobbying, and the geography of the school's area of primary responsibility (in effect, its "catchment") was somewhat odd in appearance relative to the school's location (Figure 5.4).

When public (in the U.S., public-funded, sense) schools are oversubscribed in Britain, local authorities frequently use distance measures to allocate places to children who live nearer to the school. Bristol City Council almost totally failed to understand and anticipate demand for what was, in effect, a new public service, and in the first year that the new school buildings were open, almost all the places were allocated to prospective pupils living within 1.4 km of the school gates. Much less than half of the school's catchment area was served by the new school.

Responding to complaints from local parents, the UK Local Government Ombudsman ruled that the Council had given parents an "unrealistic expectation" of securing a place at the school and ordered the school to open its gates to appellants who had been denied access. While adhering to the Ombudsman's ruling, in the longer term the Council had to adapt to unforeseen circumstances by redefining the community for which the school was intended. It has considered a number of options, including redefining the catchment area to serve a much more geographically localized community of winners in its publicly funded schools lottery.

There is no such thing as a natural area for a school catchment. This being the case, GI systems can assist in defining a functional zone that is fit for purpose by deriving measures of demand and fair access to the facility—using socioeconomic data such as numbers of children of school age, travel time, and so on. Bristol City Council failed to do this. It was able to use ArcGIS measures of distance for the operational task of rationing places once the conceptual failure had become apparent, but this could not compensate for the failure to use any of the analytical functions of GI systems in any more strategic sense to anticipate demand for the new facility at the planning stage (see Chapter 17). The conception, and hence definition, of the school's catchment area was wholly inadequate for the community function envisaged for the school when public funds were initially committed to it.

**Figure 5.4** The original catchment area of Bristol's Redland Green School, and the subarea within which offers of places were originally made.

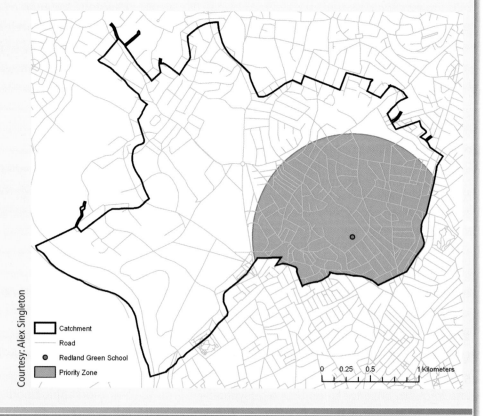

Courtesy: Alex Singleton

☐ Catchment
— Road
◉ Redland Green School
▓ Priority Zone

0    0.25    0.5          1 Kilometers

**Figure 5.5** A local assemblage of oak trees or part of an "oak woodland" on Ragged Boys Hill, in the New Forest, UK. The Ordnance Survey map of the area suggests that there are trees beyond the perimeter of the "woodland" area, whereas the tree symbology suggests that the woods are characterized by varying proportions of deciduous and coniferous trees.

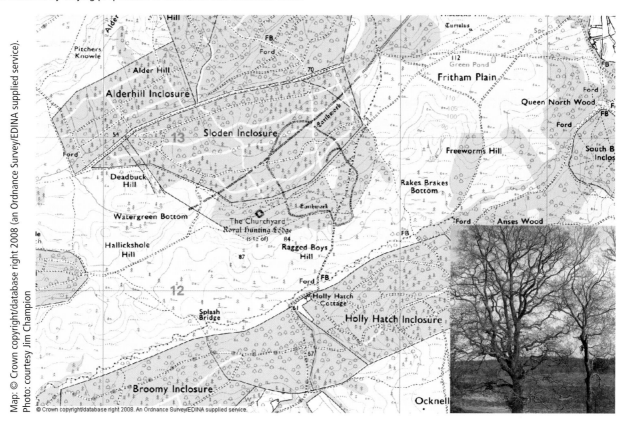

Map: © Crown copyright/database right 2008 (an Ordnance Survey/EDINA supplied service).
Photo: courtesy Jim Champion

© Crown copyright/database right 2008. An Ordnance Survey/EDINA supplied service.

regions, and the vagaries inherent in tracing lines across maps can have important political or economic implications—as, for example, with the drawing of the Dayton Agreement lines to partition Bosnia and Herzegovina in 1995, or in defining the geographical areas to which political decision making should be devolved. The data processing power of GI systems may be harnessed through *geocomputation*

(Section 15.1) to build regions from atoms of geographic information. Box 5.2 describes how the geography of Anglo-Saxon family names can be used to regionalize Great Britain.

**Many English-language terms used to convey geographic information are inherently ambiguous.**

### Vagueness, Ambiguity, and the Geographies of Family Names

In Great Britain, family names ("surnames") entered common parlance in the Thirteenth Century. Most such names are toponyms (denoting landscape features or places, such as Castle or London), metonyms (denoting occupations, such as Smith or Baker), or diminutives (denoting family linkage, such as "William's son," Williamson or the abbreviated form Williams). Family names characteristic of unique places remain concentrated near the places where they were first coined—you are on average 53 times more likely to meet someone called Rossall in Blackpool, England, than in a randomly selected location, for example, because Blackpool is very close to the settlement of that name. Other broader types of names exhibit strong regional concentrations—as with the widespread use of the diminutive "-s" suffix in Wales (e.g., Jones or Williams).

▶

Analysis of family names tells us a lot about the enduring and unique human geographies of places, for the very good reason that throughout history most people have not moved very far from the places where they were brought up. The concentrations of individual names and types of names in places provide us with an interesting indicator of the distinctiveness of places and a basis to compare the shared characteristics of their populations. Using GI systems, we can do much more than map a single family name or a single family name type (see Box 1.2). Using geocomputational clustering techniques (see Section 15.1), we can identify the degree to which the mix of names in a particular place is distinctive.

Yet "distinctive" is a subjective, hence ambiguous, term. The lower the threshold that we adopt for defining a distinctive region, the more regions we will identify. Figure 5.6 shows how geographer James Cheshire has used names to partition Great Britain into between two and seven regions. Note how the distinctiveness of Scottish and Welsh names dominates the first two maps, successively followed by an emergent "north–south divide" in England, the separation of London from the rest of the south, further partitioning of the north, and finally the separation of the urban conurbations of northwest England.

These regions are much better than vague best guesses, as they are rooted in the naming conventions of a bygone age. But why stop at seven regions? Check out James Cheshire's regional geographies of Britain at www.spatialanalysis.co.uk/surnames.

Naming conventions can also help us resolve the ambiguities inherent in mapping the local geographies of ethnic minority populations, which is important for a wide range of policy applications. Data on ethnicity are collected in many population censuses but are vulnerable to the ambiguities arising from the ways in which individuals assign themselves to ethnic groups. (Would the third-generation descendant of a Polish immigrant to the U.S.

describe himself or herself as "Polish," for example?). A different approach, which seeks neatly to circumvent these issues, begins with the adage that "a name is a statement." Research at University College London has used techniques of cluster analysis to classify names into more than 160 cultural, ethnic, and linguistic groups. The resulting geocomputational classification is the result of inductive classification of over 600 million names and enables the somewhat crude groups used in official statistics to be compared with a truly multicultural atlas of the UK. Figures 5.7A and B compare the official statistics classification with the names classification.

Check whether the classification assigns your name to the correct group at www.onomap.org and look at its global distribution at worldnames.publicprofiler.org.

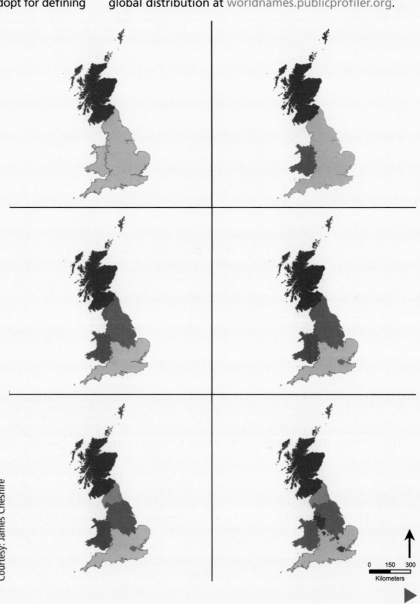

Courtesy: James Cheshire

Figure 5.6 The use of family names to regionalize Great Britain.

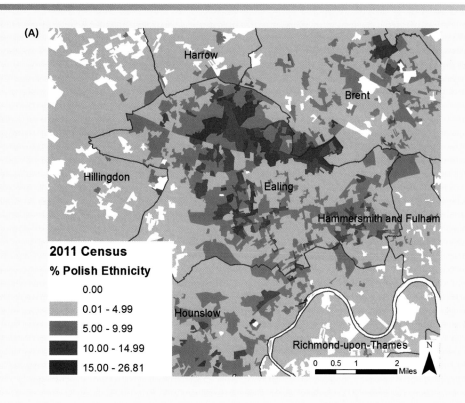

(A)

**2011 Census**

**% Polish Ethnicity**

- 0.00
- 0.01 - 4.99
- 5.00 - 9.99
- 10.00 - 14.99
- 15.00 - 26.81

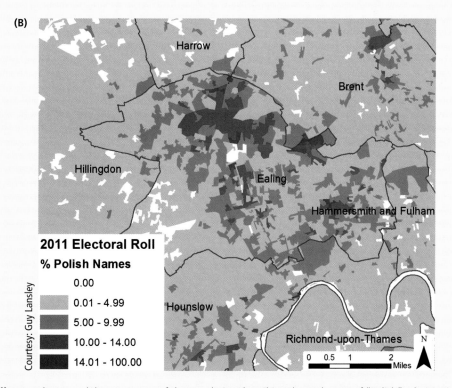

(B)

**2011 Electoral Roll**

**% Polish Names**

- 0.00
- 0.01 - 4.99
- 5.00 - 9.99
- 10.00 - 14.00
- 14.01 - 100.00

Courtesy: Guy Lansley

**Figure 5.7** Differences between (A) percentages of the population describing themselves as of "Polish" ethnicity in 2011 UK Census for part of London, compared with (B) the percentages of people identified as having Polish names by applying names classification software to an enhanced version of the public Electoral Roll for the same year. Each definition and classification procedure has its own inherent uncertainties.

These questions have statistical implications (can we put numbers on the confidence associated with boundaries or labels?), cartographic implications (how can we convey the meaning of vague boundaries and labels through appropriate symbols on maps and displays of GI?) and cognitive implications (how do people subconsciously attempt to force things into categories and boundaries to satisfy a deep need to simplify the world?).

#### 5.2.2.2 Ambiguity

Many objects are assigned different labels by different national or cultural groups, and such groups may share different spatial perceptions. Geographic prepositions in the English language such as *across*, *over*, and *in* do not have simple correspondences with terms in other languages. Object names and the topological relations between them may thus be inherently *ambiguous*. Perception, behavior, language, and cognition all play a part in the conception of real-world entities and the relationships between them. GI systems cannot provide the magic bullet of a value-neutral evidence base for decision making, and GI systems can be used to systematically privilege some worldviews over others. GI systems can also provide a formal framework for the reconciliation of different representations (see Section 3.3).

Ambiguity also arises in the conception and construction of *indicators* (see also Section 15.2.1). *Direct* indicators are deemed to bear a clear correspondence with a mapped phenomenon. Detailed household income figures, for example, can provide a direct indicator of the likely geography of expenditure and demand for goods and services; tree diameter at breast height can be used to estimate stand value; and field nutrient measures can be used to estimate agronomic yield. *Indirect* indicators are used when the best available measure is likely only to have surrogate link with the phenomenon of interest. Thus the incidence of central heating among households, or rates of multiple car ownership, might provide a surrogate for household income data if such data are not available, whereas local atmospheric measurements of nitrogen oxide can provide an indirect indicator of environmental health. Box 5.2 describes how people's names provide a direct indicator of their ethnicity and how this information may be used to improve on the detail, quality, and timeliness of ethnicity data collected in censuses of population.

Conception of the (direct or indirect) linkage between any indicator and the phenomenon of interest is subjective and hence ambiguous. Such measures will create errors of measurement if the correspondence between the two is imperfect, and these

errors may be systematic. So, for example, differences in the conception of what hardship or deprivation entail can lead to specification of different composite indicators, whereas different geodemographic systems can include correspondingly varied cocktails of census variables. With regard to the natural environment, conception of critical defining properties of soils can lead to inherent ambiguity in their classification (see Section 5.2.3).

**Ambiguity is introduced when imperfect indicators of phenomena are used instead of the phenomena themselves.**

Fundamentally, GI systems have upgraded our abilities to generalize about spatial distributions. Yet our abilities to do so may remain constrained by the different taxonomies that are conceived and used by data-collecting organizations within our overall study area. A study of wetland classification in the United States found no fewer than six agencies engaged in mapping the same phenomena over the same geographic areas, and each with its own definitions of wetland types (see Section 1.3). If wetland maps are to be used in regulating the use of land, as they are in many areas, then uncertainty in mapping clearly exposes regulatory agencies to potentially damaging and costly lawsuits. How might soils data classified according to the UK national classification be assimilated within a pan-European soils map, which uses a classification honed to the full range and diversity of soils found across the Europe rather than those just on an assemblage of offshore islands? How might different national geodemographic classifications be combined into a form suitable for a pan-European marketing exercise? These are all variants of the question:

How may mismatches between the categories of different classification schema be reconciled?

**Differences in definitions are a major impediment to integration of geographic data over wide areas.**

Like the process of pinning down the different nomenclatures developed in different cultural settings, the process of reconciling the semantics of different classification schema is an inherently *ambiguous* procedure. Ambiguity arises in data concatenation when we are unsure regarding the *metacategory* to which a particular class should be assigned.

### 5.2.3 Fuzzy Approaches to Attribute Classification

One way of resolving the assignment process is to adopt a probabilistic interpretation. If we take a statement like "the database indicates that this field contains wheat, but there is a 0.17 probability (or 17% chance) that it actually contains barley," there are at least two possible interpretations: (1) if 100 randomly chosen people were asked to make independent assessments of the field on the ground, 17 would determine that it contains barley, and 83 would decide it contains wheat; or (2) of 100 similar fields in the database, 17 actually contained barley when checked on the ground and 83 contained wheat. Of the two, we probably find the second more acceptable because the first implies that people cannot correctly determine the crop in the field.

But the important point is that, in conceptual terms, both of these interpretations are *frequentist* because they are based on the notion that the probability of a given outcome can be defined as the proportion of times the outcome occurs in some real or imagined experiment, when the number of tests is very large. Although this interpretation is reasonable for classic statistical experiments, like tossing coins or drawing balls from an urn, the geographic situation is different—there is only one field with precisely these characteristics, and one observer, and in order to imagine a number of tests we have to invent more than one observer, or more than one field. (The problems of imagining larger populations for some geographic samples are discussed further in Section 14.5.)

In part because of this problem, many people prefer the *subjectivist* conception of probability—that it represents a judgment about relative likelihood that is not the result of any frequentist experiment, real or imagined. Subjective probability is similar in many ways to the concept of fuzzy sets, and the latter framework will be used here to emphasize the contrast with frequentist probability.

Suppose we are asked to examine an aerial photograph to determine whether a field contains wheat, and we decide that we are not sure. However, we are able to put a number on our degree of uncertainty by putting it on a scale from 0 to 1. The more certain we are, the higher the number. Thus we might say we are 0.90 sure it is wheat, and this would reflect a greater degree of certainty than 0.80. This degree of belonging to the class *wheat* is termed the *fuzzy membership*, and it is common, though not necessary, to limit memberships to the range 0 to 1. In effect, we have changed our view of membership in classes, and we have abandoned the notion that things must either belong to classes or not belong to them. In this new world, the boundaries of classes are no longer clean and crisp, and the set of things assigned to a set can be fuzzy.

## In fuzzy logic, an object's degree of belonging to a class can be partial.

One of the major attractions of fuzzy sets is that they appear to let us deal with sets that are not precisely defined, and for which it is impossible to establish membership cleanly. Many such sets or classes are found in applications of GI, including land-use categories, neighborhood classifications, soil types, land cover classes, and vegetation types. Classes used for maps are often fuzzy, such that two people asked to classify the same location might disagree, not because of measurement error, but because the classes themselves are not perfectly defined and because opinions vary. As such, mapping is often forced to stretch the rules of scientific repeatability, which require that two observers will always agree.

Box 5.3 shows a typical extract from the legend of a soil map, and it is easy to see how two people might disagree, even though both are experts with years of experience in soil classification. Figure 5.8 shows an example of mapping classes using the fuzzy methods developed by A-Xing Zhu of the University of Wisconsin–Madison, which take both remote-sensing images and the opinions of experts as inputs. There are three classes, and each map shows the fuzzy membership values in one class, ranging from 0 (darkest) to 1 (lightest). This figure also shows the result of converting to *crisp* categories, or *hardening*—to obtain Figure 5.8D, each pixel is colored according to the class with the highest membership value.

Fuzzy approaches are attractive because they capture the uncertainty that many of us feel about the assignment of places on the ground to specific categories. But researchers have struggled with the question of whether they are more *accurate*. In a sense, if we are uncertain about which class to choose, then it is more accurate to say so, in the form of a fuzzy membership, than to be forced into assigning a class without qualification. But that does not address the question of whether the fuzzy membership value is accurate. If Class A is not well defined, it is hard to see how one person's assignment of a fuzzy membership of 0.83 in Class A can be meaningful to another person because there is no reason to believe that the two people share the same notions of what Class A means or of what 0.83 means, as distinct from 0.91 or 0.74. So although fuzzy approaches make sense at an intuitive level, it is more difficult to see how they could be helpful in the process of communication of geographic knowledge from one person to another.

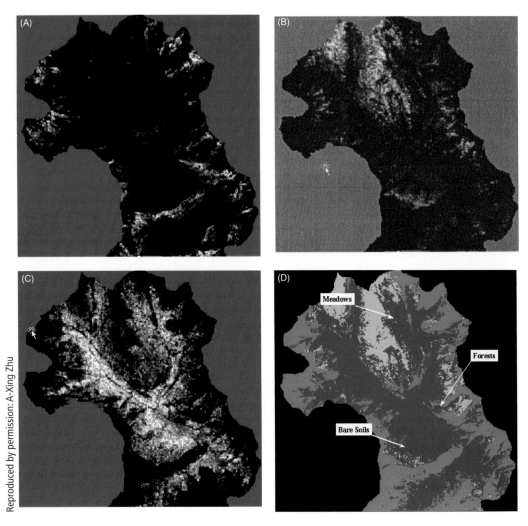

Figure 5.8 (A) Membership map for bare soils in the Upper Lake McDonald Basin, Glacier National Park. (B) Membership map for forest. (C) Membership map for alpine meadows. (D) Spatial distribution of the three cover types from hardening the membership maps.

## Fuzziness in Classification: Description of a Soil Class

The following is the description of the Limerick series of soils from New England (the type location is in Chittenden County, Vermont), as defined by the US National Cooperative Soil Survey. Note the frequent use of vague terms such as *very*, *moderate*, *about*, *typically*, and *some*. Because the definition is so loose, it is possible for many distinct soils to be lumped together in this one class—and two observers may easily disagree over whether a given soil belongs to the class, even though both are experts. The definition illustrates the extreme problems of defining soil classes with sufficient rigor to satisfy the criterion of scientific repeatability.

"The Limerick series consists of very deep, poorly drained soils on flood plains. They formed in loamy alluvium. Permeability is moderate. Slope ranges from 0 to 3 percent. Mean annual precipitation is about 34 inches and mean annual temperature is about 45 degrees F. Depth to bedrock is more than 60 inches. Reaction ranges from strongly acid to neutral in the surface layer and moderately acid to neutral in the substratum. Textures are typically silt loam or very fine sandy loam, but lenses of loamy very fine sand or very fine sand are present in some pedons. The weighted average of fine and coarser sands, in the particle-size control section, is less than 15 percent."

## 5.3 U2: Further Uncertainty in the Representation of Geographic Phenomena

As with the conception of uncertainty, it is helpful to consider the representation of uncertainty with regard to the components of geographic information—measures of places (locations), attributes, and time period (although we do not consider time in detail here). We consider both the mode of representing place (Section 5.3.1) and the accuracy and precision with which it can be measured (Section 5.3.3). We consider the measurement of attributes at the nominal, ordinal, interval, and ratio scales (see Box 2.1).

### 5.3.1 Representation of Place/Location

The conceptual models (fields and objects) that were introduced in Chapter 3 impose very different filters upon reality, and as a result, their usual corresponding representational models (raster and vector) are characterized by different uncertainties. The vector model enables a range of powerful analytical operations to be performed (see Chapters 13 through 15), yet it also requires a priori conceptualization of the nature and extent of geographic individuals and the ways in which they nest together into higher-order zones. The raster model defines individual elements as square cells, with boundaries that bear no relationship at all to natural features, but nevertheless provides a convenient and (usually) efficient structure for data handling within a GI system. However, in the absence of effective automated pattern recognition techniques, human interpretation is usually required to discriminate between real-world spatial entities as they appear in a rasterized image.

Although quite different representations of reality, both vector and raster data structures are attractive in their logical consistency, the ease with which they are able to handle spatial data, and (once the software is written) the ease with which they can be implemented in GI systems. But neither can provide any substitute for robust conception of geographic units of analysis (Section 5.2). This said, however, the conceptual distinction between fields and discrete objects is often useful in dealing with uncertainty. Figure 5.9 shows a coastline, which is often conceptualized as a discrete line object. But suppose we recognize that its position is uncertain. For example, the coastline shown on a 1:2,000,000 map is a gross generalization, in which major liberties are taken, particularly in areas where the coast is highly indented and irregular. Consequently, the 1:2,000,000 version leaves substantial uncertainty about the true location of the shoreline.

We might approach this by changing from a line to an area and mapping the area where the actual coastline lies, as shown in Figure 5.9A. But another approach would be to reconceptualize the coastline as a field by mapping a variable whose value represents the probability that a point is land. This is shown in Figure 5.9B as a raster representation. This would have far more information content and consequently much more value in many applications. But at the same time it would be difficult to find an appropriate data source for the representation—perhaps a fuzzy classification of an air photo, using one of an increasing number of techniques designed to produce representations of the uncertainty associated with objects discovered in images.

**Uncertainty can be measured differently under field and discrete object views.**

Indeed, far from offering quick fixes for eliminating or reducing uncertainty, the measurement process can actually increase it. Given that the vector and raster data models impose quite different filters on reality, it is unsurprising that they can each generate additional uncertainty in rather different ways. In field-based conceptualizations, such as those that underlie remotely sensed images expressed as rasters, spatial

Figure 5.9 The contrast between (A) discrete object and (B) field conceptualizations of an uncertain coastline.

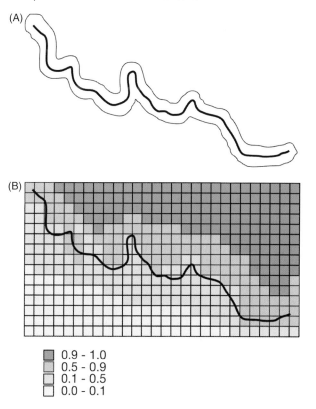

(A)

(B)

■ 0.9 - 1.0
■ 0.5 - 0.9
□ 0.1 - 0.5
□ 0.0 - 0.1

objects are not defined a priori. Instead, the classification of each cell into one or other category builds together into a representation. In remote sensing, when resolution is insufficient to detect all the detail in geographic phenomena, the term *mixel* is often used to describe raster cells that contain more than one class of land—in other words, elements in which the outcome of statistical classification suggests the occurrence of multiple land-cover categories. The total area of cells classified as mixed should decrease as the resolution of the satellite sensor increases, assuming the number of categories remains constant, yet a completely mixel-free classification is very unlikely at any level of resolution. Even where the Earth's surface is covered with perfectly homogeneous areas, such as agricultural fields growing uniform crops, the failure of real-world crop boundaries to line up with pixel edges ensures the presence of at least some mixels. Neither does finer-resolution imagery solve all problems: medium-resolution data (defined as pixel size of between 30 m × 30 m and 1000 m × 1000 m) are typically classified using between 3 and 7 bands, whereas fine-resolution data (pixel sizes 10 m × 10 m or smaller) are typically classified using between 7 and 256 bands, and this can generate much greater heterogeneity of spectral values with attendant problems for classification algorithms.

**A pixel whose area is divided among more than one class is termed a mixel.**

The vector data structure, by contrast, defines spatial entities and specifies explicit topological relations (see Section 3.6) between them. Yet this often entails transformations of the inherent characteristics of spatial objects (Chapters 13 and 14). In conceptual terms, for example, although the true individual members of a population might each be defined as point-like objects, they will often appear in a GI database only as aggregate counts for apparently *uniform* zones. Such aggregation can be driven by the need to preserve the confidentiality of individual records, or simply by the need to limit data volume. Unlike the field conceptualization of spatial phenomena, this implies that there are good reasons for partitioning space in a particular way. In practice, partitioning is often made on grounds that are principally pragmatic, yet are rarely completely random (see Section 5.4). In most socioeconomic GI applications, for example, zones that are designed to preserve the anonymity of survey respondents may often be ad hoc containers. Larger aggregations are often used for the simple reason that they permit comparisons of measures over time (see Box 5.3). They may also reflect the way that a cartographer or GI software interpolates a boundary between sampled points, as in the creation of isopleth maps (see Box 2.5).

## 5.3.2 Statistical Models of Uncertainty in Attribute Measures

Scientists have developed many widely used methods for describing errors in observations and measurements, and these methods may be applicable to GI if we are willing to think of databases as collections of measurements. For example, a digital elevation model consists of a large number of measurements of the elevation of the Earth's surface. A map of land use is also in a sense a collection of measurements because observations of the land surface have resulted in the assignment of classes to locations. Both of these are examples of observed or measured attributes, but we can also think of location as a property that is measured.

**A geographic database is a collection of measurements of phenomena on or near the Earth's surface.**

Here we consider errors in nominal class assignment, such as of types of land use and errors in continuous (interval or ratio) scales, such as elevation (see Section 3.4).

### 5.3.2.1 Nominal Case
The values of nominal data serve only to distinguish an instance of one class from an instance of another or to identify an object uniquely (Section 3.4). If classes have an inherent ranking, they are described as ordinal data, but for purposes of simplicity the ordinal case will be treated here as if it were nominal. Consider a single observation of nominal data—for example, the observation that a single parcel of land is being used for agriculture (this might be designated by giving the parcel Class A as its value of the "Land-Use Class" attribute). For some reason, perhaps related to the quality of the aerial photography being used to build the database, the class may have been recorded falsely as Class G, Grassland. A certain proportion of parcels that are truly Agriculture might be similarly recorded as Grassland, and we can think of this in terms of a probability that parcels that are truly Agriculture are falsely recorded as Grassland.

Table 5.1 shows how this might work for all of the parcels in a database. Each parcel has a true class, defined by accurate observation in the field, and a recorded class as it appears in the database. The whole table is described as a *confusion matrix*, and instances of confusion matrices are commonly encountered in applications dominated by class data, such as classifications derived from remote sensing or aerial photography. The true class might be determined by ground check, which is inherently more accurate than classification of aerial photographs but much more expensive and time consuming. Ideally,

**Table 5.1** Example of a misclassification or confusion matrix. A grand total of 304 parcels have been checked. The rows of the table correspond to the land-use class of each parcel as recorded in the database, and the columns to the class as recorded in the field. The numbers appearing on the principal diagonal of the table (from top left to bottom right) reflect correct classification.

|       | A   | B  | C  | D  | E  | Total |
|-------|-----|----|----|----|----|-------|
| A     | 80  | 4  | 0  | 15 | 7  | 106   |
| B     | 2   | 17 | 0  | 9  | 2  | 30    |
| C     | 12  | 5  | 9  | 4  | 8  | 38    |
| D     | 7   | 8  | 0  | 65 | 0  | 80    |
| E     | 3   | 2  | 1  | 6  | 38 | 50    |
| Total | 104 | 36 | 10 | 99 | 55 | 304   |

all the observations in the confusion matrix should lie along the principal diagonal, in the cells that correspond to agreement between true class and database class. But in practice certain classes are more easily confused than others, so certain cells off the diagonal will have substantial numbers of entries.

A useful way to think of the confusion matrix is as a set of rows, each defining a vector of values. The vector for any row $i$ gives the proportions of cases in which what appears to be Class $i$ is actually Class 1, 2, 3, and so on. Symbolically, this can be represented as a vector $\{p_1, p_2, \ldots, p_i, \ldots, p_n\}$, where $n$ is the number of classes and $p_i$ represents the proportion of cases for which what appears to be the class according to the database is actually Class $i$.

There are several ways of describing and summarizing the confusion matrix. If we focus on one row, then the table shows how a given class in the database falsely records what are actually different classes on the ground. For example, Row A shows that of 106 parcels recorded as Class A in the database, 80 were confirmed as Class A in the field, but 15 appeared to be truly Class D. The proportion of instances in the diagonal entries represents the proportion of correctly classified parcels, and the total of off-diagonal entries in the row is the proportion of entries in the database that appear to be of the row's class but are actually incorrectly classified. For example, there were only 9 instances of agreement between the database and the field in the case of Class D. If we look at the table's columns, the entries record the ways in which parcels that are truly of that class are actually recorded in the database. For example, of the 10 instances of Class C found in the field, 9 were recorded as such in the database and only 1 was misrecorded as Class E. The columns have been called the *producer's* perspective because the task of the producer of an accurate database is to minimize entries outside the diagonal cell in a given column; the rows have been called the *consumer's*

perspective because they record what the contents of the database actually mean on the ground—in other words, the accuracy of the database's contents.

### Users and producers of data look at misclassification in distinct ways.

For the table as a whole, the proportion of entries in diagonal cells is called the *percent correctly classified* (PCC) and is one possible way of summarizing the table. In this case 209/304 cases are on the diagonal, for a PCC of 68.8%. But this measure is misleading for at least two reasons. First, chance alone would produce some correct classifications, even in the worst circumstances, so it would be more meaningful if the scale were adjusted such that 0 represents chance. In this case, the number of chance hits on the diagonal in a random assignment is 76.2 (the sum of the row total times the column total divided by the grand total for each of the five diagonal cells). So the actual number of diagonal hits, 209, should be compared to this number, not 0. The more useful index of success is the *kappa index*, defined as

$$\kappa = \frac{\displaystyle\sum_{i=1}^{n} C_{ii} - \sum_{i=1}^{n} C_{i.}.C_{.i}/C_{..}}{C_{..} - \displaystyle\sum_{i=1}^{n} C_{i.}.C_{.i}/C_{..}}$$

where $c_{ij}$ denotes the entry in row $i$ column $j$, the dots indicate summation (e.g., $c_{i.}$ is the summation over all columns for row $i$, that is, the row $i$ total, and $c_{..}$ is the grand total), and $n$ is the number of classes. The first term in the numerator is the sum of all the diagonal entries (entries for which the row number and the column number are the same). To compute PCC, we would simply divide this term by the grand total (the first term in the denominator). For kappa, both numerator and denominator are reduced by the same amount, an estimate of the number of hits (agreements between field and database) that would occur by chance. This involves taking each diagonal cell, multiplying the row total by the column total, and dividing by the grand total. The result is summed for each diagonal cell. In this case kappa evaluates to 58.3%, a much less optimistic assessment than PCC.

The second issue with both of these measures concerns the relative abundance of different classes. In the table, Class C is much less common than Class A. The confusion matrix is a useful way of summarizing the characteristics of nominal data, but to build it there must be some source of more accurate data. Commonly, this is obtained by ground observation, and in practice the confusion matrix is created by taking samples of more accurate data, by sending observers into the field to conduct spot checks. Clearly, it makes no sense to visit every parcel, and instead a sample is taken. Because some classes are more common than others, a random sample that made every parcel equally likely to be chosen would be inefficient because too many data would be gathered on common classes, and not enough on the relatively rare ones. So, instead, samples are usually chosen such that a roughly equal number of parcels are selected in each class. Of course, these decisions must be based on the class as recorded in the database, rather than the true class. This is an instance of sampling that is *stratified* by class (see Section 2.4).

> **Sampling for accuracy assessment should pay greater attention to the classes that are rarer on the ground.**

Parcels represent a relatively easy case, if it is reasonable to assume that the land-use class of a parcel is uniform over the parcel, and class is recorded as a single attribute of each parcel object. But as we noted in Sections 2.4 and 5.2.2.1, more difficult cases arise in sampling natural areas, for example, in the case of vegetation cover class, where parcel boundaries may not exist. Figure 5.10 shows a typical vegetation cover class map and is obviously highly generalized. If we were to apply the previous strategy, then we would test each area to see if its assigned vegetation cover class checks out on the ground. But unlike the parcel case, in this example the boundaries between areas are not fixed but are themselves part of the observation process, and we need to ask whether they are correctly located. Error in this case has two forms: misallocation of an area's class and mislocation of an area's boundaries. In some cases the boundary between two areas may be fixed because it coincides with a clearly defined line on the ground; but in other cases, the boundary's location is as much a matter of judgment as the allocation of an area's class.

> **Errors in land cover maps can occur in the locations of boundaries of areas, as well as in the classification of areas.**

**Figure 5.10** An example of a vegetation cover map.

In such cases we need a different strategy that captures the influence both of mislocated boundaries and of misallocated classes. One way to deal with this is to think of error not in terms of classes assigned to areas, but in terms of classes assigned to points. In a raster dataset, the cells of the raster are a reasonable substitute for individual points. Instead of asking whether area classes are confused and estimating errors by sampling areas, we ask whether the classes assigned to raster cells are confused, and we define the confusion matrix in terms of misclassified cells. This is often called *per-pixel* or *per-point* accuracy assessment, to distinguish it from the previous strategy of *per-polygon* accuracy assessment. As before, we would want to stratify by class, to make sure that relatively rare classes were sampled in the assessment.

### 5.3.2.2 Interval/Ratio Case
The second case addresses measurements that are made on interval or ratio scales. Here, error is best thought of not as a change of class but as a change of value, such that the observed value $x'$ is equal to the true value $x$ plus some distortion $\delta x$, where $\delta x$ is hopefully small. $\delta x$ might be either positive or negative because errors are possible in both directions. For example, the measured and recorded elevation at some point might be equal to the true elevation, distorted by some small amount. If the average distortion is zero, so that positive and negative errors balance

out, the observed values are said to be *unbiased*, and the average value will be true.

### Error in measurement can produce a change of class, or a change of value, depending on the type of measurement.

Sometimes it is helpful to distinguish between *accuracy*, which has to do with the magnitude of $\delta x$, and *precision*. Unfortunately there are several ways of defining precision in this context, at least two of which are regularly encountered in the context of GI. Surveyors and others concerned with measuring instruments tend to define precision through the performance of an instrument in making repeated measurements of the same phenomenon. A measuring instrument is precise according to this definition if it repeatedly gives similar measurements, whether or not these are actually accurate. So a GPS receiver might make successive measurements of the same elevation, and if these are similar the instrument is said to be precise. Precision in this case can be measured by the variability among repeated measurements. But it is possible that all the measurements are approximately 5 m too high, in which case the measurements are said to be biased, even though they are precise, and the instrument is said to be inaccurate. Figure 5.11 illustrates this meaning of precision and its relationship to accuracy. The other definition of precision is more common in science generally. It defines precision as the number of digits

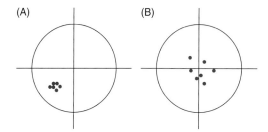

**Figure 5.11** (A) Successive measurements have similar values (they are precise). (B) Precision is lower but accuracy is higher.

used to report a measurement, and again it is not necessarily related to accuracy. For example, a GPS receiver might measure elevation as 51.3456 m. But if the receiver is in reality only accurate to the nearest 10 cm, three of those digits are spurious, with no real meaning. So, although the precision is one ten-thousandth of a meter, the accuracy is only one-tenth of a meter. Box 5.4 summarizes the rules that are used to ensure that reported measurements do not mislead by appearing to have greater accuracy than they really do.

### To most scientists, precision refers to the number of significant digits used to report a measurement, but it can also refer to a measurement's repeatability.

In the interval/ratio case, the magnitude of errors is described by the *root mean square error* (RMSE),

---

## Technical Box (5.4)

### Good Practice in Reporting Measurements

Here are some simple rules that help to ensure that people receiving measurements from others are not misled by their apparently high precision.

1. The number of digits used to report a measurement should reflect the measurement's accuracy. For example, if a measurement is accurate to 1 m, then no decimal places should be reported. The measurement 14.4 m suggests accuracy to one-tenth of a meter, as does 14.0, but 14 suggests accuracy to 1 m.

2. Excess digits should be removed by rounding. Fractions above one-half should be rounded up, whereas fractions below one-half should be rounded down. The following examples reflect rounding to two decimal places:

14.57803 rounds to 14.58

14.57397 rounds to 14.57

14.57999 rounds to 14.58

14.57499 rounds to 14.57

3. These rules are not effective to the left of the decimal place; for example, they give no basis for knowing whether 1400 is accurate to the nearest unit or to the nearest hundred units.

4. If a number is known to be exactly an integer or whole number, then it is shown with no decimal point.

defined as the square root of the average squared error, or:

$$\left[ \Sigma \delta x^2 / n \right]^{1/2}$$

where the summation is over the values of $\delta x$ for all of the $n$ observations. The RMSE is similar in a number of ways to the standard deviation of observations in a sample. Although RMSE involves taking the square root of the average squared error, it is convenient to think of it as approximately equal to the average error in each observation, whether the error is positive or negative. The U.S. Geological Survey uses RMSE as its primary measure of the accuracy of elevations in digital elevation models, and published values range up to 7 m.

Although the RMSE can be thought of as capturing the magnitude of the average error, many errors will be greater than the RMSE and many will be less. It is useful, therefore, to know how errors are *distributed* in magnitude—how many are large, how many are small. Statisticians have developed a series of models of error distributions, of which the most common and most important is the Gaussian distribution, otherwise known as the error function, the "bell curve," or the Normal distribution. Figure 5.12 shows the curve's shape. The height of the curve at any value of x gives the relative abundance of observations with that value of x. The area under the curve between any two values of x gives the probability that observations will fall in that range. If observations are unbiased, then the mean error is zero (positive and negative errors cancel each other out), and the RMSE is also the distance from the center of the distribution (zero) to the points of inflection on either side, as shown in the figure.

Let us take the example of a 7 m RMSE on elevations in a USGS digital elevation model; if error follows the Gaussian distribution, this means that some errors will be more than 7 m in magnitude, whereas some will be less, and also that the relative abundance of errors of any given size is described by the curve shown. 68% of errors will lie between +1.0 and –1.0 RMSEs, or +7 m and –7 m. In practice, many distributions of error do follow the Gaussian distribution, and there are good theoretical reasons why this should be so.

### The Gaussian distribution is used to predict the relative abundances of different magnitudes of error.

To emphasize the mathematical formality of the Gaussian distribution, its equation is shown at the end of this paragraph. The symbol $\sigma$ denotes the standard deviation, $\mu$ denotes the mean (in Figure 5.12 these values are 1 and 0, respectively), and exp is the exponential function, or "2.71828 to the power of" (also sometimes used to represent distance decay; see Section 2.5). Scientists believe that it applies very broadly and that many instances of measurement error adhere closely to the distribution because it is grounded in rigorous theory. It can be shown mathematically that the distribution arises whenever a large number of random factors contribute to error, and the effects of these factors combine additively—that is, a given effect makes the same additive contribution to error whatever the specific values of the other factors. For example, error might be introduced in the use of a steel tape measure over a large number of measurements because some observers consistently pull the tape very taut, or hold it very straight, or fastidiously keep it horizontal, or keep it cool, and others do not. If the combined effects of these considerations always contribute the same amount of error (e.g., +1 cm, or –2 cm), then this contribution to error is said to be additive.

$$f(x) = \frac{1}{\sigma \sqrt{2\pi}} \exp\left[ -\frac{(x - \mu)^2}{2\sigma^2} \right]$$

We can apply this idea to determine the inherent uncertainty in the locations of contours. The U.S. Geological Survey routinely evaluates the accuracies of its digital elevation models (DEMs) by comparing the elevations recorded in the database with those at the same locations in more accurate sources, for a sample of points. The differences are summarized in an RMSE, and in this example we will assume that errors have a Gaussian distribution with zero mean and a 7 m RMSE. Consider a measurement of 350 m. According to the error model, the truth might be as high as 360 m or as low as 340 m, and the relative frequencies of any particular error value are as predicted by the Gaussian distribution with a mean of zero and a standard

**Figure 5.12** The Gaussian or Normal distribution. The lightly shaded area (between ±1 standard deviation) encloses 68% of the area under the curve, so 68% of observations will fall between these limits.

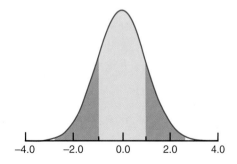

deviation of 7. If we take error into account, using the Gaussian distribution with an RMSE of 7 m, it is no longer clear that a measurement of 350 m lies exactly on the 350 m contour. Instead, the truth might be 340 m, or 360 m, or 355 m.

### 5.3.3 Statistical Models of Uncertainty in Location Measures

In the case of measurements of position, it is possible for every coordinate to be subject to error. In the two-dimensional case, a measured position $(x', y')$ would be subject to errors in both $x$ and $y$; specifically, we might write $x' = x + \delta x$, $y' = y + \delta y$, and similarly in the three-dimensional case where all three coordinates are measured, $z' = z + \delta z$. The *bivariate Gaussian distribution* describes errors in the two horizontal dimensions, and it can be generalized to the three-dimensional case. Normally, we would expect the RMSEs of $x$ and $y$ to be the same, but $z$ is often subject to errors of quite different magnitude: for example, in the case of determinations of position using GPS. The bivariate Gaussian distribution also allows for correlation between the errors in $x$ and $y$, but normally there is little reason to expect correlations.

Because it involves two variables, the bivariate Gaussian distribution has somewhat different properties from the simple (univariate) Gaussian distribution. As shown in Figure 5.12, 68% of cases lie within one standard deviation for the univariate case. But in the bivariate case with equal standard errors in $x$ and $y$, only 39% of cases lie within a circle of this radius. Similarly, 95% of cases lie within two standard deviations for the univariate distribution, but it is necessary to go to a circle of radius equal to 2.15 times the $x$ or $y$ standard deviations to enclose 90% of the bivariate distribution and 2.45 times standard deviations for 95%.

National Map Accuracy Standards often prescribe the positional errors that are allowed in databases. For example, the 1947 U.S. National Map Accuracy Standard specified that 95% of errors should fall below 1/30 inch (0.85 mm) for maps at scales of 1:20,000 and finer (more detailed), and 1/50 inch (0.51 mm) for other maps (coarser, less detailed than 1:20,000). A convenient rule of thumb is that positions measured from maps are subject to errors of up to 0.5 mm at the scale of the map. Table 5.2 shows the distance on the ground corresponding to 0.5 mm for various common map scales.

**A useful rule of thumb is that features on maps are positioned to an accuracy of about 0.5 mm.**

**Table 5.2** Positions measured from maps should be accurate to about 0.5 mm on the map. Multiplying this by the scale of the map gives the corresponding distance on the ground.

| Map scale | Ground distance corresponding to 0.5-mm map distance |
|---|---|
| 1:1250 | 52.5 cm |
| 1:2500 | 1.25 m |
| 1:5000 | 2.5 m |
| 1:10 000 | 5 m |
| 1:24 000 | 12 m |
| 1:50 000 | 25 m |
| 1:100 000 | 50 m |
| 1:250 000 | 125 m |
| 1:1 000 000 | 500 m |
| 1:10 000 000 | 5 km |

## 5.4 U3: Further Uncertainty in the Analysis of Geographic Phenomena

### 5.4.1 Internal and External Validation through Spatial Analysis

In Chapter 1 we defined a core remit of GI science as the resolution of scientific or decision-making problems through spatial analysis. Spatial analysis can be thought of as the process by which we turn raw spatial data into useful spatial information and thus far have thought of the creation of spatial information as adding value to attribute data through selectivity or preparation for purpose (see Chapters 13 and 14). A further defining characteristic is that the results of spatial analysis change when the frame or extent of the space under investigation changes. This also implies that the frame can be divided into units of analysis that are clearly defined, yet we have seen (Section 5.2.1) that there are likely to be few, if any, such units available to us. How can the outcome of spatial analysis be meaningful if it has such uncertain foundations?

Once again, this question has no easy answers, although we can begin by anticipating possible errors of positioning or the consequences of aggregating the subjects of analysis (such as individual people) into artificial geographic units of analysis (as when people are aggregated by census tracts, or disease incidences are aggregated by county). In so doing, we can illustrate how potential problems might arise, although we are unlikely to arrive at any definitive solutions—for the simple reason that the truth is inherently uncertain. The ways in which we conceive and represent geographic phenomena may distort the outcome of spatial analysis by dampening or accentuating apparent variation across space or by restricting the nature and range of questions that can meaningfully be asked.

**Good analysis cannot substitute for poor conceptions of geography or poor representation—but it can flag the likely consequences of both.**

We can deal with this risk in three ways. First, although the analyst can only rarely tackle the *source* of uncertainty (analysts are rarely empowered to collect new, completely disaggregate data, for example), analysis using GI systems can help to pinpoint the ways in which uncertainty is likely to *operate* (or *propagate*) within the GI system and identify the likely degree of distortion arising from representational expedients.

Second, although we may have to work with areally aggregated data, GI systems allow us to model within-zone spatial distributions, and this can ameliorate the worst effects of artificial zonation. This allows us to gauge the effects of scale and aggregation through simulation of different possible outcomes. This is *internal validation* of the effects of scale, point placement, and spatial partitioning.

The third way that GI systems can address uncertainty is assessing the quality of a representation with reference to other data sources, thus providing a means of *external validation* of the effects of zonal averaging. In today's advanced GI service economy, there may be other data sources that can be used to gauge the effects of aggregation on our analysis. In Section 12.2.1 we formally refine the basic scheme presented in Figure 5.1 to consider the role of geovisualization in evaluating representations, although some of the validation principles set out in this chapter also entail visual approaches.

**GI systems provide ways of validating representations, sometimes with and sometimes without reference to external data sources.**

## 5.4.2 Validation through Autocorrelation: The Spatial Structure of Errors

Understanding the spatial structure of errors is key to accommodating their likely effects on the results of spatial analysis, and hence estimates or measures of spatial autocorrelation (Section 2.3) can provide an important validation measure. This is because strong positive spatial autocorrelation will reduce the effects of uncertainty on estimates of properties such as slope or area. The cumulative effects of error, termed *error propagation*,

can nevertheless produce impacts that are surprisingly large. Some of the examples in this section have been chosen to illustrate the substantial uncertainties that can be produced by apparently innocuous data errors.

**Error propagation measures the impacts of uncertainty in data on the results of GI system operations.**

The confusion matrix, or more specifically a single row of the matrix, along with the Gaussian distribution, provide convenient ways of describing the error present in a single observation of a nominal or interval/ratio measurement, respectively. When a GI system is used to respond to a simple query, such as, "Tell me the class of soil at this point," or "What is the elevation here?" then these methods are good ways of describing the uncertainty inherent in the response. For example, a GI system might respond to the first query with the information "'Class A, with a 30% probability of Class C," and to the second query with the information "350 m, with an RMSE of 7 m." Notice how this makes it possible to describe nominal data as accurate to a percentage, but it makes no sense to describe a DEM, or any measurement on an interval/ratio scale, as accurate to a percentage. For example, we cannot meaningfully say that a DEM is "90% accurate."

However, many GI system operations involve more than the properties of single points, and this makes the analysis of error much more complex. For example, consider the query, "How far is it from this point to that point?" Suppose the two points are both subject to error of position because their positions have been measured using GPS units with mean distance errors of 50 m. If the two measurements were taken some time apart, with different combinations of satellites above the horizon, it is likely that the errors are independent of each other, such that one error might be 50 m in the direction of North, and the other 50 m in the direction of South. Depending on the locations of the two points, the error in distance might be as high as 100 m. On the other hand, if the two measurements were made close together in time, with the same satellites above the horizon, it is likely that the two errors would be similar, perhaps 50 m North and 40 m North, leading to an error of only 10 m in determining distance. The difference between these two situations can be measured in terms of the degree of *spatial autocorrelation*, or the interdependence of errors at different points in space (see Section 2.6).

**The spatial autocorrelation of errors can be as important as their magnitude in many GI system operations.**

Spatial autocorrelation is also important in analyzing probable errors in nominal data. Reconsider the agricultural field discussed in Section 5.2.3 that is known to contain a single crop, perhaps barley. When seen from above, it is possible to confuse barley with other crops, so there may be error in the crop type assigned to points in the field. But because the field has only one crop, we know that such errors are likely to be strongly correlated. Spatial autocorrelation is almost always present in errors to some degree, but very few efforts have been made to measure it systematically. As a result, it is difficult to make good estimates of the uncertainties associated with many GI system operations.

The spatial structure or autocorrelation of errors is important in many ways. DEM data are often used to estimate the slope of terrain, and this is done by comparing elevations at points a short distance apart. For example, if the elevations at two points 10 m apart are 30 m and 35 m, respectively, the slope along the line between them is 5/10, or 0.5. (A somewhat more complex method is used in practice, to estimate slope at a point in the x- and y-directions in a DEM raster, by analyzing the elevations of nine points—the point itself and its eight neighbors. The equations in Section 14.3.1 detail the procedure.) Now consider the effects of errors in these two elevation measurements on the estimate of slope. Suppose the first point (elevation 30 m) is subject to a RMSE of 2 m, and consider possible true elevations of 28 m and 32 m. Similarly, the second point might have true elevations of 33 m and 37 m. We now have four possible combinations of values, and the corresponding estimates of slope range from (33 − 32)/10 = 0.1 to (37 − 28)/10 = 0.9. In other words, a relatively small amount of error in elevation can produce wildly varying slope estimates.

> **The spatial autocorrelation between errors in geographic databases helps to minimize their impacts on many GI system operations.**

What saves us in this situation, and makes estimation of slope from DEMs a practical proposition at all, is spatial autocorrelation among the errors. In reality, although DEMs are subject to substantial errors in absolute elevation, neighboring points nevertheless tend to have similar errors, and errors tend to persist over quite large areas. Most of the sources of error in the DEM production process tend to produce this kind of persistence of error over space, including errors due to misregistration of aerial photographs. In other words, errors in DEMs exhibit strong positive spatial autocorrelation.

Another important corollary of positive spatial autocorrelation can also be illustrated using DEMs. Suppose an area of low-lying land is predicted to be submerged by sea-level rise, and our task is to estimate the area of land affected (Figure 5.13). We are asked to do this using a DEM, which is known to have an RMSE of 2 m. Suppose the data points in the DEM are 30 m apart, and preliminary analysis shows that 100 points have elevations below the flood line. We might conclude that the area flooded is the area represented by these 100 points, or 900 × 100 sq m, or 9 hectares. But because of errors, it is possible that some of this area is actually above the flood line (we will ignore the possibility that other areas outside this may also be below the flood line, also because of errors), and it is possible that *all* the area is above. Suppose the recorded elevation for each of the 100 points is 2 m below the flood line. This is one RMSE (recall that the RMSE is equal to 2 m) below the flood line, and the Gaussian distribution tells us that the chance that the true elevation is actually above the flood line is approximately 16% (see Figure 5.12).

But what is the chance that *all* 100 points are actually above the flood line? Here again the answer

Figure 5.13 The hypothetical effects of a sea-level rise of 6 m on London, viewed in the Virtual London model.

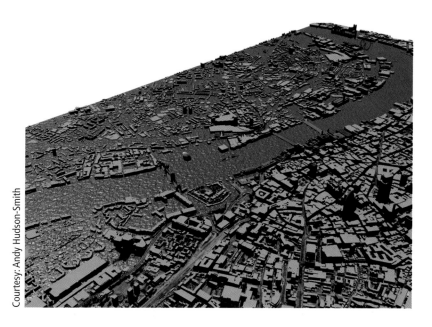

Courtesy: Andy Hudson-Smith

depends on the degree of spatial autocorrelation among the errors. If there is none, in other words, if the error at each of the 100 points is independent of the errors at its neighbors, then the answer is $(0.16)^{100}$, or 1 chance in 1 followed by roughly 70 zeroes. But if there is strong positive spatial autocorrelation, so strong that all 100 points are subject to exactly the same error, then the answer is 0.16. One way to think about this is in terms of *degrees of freedom*. If the errors are independent, they can vary in 100 independent ways, depending on the error at each point. But if they are strongly spatially autocorrelated, the effective number of degrees of freedom is much less and may be as few as 1 if all errors behave in unison. Spatial autocorrelation has the effect of reducing the number of degrees of freedom in geographic data below what may be implied by the volume of information, in this case the number of points in the DEM.

### Spatial autocorrelation acts to reduce the effective number of degrees of freedom in geographic data.

A further case concerns the accommodation of positional uncertainties in the location of the boundaries of a discrete object. Figure 5.14 shows a square approximately 100 m on each side. Suppose the square has been surveyed by determining the locations of its four corner points using GPS, and suppose the circumstances of the measurements are such that there is an RMSE of 1 m in both coordinates of all four points and that errors are independent.

Suppose our task is to determine the area of the square. A GI system can do this easily, using a standard algorithm (see Figure 14.1). Computers are precise (in the sense of Box 5.4) and capable of working to many significant digits, so the calculation might be reported by printing out a number to eight digits, such as 10,014.603 sq m, or even more. But

the number of significant digits will have been determined by the precision of the machine, and not by the accuracy of the determination. Box 5.4 summarized some simple rules for ensuring that the precision used to report a measurement reflects as far as possible its accuracy, and clearly those rules will have been violated if the area is reported to eight digits. But what is the appropriate precision?

In this case we can determine exactly how positional accuracy affects the estimate of area. It turns out that area has an error distribution that is Gaussian, with a standard deviation (RMSE), which in our particular case is 200 sq m. In other words, each attempt to measure the area will give a different result, the variation between them having a standard deviation of 200 sq m. This means that the five rightmost digits in the estimate are spurious, including two digits to the left of the decimal point. So if we were to follow the rules of Box 5.4, we would print 10,000 rather than 10,014.603. (Note the problem with standard notation here, which does not let us omit digits to the left of the decimal point even if they are spurious and so leaves some uncertainty about whether or not the tens and units digits are certain—and note also the danger that if the number is printed as an integer it may be interpreted as exactly the whole number.) We can also turn the question around and ask how accurately the points would have to be measured to justify eight digits, and the answer is approximately 0.01 mm, far beyond the capabilities of normal surveying practice.

A useful way of visualizing spatial autocorrelation and interdependence is through animation. Each frame in the animation is a single possible map, or *realization* of the error process. If a point is subject to uncertainty, each realization will show the point in a different possible location, and a sequence of images will show the point shaking around its mean position. If two points have perfectly correlated positional errors, then they will appear to shake in unison, as if they were at the ends of a stiff rod. If errors are only partially correlated, then the system behaves as if the connecting rod were somewhat elastic.

The inherent difficulties in accommodating spatially autocorrelated errors have led many researchers to explore a more general strategy of simulation to evaluate the impacts of uncertainty on the results of spatial analysis. In essence, simulation requires the generation of a series of realizations, as defined earlier. This is often called Monte Carlo simulation in reference to the realizations that occur when dice are tossed or cards are dealt in various games of chance. For example, we could simulate error in a single measurement from a DEM by generating a series of numbers with a mean equal to the measured elevation, and a standard deviation equal to the known RMSE and a Gaussian distribution. Simulation uses everything that is known

---

**Figure 5.14** Error in the measurement of the area of a square 100 m on each side.

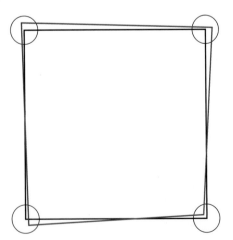

about a situation, so if any additional information is available we would incorporate it in the simulation. We might assume that elevations must be whole numbers of meters, and we would simulate this by rounding the numbers obtained from the Gaussian distribution. With a mean of 350 m and an RMSE of 7 m the results of the simulation might, for example, be 341, 352, 356, 339, 349, 348, 355, 350, . . .

**Simulation is an intuitively simple way of getting the uncertainty message across.**

Because of spatial autocorrelation, it is impossible in most circumstances to think of databases as decomposable into component parts, each of which can be independently disturbed to create alternative realizations, as in the previous example. Instead, we have to think of the entire database as a realization and create alternative realizations of the database's contents that preserve spatial autocorrelation. Figure 5.15 shows an example, simulating the effects of uncertainty on a digital elevation model. Each of the three realizations is a complete map, and the

**Figure 5.15** Three realizations of a model simulating the effects of error on a digital elevation model.

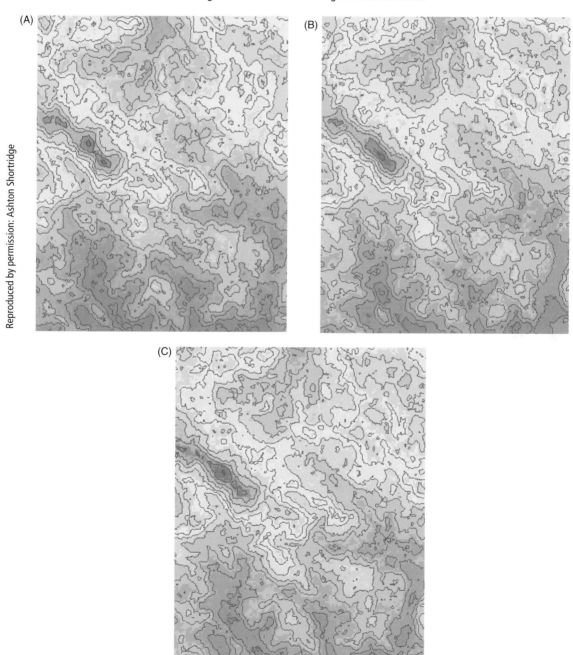

simulation process has faithfully replicated the strong correlations present in errors across the DEM.

## 5.4.3 Validation through Investigating the Effects of Aggregation and Scale

We have already seen that a fundamental difference between geography and other scientific disciplines is that the definition of its objects of study is only rarely unambiguous and, in practice, rarely precedes our attempts to measure their characteristics. In socio-economic analysis these objects of study (geographic individuals) are usually aggregations because the spaces that human individuals occupy are geographically unique, and confidentiality restrictions usually dictate that uniquely attributable information must be anonymized in some way. Even in natural-environment applications, the nature of sampling in the data collection process (Section 2.4) often makes it expedient to collect data pertaining to aggregations of one kind or another. Thus geographic individuals are likely to be defined as areal aggregations of the units of study. Moreover, in cases where data are collected to serve a range of end uses (as with general population surveys, or natural-resource inventories), the zonal systems are unlikely to be determined with the end point of particular spatial analysis applications in mind.

As a consequence, we cannot be certain in ascribing even dominant characteristics *of* areas to true individuals or point locations *in* those areas. This source of uncertainty is known as the *ecological fallacy* and has long bedeviled the analysis of spatial distributions. (The opposite of ecological fallacy is atomistic fallacy, in which the individual is considered in isolation from his or her environment. This use of the term *ecology* has nothing to do with the way that the term is commonly used today.) The ecological fallacy problem is a consequence of aggregation into the basic units of analysis and is illustrated in Figure 5.16.

**Inappropriate inference from aggregate data about the characteristics of individuals is termed the ecological fallacy.**

The likelihood of committing ecological fallacy in GI analysis depends on the nature of the aggregation being used. If the members of a set of zones are all perfectly uniform and homogeneous (Section 5.2.1), then the only geographic variation (heterogeneity) that occurs will be between zones. However, nearly all zones are internally heterogeneous to some degree, and greater heterogeneity increases the likelihood and severity of the ecological fallacy problem. That said, it is important to be aware that there are few documented case studies that demonstrate the occurrence of ecological fallacy in practice, and many

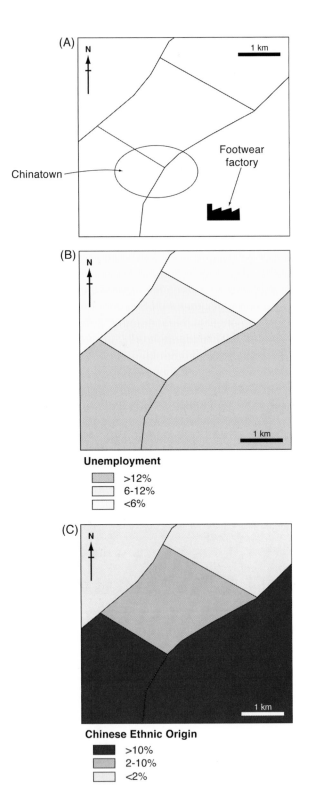

**Figure 5.16** A hypothetical illustration of the problem of ecological fallacy. (A) Before it closed down, the footwear factory drew its labor from its local neighborhood and a jurisdiction to the west. The closure caused high unemployment, but not among the service sector workers of Chinatown. Yet comparison of choropleth maps (B) and (C) indicates a spurious relationship between Chinese ethnicity and unemployment.

general-purpose zoning systems in socioeconomic GI are designed to maximize within-zone homogeneity in the most salient population characteristics.

The potential of *aggregation* to create problems in GI analysis is compounded by problems arising from the different *scales* at which zonal systems may be defined. This was demonstrated more than half a century ago in a classic paper by Yule and Kendall, who used data for wheat and potato yields from the (then) 48 counties of England to demonstrate that correlation coefficients tend to increase with scale. They aggregated the 48-county data into zones so that there were first 24, then 12, then 6, and finally just 3 zones. Table 5.3 presents the range of their results, from near zero (no correlation) to over 0.99 (almost perfect positive correlation), although subsequent research has suggested that this range of values is atypical.

**Relationships typically grow stronger when based on larger geographic units.**

Scale turns out to be important because of the property of spatial autocorrelation outlined in Section 4.3. A succession of subsequent research papers has reaffirmed the existence of similar effects in multivariate analysis. However, rather discouragingly, scale effects in multivariate cases do not follow any consistent or predictable trends.

Further *aggregation* effects can arise in GI analysis because there is a multitude of ways in which the mosaic of basic areal units (the geographic individuals) can be assembled together into zones, and the requirement that zones be made up of spatially contiguous elements presents only a weak constraint on the huge combinatorial range. This gives rise to the related *aggregation* or *zonation* problem, in which different combinations of a given number of geographic individuals into coarser-scale areal units can yield widely different results.

In a classic 1984 study, geographer Stan Openshaw applied correlation and regression analysis to the attributes of a succession of zoning schemes. He demonstrated that the constellation of elemental

zones within aggregated areal units could be used to manipulate the results of spatial analysis to a wide range of quite different prespecified outcomes. These numerical experiments have some sinister counterparts in the real world, the most notorious example of which is the political gerrymander of 1812 (see Section 14.1.2). Spatial distributions can be designed (with or without GI systems) that are unrepresentative of the scale and configuration of real-world geographic phenomena; such outcomes may even emerge by chance. The outcome of multivariate spatial analysis is also similarly sensitive to the particular zonal scheme that is used.

Taken together, the effects of scale and aggregation are generally known as the *Modifiable Areal Unit Problem* (MAUP). The ecological fallacy and the MAUP have long been recognized as problems in applied spatial analysis, and through the concept of spatial autocorrelation (Section 2.3), they are also understood to be related problems. Increased technical capacity for numerical processing and innovations in scientific visualization have refined the quantification and mapping of these measurement effects, and have also focused interest on the effects of within-area spatial distributions upon analysis.

## 5.4.4 Validation with Reference to External Sources: Data Integration and Shared Lineage

The term concatenation is used to describe the integration of two or more different data sources, such that the contents of each are accessible in the product. The polygon overlay operation that will be discussed in Section 13.2.4, and its field-view counterpart, is one simple form of concatenation. The term concatenation is used to describe the range of functions that attempt to overcome differences between datasets, or to merge their contents (as with rubber-sheeting: see Section 8.3.2.2). Conflation thus attempts to replace two or more versions of the same information with a single version that reflects the pooling, or weighted averaging, of the sources. The individual items of information in a single geographic dataset often share lineage, in the sense that more than one item is affected by the same error. This happens, for example, when a map or photograph is registered poorly because all the data derived from it will have the same error. One indicator of shared lineage is the persistence of error—because all points derived from the same misregistration will be displaced by the same, or a similar, amount. Because neighboring points are more likely to share lineage than distant points, errors tend to exhibit strong positive spatial autocorrelation.

Table 5.3 1950 Yule and Kendall's data for wheat and potato yields from 48 English counties.

| No. of geographic areas | Correlation |
| --- | --- |
| 48 | 0.2189 |
| 24 | 0.2963 |
| 12 | 0.5757 |
| 6 | 0.7649 |
| 3 | 0.9902 |

**Conflation combines the information from two data sources into a single source.**

When two datasets that share no common lineage are concatenated (for example, they have not been subject to the same misregistration), then the relative positions of objects inherit the absolute positional errors of both, even over the shortest distances. Although the shapes of objects in each dataset may be accurate, the relative locations of pairs of neighboring objects may be wildly inaccurate when drawn from different datasets. The anecdotal history of GI is full of examples of datasets that were perfectly adequate for one application, but failed completely when an application required that they be merged with some new dataset that had no common lineage. For example, merging GPS measurements of point positions with streets derived from the U.S. Bureau of the Census TIGER files may lead to surprises where points appear on the wrong sides of streets. If the absolute positional accuracy of a dataset is 50 m, as it was with parts of earlier versions of the TIGER database, points located less than 50 m from the nearest street will frequently appear to be misregistered.

**Datasets with different lineages often reveal unsuspected errors when overlaid.**

Figure 5.17 shows an example of the consequences of overlaying data with different lineages. In this case, two datasets of streets (shown in green and red) produced by different commercial vendors using their own process fail to match in position by amounts of up to 100 m; they also fail to match in the names of many streets, and even the existence of streets. The integrative functionality of GI systems makes it an attractive possibility to generate multivariate indicators from diverse sources. Such data are likely to have been collected at a range of different scales, and for a range of areal units as diverse as census

**Figure 5.17** Overlay of two street databases for part of Goleta, California (horizontal extent of image c. 5km).

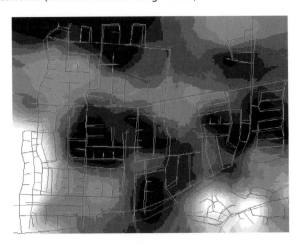

tracts, river catchments, land ownership parcels, travel-to-work areas, and market research surveys. The uncertainties arising out of overlay, as well as the sources and operation of many other forms of uncertainty, have been investigated by GI scientist Nicholas Chrisman (Box 5.5).

Established procedures of statistical inference can only be used to reason from representative samples to the populations from which they were drawn. Yet these procedures do not regulate the assignment of inferred values to other (usually smaller) zones, or their apportionment to ad hoc regions. There are emergent tensions within socioeconomic analysis, for there is a limit to the usefulness of inferences drawn from conventional, scientifically valid data sources that may be out-of-date, zonally coarse, and irrelevant to the problem under investigation.

Yet the alternative of using new rich sources of crowd-sourced VGI (see Sections 3.8.3 and 10.2), social media, or marketing data may be profoundly unscientific in its inferential procedures.

## 5.4.5 Internal and External Validation; Induction and Deduction

Reformulation of the MAUP into a *geocomputational* (Section 15.1) approach to zone design amounts to inductive use of GI systems to seek patterns through repeated scaling and aggregation experiments, alongside much better deductive *external* validation using any of the multitude of new datasets that are a hallmark of the information age. Neither of these approaches, used in isolation, is likely to resolve the uncertainties inherent in spatial analysis. Zone-design experiments are merely playing with the MAUP, and most of the new sources of external validation are unlikely to sustain full scientific scrutiny, particularly if they were assembled through nonrigorous survey designs or self-selection of respondents supplying observations.

**There is no solution to the Modifiable Areal Unit Problem, but simulation of large numbers of alternative zoning schemes can gauge its likely effects.**

The conception and measurement of elemental zones, the geographic individuals, may be ad hoc, but it is rarely wholly random either. Can our recognition and understanding of the empirical effects of the MAUP help us to neutralize its effects? Not really. In measuring the distribution of all possible zonally averaged outcomes ("simple random zoning" in analogy to simple random sampling in Section 2.4), there is no tenable analogy with the established procedures of statistical inference and its concepts of precision and error. Even if there were, as we will see

## Nicholas Chrisman, Globe-Trotting GI Scientist

Nicholas Chrisman (Figure 5.18) joined the geography discipline at what might later be seen as the high tide of the quantitative movement, graduating from the University of Massachusetts in 1972. Following graduation, he took an offer of a one-year position at the Laboratory for Computer Graphics and Spatial Analysis at Harvard University, and in the event stayed there for ten years. His early work entailed programming the topological relations between spatial objects and developing fuzzy tolerance procedures for overlay analysis (Section 13.2.4). He then moved away from software development to

Courtesy: Linda Chrisman

**Figure 5.18** Nicholas Chrisman, international GI scientist.

attend Bristol University in the UK for a PhD focused on error and data quality, studying there at the same time as two of the authors of this book (Paul Longley and David Maguire). Returning to the United States, he next joined an interdisciplinary team at the University of Wisconsin–Madison, working on the development of GI applications to understand soil erosion. This process extended his understanding of the social and institutional components of GI systems, work that continued after he moved to become Professor of Geography at the University of Washington from 1987–2004.

Nick's next move was to Canada as Scientific Director of the GEOIDE Network, linking researchers and students at 34 universities in a dozen disciplines. During this period, he taught *sciences géomatiques* in French at Université Laval. Between 2013–14 he worked as Discipline Head of Geospatial Sciences at RMIT University in Melbourne, Australia, before moving back to Washington for semi-retirement.

Within this close connection to France over the decades, Chrisman has worked in five countries in three continents. Reflecting on such a variegated career and the development and application of GI systems in his diverse career destinations, Nick remains very mindful that each has shared common scientific endeavors, yet each also has specific local requirements. Or, as he wrote in his classic textbook: "Remember that many marvelous technical advances will not save a system that was not supported by a community that really wanted it" (Chrisman, 2001, *Exploring GIS,* Wiley, p. 243). Look out for a thoroughly revised version of this classic book, infused with these personal experiences, in 2015.

in Section 14.5, there are limits to the application of classic statistical inference to spatial data.

### Zoning seems similar to sampling, but its effects are very different.

The way forward seems to entail a threefold response: first, to conduct analysis in response to specific, clearly formulated hypotheses; second, to use GI systems to customize zoning schemes to correspond with these hypotheses; and third, to undertake validation of data with respect to external sources, particularly those that might confirm or refute our assumptions about the likely level of within-zone heterogeneity within our aggregated data. In this way, the MAUP will dissipate if analysts understand the particular areal units that they wish to study.

There is also a sense here that resolution of the MAUP requires acknowledgment of the uniqueness of places. The time dimension is also important: the areal objects of study are ever-changing, and our perceptions of what constitutes an appropriate areal schema should be subject to change. Indeed, infusing the time dimension into GI arguably creates the need for new overarching conceptions and paradigms (meta theories). And finally, within the socioeconomic realm, the act of defining zones can also be self-validating if the allocation of individuals affects the interventions that are subsequently made, be they a mail-shot about a shopping opportunity or deployment of aid under policies to alleviate hardship or deprivation. Spatial discrimination affects spatial behavior, and so the principles of zone design are of much more than academic interest.

# 5.5 Consolidation

Uncertainty is much more than error. Just as the amount of available digital data and our abilities to process them have developed, so our understanding of the quality of digital depictions of reality has broadened. It is one of the supreme ironies of analysis using contemporary GI systems that as we accrue more and better data and have more computational power at our disposal, we seem to become more uncertain about the quality of our digital representations and the adequacy of our areal units of analysis. Richness of representation provided by Big Data and greater computational power serve to make us more aware of the range and variety of established uncertainties and challenge us to integrate new ones—not least understanding the sources and operation of bias that arises in the assembly of social media datasets.

The only route beyond this impasse is to continue to advance hypotheses about the likely generalized structure of spatial data, albeit in a spirit of humility rather than of conviction. Hypothesis generation requires more than the brute force of high-power computing, and so progress requires greater a priori understanding about the structure in spatial as well as attribute data. There are some general rules to guide us here, and spatial autocorrelation measures provide further structural clues (Section 2.3). In Section 13.2.1 we discuss how context-sensitive spatial analysis techniques, such as geographically weighted regression, provide a bridge between general statistics and the case-study approach.

Geocomputation helps too, by allowing us to gauge the sensitivity of outputs to inputs but, unaided, is unlikely to present unequivocal best solutions. The fathoming of uncertainty requires a combination of the cumulative development of a priori knowledge (we should expect scientific research to be cumulative in its findings), external validation of data sources, and inductive generalization in the fluid, eclectic data-handling environment that is contemporary GI science.

What does all this mean in practice? Here are some rules for how to live with uncertainty. First, because there can be no such thing as perfectly accurate analysis, it is essential to acknowledge that uncertainty is inevitable. It is better to take a positive approach by learning what one can about uncertainty, rather than to pretend that it does not exist. To behave otherwise is unconscionable and can also be very expensive in terms of lawsuits, bad decisions, and the unintended consequences of actions (see Chapter 17).

Second, most spatial analysts often have to rely on others to provide data. Historically, spatial framework data were supplied through government-sponsored mapping programs (e.g., those of the U.S. Geological Survey or Great Britain's Ordnance Survey), but commercial sources and volunteered geographic information (see Section 10.2) have come increasingly to the fore. Many Big Data, such as those from social media sources, are also sometimes georeferenced, but little is known about the degree to which they are representative of any clearly defined population (see Section 2.4). Data should never be taken as the truth; instead, it is essential to assemble all that is known about their quality and to use this knowledge to assess whether the data are fit for use. Metadata (Section 10.2) are designed specifically for this purpose and will often include assessments of quality. When these are not present, it is worth spending the extra effort to contact the creators of the data, or other people who have tried to use them, for advice on quality. Never trust data that have not been assessed for quality, or data from sources that do not have good reputations for quality.

Third, the uncertainties in the outputs of GI analysis are often much greater than one might expect, given knowledge of input uncertainties, because many GI system processes are highly nonlinear. Yet some spatial processes dampen uncertainty rather than amplify it. Given this condition, it is important to gain some impression of the likely impacts of uncertain inputs to GI systems upon outputs.

Fourth, rely on multiple sources of data whenever possible in order to facilitate external validation. It may be possible to obtain maps of an area at several different scales, for example, or to conflate several different open-source databases. Raster and vector datasets are often complementary (e.g., when combining a remotely sensed image with a topographic map). Digital elevation models can often be augmented with spot elevations, or GPS measurements.

**38.74376% of all statistics are made up (including this one). Avoid spurious precision, and evaluate the provenance of all of the data that you use.**

Finally, be honest and informative in reporting the results of GI analysis. Recognize that uncertainty is never likely to be eliminated, but that it can be managed as part of good scientific practice. Input data sources may be presented with more apparent precision than is justified by their actual accuracy on the ground, and lines may have been drawn on maps with widths that reflect relative importance, rather than uncertainty of position. It is up to the users to redress this imbalance by finding ways of communicating what they know about accuracy, rather than relying on the GI system to do so. It is wise to include plenty of caveats into reported results, so that they reflect what we believe to be true, rather than a narrow and literal interpretation of what the system appears to be saying.

## Questions for Further Study

1. What tools do GI system designers build into their products to help users deal with uncertainty? Take a look at your favorite GI system from this perspective. Does it allow you to associate metadata about data quality with datasets? Is there any support to accommodate propagation of uncertainty? How does it determine the number of significant digits when it prints numbers? What are the pros and cons of including such tools?

2. Using aggregate data for Iowa counties, Stan Openshaw found a strong positive correlation between the proportion of people over 65 and the proportion who were registered voters for the Republican Party. What, if anything, does this tell us about the tendency for older people to register as Republicans?

3. Where is the flattest place on Earth? Search the Web for appropriate documents and give reasons for your answer (Figure 5.19).

4. You are a senior retail analyst for Safemart, which is contemplating expansion from its home state to three others in the United States. Assess the relative merits of your own company's store loyalty card data (which you can assume are similar to those collected by any retail chain with which you are familiar) and of data from the most recent Census in planning this strategic initiative. Pay particular attention to issues of survey content, the representativeness of population characteristics, and problems of scale and aggregation. Suggest ways in which the two data sources might complement one another in an integrated analysis.

Courtesy: NASA

**Figure 5.19** Salt "flats"?

## Further Reading

Burrough, P. A. and Frank, A. U. (eds.). 1996. *Geographic Objects with Indeterminate Boundaries*. London: Taylor and Francis.

Fisher, P. F. 2005. Models of uncertainty in spatial data. In Longley, P. A., Goodchild, M. F., Maguire, D. J., and Rhind, D. W. (eds.). *Geographical Information Systems: Principles, Techniques, Management and Applications (Abridged Edition)*. New York: Wiley, pp. 191–205.

Heuvelink, G. B. M. 1998. *Error Propagation in Environmental Modelling with GIS*. London: Taylor and Francis.

Openshaw, S. and Alvanides, S. 2005. Applying geocomputation to the analysis of spatial distributions. In Longley, P. A., Goodchild, M. F., Maguire, D. J., and Rhind, D. W. (eds.). *Geographical Information Systems: Principles, Techniques, Management and Applications (Abridged Edition)*. New York: Wiley, pp. 267–282.

Zhang, J. X. and Goodchild, M. F. 2002. *Uncertainty in Geographical Information*. New York: Taylor and Francis.

Smith, B. and Varzi, A. C. 2000. Fiat and bona fide boundaries. *Philosophy and Phenomenological Research* 60(2), pp. 401–420.

# 6 GI System Software

At the core of an operational geographic information system lies some computer software that forms the basis of the processing engine. A geographic information (GI) system comprises an integrated collection of computer programs that implement geographic storage, processing, and display functions. The three key parts of any GI system are the user interface, the tools (functions), and the data manager. All three parts may be located on a single computer, or they may be spread over multiple machines in a departmental or an enterprise system configuration. Three main types of architecture are used to build operational GI system implementations: desktop, client server, and Cloud. There are many different types of GI system, and this chapter organizes the discussion around the main categories: desktop, Web mapping, server, virtual globe, developer, mobile, and other. Software may be licensed using a commercial or open-source model. Examples of the main types of software products from the leading commercial developers include Autodesk AutoCAD, Bentley Map, Esri Arc-GIS, and Intergraph GeoMedia. There is also a significant and growing open-source GI software community, of which Quantum GIS, Map Server, and the OpenGeo Suite are well-known examples.

## LEARNING OBJECTIVES

After studying this chapter you will be able to:

- Understand the architecture of GI systems, specifically
  - Organization by project, department, and enterprise
  - The three-tier architecture of software systems (graphical user interface, tools, and data access).
- Outline how to customize GI system products.
- Describe the main types of commercial software:
  - Desktop
  - Web mapping
  - Server
  - Virtual globe
  - Developer
  - Mobile
  - Other.
- Describe the key characteristics of the main types of commercial GI system products currently available.

## 6.1 Introduction

In Chapter 1, the technical parts of a GI system were defined as the network, hardware, software, and data, which together functioned with reference to people and the procedural frameworks within which people operate (Figure 1.12). This chapter is concerned with software: the geographic storage, processing, and display engines of a complete, working GI system. The functionality or capabilities of GI systems will be discussed later in this book (especially in Chapters 8–15). The focus here is on the different ways in which these capabilities are realized in software products and implemented in operational GI systems.

This chapter takes a fairly narrow view of GI systems, concentrating on a range of generic capabilities to collect, store, manage, query, analyze, and present geographic information. It excludes atlases, simple graphics and mapping systems, route-finding software, simple location-based services, image-processing systems, and spatial extensions to database management systems (DBMS), which are not true GI systems as defined here. The discussion is also restricted to GI system products—well-defined collections of software and accompanying documentation, install scripts, and so on—that are subject to multiversioned release control (that is to say, the software is made available in a controlled way on a periodic basis). By definition, it excludes specific-purpose utilities, unsupported routines, and ephemeral codebases.

Earlier chapters, especially Chapter 3, introduced several fundamental computer concepts, including digital representations, data, and information. Two further concepts need to be introduced here. Programs are collections of instructions that are used to manipulate digital data in a computer. System software, such as a computer operating system, is used to support application software—the programs with which end users interact. Integrated collections of application programs are referred to as software packages or systems (or just software for short).

Software can be distributed to the market in several different ways. Increasingly, download from the Internet is the main distribution mechanism, although hard-copy media (DVDs) are still widely used. GI system products usually comprise an integrated collection of software programs, and a selection of additional elements such as an install script, online help files, sample data and maps, documentation, a license, and an associated Web site. Alternative distribution models that are becoming increasingly prevalent for smaller products include *shareware* (usually intended for sale after a trial period), *liteware* (shareware with some capabilities disabled), *freeware* (free software but with copyright restrictions), *public domain software* (free with no restrictions), and *open-source* software (the *source code* is provided under an open-source license arrangement, and users agree not to limit the distribution of improvements). In the early days open-source software products provided only rather simple, poorly engineered tools with no user support. Today there are several high-quality, feature-rich software products, a number of which will be discussed in later sections.

**GI system software packages provide a unified approach to working with geographic information.**

GI system software is built on top of basic computer operating system capabilities that deal with such things as security, file management, peripheral drivers (controllers), Web interaction, printing, and display management. GI system software is constructed on these foundations to provide a controlled environment for GI collection, management, analysis, display, and interpretation. The unified architecture and consistent approach to representing and working with geographic information in a GI system both aim to provide users with a standardized way of working with geographic information.

## 6.2 The Evolution of GI System Software

In its formative years, GI system software consisted simply of collections of computer routines that a skilled programmer could use to build an operational GI system. During this period, each and every GI system was unique in terms of its capabilities, and significant levels of resource were required to create a working system. As software engineering techniques advanced and the GI system market grew in the 1970s and 1980s, demand increased for higher-level applications with a standard user interface. In the late 1970s and early 1980s, the standard means of communicating with a GI system was to type in command lines and wait for a response from the system. User interaction with a GI system entailed typing instructions, for example, to draw a topographic map, query the attributes of a forest stand object, or summarize the length of highways in a project area. Essentially, a GI system was a toolbox of geoprocessing operators or commands that could be applied to datasets to create new derivative datasets. For example, three polygon-based data layers of the same geographic area—*Soil*, *Slope*, and *Vegetation*—could be combined using an overlay processing function (see Section 13.2.4) to create an *IntegratedTerrainUnit* dataset.

Two key developments in the late 1980s and 1990s made the software easier to use and more generic. First, command line interfaces were supplemented and eventually largely replaced by graphical user interfaces (GUIs). These menu-driven, form-based interfaces greatly simplify user interaction with a GI system (see Section 12.2.1). Second, a customization capability was added to allow specific-purpose applications to be created from the generic toolboxes. Software developers and advanced technical users could make calls using a high-level programming language (such as Visual Basic, C, or Java) to published application programming interfaces (APIs) that exposed key functions. Together these stimulated enormous interest in GI systems and led to much wider adoption and expansion into new areas.

In particular, the ability to create custom application solutions allowed developers to build focused applications for end users in specific market areas. This led to the creation of GI applications specifically tailored to the needs of major markets (e.g., government, utilities, military, and environment). New terms were developed to distinguish these subtypes of GI system: planning information systems, automated mapping/facility management (AM/FM) systems, land information systems, and more recently, location-based services.

In the past few years a new method of software interaction has evolved that allows systems to communicate over the Web using a Web services paradigm. A Web service is an application that exposes its functions via a well-defined published interface that can be accessed over the Web from another program or Web service. This new software interaction paradigm allows geographically distributed GI system functions to be linked to create complete GI applications (Box 6.1). Figure 6.1 shows the program code used to draw a map of Sydney in Google Maps. For example, a market analyst who wants to determine the suitability of a particular site for locating a new store can start a small interactive program on his or her smartphone or tablet device that links to remote services over the Web that provide access to the latest population census and

---

## Technical Box (6.1)

### Cloud GI Web Services: Requesting a Google Map

GI applications that operate over the Web often use data and functionality hosted in servers managed in a Cloud computing environment. The geographic data and processing functions are accessible via APIs. Comparatively lightweight applications running on devices such as phones and tablets, or within browsers, can make calls to an API to perform actions such as request a map, geocode an address, add new data points, and route between locations. JavaScript is a widely used simple programming and API environment for mapping and analysis (and other) applications. Figure 6.1 shows a script running inside a browser that requests a simple map from the Google Maps Web service that is hosted in Google's Cloud (a set of servers in a data center accessible over the Web).

JavaScript is an XML-based language that encodes instructions as tags (pairs of instructions within angled brackets <>).

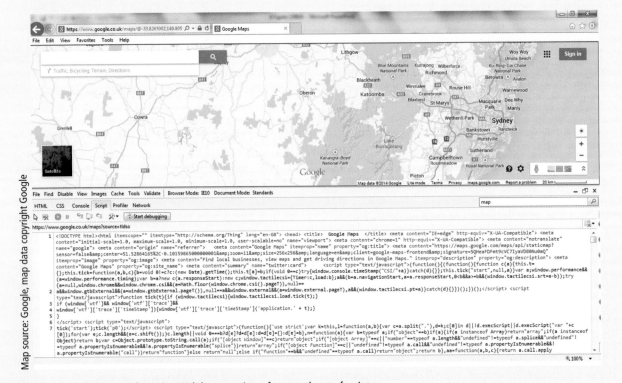

Figure 6.1 Google Maps Web service: Web browser view of map and part of script.

Map source: Google, map data copyright Google

geodemographics data, as well as analytical models. Although these data and programs are remotely hosted and maintained, they can be used for site suitability analysis as though they were resident on the market analyst's device. Chapter 10 explores Web services in more depth in the context of distributed GI systems. This software architecture for Web services is one of the key technology building blocks of Cloud GI systems (GI services hosted in data centers that can be accessed over the Web).

> **Devices connected to Cloud GI systems are becoming increasingly popular as an implementation architecture for GI technology.**

# 6.3 Architecture of GI System Software

## 6.3.1 Project, Departmental, and Enterprise GI Systems

Major GI systems can be implemented using three main scales of software implementation: project, departmental, and enterprise.

In project implementations the technical components (network, hardware, software, and data) of an operational GI system are assembled for the duration of a project, which may be from several days, to months or a few years. Data are collected or downloaded from the Web specifically for the project, and little thought is usually given to reuse of software, data, and human knowledge. In larger organizations, multiple projects may run one after another or even in parallel. The "one off" nature of the projects, coupled with an absence of organizational vision, often leads to duplication, as each project develops using different hardware, software, data, people, and procedures. Sharing data and experience is usually a low priority.

As interest in GI systems grows, to save money and encourage sharing and resource reuse, several projects in the same department may be amalgamated. This often leads to the creation of common standards, development of a focused GI team, and procurement of new GI system capabilities. Yet it is also quite common for different departments to have different GI systems and standards.

As GI systems become more pervasive, organizations learn more about them and begin to become dependent on them. This leads to the realization that GI systems are a useful way to organize many of an organization's assets, processes, and workflows. Through a process of natural growth, and possibly further major procurement (e.g., purchase of upgraded hardware, software, and data), GI systems gradually

become accepted as important enterprise-wide information systems. At this point, GI standards are accepted across multiple departments, and resources to support and manage the GI systems are often centrally funded and managed. Of course, things do not always proceed as smoothly as this and in such a linear way.

## 6.3.2 The Three-Tier Architecture

From an information systems perspective a GI system has three key parts: the user interface, the tools, and the data management system (Figure 6.2). The user's interaction with the system is via a graphical user interface (GUI; see also Section 12.2.1), an integrated collection of menus, tool bars, and other controls. The GUI provides organized access to tools for handling GI. The tool set defines the capabilities or functions that the GI system has available for processing geographic data. The data are stored as files, databases, or Web services and are organized by data management software. In standard information-system terminology this is a three-tier architecture, with the three tiers being

Figure 6.2 Classical three-tier architecture of a GI system.

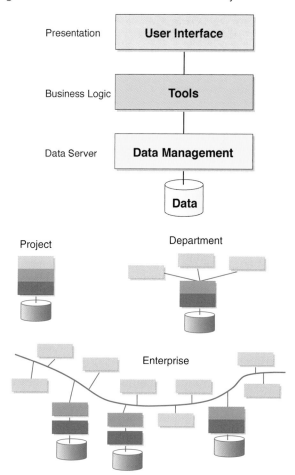

called presentation, business logic, and data server. Each of these software tiers is required to perform different types of independent tasks. The presentation tier must be adept at collecting user inputs, rendering (displaying) data, and interacting with graphic objects. The business logic tier is responsible for performing all operations such as network routing, data overlay processing, and raster analysis. It is here also that the GI data-model logic is implemented (see Section 6.3.3 for discussion of data models). The data-server tier must import and export data and service requests for subsets of data (queries) from a database or file system. In order to maximize system performance, it is useful to optimize hardware and operating system settings differently for each type of task. For example, rendering maps requires large amounts of memory and fast CPU speeds, whereas database queries need fast disks and buses (interfaces between devices) for moving large numbers of data around. By placing each tier on a separate computer, some tasks can be performed in parallel, and greater overall system performance, scalability, and resilience can be achieved.

### GI systems deal with user interfaces, tools, and data management.

Three main types of computer system architecture are used to build operational GI systems: desktop, client–server, and Cloud. In the simplest desktop configuration, as used in a single-user project, the three software tiers are all installed on a single piece of hardware (most commonly a desktop PC) in the form of a desktop GI system, and users are usually unaware of their separate identity (Figure 6.3A). In a variation

on this theme, data files are held on a PC (or sometimes a Windows, Linux, or Unix machine) centralized file server, but the data–server functionality is still part of the desktop GI system. This means that the entire contents of any accessed file must be pulled across the network, even if only a small amount of it is required (Figure 6.3B).

In larger and more advanced multiuser workgroup or departmental GI systems, the three tiers can be installed on multiple machines to improve flexibility and performance (Figure 6.4). In this type of configuration, the users in a single department (for example, the planning or public works department in a typical local government) still interact with a GUI (presentation layer) on their desktop computers, which also contain all the business logic, but the database management software (data server layer) and data may be located on another machine connected over a network. This type of computing architecture is usually referred to as client–server because clients request data or processing services from servers that perform work to satisfy client requests. The data server has local processing capabilities and is able to query and process data and thus return part of the whole database to the client, which is much more efficient than sending back the whole dataset for client-side query. Clients and servers can communicate over a local area network (LAN), wide area network (WAN), or Internet network, but given the large amount of data communication between the client and server, faster LANs are most widely used.

Both the desktop and client–server architecture configurations have significant amounts of functionality on the desktop, which is said to be a thick client (see also Section 6.3.4).

### In a client–server GI system, clients request data or processing services from servers that perform work to satisfy client requests.

---

Figure 6.3 Desktop GI system architecture: (A) standalone desktop system on PCs each with own local files; (B) Desktop system on PCs sharing files on a PC file server over a LAN.

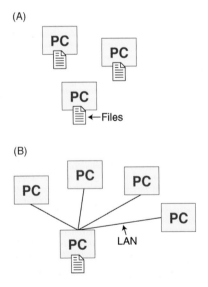

(A)

(B)

---

Figure 6.4 Client–server system: desktop software and DBMS data server in a workgroup or departmental configuration.

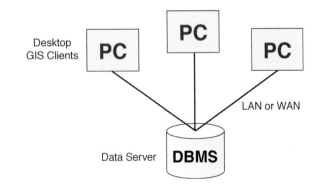

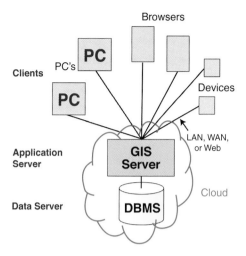

Figure 6.5 Centralized server system as used in advanced departmental and enterprise implementations. See Figure 6.2 for explanation of color-coding.

In recent years Cloud GI systems have become popular largely because of the ease of use and cost-effective nature of this architecture. The key feature of a Cloud implementation is that almost everything runs on servers that are hosted remotely in data centers that are accessible over the Web (Figure 6.5). In this architectural pattern all computing and geographic data and processing resources run on third-party hardware and software. Figure 6.6 shows the computer hardware in a typical data center. Such computing resources are usually purchased accordingly to usage (e.g., CPU and disk space utilized per month) and can be quickly provisioned (scaled up or down) depending on usage requirements.

In the Cloud model, users access business logic and data services from devices (e.g., phones or tablets), which run lightweight clients (thin clients),

Figure 6.6 Computer hardware in a typical data center.

© Baran Özdemir/Vetta/Getty Images

or from a range of heavier-weight clients, including full-functionality desktop GI systems (thick clients). In the case of thin-client access, the presentation tier (user interface) also runs in the Cloud on the server (although technically it is still presented on the desktop). The current trend is to add local processing capabilities to thin clients by supporting client-side scripting and data caching. A second server machine is usually employed to run data management software (DBMS). This type of implementation is the standard architecture for Web-based GI systems and is common in enterprise GI systems. Large, enterprise GI systems may involve more than 10 servers and hundreds or even thousands of clients that are widely dispersed geographically and nowadays are connected using the Web. The Cloud computing architecture forms the platform for many Web-based services such as the e-mail and search systems from Google, Microsoft, and Yahoo!, as well as various virtual-globe implementations (see Section 6.6.4).

**Cloud computing is the standard architecture used in Web-based and enterprise GI systems.**

Although organizations often standardize on either a project, a departmental, or an enterprise system, it is also common for large organizations to have all three configurations operating in parallel or as subparts of a full-scale system.

## 6.3.3 Software Data Models and Customization

In addition to the three-tier model, two further topics are relevant to an understanding of software architecture: data models and customization.

GI data models will be discussed in detail in Chapter 7, and so the discussion here will be brief. From a software perspective, a data model defines how the real world is represented in a GI database. It will also affect the type of software tools (functions or operators) that are available, how they are organized, and their mode of operation. A data model defines how the different tools are grouped together, how they can be used, and how they interact with data. Although such issues are largely transparent to end users whose interaction with a GI system is via a user interface, they become very important to software developers that are interested in customizing or extending software. The data model affects the capability, flexibility, and extensibility of GI system software.

Customization is the process of modifying a GI system to, for example, add new functionality to applications, embed GI functions in other applications, or create specific-purpose applications. It can be as simple as deleting unwanted controls (for example, menu choices or buttons) from a GUI, or as sophisticated as adding a major new extension to a software package for such things as network analysis, high-quality cartographic production, or land parcel management.

To facilitate customization, GI software must provide access to the data model and expose capabilities to use, modify, and supplement existing functions. In the late 1980s when customization options were first added to GI software products, each vendor had to provide a proprietary customization capability simply because no standard customization systems existed. Nowadays, with the widespread adoption of the .Net and Java frameworks, a number of industry-standard approaches and programming languages (such as Visual Basic, C/C++/C#, Java, and Python) are available for customizing GI systems. Simpler scripting languages such as JavaScript are also becoming widely used, especially for Web services, because of their high-level, interpreted nature and because a number of different libraries (e.g., for charting, database query, and search) can easily be integrated into the same end-user applications.

Modern programming languages are one component of larger developer-oriented packages called integrated development environments (IDEs). This term refers to the combination of several software development tools, including a visual programming language, an editor, a debugger, and a profiler. Many of the so-called visual programming languages, such as C++/C#, Visual Basic, and Java, support the development of windows-based GUIs containing forms, dialogs, buttons, and other controls. Program code can be entered and attached to the GUI elements using the integrated code editor. An interactive debugger will help identify syntactic problems in the code, for example, misspelled commands and missing instructions. Finally, there are also tools to support profiling programs. These show where resources are being consumed and how programs can be speeded up or improved in other ways.

**Contemporary GI systems typically use an industry-standard programming language like Visual Basic, C++/C#, or Java for customization.**

To support customization using open, industry-standard IDEs, GI system functions must be exposed in a standard way. In modern GI systems the functionality is developed as software components—self-contained software modules that implement a coherent collection of functionality. A key feature of such software components is that they have well-defined application programming interfaces (APIs) that allow the functionality to be called by the programming tools in an IDE or scripting environment.

Three predominant technology standards are widely used for defining and reusing software components. For building high-performance, interactive desktop applications, Microsoft's .Net framework is the de facto standard. It uses fine-grained components (that is, a large number of small functionality blocks). For server-centric GI systems, both .Net and the equivalent Java frameworks are widely deployed in operational applications. Although both .Net and Java work very well for building fine-grained client or server applications, they are less well suited for building applications that need to communicate over the Web. Because of the loosely coupled, comparatively slow, heterogeneous nature of Web networks and applications, fine-grained programming models have limitations. As a consequence, coarse-grained component and messaging systems have been built on top of the fine-grained .Net and Java applications using SOAP/XML (simple object access protocol/extensible markup language) and JavaScript/REST (representation state transfer protocol) that allow applications with Web services interfaces to interact over the Web. Microsoft's Silverlight and Adobe's Flex are also being widely used to build Web applications. This higher-level approach makes it easier to create custom Web applications that use Web resources in a more efficient way by minimizing network traffic.

Components are important to software developers because they are the mechanism by which reusable, self-contained, software building blocks are created and used. They allow many programmers to work together to develop a large software system incrementally. The standard open (published) format of components means that they can easily be assembled into larger systems. In addition, because the functionality within components is exposed through interfaces, developers can reuse them in many different ways, even supplementing or replacing functions if they so wish. Users also benefit from this approach because GI system software can evolve incrementally and support multiple third-party extensions. In the case of GI systems this includes, for example, tools for charting, reporting, and data table management.

Box 6.2 describes the work of one open-source software developer, Gabriel Roldan (Figure 6.7).

### 6.3.4 GI Systems on the Desktop, on the Web, and in the Cloud

Today mainstream, high-end GI system users work primarily with software that runs either on the desktop or over the Web and in the Cloud. In the desktop case, a PC (personal computer) is the main hardware platform, and Microsoft Windows remains the dominant

### Gabriel Roldan, Open-Source Software Developer

Gabriel has been an open-source GI system developer for over a decade and is a core OpenGeo developer. OpenGeo is the commercial developer of the OpenGeo Suite of open-source geospatial software. This is a geospatial Web services stack for deploying solutions for Web mapping, transportation, telecommunications, open government, and a huge range of other solutions. The OpenGeo Suite is a continually updated geo Web services platform along with maintenance agreements, which include support and training. OpenGeo is funded by leading investment firms Vanedge Capital and IQT.

Gabriel began in the GIS and Remote Sensing Department in the Instituto Politecnico Superior in Rosario, Argentina, and since then he has worked at various companies in Argentina and Spain on route finding, vehicle tracking, and Web mapping, improving corporate GI and spatial data infrastructure software. Working with local governments, he has introduced GeoTools and GeoServer and helped with their adoption as core components of GI software infrastructures.

Gabriel joined the open-source community in 2003 and has developed the ArcSDE datastore and Web Mapping Service for GeoServer as well as the first ArcSDE data reader, which he still maintains. At OpenGeo, he rearchitected GeoTools' ArcSDE raster support to match ArcSDE's capabilities and became a core GeoServer and GeoWebCache developer. Since 2011 he has been the technical lead for the Geogit project, which is implementing database versioning in GeoServer.

"Working in professional open-source is an incredible opportunity for continuous learning and improvement, both technically and as a human being," says Gabriel.

**Figure 6.7** Gabriel Roldan, open-source software developer.

operating system (Figure 6.8 and Table 6.1). In the desktop paradigm, clients tend to be functionally rich and substantial in size and are often referred to as thick or fat clients. Using the Windows standard facilitates interoperability (interaction) with other desktop applications, such as word processors, spreadsheets, and databases. As noted earlier (Section 6.3.2), most sophisticated and mature GI workgroups have adopted the client–server implementation approach by adding either a thin or thick server application running on the Windows, Linux, or Unix operating system. The terms *thin* and *thick* are less widely used in the context of

**Figure 6.8** Desktop and network GI system paradigms. See Figure 6.2 for explanation of color coding.

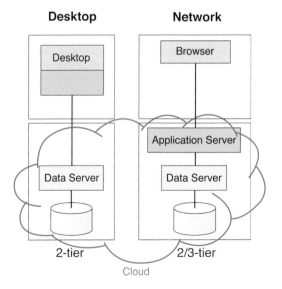

**Table 6.1** Comparison of desktop and network GI systems.

| Feature | Desktop | Network |
|---|---|---|
| **Client size** | Thick | Thin |
| **Client platform** | Windows | Browser or Device |
| **Server size** | Thin/thick | Thick |
| **Server platform** | Windows/Unix/ Linux | Windows/ Unix/Linux |
| **Component standard** | .Net | .Net/Java |
| **Network** | LAN/WAN | LAN/WAN/ Internet |

servers, but they mean essentially the same as when applied to clients. Thin servers perform relatively simple tasks, such as serving data from files or databases, whereas thick servers also offer more extensive analytical capabilities such as geocoding, routing, mapping, and spatial analysis. In desktop implementations, LANs and WANs tend to be used for client–server communication. It is natural for developers to select Microsoft's .Net technology framework to build the underlying components making up these systems given the preponderance of the Windows operating system, although other component standards could also be used. The Windows-based client–server architecture is a good platform for hosting interactive, high-performance GI applications. Examples of applications well suited to this platform include those involving geographic data editing, map production, 2-D and 3-D visualization, spatial analysis, and modeling. It is currently the most practical platform for general-purpose systems because of its wide availability, good performance for a given price, and common usage in business, education, and government.

**GI system users are standardizing their systems on the desktop and Web implementation models.**

In the past few years there has been increasing interest in harnessing the power of the Web for GI systems and for hosting applications in a cloud computing environment. Although desktop GI systems have been and continue to be very successful, users are constantly looking for lower costs of ownership and improved access to geographic information. Network-based (sometimes called distributed) GI systems allow previously inaccessible information resources to be made more widely available. A particularly interesting type of network GI system is Web-based, using the Web as a network and Cloud infrastructure for managing resources. The network model intrigues many organizations because it is based on centralized software and data management, which can dramatically reduce initial implementation and ongoing support and maintenance costs. It also provides the opportunity to link nodes of distributed user and processing resources using the medium of the Internet. The continued rise in network GI systems will not signal the end of desktop GI systems—indeed quite the reverse is true because it is likely to stimulate the demand for content and professional skills in geographic database automation and administration and applications that are well suited to running on the desktop.

In contrast to desktop systems, networked GI systems can use cross-platform Web browsers or devices like smartphones and tablets to host the viewer user interface. Currently, clients are typically very thin, often with simple display and query capabilities, although

there is an increasing trend for them to become more functionally rich. Server-side functionality may be encapsulated on a single server, although in medium and large systems it is more common to have two servers, one containing the business logic (a middleware application server), the other the data manager (data server). The server applications typically contain all the business logic, are comparatively thick, and may run on a Windows, Unix, or Linux platform.

Recently, there has been a move to combine the best elements of the desktop and network paradigms to create so-called rich clients. These are stored and managed on the server and are dynamically downloaded to a client computer according to user demand. The business logic and presentation layers run on the server, with the client hardware simply used to render the data and manage user interaction. The new software capabilities in recent editions of the .Net and Java software development kits allow the development of applications with extensive user interaction that closely emulate the user experience of working with desktop software. Microsoft's Bing Maps 3D is an example of a rich client. When a user visits the Bing Web site, he or she is given the opportunity to download the Bing Maps 3D client. Upon installation, the client talks to the Bing Maps 3D server that holds vast databases of geographic information for the Earth and provides server-side query and analysis features.

## 6.4 Building GI Software Systems

Widely used GI systems are built and released as software products by development and product teams that may operate according to commercial or open-source models. Mature software products are subject to carefully planned versioned release cycles that incrementally enhance and extend the pool of capabilities. The key parts of a GI system architecture—the user interface, business logic (tools), data manager, data model, and customization environment—were outlined in the previous section.

GI product teams start with a formal design for a software system and then build each part or component separately before assembling the whole system. Typically, development will be iterative with creation of an initial prototype framework containing a small number of partially functioning parts, followed by increasing refinement of the system. Core software systems are usually written in a modern programming language like Visual C++/C# or Java, with Visual Basic or another scripting language such as Python used for operations that do not involve significant amounts of computer processing like GUI interaction.

As standards for software development become more widely adopted, so the prospect of reusing software components becomes a reality. A key choice that then faces all software developers or customizers is whether to design a system by reusing third-party components or libraries (that might be licensed commercially or using an open-source distribution model), or to build one more or less from scratch. All three options have advantages and disadvantages: building components gives greater control over capabilities and enables specific-purpose optimization, but can be complex and slow; buying components can save time, but can be expensive and inflexible; the open-source model provides low-cost access to functionality and the ability to view and control source code, but often lacks ongoing support.

**A key GI system implementation issue is whether to buy a system or to build one.**

A modern GI system comprises an integrated suite of software components of three basic types: data management software for controlling access to data (Chapters 7, 8, and 9); mapping software for display and interaction with maps and other geographic visualizations (Chapters 11 and 12); and spatial analysis and modeling software for transforming geographic data using operators (Chapters 13, 14, and 15). The components for these parts may reside on the same computer or can be distributed widely (Chapter 10) over a network.

## 6.5 GI Software Vendors

The GI industry is fortunate to have a diverse range of significant and well-established software vendors. Each of the major vendors brings its own heritage and experience and has created unique elements and interpretations of core capabilities in the form of software that is constantly evolving in response to ongoing IT changes and expanded user requirements. The following sections briefly describe the main commercial GI software vendors. Space does not permit similar treatment of smaller open-source development teams, although the work of one open-source developer is discussed in Box 6.2.

### 6.5.1 Autodesk

Autodesk is a large and well-known publicly traded company with headquarters in San Rafael, California. It is one of the world's leading digital design and content companies and serves customers in markets where design is critical to success: building, manufacturing, infrastructure, digital media, and location

services. Autodesk is best known for its AutoCAD product family, which is used worldwide by more than 4 million customers. The company was founded more than 30 years ago and has grown to become a $2 billion entity employing over 7,500 staff.

Autodesk's product family for the GI marketplace features several products including AutoCAD Map 3D, Autodesk Infrastructure Map Server, and Autodesk Infrastructure Design Server. AutoCAD Map 3D is a desktop product that enables organizations to create, maintain, and visualize geographic data and is especially adept at bridging the gap between engineering and GI teams on the one hand, and the rest of an organization on the other. It is very widely used for data capture in architecture, engineering, and construction firms, as well as local-government engineering departments. Autodesk Infrastructure Map Server is used to publish and share CAD, GI, and asset information with Web-based mapping software. It has capabilities to access, visualize, and coordinate enterprise data with workers in the field, operations and customer service personnel, and the public by using Web-based mapping dashboards and reports. Autodesk Infrastructure Design Suite is a design solution that combines intelligent, model-based tools to gain accurate, accessible, and actionable insight in transportation, land, and water projects.

### 6.5.2 Bentley

Bentley, a privately held developer of software solutions for the infrastructure lifecycle market, is headquartered in Exton, Pennsylvania. It has applications that help engineers, architects, contractors, governments, institutions, utilities, and owner–operators design, build, and operate productively, collaborate globally, and deliver infrastructure assets that perform sustainably. Bentley has total annual revenues of $500 million and employs over 3,000 staff worldwide; over $200 million in revenue is derived from Bentley's GI business.

Bentley's flagship software product, MicroStation, supports a number of applications and serves the needs of a range of communities. Bentley Map is a system designed to address the needs of organizations that map, plan, design, build, and operate the world's infrastructure. It enhances the underlying MicroStation platform with capabilities for geospatial data creation, maintenance, and analysis. With Bentley Map, users can integrate data from a wide variety of sources into engineering and mapping workflows. Because Bentley Map is tightly integrated with MicroStation, it allows simultaneous manipulation of raster and vector data, which are standard features of the core platform. An interoperability environment makes it easy to work with many spatial data formats. Bentley's GI system can be implemented with any database connection supported by MicroStation (e.g., Oracle Spatial or ArcGIS Server).

## 6.5.3 Esri

Esri is a privately held company founded in 1969 by Jack and Laura Dangermond. Headquartered in Redlands, California, Esri employs over 6,000 people worldwide and has annual revenues of over $1 billion. Today it serves more than 300,000 organizations and more than 2 million users. Esri focuses solely on the GI systems market, primarily as a software product company, but it also generates about a sixth of its revenue from project work such as advising clients on how to implement GI systems. Esri started building commercial software products in the late 1970s. Today Esri's product strategy is centered on an integrated family of products called ArcGIS (Box 6.3). The ArcGIS platform is aimed at both end users

## Application Box (6.3)

### Desktop GI System: Esri ArcGIS Desktop

ArcGIS for Desktop (Figure 6.9) is the desktop part of Esri's ArcGIS platform, an integrated suite of products for all key IT environments (desktop, server, Cloud, and mobile). It supports the full range of GI system functions including data collection and import; editing, restructuring, and transformation; display; query; and analysis.

A collection of analytical extensions is available for 3-D analysis, network routing, geostatistics, and spatial (raster) analysis, among others. ArcGIS for Desktop is available in three functionality levels: Basic, Standard, and Advanced. Its strengths include a comprehensive portfolio of capabilities, high-quality cartography and

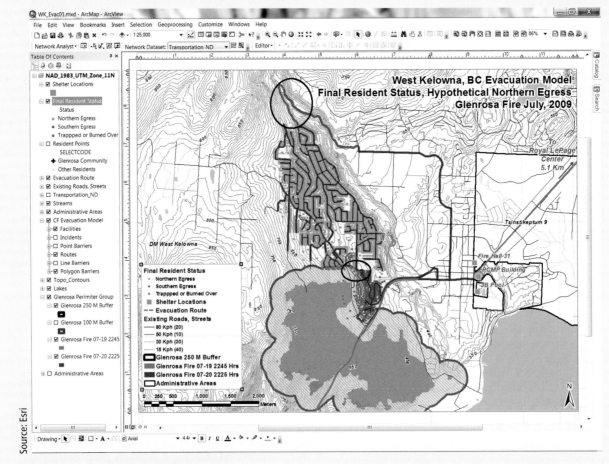

Source: Esri

**Figure 6.9** Screenshot of Esri ArcGIS for Desktop.

display tools, analysis functions, extensive customization options, strong data management (using ArcGIS Server), and a vast array of third-party tools and interfaces.

ArcGIS for Desktop comprises a series of integrated menu-driven, end-user applications: ArcMap—a map-centric application supporting integrated editing and viewing; ArcCatalog—a data-centric application for browsing and managing geographic data in files and databases; and ArcToolbox—a tool-oriented application for performing geoprocessing tasks such as proximity analysis, map overlay, and data conversion. ArcGIS is customizable using any Microsoft .Net-compliant programming language such as Visual C$^{++}$ and C#. The software is also notable because of the ability to store and manage all data (geographic and attribute) in standard commercial off-the-shelf DBMS (e.g., DB2, Informix, SQL Server, Oracle, and Postgres).

ArcGIS is the standard against which other products are often compared. It is functionally very rich, but at times can appear complex to new users. It has evolved considerably over the past 30 years, always being reinvented for new technology generations and application requirements.

(especially professional or highly technical users) and technical developers and includes products that run on handheld devices, desktop personal computers (Figure 6.9), servers, and over the Web. Most recently, the company has released an advanced Cloud GI system called ArcGIS Online. ArcGIS Explorer is Esri's virtual globe and is not unlike Google Earth and Microsoft Bing Maps.

Esri is the classic high-end GI system vendor. It has a wide range of mainstream products covering all the main technical and industry markets. Esri is a technically led geographic company focused squarely on the needs of hard-core users; in recent years it has started to address the needs of less technical mapping users. As a company, Esri is also very much concerned with the practical applications of GI, producing everything from data to data models and Web mapping services.

## 6.5.4 Intergraph

Like Esri, Intergraph was also founded in 1969 as a private company. The initial focus from their Huntsville, Alabama, offices was the development of computer graphics systems. After going public in 1981, Intergraph grew rapidly and diversified into a range of graphics areas including CAD and mapping software, consulting services, and hardware. Today Intergraph is part of Hexagon, a global provider of design, measurement, and visualization technologies.

Intergraph's geospatial products are able to work with content collected from a variety of sources. The portfolio includes the desktop GI system (Figure 6.10),

**Figure 6.10** Screenshot of Intergraph GeoMedia desktop system.

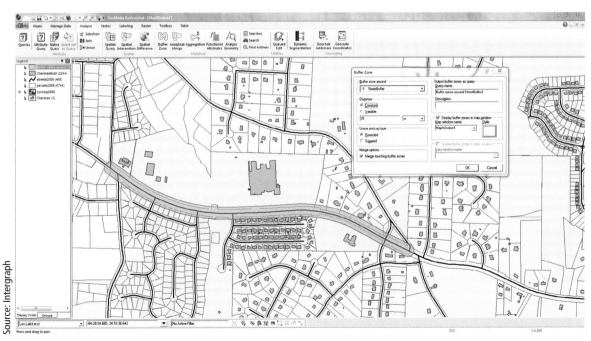

Source: Intergraph

## Desktop GI System: Intergraph GeoMedia

GeoMedia is an archetypal example of a mainstream commercial desktop product (Figure 6.10). First released in the late 1990s, it was created from the ground up to run on the Windows desktop operating system. Like other products in the desktop category, it is primarily designed with the end user in mind (rather than the technical user). It has a Windows-based graphical user interface and many tools for editing, querying, mapping, and spatial analysis. Data can be stored in proprietary GeoMedia files or in a DBMS such as Oracle, Microsoft Access, or SQL Server. GeoMedia enables data from multiple disparate databases to be brought into a single environment for viewing, analysis, and presentation. The data are read and translated on the fly directly into memory. GeoMedia provides access to all major geospatial/CAD data file formats and to industry-standard relational databases.

GeoMedia is built as a collection of software components. These underlying objects are exposed to developers who can customize the software using a programming language such as Visual Basic or C#. It offers a suite of analysis tools, including attribute and spatial query, buffer zones, spatial overlays, and thematic analysis. The product's layout tools provide the flexibility to design a range of types of maps that can be distributed on the Web, printed, or exported as files.

GeoMedia is available in three functionality levels: Essentials, Advantage, and Professional. Other members of the product family include GeoMedia Viewer (a free data viewer) and Geo-Media WebMap (for Internet publishing).

All in all, the GeoMedia product family offers a wide range of capabilities for core GI activities in the mainstream markets of government, education, and private companies. It is a modern and integrated product line with strengths in the areas of data access and user productivity.

remote sensing, and photogrammetry software, as well as the synthesis of these technologies in server-based products specializing in data management, spatial data infrastructure, workflow optimization and Web editing, and Web mapping.

From a GI perspective the principal product family is GeoMedia, which spans the desktop and Web-based GI system markets across a range of application domains. Box 6.4 describes Intergraph GeoMedia.

## 6.6 Types of GI Systems

Over 100 commercial software products claim to have mapping and GI system capabilities. The main categories of generic GI software that dominate today are desktop, Web mapping, server, virtual globe, developer, and mobile. The distinction between the Web mapping, server, and virtual globe categories is sometimes blurred, not least because increasingly they are being deployed in Cloud computing environments. Web mapping systems are those that integrate software and data to create a unified online mapping service; servers typically have a wider range of more advanced functionality and can work with multiple data sources; and virtual globes provide 3-D visualization, query, and some analysis capabilities. In this section, the categories will be discussed followed by a brief summary of other types of software. Reviews of currently popular software packages can be found in the various GI magazines on the Web.

### 6.6.1 Desktop Systems

Since the mid-1990s, desktop GI systems have been the mainstay of the majority of operational implementations. Desktop software owes its origins to the personal computer and the Microsoft Windows operating system. Desktop software provides personal productivity tools for a wide variety of users across a broad cross section of industries. PCs are widely available, are relatively inexpensive, and offer a large collection of user-oriented tools, including databases, word processors, and spreadsheets. The desktop software category includes a range of options from simple viewers (such as Esri ArcGIS ArcReader, Intergraph GeoMedia Viewer, Pitney Bowes MapInfo ProViewer, as well as a growing number of open-source products) to desktop mapping and GI system software (such as Autodesk AutoCAD Map 3D, Clark Labs Idrisi, Esri ArcGIs; see Box 6.3, Figure 6.9), open-source GRASS, Intergraph GeoMedia (see Box 6.4, Figure 6.10), the Manifold System (Figure 6.11), Pitney Bowes MapInfo Professional, Supermap, Smallworld Spatial Intelligence, and open-source Quantum GIS (see Figure 6.12).

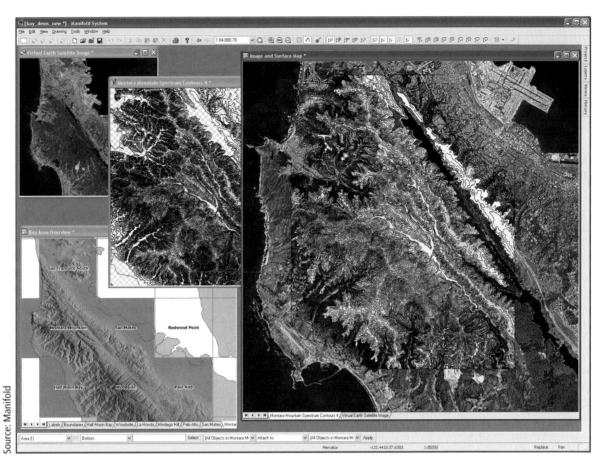

**Figure 6.11** Screenshot of Manifold desktop system.

**Figure 6.12** Screenshot of Quantum GIS desktop system.

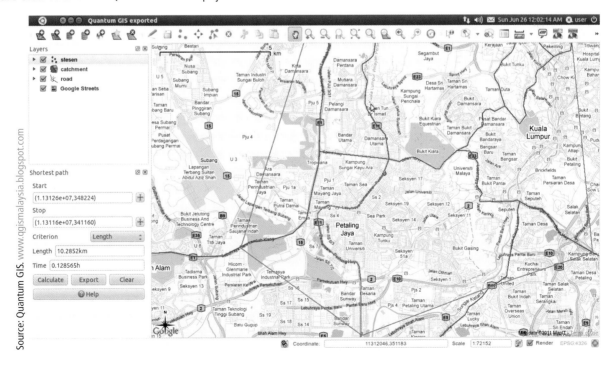

**Desktop GI systems are the mainstream workhorses of today.**

In the late 1990s, a number of vendors released free GI viewers that are able to display and query popular file formats. Today, the GI viewer has developed into a significant product subcategory. The rationale behind the development of these products from the commercial vendor point of view is that they help to establish market share and can create de facto standards for specific vendor terminology and data formats. A number of open-source products have also been developed to provide free and easy access to GI and tools. Users often work with viewers on a casual basis and use them in conjunction with more sophisticated software products. Viewers have limited functional capabilities restricted to display, query, and simple mapping. They do not support editing, sophisticated analysis, modeling, or customization.

With their focus on data use rather than data creation and their excellent tools for making maps, reports, and charts, desktop mapping and GI systems represent most people's experience of high-end systems today. Many of the successful systems have adopted Microsoft standards for interoperability and user-interface style (although some open-source systems are based on Java). Users often see a desktop mapping package and a GI system as simply a tool to enable them to do their full-time job faster, more easily, or more cheaply. Desktop mapping and GI system users work in planning, engineering, teaching, the army, marketing, and other similar professions; they are often not advanced technical users. Desktop GI system prices typically range from $400 to $1,500 (these and other prices mentioned later typically have discounts for multiple purchases).

The most advanced and functionally rich desktop software packages include data collection and editing, database administration, advanced geoprocessing and analysis, and other specialist tools. Such professional software packages offer a superset of the functionality of the systems in other classes. The people who use these systems typically are technically literate and think of themselves as GI professionals (career GI staff); they have degrees and, in many cases, advanced degrees in GI science or related disciplines. Prices for professional GI systems are typically in the $5,000–$15,000 range per user.

**Professional GI systems are high-end and fully functional systems.**

## 6.6.2 Web Mapping Systems

GI systems on the Web have a comparatively long history that goes back to 1993 when researchers at Xerox PARC in California created the first map server that could be accessed over the Internet. Since then several major milestones have been reached in developmental and organizational terms: the popularity of the MapQuest site and subsequent acquisition of the company by AOL for over $1.1 billion; the introduction of major mapping sites by Google, Yahoo!, and Microsoft (among others); and the rise of computer mashups and neogeography (see Section 6.6.3 and Chapter 10), to name but a few.

Here the term *Web mapping* is taken to mean integrated Web-accessible software, a 2-D database (3-D base maps are covered later in Section 6.6.4 on virtual globes) comprising one or more base maps, and an associated collection of services. Web access is provided via easily accessible, open interfaces. Typically, image tiles—each containing a fragment of a map—are returned in response to a user request for a map of a given area. Requests can be issued from, and responses processed and visualized in, standard Web browsers and Web-accessible devices (e.g., phones and tablets), and Web mapping sites have proven to be very popular (being consistently rated among the top 10 in terms of Web site traffic). In addition to providing maps, a range of other useful services are normally closely coupled, for example, a gazetteer to find places of interest, multipoint driving directions, a choice of image and street map base maps, and the ability to overlay, or mash up, many other datasets.

Some of the success of Web mapping sites is also due to the fact that they can be easily accessed programmatically via well-defined APIs. The quintessential example here is KML and the associated API for interacting with Google Maps (see Box 6.1). Web APIs have spawned a large number of add-ons and have allowed Web maps to be integrated or mashed up with many other Web services. Figure 6.13 shows multiple Web services overlaid (mashed up) on a topographic base map; see Section 10.4.1 for a further example.

OpenStreetMap is a base-map service collected by volunteers and published on the Web using a copyright-free license. The software used to edit, access, and view the data is open source (including Mapnik for map rendering, JOSM and Potlatch for editing, and PostGIS for data management). In certain cities, OpenStreetMap is now beginning to rival and even exceed the content and coverage of commercial online base-map databases.

## 6.6.3 Server GI Systems

In simple terms, a server GI system is a system that runs on a computer server and can handle concurrent processing requests from a range of networked

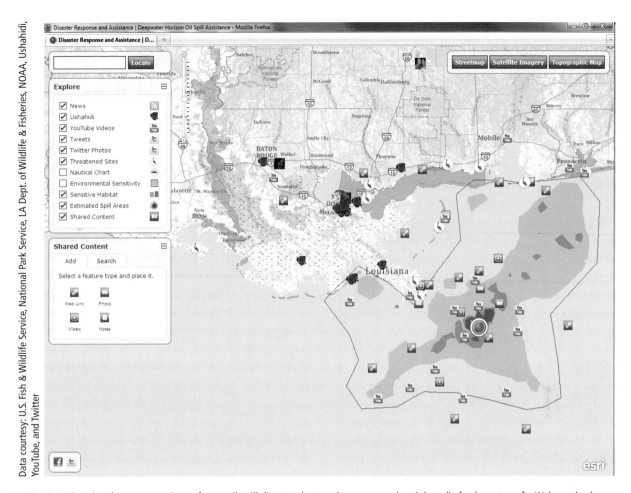

**Figure 6.13** A situational awareness viewer for an oil spill disaster that mashes up several social media feeds on top of a Web-service base map.

clients. Server GI systems offer a wider range of functions than Web mapping systems, which focus solely on mapping and closely related services and can work with any base map. Server GI system products have the potential for the largest user base and lowest cost per user of all software system types. Stimulated by advances in server hardware and networks, the widespread availability of the Internet, and market demand for greater access to geographic information, GI system vendors have been quick to release server-based products that are accessible over the Web. Examples of server systems include Esri ArcGIS Server, Intergraph GeoMedia Webmap, MapGuide and MapServer (both open source), Pitney Bowes MapInfo MapXtreme, and Smallworld Spatial Application Server. The cost of commercial server products varies from around $5,000 to $25,000, for small to medium-sized systems, to well beyond for large multifunction, multiuser systems. Box 6.5 highlights the capabilities of MapGuide.

**Server GI systems are growing in importance as capabilities increase and organizations shift from desktop to network implementations.**

The most popular open-source GI system today is probably the MapServer Web mapping system (Figure 6.14). Originally developed at the University of Minnesota, it now has community ownership. The primary purpose of MapServer is to display and interact with dynamic maps over the Internet. It can work with hundreds of raster, vector, and database formats and can be accessed using popular scripting languages and development environments. It supports state-of-the-art functionality such as on-the-fly projections, high-quality map rendering, and map navigation and queries. MapServer has been used quite widely as the core of many spatial data infrastructure projects to render maps and respond to queries for data download.

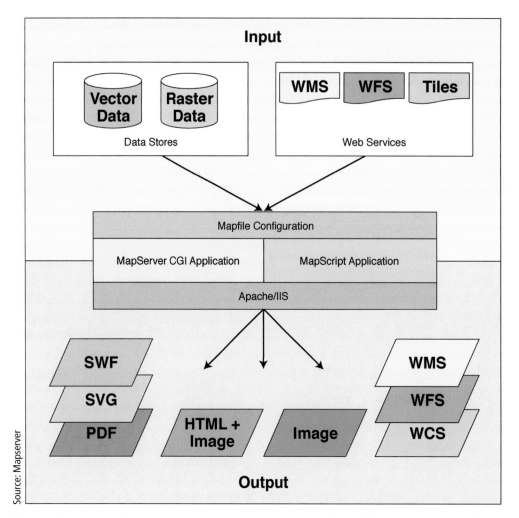

**Figure 6.14** Architecture of MapServer. The central boxes are the software components that read data, render maps, and fulfill other client requests. A Web server (open-source Apache or Microsoft IIS) is used to communicate with browser clients.

## Application Box (6.5)

### Server System: MapGuide Open Source

MapGuide Open Source is used to develop and deploy Web mapping applications and Web services. It is licensed under the open-source GNU Lesser General Public License. MapGuide is an important system for the many users who have spent considerable amounts of time and money creating valuable databases and who want to make them available to other users inside or outside their organization. It allows users to leverage their existing GI investment by publishing dynamic, intelligent maps at the point at which they are most valuable—in the field, at the job site, or on the desks of colleagues, clients, and the public.

Figure 6.15 is a screenshot of a MapGuide application that provides tourists and residents an easy way to locate streets and businesses in or around San Miguel de Allende, Guanajuato, Mexico. Using this application, community members are encouraged to include small local businesses that would not otherwise have an Internet presence.

There are three key components of MapGuide: the viewer—a relatively easy-to-use Web application with a browser-style interface; the author—a menu-driven authoring environment used to create and publish a site for client access; and the server—the administrative software that monitors site usage and manages requests from multiple clients and to external databases. MapGuide works directly with Web browsers and servers and uses the HTTP (Internet protocol) for communication.

▶

It makes good use of standard Internet tools like HTML (hypertext markup language) and JavaScript for building client applications. Typical features of MapGuide sites include the display of raster and vector maps, map navigation (pan and zoom), geographic and attribute queries, buffering, report generation, and printing. Like other advanced Web-based server systems, MapGuide has tools for redlining (drawing on maps) and basic editing of geographic objects.

To date, MapGuide has been used most widely in existing mature GI sites that want to publish their data internally or externally and in new sites that want a way to publish dynamic maps quickly to a widely dispersed collection of users (for example, maps showing election results or transportation network status). MapGuide can be used to serve maps (using the OGC WMS protocol) and features (using the OGC WFS protocol). Configured as a mapping service, MapGuide supports a client/server environment. It can retrieve geospatial data from WFS and WMS sites, enabling the use of data from other organizations that share their geospatial data. As a further service, MapGuide allows organizations to share their data, in vector form, with authorized outside organizations.

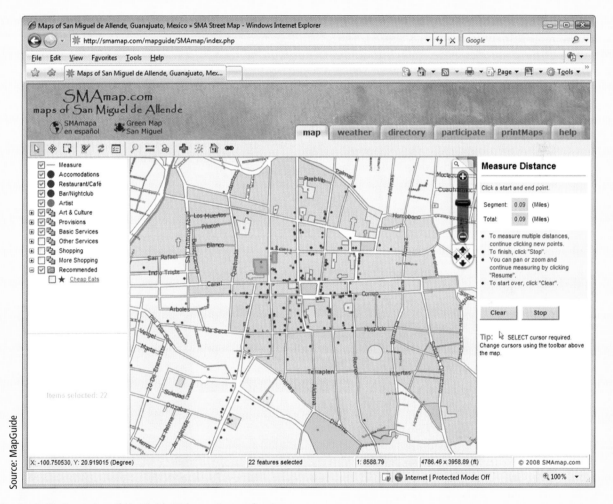

Source: MapGuide

**Figure 6.15** Screenshot of MapGuide Web mapping application.

Initially, server GI systems were nothing more than ports of desktop products, but second-generation systems were subsequently built using a multiuser services-based architecture that allows them to run unattended and to handle many concurrent requests from remote networked users. These systems often focus on display and query applications—making things simple and cost effective—with more advanced applications available at a higher price. Server GI systems are especially adapted for well-defined tasks that need to be performed repeatedly (ad hoc tasks largely remain the preserve of desktop systems, at least for now). Today, it is routinely possible to perform standard operations like making maps, network routing, geocoding, editing, and publishing many types of thematic and topographic data. A number of Web GI services are indexed using geoportals—gateways to GI resources—for example, the U.S. federal government geoportal www.data.gov. These are built using server software. Many Cloud GI systems are underpinned by server software.

A second generation of Web-based server products is becoming increasingly prevalent. They exploit the unique characteristics of the Web and integrate GI technology with Web browsers and servers. Initially, these new systems had limited functionality, but now there is a new breed of true server products that offer complete GI system functionality in a multiuser, Web-based server environment (e.g., ArcGIS Server). These server products have functions for distributed editing, mapping, data management, and spatial analysis, and support state of-the-art customization.

In conclusion, server products are growing in importance. Their cost-effective nature, ability to be centrally managed, and focus on ease of use will help to disseminate GI even more widely and will introduce many new users to the field. Server systems are the heart of many Cloud GI implementations.

## 6.6.4 Virtual Globes

One of the most exciting recent developments in the field has been the advent of virtual globes—3-D Web services hosted on Web-based systems that publish global 3-D databases and associated services for use over the Web. Virtual globes allow users to visualize geographic information on top of 3-D global base maps. Users can fly over a comparatively fine-resolution virtual globe with thematic data overlaid. Google Earth, released in 2005, set the standard for user interaction and the quality and quantity of data and established in a matter of months an engaged user community. Figure 6.16 shows information about the Millennium Development Goals (MDGs) for the Lao People's Democratic Republic as represented in Google Earth. Since then, a number of other vendors have released comparable globes, not least of which is Microsoft Bing Maps 3D. Figure 6.17 shows the center of Paris in the Bing Maps 3D mode.

Virtual globes have gained considerable traction in the wider community because, comparatively speaking, they are low cost (basic versions are free), they have high-quality image and vector databases for the Earth (and other planets), it is simple to overlay user data on a globe, and they are easy to use.

**Figure 6.16** Screenshot of Google Earth (earth.google.com).

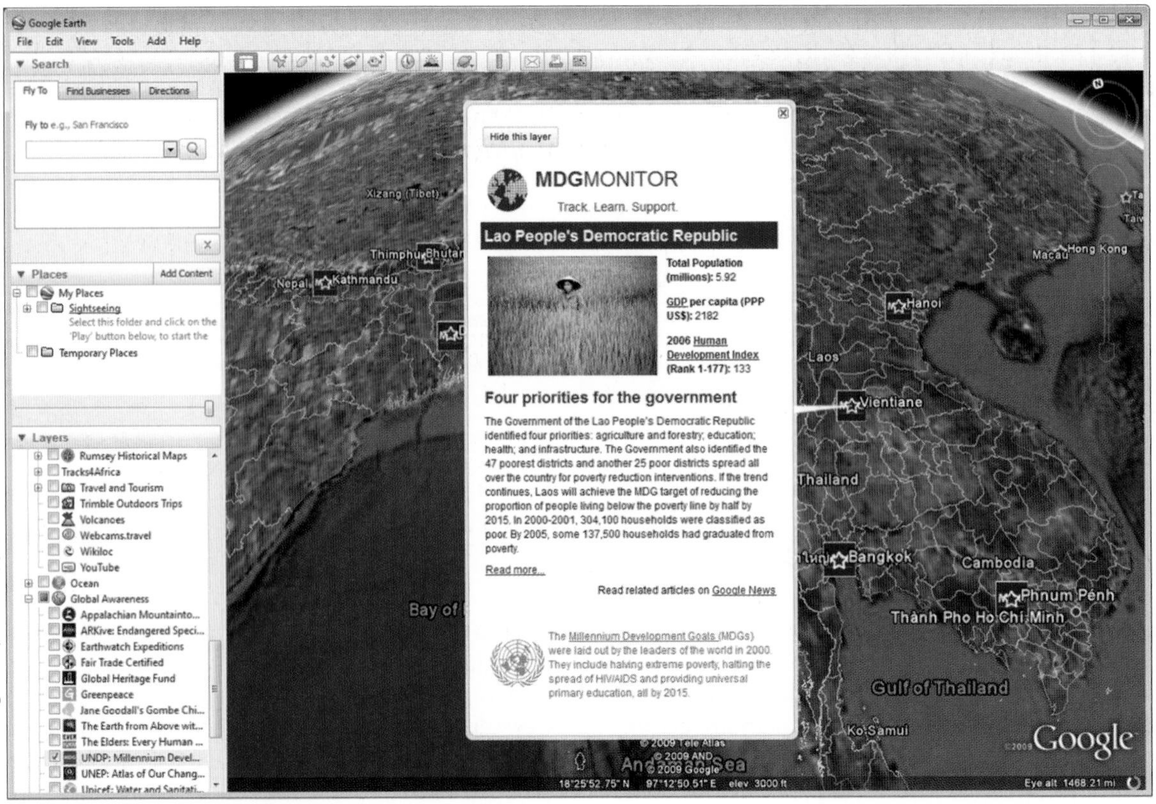

Source: Google, map data copyright Google, AND, TeleAtlas

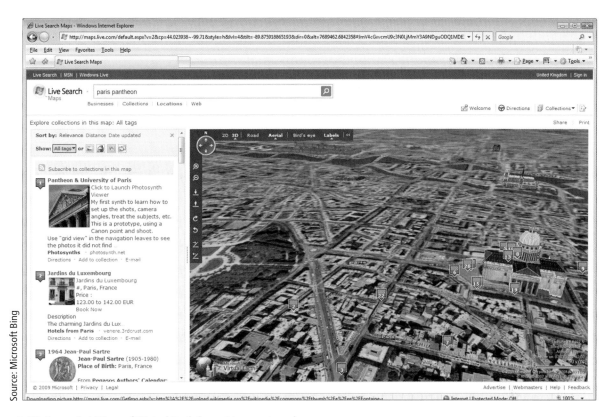

**Figure 6.17** Screenshot Microsoft Virtual Earth (www.bing.com/maps).

They have stimulated in no small measure two whole new subfields called neogeography (see Section 1.5.5) and volunteered geographic information (VGI; see Sections 1.5.6 and 10.2). Neogeography is the "new" geography that among other things includes the overlay or mashing up of two or more sources of geographic information (for example, webcams from Caltrans [California Department of Transportation] on top of a Yahoo! base map). VGI focuses on the fact that humans are acting as sensors and are building and publishing content from the ground up, often using virtual globes as base maps. The nonauthoritative and sometimes transient and dynamic nature of this information provides new geographic challenges and opportunities. These issues are explored in more depth in Chapter 10.

## 6.6.5 Developer GI Systems

With the advent of component-based software development (see Section 6.4), a number of collections of software components oriented toward the needs of developers have been released. These are really tool kits of GI system functions (components) that a reasonably knowledgeable programmer can use to build a specific-purpose application. They are of interest to developers because such components can be used to create highly customized and optimized applications that can either stand alone or can be embedded within other systems. Typically, component packages offer strong display and query capabilities, but only limited editing and analysis tools, mainly because there is greatest demand for products with such data exploration and visualization functionality. However, this is beginning to change as products mature. In the past few years there has been a surge in Web services that can be customized and embedded using published APIs (see Box 6.1).

**Developer products are collections of components used by developers to create focused GI applications.**

Esri ArcGIS Engine is a good example of a desktop development product that can be used in .Net, and Java environments. Java-based tool kits include ObjectFX SpatialFX and IBM ILOG JViews Maps (see Figure 6.18). The typical cost for a commercial developer product is $1,000–5,000 for the developer kit and $50–500 per deployed application. The people who use deployed applications may not even realize that they are using a GI system because often the run-time deployment is embedded in other applications (e.g., customer care systems, routing systems, or interactive atlases).

**Figure 6.18** Screenshot of a Web application written using the IBM ILOG JViews Map component tool kit.

There are a number of open-source developer toolkits of which GeoTools and GDAL are two of the most widely used. GeoTools is a Java-based code library that provides OGC standards-compliant methods for the manipulation of geospatial data. It has been used in a number of GI systems including GeoServer and is a key part of the GeoAPI set of tools. GDAL/ODR is a software library for reading and writing raster and vector data formats. It is widely used in commercial and open-source products.

## 6.6.6 Mobile GI Systems

As hardware design and miniaturization have progressed dramatically over the past few years, so it has become possible to develop software for mobile and personal use on handheld systems. The development of low-cost, lightweight location positioning technologies (primarily based on the Global Positioning System; see Section 4.9) and wireless networking has further stimulated this market. With capabilities similar to the desktop systems of just

a few years ago, mobile smartphones, tablets, and specialist devices can support many display, query, and simple analytical applications, even on displays as small as 320 by 240 pixels (although today much finer-resolution screens are available). An interesting characteristic of these systems is that all programs and data are held in local memory because of the lack of a hard disk. This provides fast access, but because of the cost of memory compared to disk systems, designers have had to develop compact data storage structures. Esri's ArcPad product is one of a number of products in this space (Figure 6.19). It specializes in field data collection and mobile mapping. When linked to Esri ArcGIS Server, it can act as an enterprise field client.

**Mobile GI systems are lightweight and designed for handheld field use.**

One of the most innovative areas at present is the development of software for smartphones and related devices. Despite their compact size, they can deal with comparatively large amounts of data (16 GB and more) and handle surprisingly sophisticated

**Figure 6.19**  Esri ArcPad running on a rugged handheld field portable device.

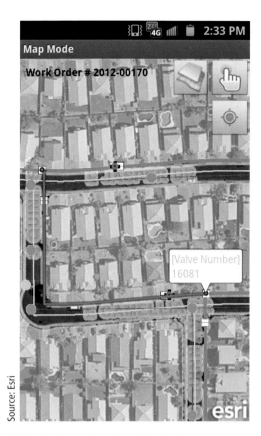

**Figure 6.20**  A utility work order application that runs on a smartphone device.

applications. The systems usually operate in a mixed connected/disconnected environment and so can make active use of data and software applications held on a server (see earlier discussion of server GI systems in Section 6.6.3) and accessed over a wireless telephone or WiFi network. Many applications can run on smartphones, tablets, and other specialist devices and can deliver highly productive specific-purpose GI applications (Figure 6.20).

## 6.6.7 Other types of GI Software

The previous section focused on mainstream software from the major commercial vendors. There are many other types of commercial and noncommercial software that provide valuable GI capabilities. This section briefly reviews some of the main types.

Raster-based GI systems, as the name suggests, focus primarily on raster (image) data and raster analysis. Chapters 2, 3, and 7 provide a discussion of the principles and techniques associated with raster and other data models, whereas Chapters 13 and 14 review their specific capabilities. Just as many vector-based systems have raster analysis capabilities (for example, Esri ArcGIS has Spatial Analyst, and Intergraph GeoMedia has Image and Grid), in recent years raster systems have added vector capabilities (for example, ERDAS Imagine and Clark Labs Idrisi now

have vector capabilities built in; see Figure 6.21). As a result, the distinction between raster-based and other system categories is becoming increasingly blurred. The users of raster-based systems are primarily interested in working with imagery and undertaking spatial analysis and modeling activities. The prices for raster-based systems range from $500 to $10,000.

CAD (or computer-aided design)-based systems started life as CAD packages and then had GI functions added. Typically, a CAD system is supplemented with database, spatial analysis, and cartography capabilities. Not surprisingly, these systems appeal mainly to users whose primary focus is in typical CAD application areas such as architecture, engineering, and construction, but who also want to use GI and geographic analysis in their projects. These systems are often used in data collection and mapping applications. The best-known examples of CAD-based systems are Autodesk Map 3D and Bentley Map (see Section 6.5). CAD-based GI systems typically cost from $3,000 to $5,000.

To assist in managing data in standard DBMS, some vendors—notably IBM, Microsoft, and Oracle—have developed technology to extend their DBMS servers so that they are able to store and process GI efficiently. Although not strictly GI systems in their own right (because of the absence of editing,

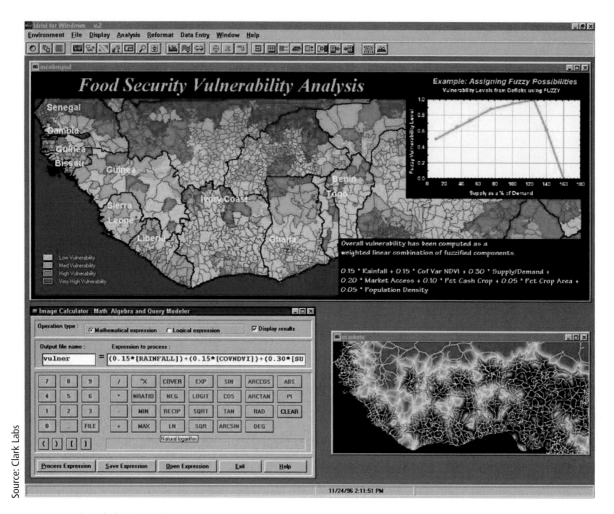

Figure 6.21 Screenshot of Clark Labs Idrisi desktop system.

mapping, and analysis tools), they are included here for completeness. Box 9.1 provides an overview of Oracle's Spatial DBMS extension.

Some noteworthy examples of open-source systems not already mentioned in earlier sections include GRASS (desktop), GeoDA for spatial analysis and visualization, gvSIG for desktop mapping, and PostGIS for storing data in a DBMS.

The Open Geospatial Consortium, although not itself a software vendor or software-development organization, has played a very important role in the evolution of GI software. OGC's role has been to encourage the development of standards that facilitate the sharing of GI and the interaction of geographic software. Perhaps the most significant progress has been in the area of interoperability of Web services, and a series of OGC standards now allow images, features, coverages, and metadata to be integrated together over the Web.

## 6.7 Conclusion

Software is a fundamental and critical part of any operational GI system. The software employed in a project has a controlling impact on the type of studies that can be undertaken and the results that can be obtained. There are also far-reaching implications for user productivity and project costs. Today, there are many types of software to choose from and a number of ways to configure implementations. One of the exciting and at times unnerving characteristics of GI software is its very rapid rate of development, not least in the areas of Web and open-source systems. This trend seems set to continue as software developers push ahead with significant research and development efforts. The following chapters explore in more detail the functionality of GI software and how it can be applied in real-world contexts.

## Questions for Further Study

1. Design a GI system architecture that 25 users in three cities could access to create an inventory of recreation facilities.

2. Discuss the role of each of the three tiers of software architecture in an enterprise GI implementation.

3. With reference to a large organization that is familiar to you, describe the ways in which its staff might use a GI system, and evaluate the different types of systems that might be implemented to fulfill these needs.

4. Go to the Web sites of the main GI system vendors, and compare their product strategies with those of open-source GI products such as OpenGeo. In what ways are they different?
   - Autodesk: www.autodesk.com
   - Bentley: www.bentley.com
   - Esri: www.esri.com
   - Intergraph: www.intergraph.com
   - OpenGeo: opengeo.org

## Further Reading

Coleman, D. J. 2005. GIS in networked environments. In Longley, P. A., Goodchild, M. F., Maguire, D. J., and Rhind, D. W. (eds.). *Geographical Information Systems: Principles, Techniques, Management and Applications* (abridged edition). Hoboken, NJ: Wiley, 317–329.

Fu, P. and Sun, J. 2011. *Web GIS: Principles and Applications*. Redlands, Esri Press.

Maguire, D. J. 2005. GIS customization. In Longley, P. A., Goodchild, M. F., Maguire, D. J., and Rhind, D. W. (eds.). *Geographical Information Systems: Principles, Techniques, Management and Applications* (abridged edition). Hoboken, NJ: Wiley, 359–369.

Tsou, M.-H., and Smith, J. 2011 *Free and open source software for education*. National Geospatial Technology Centre of Excellence geoinfo.sdsu.edu/hightech/WhitePaper/tsou_free-GIS-for-educators-whitepaper.pdf

# 7 Geographic Data Modeling

This chapter discusses the technical issues involved in modeling the real world in a geographic information (GI) database. It describes the process of data modeling and the various data models that have been used. A data model is a set of constructs for describing and representing parts of the real world in a digital computer system. Data models are vitally important because they control the way that data are stored and have a major impact on the type of analytical operations that can be performed. Early GI databases were based on extended CAD, simple graphical, and image data models. In the 1980s and 1990s, the hybrid georelational model came to dominate. More recently, major software systems have been developed on more advanced and standards-based geographic object models that include elements of all earlier models.

## LEARNING OBJECTIVES

After studying this chapter you will be able to:

- Define what geographic data models are and discuss their importance.

- Understand how to undertake GI data modeling.

- Outline the main geographic models used in GI databases and their strengths and weaknesses.

- Understand key topology concepts and why topology is useful for data validation, analysis, and editing.

- Read data model notation.

- Describe how to model the world and create a useful geographic database.

## 7.1 Introduction

This chapter builds on the material on geographic representation presented in Chapter 3. By way of introduction, it should be noted that the terms *representation* and *model* overlap considerably (we will return to a more detailed discussion of models in Chapter 15). *Representation* is typically used in conceptual and scientific discussions, whereas *model* is used in practical and database circles and is preferred in this chapter. The focus here is on how geographic reality is modeled (that is to say, abstracted or simplified) in GI databases, with particular emphasis on the different types of data models that have been developed. A data model is an essential ingredient of any operational GI system and, as the discussion will show, has important implications for the types of operations that can be performed and the results that can be obtained.

### 7.1.1 Data Model Overview

The heart of any GI database is the data model, which is a set of constructs for representing objects and processes in the digital environment of the computer (Figure 7.1). People (GI system users) interact with operational GI systems in order to undertake tasks such as making maps, querying databases, and performing site suitability analyses. Because the types of analyses that can be undertaken are strongly influenced by the way the real world is modeled, decisions about the type of data model to be adopted are vital to the success of a GI project.

**A data model is a set of constructs for describing and representing selected aspects of the real world in a computer.**

As described in Chapters 2 and 3, geographic reality is continuous and of seemingly infinite complexity, but computers are finite, are comparatively

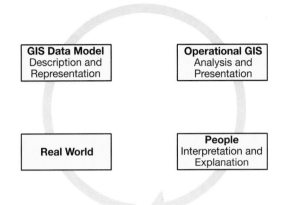

**Figure 7.1** The role of a data model in GI systems.

simple, and can only work with digital data. Therefore, difficult choices have to be made about what things are modeled in a GI database and how they are represented. Because different types of people use GI systems for different purposes, and the phenomena these people study have different characteristics, there is no single type of all-encompassing data model that is best for all circumstances. Modern GI systems are able to incorporate multiple data models and so can be applied to a wide range of different application areas.

## 7.1.2 Levels of Data Model Abstraction

When representing the real world in a computer, it is helpful to think in terms of four different levels of abstraction (levels of generalization or simplification); these are shown in Figure 7.2. First, *reality* is made up of real-world phenomena (buildings, streets, wells, lakes, people, etc.) and includes all aspects that may or may not be perceived by individuals or deemed relevant to a particular application. Second, the *conceptual model* is a human-oriented, often partially structured, model of selected objects and

**Figure 7.2** Levels of GI data model abstraction.

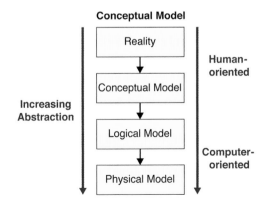

processes that are thought relevant to a particular problem domain. Third, the *logical model* is an implementation-oriented representation of reality that is often expressed in the form of diagrams and lists. Last, the *physical model* portrays the actual implementation in a GI system and often comprises tables stored as files or databases (see Chapter 9). Use of the term *physical* here is actually misleading because the models are not physical and only exist digitally in computers, but this is the generally accepted use of the term. Relating back to the discussion of uncertainty in Chapter 5 (Figure 5.1), the conceptual and logical models are found beyond the U1 filter, and the physical model is that on which analysis may be performed (beyond the U2 filter).

In data modeling, users and system developers participate in a process that successively engages with each of these levels. The first phase of modeling begins with the definition of the main types of objects to be represented in the GI database and concludes with a conceptual description of the main types of objects and relationships between them. Once this phase is complete, further work will lead to the creation of diagrams and lists describing the names of objects and what they look like, their behavior (how they act), and the type of interaction between objects. This type of logical data model is very valuable for defining what a GI system will be able to do and the type of domain over which it will extend. Logical models are implementation-independent and in theory can be created in any GI system with appropriate capabilities. The final data modeling phase involves creating a model showing how the objects under study can be digitally implemented in a GI database. Physical models describe the exact files or database tables used to store data, the relationships between object types, and the precise operations that can be performed. For more details about the practical steps involved in data modeling, see Sections 7.3 and 7.4.

A data model provides system developers and users with a common understanding and reference point. For developers, a data model is the means to represent an application domain in terms that may be translated into a design and then implemented in a system. For users, it provides a description of the structure of the system, independent of specific items of data or details of the particular application. A data model controls the types of things that a GI system can handle and the range of operations that can be performed on them.

The discussion of geographic representation in Chapter 3 introduced discrete objects and fields, the two fundamental conceptual models for representing real-world things geographically. In the same chapter the raster and vector logical models were also

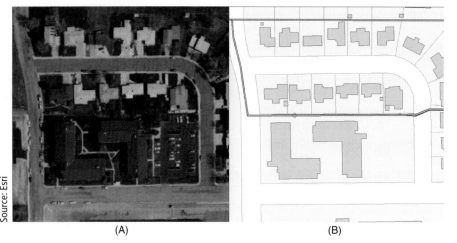

(A)  (B)

**Figure 7.3** Different representational models of the same area in Colorado, USA: (A) aerial photograph; and (B) vector objects, some digitized from the photograph.

introduced. Figure 7.3 shows two representations of the same area in a GI database, one raster and the other vector. Notice the difference in the objects represented. The roads and buildings can be clearly seen on both figures, but the vector representation (Figure 7.3B) also shows the lot (property) boundaries and water mains and valves. Cars and other surface texture ephemera can be observed on the raster representation (Figure 7.3A). The next sections in this chapter focus on the logical and physical representation of raster, vector, and related models in GI systems.

## 7.2 GI Data Models

In the past half-century, many GI data models have been developed and deployed in GI systems. The key types of geographic data models and their main areas of application are listed in Table 7.1. All are based in some way on the conceptual discrete object/field and logical vector/raster geographic data models. All GI systems include a core data model that is built on one or more of these GI data models. In practice, any modern comprehensive GI system supports at least some elements of all these models. As discussed earlier, the GI core system data model is the means to represent geographic aspects of the real world and defines the type of geographic operations that can be performed. It is the responsibility of the implementation team to populate this generic model with information about a particular problem (e.g., utility outage management, military mapping, or natural resource planning). Some GI software packages come with a fixed data model, whereas others have models that can be extended by adding new object types and relationships. Those that can easily be extended are better able to model the richness of geographic domains and in general are the easiest to use and the most productive systems.

**Table 7.1** Some commonly used geographic data models.

| Data model | Example application |
| --- | --- |
| Computer-aided design (CAD) | Automating engineering design and drafting |
| Graphical (nontopological) | Simple mapping and graphic arts |
| Image | Image processing and simple grid analysis |
| Raster/grid | Spatial analysis and modeling, especially in environmental and natural resources applications |
| Vector/georelational topological | Many operations on geometric features in cartography, socioeconomic, and resource analysis and modeling |
| Network | Network analysis in transportation and utilities |
| Triangulated irregular network (TIN) | Surface/terrain visualization, analysis and modeling |
| Object | Many operations on all types of entities (raster/vector/TIN, etc.) in all types of applications |

When modeling the real world for representation inside a GI database, it is convenient to group entities of the same geometric type together (for example, all point entities such as lights, garbage cans, or dumpsters might be stored together). A collection of entities of the same geometric type (dimensionality) is referred to as a class or layer. It should also be noted that the term *layer* is quite widely used in GI systems as a general term for a specific dataset. It is derived from the process of entering different types of data into a GI system from paper maps, which was undertaken one plate at a time. (All entities of the same type were represented in the same color and, using printing technology, were reproduced together on film or printing plates.) Grouping entities of the same geographic type together makes the storage of geographic databases more efficient (for further discussion, see Section 9.3). It also makes it much easier to implement rules for validating edit operations (for example, the addition of a new building or census administrative area) and for building relationships between entities. All the data models discussed in this chapter use layers in some way to handle geographic entities.

**A layer is a collection of geographic entities of the same geometric type (e.g., points, lines, or areas). Grouped layers may combine layers of different geometric types.**

## 7.2.1 CAD, Graphical, and Image Data Models

The earliest GI systems were based on very simple data models derived from work in the fields of CAD (computer-aided design and drafting), computer cartography, and image analysis. In a CAD system, real-world entities are represented symbolically as simple point, line, and area vectors. This basic CAD data model never became widely popular for GI because of three severe problems for most applications at geographic scales. First, because CAD models typically use local drawing coordinates instead of real-world coordinates for representing objects, they are of little use for map-centric applications. Second, because individual objects do not have unique identifiers, it is difficult to tag them with attributes. As the following discussion shows, this is a key requirement for GI applications. Third, because CAD data models focus on graphical representation of objects, they cannot store details of any relationships between objects (e.g., topology or networks), yet this type of information is essential in many spatial analytical operations.

A second type of simple GI geometry model was derived from work in the field of computer cartography.

The main requirement for this field in the 1960s was the automated reproduction of paper topographic maps and the creation of simple thematic maps. Techniques were developed to digitize maps and store them in a computer for subsequent plotting and printing. All paper map entities were stored as points, lines, and areas, with annotation used for placenames. Like CAD systems, there was no requirement to tag objects with attributes or to work with object relationships.

At about the same time that CAD and computer cartography systems were being developed, a third type of data model emerged in the field of image processing. Because the main data sources for geographic image processing are scanned aerial photographs and digital satellite images, it was natural that these systems would use rasters or grids to represent the patterning of real-world objects on the Earth's surface. The image data model is also well suited to working with pictures of real-world objects, such as photographs of water valves and scanned building floor plans that are held as attributes of geographically referenced entities in a database (Figure 7.4).

Despite their many limitations, some of which have been overcome through data model extensions, GI systems still exist based on these simple data models, although the number is in significant decline. This is partly for historical reasons—the GI system may have been built before newer, more advanced models became available—but also because of lack of knowledge about the newer approaches described in this chapter.

## 7.2.2 Raster Data Model

The raster data model uses an array of cells, or pixels, to represent real-world objects (see Figure 3.10). The cells can hold attribute values based on one of several encoding schemes, including categories, and integer and floating-point numbers (see Box 3.2 for details). In the simplest case a binary representation is used (for example, presence or absence of vegetation), but in more advanced cases floating-point values are preferred (for example, height of terrain above sea level in meters). In some systems, multiple attributes can be stored for each cell in a type of value attribute table where each column is an attribute and each row either a pixel or a pixel class (Figure 7.5).

Raster data are usually stored as an array of grid values, with metadata (data about data: see Section 10.2) about the array held in a file header. Typical metadata include the geographic coordinate of the upper-left corner of the grid, the cell size, the number of row and column elements, and the projection. The data array itself is usually stored as a compressed file or record in a

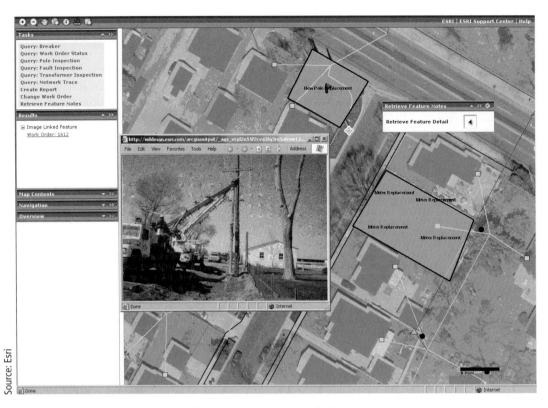

**Figure 7.4** A photographic image used as a building object attribute in an electric facility system.

**Figure 7.5** Raster data of the Olympic Peninsula, Washington State, with associated value attribute table. Bands 4,3,2 from Landsat 5 satellite with land cover classification overlaid.

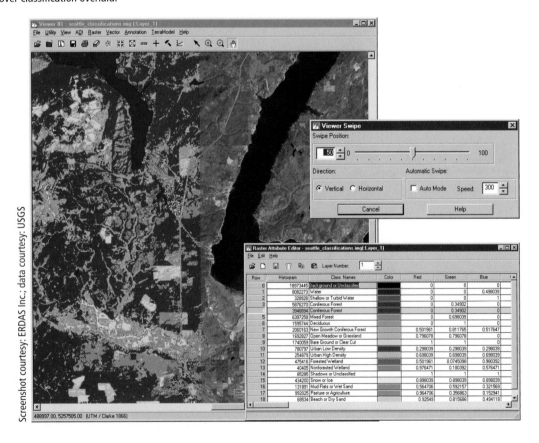

database management system (see Section 9.3). Techniques for compressing rasters are described in Box 7.1.

Datasets encoded using the raster data model are particularly useful as a backdrop map display because they look like conventional maps and can communicate a lot of information quickly. They are also widely used for analytical applications such as disease dispersion modeling, surface-water flow analysis, and store location modeling.

## 7.2.3 Vector Data Model

The raster data model discussed earlier is most commonly associated with the field conceptual data

---

## Technical Box (7.1)

### Technical Raster Compression Techniques

Although the raster data model has many uses in GIS, one of the main operational problems associated with it is the sheer amount of raw data that must be stored. To improve storage efficiency, many types of raster compression techniques have been developed such as run-length encoding, block encoding, wavelet compression, and quadtrees (see Section 9.7.2.2 for another use of quadtrees as a means of indexing geographic data). Table 7.2 presents a comparison of file sizes and compression rates for three compression techniques based on the image in Figure 7.6. It can be seen that even the

**Table 7.2** Comparison of file sizes and compression rates for selected raster compression techniques (using image shown in Figure 7.6).

| Compression technique | File size (MB) | Compression rate |
|---|---|---|
| Uncompressed original | 80.5 | — |
| Run-length | 17.7 | 5.1 |
| Wavelet | 2.3 | 38.3 |

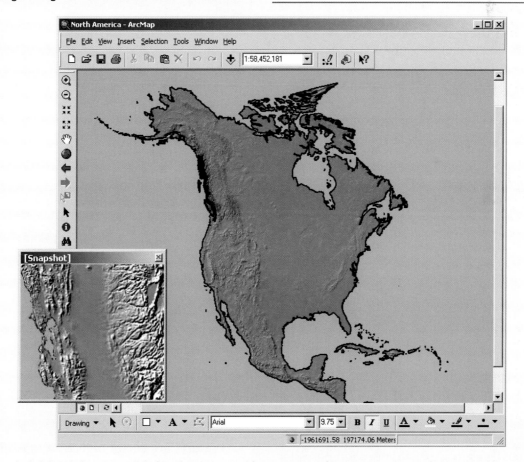

**Figure 7.6** Shaded digital elevation model of North America used for comparison of image compression techniques in Table 7.2. Original image is 8,726 by 10,618 pixels, 8 bits per pixel. The inset shows part of the image at a zoom factor of 1,000 for the San Francisco Bay area.

comparatively simple run-length encoding technique compresses the file size by a factor of 5. The more sophisticated wavelet compression technique results in a compression rate of almost 40, reducing the file from 80.5 to 2.3 MB.

### Run-Length Encoding

Run-length encoding is perhaps the simplest compression method and is very widely used. It involves encoding adjacent row cells that have the same value, with a pair of values indicating the number of cells with the same value and the actual value.

### Block Encoding

Block encoding is a two-dimensional version of run-length encoding in which areas of common cell values are represented with a single value. An array is defined as a series of square blocks of the largest size possible. Recursively, the array is divided using blocks of smaller and smaller sizes. It is sometimes described as a quadtree data structure (see also Section 9.7.2.2).

### Wavelet

Wavelet compression techniques invoke principles similar to those discussed in the treatment of fractals (see Section 2.8). They remove information by recursively examining patterns in datasets at different scales, always trying to reproduce a faithful representation of the original. A useful by-product of this for geographic applications is that wavelet-compressed raster layers can be quickly viewed at different scales with appropriate amounts of detail. The JPEG 2000 standard defines wavelet image compression.

Run-length and block encoding both result in lossless compression of raster layers; that is, a layer can be compressed and decompressed without degradation of information. In contrast, the wavelet compression technique is *lossy* because information is irrevocably discarded during compression. Although wavelet compression results in very high compression ratios, because information is lost its use is limited to applications that do not need to use the raw digital numbers for processing or analysis. It is not appropriate for compressing DEMs, for example, but many organizations use it to compress scanned maps and aerial photographs when access to the original data is not necessary.

model. The vector data model, on the other hand, is closely linked with the discrete object view. Both of these conceptual perspectives were introduced in Section 3.5. The vector data model is used in GI databases because of the precise nature of its representation method, its storage efficiency, the quality of its cartographic output, and the wide availability of functional tools for operations like map projection, overlay processing, and cartographic analysis.

In the vector data model, each object in the real world is first classified into a geometric type; in the 2-D case it is point, line, or area (Figure 7.7). Points (e.g., wells, soil pits, and retail stores) are recoded as single coordinate pairs, lines (e.g., roads, streams, and geologic faults) as a series of ordered coordinate pairs (also called polylines—Section 3.6.2), and areas (e.g., census tracts, soil areas, and oil license zones) as one or more line segments that close to form a polygonal area. The coordinates that define the geometry of each object may have 2, 3, or 4 dimensions: 2 ($x$, $y$: row and column, or latitude and longitude), 3 ($x$, $y$, $z$: the addition of a height value), or 4 ($x$, $y$, $z$, $m$: the addition of another value to represent time or some other property—perhaps the offset of road signs from a road centerline, or an attribute).

For completeness, it should also be said that in some data models linear features can be represented not only as a series of ordered coordinates, but also as curves defined by a mathematical function (e.g., a spline or Bézier curve). These are particularly useful for representing built-environment entities like road curbs and some buildings.

#### 7.2.3.1 Simple Features

Geographic entities encoded using the vector data model are usually called features, and this will be the convention adopted here. Features of the same geometric type are stored in a GI database as a feature class, or when speaking about the physical (database) representation the term *feature table* is preferred. Here each feature occupies a row, and each property of the feature occupies a column. GI databases commonly deal with two types of feature: simple and topological. The structure of simple feature polyline and polygon datasets is sometimes called *spaghetti* because, like a plate of cooked spaghetti, lines (strands of spaghetti) and polygons (spaghetti hoops) can overlap, and there are no stored relationships between any of the objects.

**Features are vector objects of type point, polyline, or polygon.**

Simple feature datasets are useful in GI applications because they are easy to create and store, and because they can be retrieved and rendered on

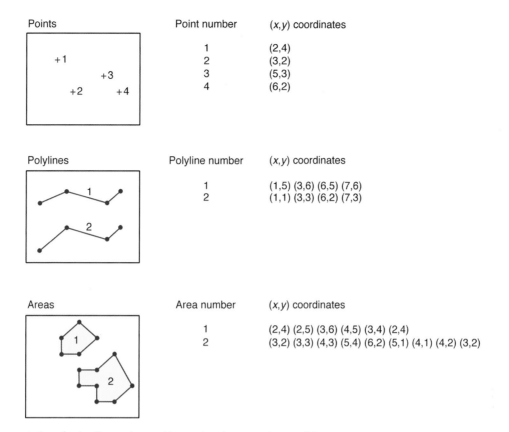

| Points | Point number | (x,y) coordinates |
|---|---|---|
| | 1 | (2,4) |
| | 2 | (3,2) |
| | 3 | (5,3) |
| | 4 | (6,2) |

| Polylines | Polyline number | (x,y) coordinates |
|---|---|---|
| | 1 | (1,5) (3,6) (6,5) (7,6) |
| | 2 | (1,1) (3,3) (6,2) (7,3) |

| Areas | Area number | (x,y) coordinates |
|---|---|---|
| | 1 | (2,4) (2,5) (3,6) (4,5) (3,4) (2,4) |
| | 2 | (3,2) (3,3) (4,3) (5,4) (6,2) (5,1) (4,1) (4,2) (3,2) |

**Figure 7.7** Representation of point, line, and area objects using the vector data model.

screen very quickly. However, because simple features lack more advanced data structure characteristics, such as topology (see next section), operations such as shortest-path network analysis and querying polygon adjacency cannot be performed without additional calculations.

### 7.2.3.2 Topological Features

Topological features are essentially simple features structured using topological rules. Topology is the mathematics and science of geometrical relationships. Topological relationships are nonmetric (qualitative) properties of geographic objects that remain constant when the geographic space of objects is distorted. For example, when a map is stretched, properties such as distance and angle change, whereas topological properties such as adjacency and containment do not change. Topological structuring of vector layers introduces some interesting and very useful properties, especially for polyline (also called 1-cell, arc, edge, line, and link) and area (also called 2-cell, polygon, and face) data. Topological structuring of line layers forces all line ends that are within a user-defined distance to be snapped together so that they are given exactly the same coordinate value. A node is placed wherever the ends of lines meet or cross. Following on from the earlier analogy, this type of data model is sometimes

referred to as spaghetti with meatballs (the nodes being the meatballs on the spaghetti lines). Topology is important in GI systems because of its role in data validation, modeling integrated feature behavior, editing, and query optimization.

> **Topology is the science and mathematics of relationships used to validate the geometry of vector entities and for operations such as network tracing and tests of polygon adjacency.**

***Data validation*** Many of the geographic data collected from basic digitizing, field data collection devices, photogrammetric workstations, and CAD systems comprise simple features of type point, polyline, or polygon, with limited structural intelligence (for example, no topology). Testing the topological integrity of a dataset is a useful way to validate the geometric quality of the data and to assess its suitability for geographic analysis. Some useful data validation topology tests include the following:

- Network connectivity—do all network elements connect to form a graph (i.e., are all the pipes in a water network joined to form a wastewater system)? Network elements that connect must be "snapped" together (that is, given the same coordinate value) at junctions (intersections).

- Line intersection—are there junctions at intersecting polylines, but not at crossing polylines? It is, for example, perfectly possible for roads to cross in planimetric 2-D view, but not intersect in 3-D (for example, at a bridge or underpass).

- Overlap—do adjacent polygons overlap? In many applications (e.g., land ownership), it is important to build databases free from overlaps and gaps so that ownership is unambiguous.

- Duplicate lines—are there multiple copies of network elements or polygons? Duplicate polylines often occur during data capture. During the topological creation process, it is necessary to detect and remove duplicate polylines to ensure that topology can be built for a dataset.

### Modeling the integrated behavior of different feature types

In the real world, many objects share common locations and partial identities. For example, water distribution areas often coincide with water catchments, electric distribution areas often share common boundaries with building subdivisions, and multiple telecommunications fibers are often run down the same conduit. The data modeling challenge is how to model such integrated features effectively in a GI database. These situations can be modeled as either objects with shared geometries or as objects with separate geometries integrated for editing, analysis, and representation. There are advantages and disadvantages to both approaches. In the former, there are fewer coordinates to store, and feature integrity is inherent in the data. In the case of the latter, multiple objects with separate geometries are easier to implement in commercially available database management systems. If one feature is moved during editing, then logically both features should move. This is achieved by storing both objects separately in the database, each with its own geometry, but integrating them inside the GI editor application so that they are treated as single features. When the geometry of one is moved, the other automatically moves with it.

### Editing productivity

Topology improves editor productivity by simplifying the editing process and providing additional capabilities to manipulate feature geometries. Editing requires both topological data structuring and a set of topologically aware tools. The productivity of editors can be improved in several ways:

- Topology provides editors with the ability to manipulate common, shared polylines and nodes as single geometric objects to ensure that no differences are introduced into the common geometries.

- Rubberbanding is the process of moving a node, polyline, or polygon boundary and receiving interactive feedback on-screen about the location of all topologically connected geometry.

- Snapping (e.g., forcing the end of two polyline features to have the same coordinate) is a useful technique to both speed up editing and maintain a high standard of data quality.

- Autoclosure is the process of completing a polygon by snapping the last point to the first digitized point.

- Tracing is a type of network analysis technique that is used, especially in utility applications, to test the connectivity of linear features (e.g., is the newly designed service going to receive power?).

### Optimizing queries

Many GI system queries can be optimized by precomputing and storing information about topological relationships. Some common examples include the following:

- Network tracing (e.g., find all connected water pipes and fittings)

- Polygon adjacency (e.g., determine who owns the parcels adjoining those owned by a specific owner)

- Containment (e.g., find out which manholes lie within the pavement area of a given street)

- Intersection (e.g., examine which census tracts intersect with a set of health areas)

The remainder of this section focuses on a conceptual understanding of GI topology, emphasizing the polygon case. The next section considers the network case; the relative merits and implementations of two approaches to GI topology are discussed in Section 9.7.1. These two implementation approaches differ because in one case relationships are batch built and stored along with feature geometry, and in the other relationships are calculated interactively when they are needed.

Conceptually speaking, in a topologically structured polygon data layer, each polygon is defined as a collection of polylines that in turn are made up of an ordered list of coordinates (vertices). Figure 7.8 shows an example of a polygon dataset comprising six polygons (including the "outside world": Polygon 1). A number in a circle identifies a polygon. The lines that make up the polygons are shown in the area-polyline list. For example, Polygon 2 can be assembled from lines 4, 6, 7, 10, and 8. In this particular implementation example, the 0 before the 8 is used to indicate that Line 8 actually defines an "island" inside Polygon 2. The list of coordinates for each line is also shown in Figure 7.8. For example, Line 6 begins and ends with coordinates (7,4) and (6,3)—other coordinates have been omitted for brevity. A line may appear in the area-polyline list more than once (for example, Line 6 is used in the definition of both Polygons 2 and 5), but the actual coordinates for each polyline are only stored once in the polyline-coordinate

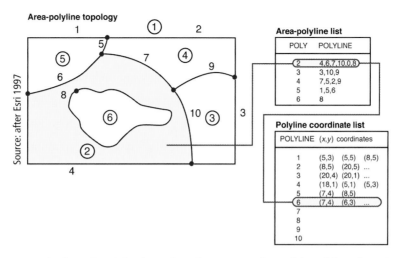

**Figure 7.8** A topologically structured polygon (area) data layer. The polygons are made up of the polylines shown in the area polyline list. The lines are made up of the coordinates shown in the line coordinate list.

list. Storing common boundaries between adjacent polygons avoids the potential problems of gaps (slivers) or overlaps between adjacent polygons. It has the added bonus that there are fewer coordinates in a topologically structured polygon feature layer compared with a simple feature layer representation of the same entities. The downside, however, is that drawing a polygon requires that multiple polylines must be retrieved from the database and then assembled into a boundary. This process can be time consuming when repeated for each polygon in a large dataset.

Planar enforcement is a very important property of topologically structured polygons. In simple terms,

planar enforcement means that all the space on a map must be filled and that any point must fall in one polygon alone; that is, polygons must not overlap. Planar enforcement implies that the phenomenon being represented is conceptualized as a field.

The contiguity (adjacency) relationship between polygons is also defined during the process of topological structuring. This information is used to define the polygons on the left- and right-hand sides of each polyline, in the direction defined by the list of coordinates (Figure 7.9). In Figure 7.9, Polygon 2 is on the left of Polyline 6 and Polygon 5 is on the right (polygon identifiers are in circles). Thus from a simple

**Figure 7.9** The contiguity of a topologically structured polygon data layer. For each polyline the left and right polygons are stored with the geometry data.

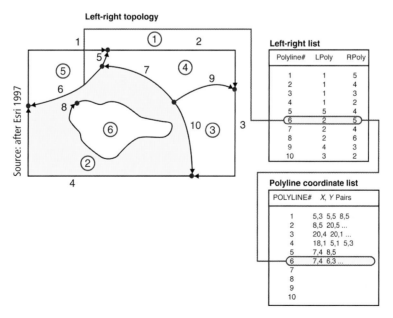

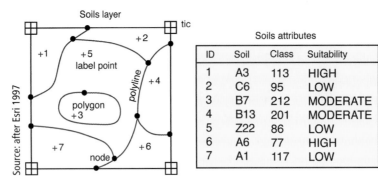

**Figure 7.10** An example of a georelational polygon dataset. Each of the polygons is linked to a row in an RDBMS table. The table has multiple attributes, one in each column.

look-up operation, we can deduce that Polygons 2 and 5 are adjacent.

Software systems based on the vector topological data model have become popular over the years. A special case of the vector topological model is the *georelational* model. In this model derivative, the feature geometries and associated topological information are stored in regular computer files, whereas the associated attribute information is held in relational database management system (RDBMS) tables. The GI system maintains the intimate linkage between the geometry, topology, and attribute information. This hybrid data management solution was developed to take advantage of RDBMS to store and manipulate attribute information. Geometry and topology were not placed in RDBMSs because, until a decade ago, RDBMSs were unable to store and retrieve geographic data efficiently. Figure 7.10 is an example of a georelational model as implemented in Esri's original ARC/INFO coverage polygon dataset. It shows file-based geometry and topology information linked to attributes in an RDBMS table. The ID (identifier) of the polygon, the label point, is linked (related or joined) to the ID column in the attribute table. Thus, in this soils dataset Polygon 3 is soil B7, of Class 212, and its suitability is MODERATE.

The topological feature geographic data model has been extensively used in GI applications over the last 25 years, especially in government and natural resources applications based on polygon representations. Typical government applications include cadastral management, tax assessment, land/property parcel management, land-use zoning, planning, and building control. In the areas of natural resources and environment, key applications include site suitability analysis, integrated land use modeling, license mapping, natural resource management, and conservation. Tax appraisal is an example of a GI application based on the topological feature data model; the topological data model is chosen in order to avoid

overlaps and gaps in tax parcels (polygons), to ensure that all parcel boundaries closed (were validated), and to store data in an efficient way. This is despite the fact that there is an overhead in creating and maintaining parcel topology, as well as degradation in draw and query performance for large databases.

### 7.2.3.3 Network Data Model

The network data model is a special type of the topological feature model. It is discussed here separately because it raises several new issues and has been widely applied in GI projects.

Networks can be used to model the flow of goods and services. There are two primary types of networks: *radial* and *looped*. In radial or tree networks, flow always has an upstream and downstream direction. Stream and storm drainage systems are examples of radial networks. In looped networks, self-intersections are common occurrences. Water distribution networks are looped by design to ensure that service interruptions affect the fewest customers.

In GI databases, networks are modeled as points (for example, street intersections, fuses, switches, water valves, and the confluence of stream reaches, usually referred to as nodes in topological models) and lines (for example, streets, transmission lines, pipes, and stream reaches). Network topological relationships define how lines connect with each other at nodes. For the purpose of network analysis, it is also useful to define rules about how flows can move through a network. For example, in a sewer network, flow is directional from a customer (source) to a treatment plant (sink), but in a pressurized gas network flow can be in any direction. The rate of flow is modeled as impedances (weights) on the nodes and lines. Figure 7.11 shows an example of a street network. The network comprises a collection of nodes (types of street intersection) and lines (types of street), as well as the topological relationships between them. The topological information makes it possible, for

**Figure 7.11** An example of a street network.

example, to trace the flow of traffic through the network and to examine the impact of street closures. An impedance defined on the intersections and streets determines the speed at which traffic flows. Typically, the rate of flow is proportional to the street speed limit, the number of lanes, and the timing of stoplights at intersections. Although this example relates to streets, the same basic principles also apply to, for example, electric, water, and railroad networks.

In georelational implementations of the topological network feature model, the geometry and topology information is typically held in ordinary computer files, and the attributes are stored in a linked database. The GI software tools are responsible for creating and maintaining the topological information each time a change in the feature geometry takes place. In more modern object models the geometry, attributes, and topology may be stored together in a DBMS, or topology may be computed on the fly.

Many applications utilize networks. Prominent examples include calculating power load drops over an electricity network, routing emergency response vehicles over a street network, optimizing the route of mail deliveries over a street network, and tracing pollution upstream to a source over a stream network.

Network data models are also used to support another data model variant called *linear referencing* (see Section 4.5). The basic principle of linear referencing is quite simple. Instead of recording the locations of geographic entities as explicit $x$, $y$, $z$ coordinates, they are recorded as distances along a network (called a route system) from a point of origin. This is a very efficient way of storing information such as road pavement (surface) wear characteristics (e.g., the location of potholes and degraded asphalt), geological seismic data (e.g., shockwave measurements at sensors along seismic lines), and pipeline corrosion data.

An interesting aspect of this is that a two-dimensional network is reduced to a one-dimensional linear route list. The location of each entity (often called an event) is simply a distance along the route from the origin. Offsets are also often stored to indicate the distance from a network centerline. For example, when recording the surface characteristics of a multicarriageway road, several readings may be taken for each carriageway at the same linear distance along the route. The offset value will allow the data to be related to the correct carriageway. Dynamic segmentation is a special type of linear referencing. The term derives from the fact that event data values are held separately from the actual network route in database tables (still as linear distances and offsets) and then dynamically added to the route (segmented) each time the user queries the database. This approach is especially useful in situations in which the event data change frequently and need to be stored in a database due to access from other applications (e.g., traffic volumes or rate of pipe corrosion).

### 7.2.3.4 TIN Data Model

The geographic data models discussed so far have concentrated on one- and two-dimensional data. There are several ways to model three-dimensional data, such as terrain models, sales cost surfaces, and geological strata. The term *2.5-D* is sometimes used to describe surface structures because they have dimensional properties between 2-D and 3-D. A true 3-D structure will contain multiple $z$ values at the same $(x,y)$ location and thus is able to model overhangs and tunnels, as well as support accurate volumetric calculations such as cut and fill (a term derived from civil-engineering applications that describes cutting earth from high areas and placing it in low areas to construct a flat surface, as is required in, for example, railroad construction). Both grids and triangulated irregular networks (TINs) are used to create and represent surfaces in GI databases. A regular grid surface is really a type of raster dataset, as discussed earlier in Section 7.2.2. Each grid cell stores the height of the surface at a given location. The TIN structure, as the name suggests, represents a surface as contiguous nonoverlapping triangular elements (Figure 7.12). A TIN is created from a set of points with $x$, $y$, and $z$ coordinate values. A key advantage

of the TIN structure is that the density of sampled points, and therefore the size of triangles, can be adjusted to reflect the relief of the surface being modeled, with more points sampled in areas of variable relief (see Section 2.4). TIN surfaces can be created by performing what is called a Delaunay triangulation (Figure 7.13; Section 13.3.6.1).

A TIN is a topological data structure that manages information about the nodes comprising each triangle and the neighbors of each triangle. Figure 7.13 shows the topology of a simple TIN. As with other topological data structures, information about a TIN may be conveniently stored in a file or database table, or computed on the fly. TINs offer many advantages for surface analysis. First, they incorporate the original sample points, providing a useful check on the accuracy of the model. Second, the variable density of triangles means that a TIN is an efficient way of storing surface representations such as terrains that have substantial variations in topography. Third, the data structure makes it easy to calculate elevation, slope, aspect, and line-of-sight between points. The combination of these factors has led to the widespread use of the TIN data structure in applications such as volumetric calculations for roadway design, drainage studies for land development, and visualization of urban forms. Figure 7.14 shows two example applications of TINs. Figure 7.14A is a shaded landslide risk TIN of the Pisa district in Italy, with building objects draped on top to give a sense of landscape. Figure 7.14B is a TIN of the Yangtse River, China, greatly exaggerated in the $z$ dimension. It shows how TINs draped with images can provide photorealistic views of landscapes.

Like all 2.5-D and 3-D models, TINs are only as good as the input sample data (see Section 2.4). They are especially susceptible to extreme high and low values because there is no smoothing of the original data. Other limitations of TINs include their inability to deal with discontinuity of slope across triangle boundaries, the difficulty of calculating optimum routes, and the need to ensure that peaks, pits, ridges, and channels are captured if a drainage network TIN is to be accurate.

## 7.2.4 Object Data Model

All the geographic data models described so far are geometry-centric; that is, they model the world as collections of points, lines, areas, TINs, or rasters. Any operations to be performed on the geometry (and, in some cases, associated topology) are created as procedures (programs or scripts) that are separated from the data. Unfortunately, this approach can present several limitations for modeling geographic systems. All but the simplest of geographic systems

**Figure 7.12** TIN surface of Death Valley, California: (A) "wireframe" showing all triangles; (B) shaded by elevation; and (C) draped with satellite image.

(A)

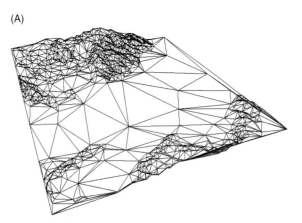

(B)

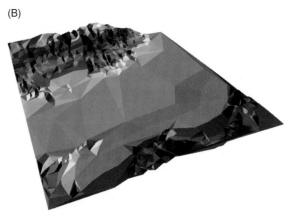

(C)

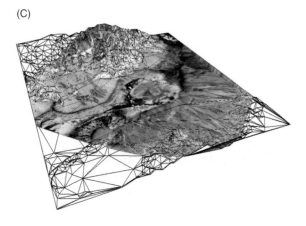

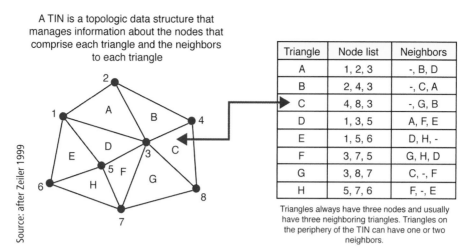

A TIN is a topologic data structure that manages information about the nodes that comprise each triangle and the neighbors to each triangle

| Triangle | Node list | Neighbors |
|----------|-----------|-----------|
| A | 1, 2, 3 | -, B, D |
| B | 2, 4, 3 | -, C, A |
| C | 4, 8, 3 | -, G, B |
| D | 1, 3, 5 | A, F, E |
| E | 1, 5, 6 | D, H, - |
| F | 3, 7, 5 | G, H, D |
| G | 3, 8, 7 | C, -, F |
| H | 5, 7, 6 | F, -, E |

Triangles always have three nodes and usually have three neighboring triangles. Triangles on the periphery of the TIN can have one or two neighbors.

Source: after Zeiler 1999

**Figure 7.13** The topology of a TIN.

**Figure 7.14** Examples of applications that use the TIN data model: (A) Landslide risk map for Pisa, Italy; (B) Yangtse River, China.

(A)

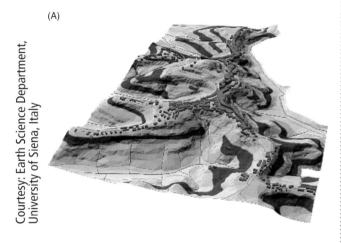

Courtesy: Earth Science Department, University of Siena, Italy

(B)

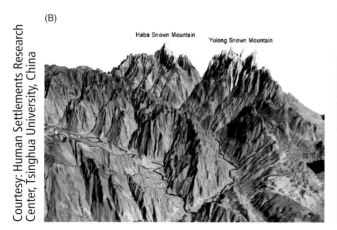

Courtesy: Human Settlements Research Center, Tsinghua University, China

contain many entities with large numbers of properties, complex relationships, and sophisticated behavior. Modeling such entities as simple geometry types is overly simplistic and does not easily support the sophisticated characteristics required for modern analysis. In addition, separating the state of an entity (attributes or properties defining what it *is*) from the behavior of an entity (methods defining what it *does*) makes software and database development tedious, time-consuming, and error prone. To try to address these problems, geographic object data models were developed; they allow the full richness of geographic systems to be modeled in an integrated way in GI databases.

The central focus of a GI object data model is the collection of geographic objects and the relationships between the objects (see Box 7.2). Each geographic object is an integrated package of geometry, properties, and methods. In the object data model, geometry is treated like any other attribute of the object and not as its primary characteristic (although clearly from an applications perspective it is often the major property of interest). Geographic objects of the same type are grouped together as object classes, with individual objects in the class referred to as "instances." In many GI systems each object class is stored physically as a database table, with each row an object and each object property a column. The methods that apply are attached to the object instances when they are created in memory for use in the application.

**An object is the basic atomic unit in an object data model and comprises all the properties that define the state of an object, together with the methods that define its behavior.**

## Object-Oriented Concepts

An *object* is a self-contained package of information describing the characteristics and capabilities of an entity (a thing of interest) under study. An interaction between two objects is called a *relationship*. In a geographic object data model, the real world is modeled as a collection of objects and the relationships between the objects. Each entity in the real world to be included in the GI database is an object. A collection of objects of the same type is called a *class*. In fact, classes are a more central concept than objects from the implementation point of view because many of the object-oriented characteristics are built at the class level. A class can be thought of as a template for objects. When creating an object data model the designer specifies classes and the relationships between classes. Only when the data model is used to create a database are objects (instances or examples of classes) actually created.

Examples of objects include oil wells, soil bodies, stream catchments, and aircraft flight paths. In the case of an oil-well class, each oil-well object might include *properties* defining its *state*—annual production, owner name, date of construction, and type of geometry used for representation at a given scale (perhaps a point on a coarse-scale map and a polygon on a fine-scale one). The oil-well class could have connectivity relationships with a pipeline class that represents the pipeline used to transfer oil to a refinery. There could also be a relationship defining the fact that each well must be located on a drilling platform. Finally, each oil-well object might also have *methods* defining the *behavior* or what it can do. Example behavior might include how objects draw themselves on a computer screen, how objects can be created and deleted, and editing rules about how oil wells snap to pipelines.

Three key facets of object data models make them especially good for modeling geographic systems: encapsulation, inheritance, and polymorphism.

*Encapsulation* describes the fact that each object packages together a description of its state and behavior. The state of an object can be thought of as

its properties or attributes (e.g., for a forest object, it could be the dominant tree type, average tree age, and soil pH). The behavior is the sum of the methods or operations that can be performed on an object (for a forest object these could be create, delete, draw, query, split, and merge). For example, when splitting a forest polygon into two parts, perhaps following a part sale, it is useful to get the GI system to automatically calculate the areas of the two new parts. Combining the state and behavior of an object together in a single package is a natural way to think of geographic entities and a useful way to support the reuse of objects.

*Inheritance* is the ability to reuse some or all of the characteristics of one object in another object. For example, in a gas facility system a new type of gas valve could easily be created by overwriting or adding a few properties or methods to a similar existing type of valve. Inheritance provides an efficient way to create models of geographic systems by reusing objects and also a mechanism to extend models easily. New object classes can be built to reuse parts of one or more existing object classes and add some new unique properties and methods. The example described in Section 7.3 shows how inheritance and other object characteristics can be used in practice.

*Polymorphism* describes the process whereby each object has its own specific implementation for operations like draw, create, and delete. One example of the benefit of polymorphism is that a geographic database can have a generic object-creation component that issues requests to be processed in a specific way by each type of object class. A utility system's editor software can send a generic create request to all objects (e.g., gas pipes, valves, and service lines), each of which has specific create algorithms. If a new object class is added to the system (e.g., landbase), then this mechanism will work because the new class is responsible for implementing the create method. Polymorphism is essential for isolating parts of software as self-contained components (see Chapter 6).

All geographic objects have some type of relationship to other objects in the same object class and, possibly, to objects in other object classes. Some of these relationships are inherent in the class definition (for example, some GI systems remove overlapping polygons), whereas other interclass relationships are user-definable. Three types of relationships are

commonly used in geographic object data models: topological, geographic, and general.

**A class is a template for creating objects.**

Generally, topological relationships are built into the class definition. For example, modeling real-world entities as a network class will cause network topology to be built for the nodes and lines participating in the network. Similarly, real-world entities modeled as topological polygon classes will be structured using the node–polyline model described in Section 7.2.3.2.

Geographic relationships between object classes are based on geographic operators (such as overlap, adjacency, inside, and touching) that determine the interactions between objects. In a model of an agricultural system, for example, it might be useful to ensure that all farm buildings are within a farm boundary using a test for geographic containment.

General relationships are useful to define other types of relationships between objects. In a tax assessment system, for example, it is advantageous to define a relationship between land parcels (polygons) and ownership data that is stored in an associated DBMS table. Similarly, an electric distribution system relating light poles (points) to text strings (called annotation) allows depiction of pole height and material of construction on a map display. This type of information is very valuable for creating work orders (requests for change) that alter the facilities. Establishing relationships between objects in this way is useful because if one object is moved, then the other will move as well, or if one is deleted, then the other is also deleted. This makes maintaining databases much easier and safer.

In addition to supporting relationships between objects (strictly speaking, between object classes), object data models also allow several types of rules to be defined. Rules are a valuable means of maintaining database integrity during editing tasks. The most popular types of rules used in object data models are attribute, connectivity, relationship, and geographic.

Attribute rules are used to define the possible attribute values that can be entered for any object. Both range and coded-value attribute rules are widely employed. A range attribute rule defines the range of valid values that can be entered. Examples of range rules include the following: highway traffic speed must be in the range 25–70 miles per hour; forest compartment average tree height must be in the range of 0–50 meters. Coded attribute rules are used for categorical data types. For example, land use must be of type commercial, residential, park, or other; or pipe material must be of type steel, copper, lead, or concrete.

Connectivity rules are based on the specification of valid combinations of features, derived from the geometry, topology, and attribute properties. For example, in an electric distribution system a 28.8-kV conductor can only connect to a 14.4-kV conductor via a transformer. Similarly, in a gas distribution system it should not be possible to add pipes with free ends (that is, with no fitting or cap) to a database.

Geographic rules define what happens to the properties of objects when an editor splits or merges them (Figure 7.15). In the case of a land parcel split following the sale of part of the parcel, it is useful to define rules to determine the impact on properties such as area, land-use code, and owner. In this example, the original parcel area value should be divided in proportion to the size of the two new parcels, the land-use code should be transferred to both parcels, and the owner name should remain for one parcel, but a new one should be added for the part that was sold off. In the case of a merge of two adjacent water pipes, decisions need to be made about what happens to attributes such as material, length, and corrosion rate. In this example, the two pipe materials should be the same, the lengths should be summed, and the new corrosion rate determined by a weighted average of both pipes.

**Figure 7.15** Example of split and merge rules for parcel objects: (A) split; and (B) merge.

(A)

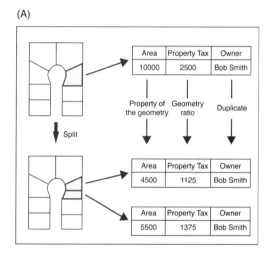

(B)

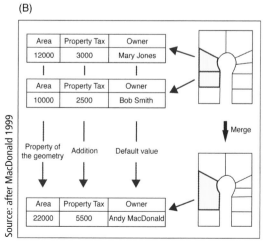

Source: after MacDonald 1999

## 7.3 Example of a Water-Facility Object Data Model

This section presents an example of a geographic object model and discusses how many of the concepts introduced earlier in this chapter are used in practice. The example selected is that of an urban water facility model. The types of issues raised in this example apply to all geographic object models, although of course the actual objects, object classes, and relationships under consideration will differ. The role of data modeling, as discussed in Section 7.1, is to represent the key aspects of the real world inside the digital computer for management, analysis, and display purposes.

Figure 7.16 is a diagram of part of a water distribution system, a type of pressurized network controlled by several devices. A pump is responsible for moving water through pipes (mains and laterals) connected together by fittings. Meters measure the rate of water consumption at houses. Valves and hydrants control the flow of water.

The goal of the example object model is to support asset management, mapping, and network analysis applications. Based on this goal it is useful to classify the objects into two types: the landbase and the water facilities. Landbase is a general term for objects such as houses and streets that provide geographic context but are not used in network analysis. The landbase object types are Pump House, House, and Street. The water-facilities object types are Main, Lateral (a smaller type of *WaterLine*), Fitting (water line connectors), Meter, Valve, and Hydrant. All these object types need to be modeled as a network in order to support network analysis operations such as network isolation traces and flow prediction. A network isolation trace is used to find all parts of a network that are unconnected (isolated). By using the topological connectivity of the network and information about whether pipes and fittings support water flow, it is possible to determine connectivity. Flow prediction is used to estimate the flow of water through

**Figure 7.16** Water distribution system water-facility object types and geographic relationships.

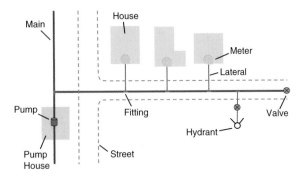

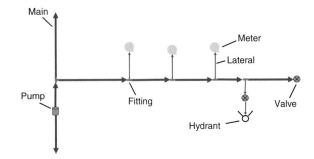

**Figure 7.17** Water distribution system network.

the network based on network connectivity and data about water availability and consumption. Figure 7.17 shows all the main object types and the implicit geographic relationships to be incorporated into the model. The arrows indicate the direction of flow in the network. When digitizing this network using a GI editor, it is useful to specify topological connectivity and attribute rules to control how objects can be connected (see Section 7.2.3.3). Before this network can be used for analysis, it will also be necessary to add flow impedances to each link (for example, pipe diameter).

Having identified the main object types, the next step is to decide how objects relate to each other and the most efficient way to implement them. Figure 7.18 shows one possible object model that uses a popular object modeling language called the Unified Modeling Language (UML) to show objects and the relationships between them. Some additional color-coding has been added to help interpret the model. In UML models each box is an object class, and the lines define how one class reuses (inherits) part of the class above it in a hierarchy. Object class names in an italic font are abstract classes; those with regular font names are used to create (instantiate) actual object instances. Abstract classes do not have instances and exist for reasons of efficiency. It is sometimes useful to have a class that implements some capabilities once, so that several other classes can then be reused. For example, Main and Lateral are both types of *Line*, as is Street. Because Main and Lateral share several things in common—such as ConstructionMaterial, Diameter, and InstallDate properties, and connectivity and draw behavior—it is efficient to implement these in a separate abstract class, called *WaterLine*. The triangles indicate that one class is a type of another class. For example, Pump House and House are types of *Building*, and Street and *WaterLine* are types of *Line*. The diamonds indicate composition. For example, a network is composed of a collection of *Line* and *Node* objects. In the water-facility object model, object classes without any geometry are colored pink. The Equipment and Operations-Record object classes have their location determined by being associated with other objects (e.g., valves and

**Figure 7.18** A water-facility object model.

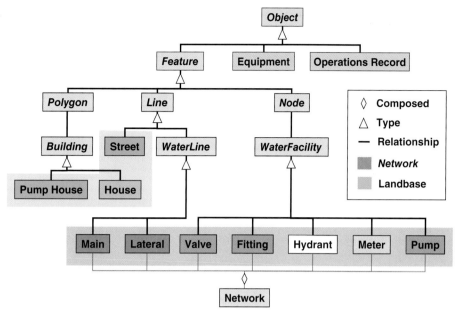

mains). The Equipment and OperationsRecord classes are useful places to store properties common to many facilities, such as EquipmentID, InstallDate, ModelNumber, and SerialNumber.

Once this logical geographic object model has been created, it can be used to generate a physical data model. One way to do this is to create the model using a computer-aided software engineering (CASE) tool. A CASE tool is a software application that has graphical tools that draw and specify a logical model (Figure 7.19). A further advantage of a CASE tool is that physical models can be generated directly from

**Figure 7.19** An example of a CASE tool (Microsoft Visio). The UML model is for a utility water system.

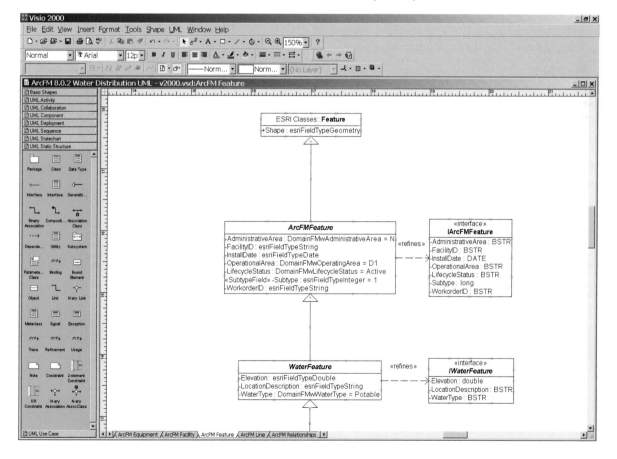

the logical models, including all the database tables and much of the supporting code for implementing behavior. Once a database structure (schema) has been created, it can be populated with objects and the intended applications put into operation.

## 7.4 Geographic Data Modeling in Practice

Geographic analysis is only as good as the geographic database on which it is based, and a geographic database is only as good as the geographic data model from which it is derived. Geographic data modeling begins with a clear definition of the project goals and progresses through an understanding of user require-

ments, a definition of the objects and relationships, formulation of a logical model, and then creation of a physical model. These steps are a prelude to database creation and, finally, database use.

No step in data modeling is more important than understanding the purpose of the data-modeling exercise. This understanding can be gained by collecting user requirements from the main users. Initially, user requirements will be vague and ill-defined, but over time they will become clearer. Project goals and user requirements should be precisely specified in a list or narrative. Peter Batty's career and experience of GI data modeling is highlighted in Box 7.3.

Formulation of a logical model necessitates identification of the objects and relationships to be modeled. Both the attributes and behavior of objects are required for an object model. A useful graphic

---

### Biographical Box 7.3

### Peter Batty: Geospatial Data Modeler

Peter Batty has served as Chief Technology Officer at several major geospatial software companies and has been involved in a number of innovations in the industry.

He studied at Oxford University, receiving a Bachelor's degree in Mathematics and a Master's degree in Computation. He began work for IBM in 1986 on their Geographic Facilities Information System (GFIS), a pioneering GI system focused on electric and gas utilities. Peter was an early advocate of using standard relational databases for GIS. He wrote an influential article, "Exploiting Relational Database Technology in GIS," in 1990[1], which was referenced by Oracle at the initial launch of Oracle Spatial (then Oracle MM) in 1995.

In 1992 Peter went to work for Smallworld Systems, at the time a small startup in the UK, which grew to become the market leader for GI systems in utilities and telecommunications during the 1990s. Peter was the first employee to move to the United States when Smallworld began operating there in 1993, and he held various technical roles before becoming Chief Technology Officer.

Smallworld introduced several innovations that had a significant impact on the industry, most notably the handling of long transactions using a new approach called version management. This eventually became a standard approach in the industry, being adopted in some form by Esri, Intergraph, and Oracle, among others. Long transactions are fundamental to handling the design and planning process in utilities.

The Smallworld data model introduced a number of new ideas that were especially good for modeling

Courtesy: Peter Batty

**Figure 7.20** Peter Batty

complex networks. One of these was the ability for a feature to have multiple spatial attributes, rather than a feature having just a single spatial type, as is still common today. Another was the notion of "internal worlds" that schematically represented the network connections inside a complex object such

▶

as a switchgear or substation, in a separate coordinate space from the main "geographic world".

In 1997–98, Peter led the development of PowerOn, a new product for electric outage management built on top of Smallworld, which quickly became the outage management market leader. A key innovation in PowerOn was a "tree key" hierarchical indexing scheme that enabled very fast updates of large numbers of electrical devices when power switched on or off.

In 2000, Smallworld was acquired by General Electric, and in 2002 Peter left to cofound a company called Ten Sails, now known as Ubisense. From 2005 to 2007 he left to serve as CTO of Intergraph, at that time the second-largest geospatial software company. In 2007 he founded a startup called Spatial Networking that worked on future location and social networking. He returned to Ubisense in 2010, and his current work is focused on applying the new generation of Web mapping systems, including Google Maps and a number of open-source software packages, to enterprise utility and telecom applications.

Peter has in recent years become quite involved with Open Source geospatial software. In 2011 he chaired the FOSS4G conference in Denver, the leading global conference for open-source geospatial software. This was the largest FOSS4G event to date, attracting 900 attendees from 42 countries. He served on the board of OSGeo, the Open Source Geospatial Foundation, from 2011 to 2013. Previously he also served on the board of the Geospatial Information and Technology Association (GITA) for a total of 5 years.

[1]www.ebatty.com/exploiting_rdb_in_gis.htm (originally published in 1990, republished in Computers and Geosciences in 1992)

tool for creating logical data models is a CASE tool, and a useful language for specifying models is UML. It is not essential that all objects and relationships be identified at the first attempt because logical models can be refined over time. The key objects and relationships for a water distribution system object model are shown in Figure 7.18.

Once an implementation-independent logical model has been created, this model can be turned into a system-dependent physical model. A physical model

Figure 7.21 Smallworld "internal worlds" screenshot. This screenshot shows the concept of "internal worlds" that can be used to model complex network devices in Smallworld. The main map shows an electric substation, represented by a polygon geometry. This contains a number of switches and other electric devices that need to be independently modeled in the GI system. The top right window shows a schematic representation of the devices inside the substation. These are stored in an independent "world" with its own (nongeographic) coordinate system. This world is owned by the substation. The highlighted cables on the map show an electric network trace that starts at the top right, then when it reaches the substation it jumps into the internal world and continues tracing there, looking at which switches are open and closed, then it jumps back out to the geographic world and continues tracing along two of the circuits that leave the substation.

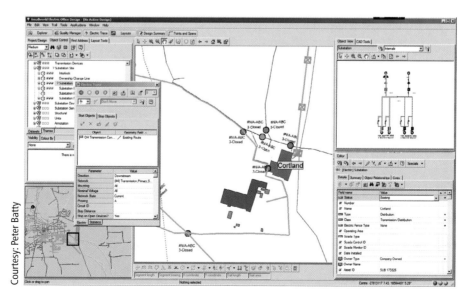

Courtesy: Peter Batty

will result in an empty database schema—a collection of database tables and the relationships between them. Sometimes, for performance optimization reasons or because of changing requirements, it is necessary to alter the physical data model. Even at this relatively late stage in the process, flexibility is still necessary.

It is important to realize that there is no such thing as the correct geographic data model. Every problem can be represented with many possible data models. Each data model is designed with a specific purpose in mind and is suboptimal for other purposes. A classic dilemma is whether to define a general-purpose data model that has wide applicability, but that can, potentially, be complex and inefficient, or to focus on a narrower highly optimized model that will yield better performance. A small prototype can often help resolve some of these issues.

Geographic data modeling is both an art and a science. It requires a scientific understanding of the key geographic characteristics of real-world systems, including the state and behavior of objects, and the relationships between them. Geographic data models are of critical importance because they have a controlling influence over the type of data that can be represented and the operations that can be performed. As we have seen, object models are the best type of data model for representing rich object types and relationships in facility systems, whereas simple feature models are sufficient for elementary applications such as a map of the body. In a similar vein, so to speak, raster models are good for data represented as fields such as soils, vegetation, pollution, and population counts.

## Questions for Further Study

1. Figure 7.22 is an oblique aerial photograph of an area of northern Italy. Take 10 minutes to list the main five object classes (including their attributes and behaviors) and the relationships between the classes that you can see in this picture that would be appropriate for a facilities information management system study.

2. Why is it useful to include the conceptual, logical, and physical levels in geographic data modeling?

3. Describe, with examples, five key differences between the topological vector and raster geographic data models. It may be useful to consult Figure 7.3 and Chapters 2 and 3.

4. Review the terms *encapsulation, inheritance,* and *polymorphism,* and explain with geographic examples why they make object data models superior for representing geographic systems.

Figure 7.22 Oblique view northern Italy

## Further Reading

Arctur, D. and Zeiler, M. 2004. *Designing Geodatabases: Case Studies in GIS Data Modeling.* Redlands, CA: Esri Press.

Esri. 1999. *Understanding GIS: The ArcInfo Method.* Redlands, CA: Esri Press.

Perencsik, A., Woo, S., Booth, B., Crosier S., Clark, J., and MacDonald, A. 2004. *Building a Geodatabase.* Redlands, CA: Esri Press.

Worboys, M. F. and Duckham, M. 2004. *GIS: A Computing Perspective* (2nd ed.). Boca Raton, FL: CRC Press.

Zeiler, M. and Murphy, J. 2010. *Modeling Our World: The Esri Guide to Geodatabase Concepts* (2nd ed). Redlands, CA: Esri Press.

# 8 Data Collection

Data collection is one of the most time-consuming and expensive, yet important, tasks associated with geographic information (GI). There are many diverse sources of GI, and many methods are available to enter them into a GI system. The two main methods of data collection are data capture and data transfer. It is useful to distinguish between primary (direct measurement) and secondary (derivation from other sources) data capture for both raster and vector data types. Data transfer involves importing digital data from other sources such as geoportals. Planning and executing an effective data collection plan touches on many practical issues. This chapter reviews the main methods of data capture and transfer and introduces key practical management issues.

## LEARNING OBJECTIVES

After studying this chapter you will be able to:

- Describe data collection workflows.
- Understand the primary techniques of data capture in remote sensing and surveying.
- Discuss secondary techniques of data capture, including scanning, digitizing, vectorization, photogrammetry, and COGO (a contraction of the term *coordinate geometry*) feature construction.
- Understand the principles of data transfer, where to search for digital GI, and GI formats.
- Analyze practical issues associated with managing data capture projects.

## 8.1 Introduction

GI systems can contain a wide variety of geographic data types originating from many diverse sources. Indeed, one of the key defining characteristics of GI systems is their ability to integrate data about places and spaces from many different sources. Data collection activities for the purposes of organizing the material in this chapter are split into data capture (direct data input) and data transfer (input of data from other systems). From the perspective of creating geographic databases, it is convenient to classify both raster and vector geographic data as either primary or secondary (Table 8.1). Primary data sources are those collected in digital format specifically for use in a GI project.

Typical examples of primary GI sources include raster SPOT and Quickbird Earth satellite images and vector building-survey measurements captured using a total survey station (see Section 8.2 for discussion of these terms). Secondary sources are digital and analog datasets that may have been originally captured for another purpose and need to be converted into a suitable digital format for use in a GI project. Typical secondary sources include raster-scanned color aerial photographs of urban areas, or United States Geological Survey (USGS) or Institut Géographique National (IGN, France) paper maps that can be scanned and vectorized. There is a new trend for citizens to collect geographic data in the field directly into a GI system. Some issues surrounding volunteered geographic information are discussed in Section 2.5.

This classification scheme provides a useful organizing framework for this chapter, and more important, it highlights the number of processing transformations that a dataset goes through, and therefore the opportunities for errors to be introduced. However, the distinctions between primary and secondary and between raster and vector are not always easy to determine. For example, are digital satellite

**Table 8.1** Classification of geographic data for data collection purposes with examples of each type.

| | Raster | Vector |
|---|---|---|
| **Primary** | Digital satellite remote-sensing images | GPS measurements |
| | Digital aerial photographs | Field survey measurements |
| **Secondary** | Scanned maps or photographs | Topographic maps |
| | Digital elevation models from topographic map contours | Toponymy (place-name) databases |

remote-sensing data obtained on a DVD primary or secondary? Clearly, the commercial satellite sensor feeds do not run straight into GI databases, but to ground stations where the data are preprocessed and then written onto digital media. Here they are considered primary because the data have usually undergone only minimal transformation since being collected by the satellite sensors and because the characteristics of the data make them suitable for direct use in GI projects.

> **Primary geographic data sources are captured by direct measurement specifically for use in GI systems. Secondary sources are reused from earlier studies or obtained from other systems.**

Both primary and secondary geographic data may be obtained in either digital or analog format (see Section 3.7 for a definition of analog), although digital data are becoming much more common. Analog data must always be digitized before being added to a geographic database. Analog-to-digital transformation may involve scanning paper maps or photographs, optical character recognition (OCR) of text describing geographic object properties, or vectorization of selected features from an image. Depending on the format and characteristics of the digital data, considerable reformatting and restructuring may be required prior to importing into a GI system. Each of these transformations alters the original data and will introduce further uncertainty into the data (see Chapter 5 for a discussion of uncertainty).

This chapter describes the data sources, techniques, and workflows involved in data collection. The processes of data collection are also variously referred to as data capture, data automation, data conversion, data transfer, data translation, and digitizing. Although there are subtle differences between these terms, they essentially describe the same thing, namely, adding geographic data to a database. Here data capture refers to direct entry, and data transfer is the importing of existing digital data across a network connection (Internet, WAN, or LAN) or from physical media such as DVD or portable hard disk. This chapter focuses on the techniques of data collection; of equal, or perhaps more, importance to a real-world GI system implementation are project management, cost, legal, and organization issues. These are covered briefly in Section 8.7 as a prelude to more detailed treatment in Chapters 16 through 18.

Table 8.2 shows a breakdown of costs (in $1,000s) for two typical client–server GI system implementations: one with 10 seats (systems) and the other with 100. The hardware costs include desktop clients and servers only (i.e., not network infrastructure). The data costs assume the purchase of a landbase (e.g., streets, parcels, and landmarks) and digitizing assets such as pipes and fittings (water utility), conductors, and devices (electrical utility), or land and property parcels (local government). Staff costs assume that all core staff will be full time, but that users will be part time.

In the early days when geographic data were very scarce, data collection was the main project task, and typically it consumed the majority of the available resources. Even today, with the widespread availability of maps on the Internet, data collection still remains a time-consuming, tedious, and expensive process in many serious GI projects. Typically, it accounts for 15 to 50% of the total cost of a project (Table 8.2). Data capture costs can in fact be much more significant because in many organizations (especially those that are government funded) staff

**Table 8.2** Breakdown of costs (in $1,000s) for two typical client–server GI systems as estimated by the authors.

| | 10 seats | | 100 seats | |
|---|---|---|---|---|
| | $ | % | $ | % |
| **Hardware** | 30 | 3.4 | 250 | 8.8 |
| **Software** | 25 | 2.8 | 150 | 5.3 |
| **Data** | 400 | 44.7 | 450 | 15.8 |
| **Staff** | 440 | 48.2 | 2,000 | 70.2 |
| **Total** | 885 | 100.0 | 2,850 | 100.0 |

costs are often assumed to be fixed and are not used in budget accounting. Furthermore, as the majority of data capture effort and expense tends to fall at the start of projects, data capture costs often receive greater scrutiny from senior managers. If staff costs are excluded from a GI project budget, then in cash expenditure terms, data collection can be as much as 60 to 85% of costs.

**Data capture costs can account for up to 85% of the cost of a GIS.**

After an organization has completed basic data collection tasks the focus of a GI project moves on to data maintenance. Over the multiyear lifetime of a GI project, data maintenance can turn out to be a far more complex and expensive activity than initial data collection. This is because of the high volume of update transactions in many systems (for example, changes in land parcel ownership, maintenance work orders on a highway transport network, or logging military operational activities) and the need to manage multiuser access to operational databases. For more information about geographic data maintenance see Chapter 9.

### 8.1.1 Data Collection Workflow

In all but the simplest of projects, data collection involves a series of sequential stages (Figure 8.1). The workflow commences with planning, followed by preparation, digitizing/transfer (here taken to mean a range of primary and secondary techniques such as table digitizing, survey entry, scanning, and photogrammetry), editing and improvement, and finally, evaluation.

Planning is obviously important to any project, and data collection is no exception. It includes establishing user requirements, garnering resources (staff, hardware, and software), and developing a project plan. Preparation is especially important in data col-

Figure 8.1 Stages in data collection projects.

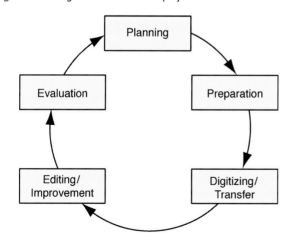

lection projects. It involves many tasks such as obtaining data, redrafting poor-quality map sources, editing scanned map images, and removing noise (unwanted data such as speckles on a scanned map image). It may also involve setting up appropriate hardware and software systems to accept data. Digitizing and transfer are the stages where the majority of the effort will be expended. It is naive to think that data capture is really just digitizing, when in fact it involves very much more as discussed later in this chapter. Editing and improvement follows digitizing/transfer. This covers many techniques designed to validate data, as well as to correct errors and to improve quality. Evaluation, as the name suggests, is the process of identifying project successes and failures; these may be qualitative or quantitative. Because all large data projects involve multiple stages, this workflow is iterative, with earlier phases (especially a first, pilot, phase) helping to improve subsequent parts of the overall project.

## 8.2 Primary Geographic Data Capture

Primary geographic capture involves the direct measurement of objects. Digital data measurements may be input directly into the GI database or may reside in a temporary file prior to input. Although the direct method is preferable as it minimizes the amount of time and the possibility of errors, close coupling of data collection devices and GI databases is not always possible. Both raster and vector primary data capture methods are available.

### 8.2.1 Raster Data Capture

The most popular form of primary raster data capture is remote sensing. Broadly speaking, remote sensing is a technique used to derive information about the physical, chemical, and biological properties of objects without direct physical contact. Information is derived from measurements of the amount of electromagnetic radiation reflected, emitted, or scattered from objects. A variety of sensors, operating throughout the electromagnetic spectrum from visible to microwave wavelengths, are commonly employed to obtain measurements (see Section 3.6.1). Passive sensors rely on reflected solar radiation or emitted terrestrial radiation; active sensors (such as synthetic aperture radar) generate their own source of electromagnetic radiation. The platforms on which these instruments are mounted are similarly diverse. Although Earth-orbiting satellites and fixed-wing aircraft are by far the most common, helicopters, balloons, masts, booms, and even handheld devices are also employed (Figure 8.2). As used here,

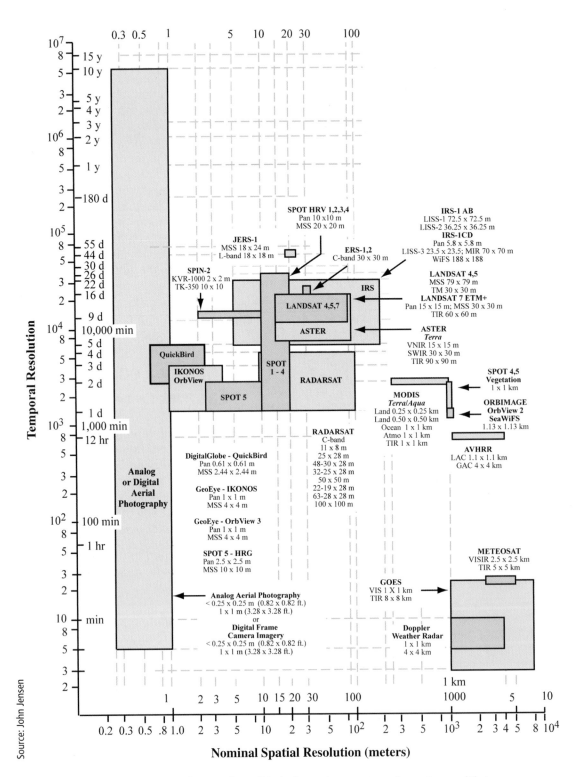

**Figure 8.2** Spatial and temporal characteristics of commonly used Earth observation remote-sensing systems and their sensors.

the term *remote sensing* subsumes the fields of satellite remote sensing and aerial photography.

**Remote sensing is the measurement of physical, chemical, and biological properties of objects without direct contact.**

From the perspective of GI, resolution is a key physical characteristic of remote-sensing systems. There are three aspects to resolution: spatial, spectral, and temporal. All sensors need to trade off spatial, spectral, and temporal properties because of storage, processing, and bandwidth considerations. For further discussion of the important topic of resolution, see also Sections 2.3, 5.3.2, and 15.1 and Box 2.3.

### Three key aspects of resolution are spatial, spectral, and temporal.

Spatial resolution refers to the size of object that can be resolved, and the most usual measure is the pixel size. Satellite remote-sensing systems typically provide data with pixel sizes in the range 0.4 m–1 km. The resolution of digital cameras used for capturing aerial photographs usually range from 0.01 m–5 m. Image (scene) sizes vary quite widely between sensors—typical ranges include 900 by 900 to 3,000 by 3,000 pixels. The total coverage of remote-sensing images is usually in the range 9 by 9 km to 200 by 200 km.

Spectral resolution refers to the parts of the electromagnetic spectrum that are measured. Because different objects emit and reflect different types and amounts of radiation, selecting which part of the electromagnetic spectrum to measure is critical for each application area. Figure 8.3 shows the spectral signatures of water, green vegetation, and dry soil. Remote-sensing systems may capture data in one part of the spectrum (referred to as a single band) or simultaneously from several parts (multiband or multispectral). The radiation values are usually normalized and resampled to give a range of integers from 0–255 for each band (part of the electromagnetic spectrum measured), for each pixel, in each image. Until recently, remote-sensing satellites typically measured a small number of bands, in the visible part of the spectrum. More recently, a number of hyperspectral systems have come into operation that measure very large numbers of bands across a much wider part of

the spectrum. Temporal resolution, or repeat cycle, describes the frequency with which images are collected for the same area. There are essentially two types of commercial remote-sensing satellite: Earth-orbiting and geostationary. Earth-orbiting satellites collect information about different parts of the Earth surface at regular intervals. To maximize utility, typically orbits are polar, at a fixed altitude and speed, and are sun synchronous.

The French SPOT (Système Probatoire d'Observation de la Terre) 6 satellite launched in 2012, for example, orbits over the poles at an altitude of 694 km, sensing the same location on the Earth surface every 1 to 3 days. The SPOT 6 platform carries multiple sensors: a panchromatic sensor measuring radiation in the visible part of the electromagnetic spectrum at a spatial resolution of 1.5 m; and a multispectral sensor measuring blue, green, red, and near infrared radiation separately at a spatial resolution of 6 m. The SPOT 6 system is also able to provide stereo images from which digital elevation models and 3-D measurements can be obtained. SPOT 6 scenes cover areas from about 60 by 60 km to 120 by 120 km. A number of digital image products are available from the ground distribution center: orthoimages (rectified to the ground), stereo pairs for production of digital elevation models, and various image mosaics.

Much of the discussion so far has focused on commercial satellite remote-sensing systems. Of equal importance, especially in GI projects that focus on smaller areas in great detail, is aerial photography. Although the data products resulting from remote-sensing satellites and digital aerial photography systems are technically very similar (i.e., they are both images), there are some significant differences in the

**Figure 8.3** Typical reflectance signatures for water, green vegetation, and dry soil.

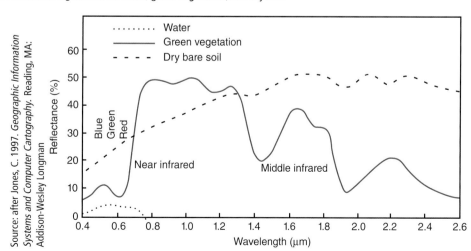

Source: after Jones, C. 1997. *Geographic Information Systems and Computer Cartography*. Reading, MA: Addison-Wesley Longman

way data are captured and can, therefore, be analyzed and interpreted. One notable difference is that older aerial photography systems use analog optical cameras and then later rasterize the photographs (e.g., scanning a film negative) to create an image. Both the quality of the optics of the camera and the mechanics of the scanning process affect the spatial and spectral characteristics of the resulting images. Modern aerial photography systems now use digital cameras with on-board position-fixing units to collect digital imagery (although the term *aerial photography* still persists when the digital cameras are carried on planes). Most aerial photographs are collected on an ad hoc basis using cameras mounted in airplanes flying at low altitudes (3,000–9,000 m) and are either panchromatic (black and white) or color, although multispectral cameras/sensors operating in the nonvisible parts of the electromagnetic spectrum are also used. Aerial photographs are very suitable for detailed surveying and mapping projects.

An important feature of satellite and aerial photography systems is that they can provide stereo imagery from overlapping pairs of images. These images are used to create a 3-D analog or digital model from which 3-D coordinates, contours, and digital elevation models can be created (see Section 8.3.2.3).

Satellite and aerial photograph data offer a number of advantages for GI projects. The consistency of the data and the availability of systematic global coverage make satellite data especially useful for large-area, very detailed projects (for example, mapping landforms and geology at the river catchment-area level) and for mapping inaccessible areas. The regular repeat cycles of commercial systems and the fact that they record radiation in many parts of the spectrum make such data especially suitable for assessing the condition of vegetation (for example, the moisture stress of wheat crops). Aerial photographs in particular are very useful for detailed surveying and mapping of, for example, urban areas and archaeological sites, especially those applications requiring 3-D data (see Chapter 11).

On the other hand, the spatial resolution of commercial satellites is too coarse for many detailed projects, and the data collection capability of many sensors is restricted by cloud cover. Some of this is changing, however, as the new generation of satellite sensors now provide data at 0.4 m spatial resolution and better, and radar data can be obtained that are not affected by cloud cover. The data volumes from both satellites and aerial cameras can be very large and create storage and processing problems for all but the most modern systems. The cost of data can also be prohibitive for a single project or organization.

Box 8.1 describes the work of Huadong Guo.

## Biographical Box (8.1)

### Huadong Guo

Huadong Guo (Figure 8.4) received his postgraduate degree in Cartography and Remote Sensing from the University of the Chinese Academy of Sciences (UCAS). He is a Professor and Director-General of the Chinese Academy of Sciences (CAS) Institute of Remote Sensing and Digital Earth (RADI) and a Member of CAS. He presently serves as President of the International Council for Science (ICSU) Committee on Data for Science and Technology (CODATA), Secretary-General of the International Society for Digital Earth (ISDE), and Editor-in-Chief of the *International Journal of Digital Earth* (IJDE), among other organizations. He has published more than 300 papers and 15 books and is the principal awardee of thirteen national and CAS prizes, one being "National Outstanding Expert," awarded by the State Council of China.

Guo has over thirty years of experience in remote sensing, specializing in radar for Earth observation and remote-sensing applications (see, for example, Figure 8.5) and has been involved in research on

Courtesy: Huadong Guo

Figure 8.4 Huadong Guo.

Digital Earth since the end of the previous century. Through his research in radar remote sensing, Guo established a nonvegetation sand dune radar backscattering model, developed radar polarimetric theory, researched the depolarization phenomena of volcanic lava flow, and revealed the relationship between radar

**Figure 8.5** Remote-sensing image of the Great Wall. SIR-C Image showing three segments of the Great Wall, of which two were built in the Ming dynasty (1474 A.D. and 1531 A.D.), and one was built in the Sui dynasty (585 A.D.) between the Ming Wall and the road.

Courtesy: Huadong Guo

phase information and vegetation parameters. He is a forerunner in the use of full polarization radar and has achieved high-accuracy crop classification mapping.

With the unceasing development of science and technology, the role of data is becoming increasingly important. Scientists are surrounded by petabyte-level and even exabyte-level data in the realm of remote sensing. Guo has established an all-weather, day and night remote-sensing monitoring system for disaster reduction. The system was used in the aftermath of the Wenchuan and Yushu earthquakes in 2008 and 2010. Data were made available to all partners who needed the acquired data. In all, 4,851 GB of data were shared with 15 government agencies.

Since the advent of the Digital Earth concept, Guo has actively promoted the vision of Digital Earth. He led a research team to build up the first Digital Earth proto-type system in China. This system comprises integrated, advanced technology both in hardware and software to conduct Earth system science at global, national, and regional scales.

Guo's research and contributions in radar remote-sensing theory, integrated technologies for mineral exploration, Earth observation for disaster mitigation, and Digital Earth have demonstrated the importance of remote sensing and its applications.

## 8.2.2 Vector Data Capture

Primary vector data capture is a major source of geo-graphic data. The two main branches of vector data capture are ground surveying and GPS (which is covered in Section 4.9), although as more surveyors use GPS rou-tinely, so the distinction between the two is becoming increasingly blurred. This is also leading to a fundamen-tal shift from measurement-based to coordinate-based

ground surveying (i.e., surveyors in the field work with geographic coordinates, rather than angle and distance measurements, which are harder to interpret).

### 8.2.2.1 Surveying

Ground surveying is based on the principle that the 3-D location of any point can be determined by mea-suring angles and distances from other known points.

Surveys begin from a benchmark point. If the coordinate system and location of this point are known, all subsequent points can be collected in this coordinate system. If they are unknown, then the survey will use a local or relative coordinate system (see Section 4.8).

Because all survey points are obtained from survey measurements, their known locations are always relative to other points. Any measurement errors need to be apportioned between multiple points in a survey. For example, when surveying a field boundary, if the last and first points are not identical in survey terms (within the tolerance employed in the survey), then errors need to be apportioned between all points that define the boundary (see Sections 5.3.2 and 5.4.2). As new measurements are obtained, these may change the locations of points.

Traditionally, surveyors used equipment such as transits and theodolites to measure angles, and tapes and chains to measure distances. Today these have been replaced by electro-optical devices called total stations that can measure both angles and distances to an accuracy of 1 mm (Figure 8.6). Total stations automatically log data, and the most sophisticated can create vector point, line, and area objects in the field, thus providing direct validation. The basic principles of surveying have changed very little in the past 100 years, although new technology has considerably improved accuracy and productivity. Two people are usually required to perform a survey, one to operate the total station and the other to hold a reflective prism that is placed at the object being measured. On some remote-controlled systems a single person can control both the total station and the prism.

Ground survey is a very time-consuming and expensive activity, but it is still the best way to obtain highly accurate point locations. Surveying is typically used for capturing buildings, land and property boundaries, manholes, and other objects that need to be located accurately. It is also used to obtain reference marks for use in other data capture projects. For example, detailed, fine-scale aerial photographs and satellite images are frequently georeferenced using points obtained from ground survey.

### 8.2.2.2 LiDAR

LiDAR (light detection and ranging, also known as airborne laser swath mapping, or ALSM) is a relatively recent technology that employs a scanning laser range finder to produce accurate topographic surveys of great detail. A LiDAR scanner is an active remote-sensing instrument; that is, it transmits electromagnetic radiation and measures the radiation that is scattered back to a receiver after interacting with the Earth's atmosphere or objects on the surface. LiDAR uses radiation in the ultraviolet, visible, or infrared region of the electromagnetic spectrum. The scanner is typically carried on a low-altitude aircraft that also has an inertial navigation system and a differential GPS to provide location. LiDAR scanners are capable of collecting extremely large quantities of very detailed information (i.e., scanning of the order of 30,000 points per second at an accuracy of around 15 cm). The data collected from a LiDAR scanner can be described as a point cloud, that is, a massive collection of independent points with $(x, y, z)$ values. After initial data capture, extensive processing is usually required to remove tree canopies, buildings, and other unwanted features and to correct errors in order to provide a "bare Earth" point dataset. The points in a LiDAR dataset are often rasterized to create a digital

**Figure 8.6** A tripod-mounted total station.

© Klubovy/iStockphoto

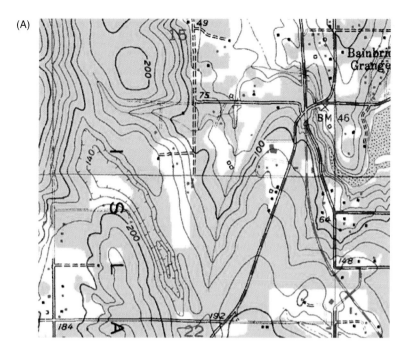

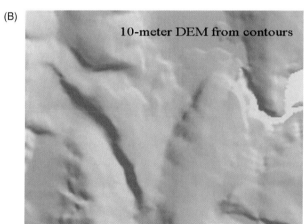

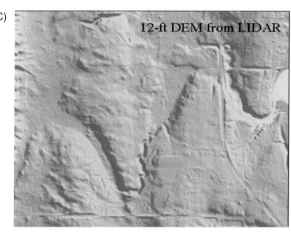

**Figure 8.7** Comparison of three datasets for 1 square mile of Bainbridge Island, Washington State: (A) scanned USGS 1:24,000 topographic map sheet; (B) 10-m digital elevation model (DEM) derived from contours digitized from a map sheet; (C) 12-ft (365-cm) resolution DEM derived from a LiDAR survey. (From pugetsoundlidar.ess.washington.edu/example2.htm)

elevation model that is smaller in size and easier to work with in a GI system. Figure 8.7 presents a comparison of three datasets for the same area, one of which is derived from LiDAR.

## 8.3 Secondary Geographic Data Capture

Geographic data capture from secondary sources is the process of creating raster and vector files and databases from maps, photographs, and other hardcopy documents. Scanning is used to capture raster data. Heads-up digitizing, stereo-photogrammetry,

and COGO data entry are the most widely used methods for capturing vector data.

### 8.3.1 Raster Data Capture Using Scanners

A scanner is a device that converts hardcopy analog media into digital images by scanning successive lines across a map or document and recording the amount of light reflected from a local data source (Figure 8.8). The differences in reflected light are normally scaled into bi-level black and white (1 bit per pixel) or multiple gray levels (8, 16, or 32 bits). Color scanners typically output data into 8-bit red, green, and blue color bands. The spatial resolution of scanners varies widely from as little as 200 dpi (8 dots per mm) to 2400 dpi (86 dots per mm) and beyond. Most GIS scanning

**Figure 8.8** A large-format roll-feed image scanner.

is in the range 400–800 dpi (16–40 dots per mm). Depending on the type of scanner and the resolution required, it can take from 30 seconds to 30 minutes or more to scan a map.

**Scanned maps and documents are used extensively as background maps and data stores.**

There are three main reasons to scan hardcopy media for use in GI systems:

- Documents, such as building plans, CAD drawings, property deeds, and equipment photographs are scanned to reduce wear and tear, to improve access, to provide integrated database storage, and to index them geographically (e.g., building plans can be attached to building objects in geographic space).

- Film and paper maps, aerial photographs, and images are scanned and georeferenced so that they provide geographic context for other data (typically vector layers). This type of unintelligent image or background geographic wallpaper is very popular in systems that manage equipment and land and property assets (Figure 8.9).

- Maps, aerial photographs, and images are scanned prior to vectorization (see Section 8.3.2.1) and sometimes as a prelude to spatial analysis.

An 8-bit (256 gray level) 400 dpi (16 dots per mm) scanner is a good choice for minimum resolution for scanning maps to be used as a background

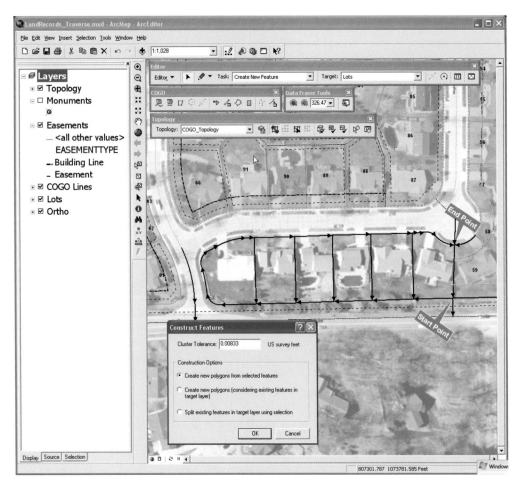

**Figure 8.9** An example of raster background data (color aerial photography) underneath vector data (land parcels) that are being digitized on-screen.

reference layer. For a color aerial photograph that is to be used for subsequent photo interpretation and analysis, a color (8 bit for each of three bands) 900 dpi (40 dots per mm) scanner provides a more appropriate minimum resolution. The quality of data output from a scanner is determined by the nature of the original source material, the quality of the scanning device, and the type of preparation prior to scanning (e.g., redrafting key features or removing unwanted marks will improve output quality).

## 8.3.2 Vector Data Capture

Secondary vector data capture involves digitizing vector objects from maps and other geographic data sources. The most popular methods are heads-up digitizing and vectorization, photogrammetry, and COGO data entry. Historically, manual digitizing from digitizing tables was the most popular way of secondary vector data capture, but its use has now almost entirely been replaced by the other more modern methods.

### 8.3.2.1 Heads-up Digitizing and Vectorization

One of the main reasons for scanning maps (see Section 8.3.1) is as a prelude to vectorization—the process of converting raster data into vector data. The simplest way to create vectors selectively from raster data is to digitize vector objects manually straight off a computer screen using a mouse or digitizing cursor. This method is called heads-up digitizing because the map is vertical and can be viewed without bending the head down. It is widely used for capturing, for example, land parcels, buildings, and utility assets.

> **Vectorization is the process of converting raster data into vector data. The opposite is called rasterization.**

Any type of image derived from either primary sources, such as satellite images or aerial photographs, or secondary sources, such as maps or other documents, can be used in heads-up digitizing and vectorization. After loading a scanned image into a GI database, it must be georeferenced (the term *georegistration* is also sometimes used synonymously: see Section 4.12) before digitizing can begin. This involves a geometric transformation process that uses well-known algorithms to convert image coordinates into database coordinates. The algorithms use both image and database coordinates from a minimum of three well-defined reference points. Once the transformation has been set up, any future coordinates digitized from the image will be in the database coordinate reference system.

Vertices defining point, polyline, and polygon objects are captured on-screen using point- or stream-digitizing methods. Point digitizing involves placing the screen cursor at the location for each object vertex and then clicking a button to record the location of the vertex. Stream-mode digitizing partially automates this process by collecting vertices automatically every time a distance or time threshold is crossed (e.g., every 0.02 inch (0.5 mm) or 0.25 second). Stream-mode digitizing is a much faster method, but it typically produces larger files with many redundant coordinates (although these can be generalized using standard algorithms).

A faster and more consistent approach is to use software to perform automated vectorization in either batch or semi-interactive mode. Batch vectorization takes an entire raster file and converts it to vector objects in a single operation. Vector objects are created using software algorithms that build simple (spaghetti) line strings from the original pixel values. The lines can then be further processed to create topologically correct polygons (Figure 8.10). A typical map will take only a few minutes to vectorize using modern hardware and software systems. See Section 9.7 for further discussion of structuring geographic data.

---

**Figure 8.10** Batch vectorization of a scanned map: (A) original raster file; (B) vectorized polygons. Adjacent raster cells with the same attribute values are aggregated. Class boundaries are then created at the intersection between adjacent classes in the form of vector lines.

(A)

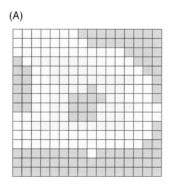

(B)

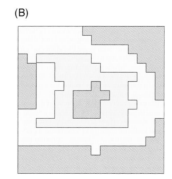

Unfortunately, batch vectorization software is far from perfect, and postvectorization editing is usually required to clean up errors. To avoid large amounts of vector editing, it is useful to undertake a little editing of the original raster file prior to vectorization in order to remove unwanted noise that may affect the vectorization process. For example, text that overlaps lines should be deleted, and dashed lines are best converted to solid lines. Following vectorization, topological relationships are usually created for the vector objects. This process may also highlight some previously unnoticed errors that require additional editing.

Batch vectorization is best suited to simple bi-level maps of, for example, contours, streams, and highways. For more complicated images and maps and where selective vectorization is required (for example, digitizing electric conductors and devices, or water mains and fittings off topographic maps), interactive vectorization (also called semiautomatic vectorization, line following, or tracing) is preferred. In interactive vectorization, software is used to automate digitizing. The operator snaps the cursor to a pixel on the screen, indicates a direction for line following, and the software then automatically digitizes lines. Typically, many parameters can be tuned to control the density of points (level of generalization), the size of gaps (blank pixels in a line) that will be jumped, and whether to pause at junctions for operator intervention or always to trace in a specific direction (most systems require that all polygons are ordered either clockwise or counterclockwise). Interactive vectorization is still quite labor intensive, but generally it results in much greater productivity than manual or heads-up digitizing. It also produces high-quality data, as software is able to represent lines more accurately and consistently than can humans. For these reasons specialized data capture groups much prefer vectorization to manual digitizing.

### 8.3.2.2 Measurement Error

Data capture, like all geographic workflows, is likely to generate both measurement and operator errors. Because digitizing is a tedious and hence error-prone practice, it presents a source of operator errors—as when the operator fails to position the cursor correctly or fails to record line segments. Figure 8.11 presents some examples of human errors that are commonly introduced in the digitizing process. These errors are overshoots and undershoots where line intersections are inexact (Figure 8.11A); invalid polygons that are topologically inconsistent because of omission of one or more lines, or omission of attribute data (Figure 8.11B); and sliver polygons, in which multiple digitizing of the common boundary between adjacent

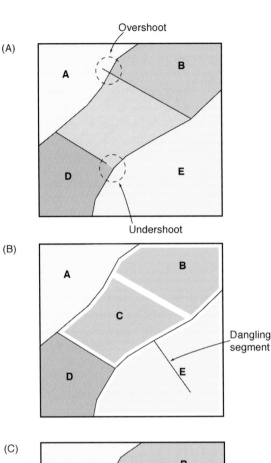

Figure 8.11 Examples of human errors in digitizing: (A) undershoots and overshoots; (B) invalid polygons; and (C) sliver polygons.

polygons leads to the creation of additional polygons (Figure 8.11C).

Most GI software packages include standard functions which can be used to restore integrity and clean (or rather obscure, depending on your viewpoint!) obvious measurement errors. Such operations are best carried out immediately after digitizing, so that omissions may be easily rectified. Data-cleaning operations require sensitive setting of threshold values, or else damage can be done to real-world features, as Figure 8.12 shows.

**Many errors in digitizing can be remedied by appropriately designed software.**

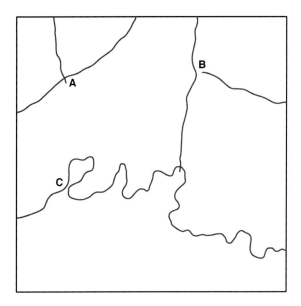

**Figure 8.12** Error induced by data cleaning. If the tolerance level is set large enough to correct the errors at A and B, the loop at C will also (incorrectly) be closed.

Further classes of problems arise when the products of digitizing adjacent map sheets are merged. Stretching of paper base maps, coupled with errors in rectifying them, give rise to the kinds of mismatches shown in Figure 8.13. Rubber sheeting is the term used to describe methods for removing such errors on the assumption that strong spatial autocorrelation exists among errors. If errors tend to be spatially autocorrelated up to a distance of $x$, say, then rubber sheeting will be successful at removing them, at least partially, provided control points can be found that are spaced less than $x$ apart. For the same reason, the

**Figure 8.13** Mismatches of adjacent spatial data sources that require rubber sheeting.

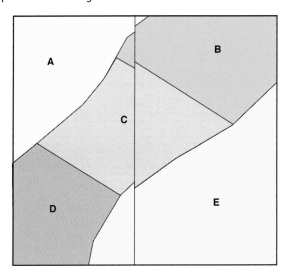

shapes of features that are less than $x$ across will tend to have little distortion, whereas very large shapes may be badly distorted. The results of calculating areas (Section 14.1.1), or other geometric operations that rely only on relative position, will be accurate as long as the areas are small, but will grow rapidly with feature size. Thus it is important for the user to know which operations depend on relative position, and over what distance; and where absolute position is important (of course, the term *absolute* simply means relative to the Earth frame, defined by the Equator and the Greenwich Meridian, or relative over a very long distance; see Section 4.7). Analogous procedures and problems characterize the rectification of raster datasets, be they scanned images of paper maps or satellite measurements of the curved Earth surface.

### 8.3.2.3 Photogrammetry

Photogrammetry is the science and technology of making measurements from pictures, aerial photographs, and images. Although in the strict sense it includes 2-D measurements taken from single aerial photographs, today it is almost exclusively concerned with capturing 2.5-D and 3-D measurements from models derived from stereopairs of photographs and images. In the case of aerial photographs, it is usual to have 60% overlap along each flight line and 30% overlap between flight lines. Similar layouts are used by remote-sensing satellites. The amount of overlap defines the area for which a 3-D model can be created.

> **Photogrammetry is used to capture measurements from photographs and other image sources.**

To obtain true georeferenced Earth coordinates from a model, it is necessary to georeference photographs using control points (the procedure is essentially analogous to that described for digitizing in Section 8.3.2.1). Control points can be defined by ground survey, or nowadays more usually with GPS (see Section 8.2.2.1 for discussion of these techniques).

Measurements are captured from overlapping pairs of images using stereoplotters. These build a model and allow 3-D measurements to be captured, edited, stored, and plotted. Stereoplotters have undergone three major generations of development: analog (optical), analytic, and digital. Mechanical analog devices are seldom used today, whereas analytical (combined mechanical and digital) and digital (entirely computer based) are much more common. It is likely that digital (softcopy) photogrammetry will eventually replace mechanical devices entirely.

There are many ways to view stereo models, including a split screen with a simple stereoscope and the use of special glasses to observe a red/green

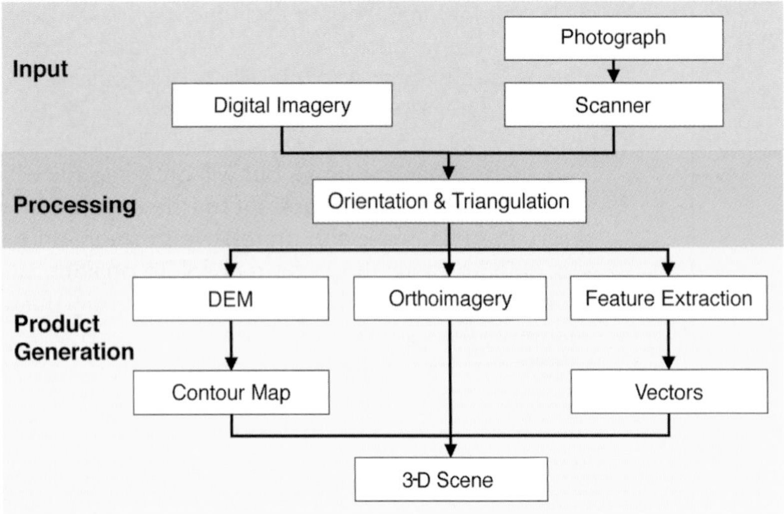

**Figure 8.14** Typical photogrammetry workflow.

display or polarized light. To manipulate 3-D cursors in the $x$, $y$, and $z$ planes, photogrammetry systems offer free-moving hand controllers, hand wheels and foot disks, and 3-D mice. The options for extracting vector objects from 3-D models are directly analogous to those available for manual digitizing as described earlier: namely, batch, interactive, and manual (Section 8.3.2.1). The obvious difference, however, is that there is a requirement for capturing $z$ (elevation) values.

Figure 8.14 shows a typical workflow in digital photogrammetry derived from the work of Vincent Tao. There are three main parts to digital photogrammetry workflows: data input, processing, and product generation. Data can be obtained directly from sensors or by scanning secondary sources. Orientation and triangulation are fundamental photogrammetry processing tasks. Orientation is the process of creating a stereo model suitable for viewing and extracting 3-D vector coordinates that describe geographic objects. Triangulation (also called block adjustment) is used to assemble a collection of images into a single model so that accurate and consistent information can be obtained from large areas.

Photogrammetry workflows yield several important product outputs, including digital elevation models (DEMs), contours, orthoimages, vector features, and 3-D scenes. DEMs—regular arrays of height values—are created by matching stereo image pairs together using a series of control points. Once a DEM has been created, it is relatively straightforward to derive contours using a choice of algorithms. Orthoimages are images corrected for variations in terrain using a DEM so as to appear as if every point was seen from vertically above. They have become popular because of their relatively low cost of creation (when compared with topographic maps) and ease of interpretation as base maps except where tall buildings and other dramatic topographic features

are present. They can also be used as accurate data sources for heads-up digitizing (see Section 8.3.2.1). Vector feature extraction is still an evolving field, and there are no widely applicable fully automated methods. The most successful methods use a combination of spectral analysis and spatial rules that define context, shape, proximity, and the like. Finally, 3-D scenes can be created by merging vector features with a DEM and an orthoimage (Figure 8.15).

In summary, photogrammetry is a very cost-effective data capture technique that is sometimes the only practical method of obtaining detailed topographic data about an area of interest. Unfortunately, the complexity and high cost of equipment have restricted its use to primary data capture projects and specialist data capture organizations where very detailed information is required.

### 8.3.2.4 COGO Data Entry

COGO, which as noted earlier is a contraction of the term *coordinate geometry*, is a method for capturing and representing geographic data. COGO uses survey-style bearings and distances to define each part of an object in much the same way as described in Section 8.2.2. Figure 8.9 shows how land-parcel features can be created using COGO tools and then formed into topologically correct polygons. Some examples of COGO object-construction tools are shown in Figure 8.16. The Construct Along tool creates a point along a curve using a distance along the curve. The Line Construct Angle Bisector tool constructs a line that bisects an angle defined by a from-point, a through-point, a to-point, and a length. The Construct Fillet tool creates a circular-arc tangent from two segments and a radius.

The COGO system is widely used in North America to represent land records and property parcels (also called lots). Coordinates can be obtained

**Figure 8.15** Automatically created 3-D model of central Philadelphia.

Source: Esri

from COGO measurements by geometric transformation (i.e., bearings and distances are converted into *x, y* coordinates). Although COGO data obtained as part of a primary data capture activity are used in some projects, it is more often the case that secondary measurements are captured from hardcopy maps and documents. Source data may be in the form of legal descriptions, records of survey, tract (housing estate) maps, or similar documents.

**COGO stands for coordinate geometry. It is a vector data structure and method of data entry.**

**Figure 8.16** Example COGO construction tools used to represent geographic features.

**Construct Along**

distance
(or ratio)

curve

**Line Construct Angle Bisector**

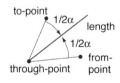

to-point

1/2α

length

1/2α

from-
point

through-point

**Construct Fillet**

segment 1

segment 2

radius

COGO data are very precise measurements and are often regarded as the only legally acceptable definition of land parcels. Measurements are usually very detailed, and data capture is often time consuming. Furthermore, commonly occurring discrepancies in the data must be manually resolved by highly qualified individuals.

## 8.4 Obtaining Data from External Sources (Data Transfer)

One major decision that needs to be faced at the start of a GI project is whether to build or buy part or all of a database. All of the preceding discussion has been concerned with techniques for building databases from primary and secondary sources. This section focuses on how to import or transfer data into a GI system that has been captured by others. Some datasets are freely available, but many of them are sold as commodities, mainly over the Web. Increasingly geographic data are being made available as direct-use services that can be applied in geographic analysis and mapping without having to import the data.

There are many sources and types of geographic data. Space does not permit a comprehensive review of all geographic data sources here, but a small selection of key sources is listed in Table 8.3. In any case, the characteristics and availability of datasets are constantly changing, so those seeking an up-to-date list should consult one of the online sources described in this section. Kerski and Clark offer a useful guide to public domain data. Chapter 17 also discusses the characteristics of geographic information and highlights several issues to bear in mind when using data collected by others.

Table 8.3 Examples of some digital data sources that can be imported into a GI system. NMO = National Mapping Organizations, USGS = U.S. Geological Survey, NGA = U.S. National Geospatial-Intelligence Agency, NASA = National Aeronautics and Space Administration, DEM = Digital Elevation Model, EPA = U.S. Environmental Protection Agency, WWF = World Wide Fund for Nature, FEMA = Federal Emergency Management Agency, Esri BIS = Esri Business Information Solutions.

| Type | Source | Details |
|---|---|---|
| **Base Maps** | | |
| Geodetic framework | Many NMOs, e.g., USGS and Ordnance Survey | Definition of framework, map projections, and geodetic transformations |
| General topographic map data | NMOs and military agencies, e.g., NGA | Many types of data at detailed to medium scales |
| Elevation | NMOs, military agencies, and several commercial providers, e.g., USGS, SPOT Image, NASA | DEMs, contours at local, regional, and global levels |
| Transportation | National governments and several commercial vendors | Highway/street centerline databases at national levels |
| Hydrology | NMOs and government agencies | National hydrological databases are available for many countries |
| Toponymy | NMOs, other government agencies, and commercial providers | Gazetteers of place-names at global and national levels |
| Satellite images | Commercial, government, and military providers, e.g., EROS Data Center, IRS, NASA, SPOT Image, i-cubed, and DigitalGlobe | See Figure 8.2 for further details |
| Aerial photographs | Many private and public agencies | Scales vary widely, typically from 1:500–1:20,000 |
| **Environmental** | | |
| Wetlands | National agencies, e.g., U.S. National Wetlands Inventory | Government wetlands inventory |
| Toxic release sites | National environmental protection agencies, e.g., EPA | Details of thousands of toxic sites |
| World ecoregions | Conservation agencies, e.g., WWF | Habitat types, threatened areas, biological distinctiveness |
| Flood zones | Many national and regional government agencies, e.g., FEMA | National flood-risk areas |
| **Socioeconomic** | | |
| Population census | National governments, with value added by commercial providers | Typically, every 10 years with annual estimates |
| Lifestyle classifications | Private agencies (e.g., CACI and Experian) | Derived from population censuses and other socioeconomic data |
| Geodemographics | Private agencies (e.g., Experian, Esri BIS) | Many types of data at many scales and prices |
| Land and property ownership | National governments | Street, property, and cadastral data |
| Administrative areas | National governments | Obtained from maps at scales of 1:5,000–1:750,000 |

The best way to find geographic data is to search the Internet. Several types of resources and technologies are available to assist searching, and these are described in detail in Section 10.2. They include specialist geographic data catalogs and stores, as well as the sites of specific geographic data vendors. These sites provide access to information about the characteristics and availability of geographic data. Some also have facilities to purchase and download data directly. Probably the most useful resources for locating geographic data are the geolibraries and geoportals (see Section 10.2.2) that have been created as part of national and global spatial data infrastructure initiatives (SDI).

> **The best way to find geographic data is to search the Internet using one of the specialist geolibraries or SDI geographic data geoportals.**

A major challenge of using data obtained from the Web is evaluation of fitness for purpose. Too often inexperienced practitioners download data from the Web and assume that its accuracy and licensing terms are adequate for use in a GI project. It is essential that the suitability of all datasets be checked before they are used. A good starting point is to examine the metadata records associated with the dataset (Section 10.2); these records should indicate age, provenance, projection, and a range of other relevant properties. Simple checks include overlay of the data on top of a base map of known and acceptable accuracy and independent verification (e.g., by fieldwork or by comparison with other datasets) of the geometric and attribute properties of a representative sample of objects. It is also best to assume that all data are proprietary until open access/use is confirmed.

## 8.4.1 Geographic Data Formats

One of the biggest problems with data obtained from external sources is that they can be encoded in many different formats. There are so many different geographic data formats because no single format is appropriate for all tasks and applications. It is not possible to design a format that supports, for example, both fast rendering in police command and control systems and sophisticated topological analysis in natural resource information systems: the two are mutually incompatible. Also, given the great diversity of geographic information, a single comprehensive format would simply be too large and cumbersome. The many different formats that are in use today have evolved in response to diverse user requirements.

Given the high cost of creating databases, many tools have been developed to move data between

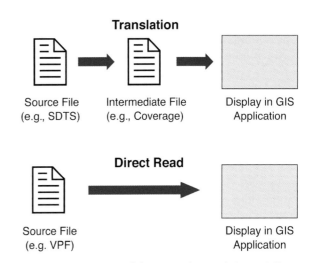

**Figure 8.17** Comparison of data access by translation and direct read.

systems and to reuse data through open application programming interfaces (APIs). In the former case, the approach has been to develop software that is able to translate data (Figure 8.17), either by a direct read into memory or via an intermediate file format. In the latter case, software developers have created open interfaces to allow access to data.

Many GI systems are now able to read directly Auto-CAD DWG and DXF, Microstation DGN, Esri Shapefile, VPF, and many image formats. Unfortunately, direct read support can only be easily provided for relatively simple product-oriented formats. Complex formats, such as SDTS (Spatial Data Transfer Standard), were designed for exchange purposes and require more advanced processing before they can be viewed (e.g., multipass read and feature assembly from several parts).

> **Data can be transferred between systems by direct read into memory or via an intermediate file format.**

More than 25 organizations are involved in the standardization of various aspects of geographic data and geoprocessing; several of them are country and domain specific. At the global level, the ISO (International Organization for Standards) is responsible for coordinating efforts through the work of technical committees TC 211 and 287. In Europe, CEN (Comité Européen de Normalisation) is engaged in geographic standardization. At the national level, there are many complementary bodies. One other standards-forming organization of particular note is OGC (Open Geospatial Consortium), a group of vendors, academics, and users interested in the interoperability of geographic systems. To date, there have been promising OGC-coordinated efforts to standardize on simple feature access (simple geometric object types), metadata catalogs, and Web access (see Chapter 10 for further details).

**The most efficient way to translate data between systems is usually via a common intermediate file format.**

Having obtained a potentially useful source of geographic information, the next task is to import it into a GI database. If the data are already in the native format of the target GI system, or the software has a direct read capability for the format in question, then this is a relatively straightforward task. If the data are not compatible with the target software, then the alternatives are to ask the data supplier to convert the data to a compatible format or to use a third-party translation software system, such as the Feature Manipulation Engine from Safe Software (www.safe.com lists over 300 supported geographic data formats) to convert the data. Geographic data translation software must address both syntactic and semantic translation issues. Syntactic translation involves converting specific digital symbols (letters and numbers) between systems, whereas semantic translation is concerned with converting the meaning inherent in geographic information. Although syntactic translation is relatively simple to encode and decode, semantic translation is much more difficult and has seldom met with much success to date.

Although the task of translating geographic information between systems was described earlier as relatively straightforward, those that have tried this in practice will realize that things on the ground are seldom quite so simple. Any number of things can (and do!) go wrong, ranging from corrupted media to incomplete data files, incompatible versions of translators, and different interpretations of a format specification, to basic user error.

Two basic strategies are used for data translation: one is direct and the other uses a neutral intermediate format. For small systems that involve the translation of a small number of formats, the first is the simplest. Directly translating data back and forth between the internal structures of two systems requires two new translators (A to B, B to A). Adding two further systems will require 12 translators to share data between all systems (A to B, A to C, A to D, B to A, B to C, B to D, C to A, C to B, C to D, D to A, D to B, and D to C). A more efficient way of solving this problem is to use the concept of a data switchyard and a common intermediate file format. Systems now need only to translate to and from the common format. The four systems use only 8 translators instead of 12 (A to Neutral, B to Neutral, C to Neutral, D to Neutral, Neutral to A, Neutral to B, Neutral to C, and Neutral to D). The more systems there are, the more efficient the result. This is one of the key principles underlying the need for common file-interchange formats.

## 8.5 Capturing Attribute Data

All geographic objects have attributes of one type or another. Although attributes can be collected at the same time as vector geometry, it is usually more cost effective to capture attributes separately. In part, this is because attribute data capture is a relatively simple task that can be undertaken by lower-cost clerical staff. It is also because attributes can be entered by direct data loggers, manual keyboard entry, optical character recognition (OCR), or increasingly, voice recognition—methods that do not require expensive hardware and software systems. By far the most common method is direct keyboard data entry into a spreadsheet or database. For some projects, a custom data-entry form with built-in validation is preferred. On small projects single entry is used, but for larger, more complex projects data are entered twice and then compared as a validation check.

An essential requirement for separate data entry is a common identifier (also called a key, or object-id) that can be used to relate object geometry and attributes following data capture (see Figure 7.10 for a diagrammatic explanation of relating geometry and attributes).

Metadata are a special type of nongeometric data that are increasingly being collected. Some metadata are derived automatically by the GI system (for example, length and area, extent of data layer, and count of features), but some must be explicitly collected (for example, owner name, quality estimate, and original source). Explicitly collected metadata can be entered in the same way as other attributes, as described earlier. For further information about metadata, see Section 10.2.

## 8.6 Citizen-Centric Web-Based Data Collection

New developments in Web technology have opened up a new vista of opportunities for distributed geographic data collection. A raft of new Web 2.0 technologies has enabled organizations and individual projects to use citizens to collect data very rapidly across a wide variety of thematic and geographic areas that represent a wide spectrum of viewpoints. It is now very simple to create a Web site with a form-based interface to collect geographic data about many types of phenomena, events, and activities. Locations can be obtained by asking the user to digitize the location on a map or to upload coordinates collected in any of the ways outlined earlier in this chapter. This type of approach to data collection by volunteers is discussed in Section 2.5. Box 8.2 presents an example of a Web 2.0, citizen-centric data collection application.

## E-Flora British Columbia, Citizen-Centric Data Collection

The British Columbia, Canada E-Flora BC project is a good example of the way data collection is changing to incorporate citizen-centric data input. E-Flora BC is an online, Web-accessible electronic atlas of plant species that allows interactive data collection, reporting, and mapping (www.eflora.bc.ca).

The public is encouraged to participate in an Invasive Alien Plant Program run by the Project by reporting suspected new occurrences of invasive plants using an interactive form-based Report-a-Weed tool. This tool (Figure 8.18) allows citizen–scientists to collect and enter pertinent information about invasive species, which is then delivered to an appropriate botanist for review and then entered into the organization's master database.

The Web has radically transformed the way that databases of this type are collected and maintained: no longer is this process the sole preserve of the official government organizations. This type of Web 2.0 data collection and public empowerment is especially good at handling local (in space and time) phenomena. There is a rapidly growing list of examples of Web projects that rely on volunteers to collect geographic information (see also Section 10.2 for further discussion).

(A)

© DND ASU Chilliwack

(B)

(C)

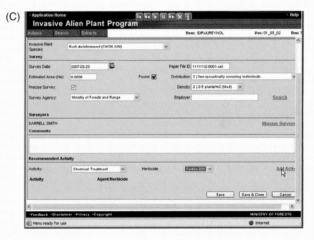

Figure 8.18 E-Flora information about *Senecio jacobaea* (the "stinking willie" or tansy ragwort): (A) photograph; (B) distribution map; and (C) citizen-centric Web data input form.

## 8.7 Managing a Data Collection Project

The subject of managing a GI project is given extensive treatment in Chapters 16–18. The management of data capture projects is discussed briefly here, both because of its critical importance and because it involves several unique issues. That said, most of the general principles for any GI project apply to data collection: the need for a clearly articulated plan, adequate resources, appropriate funding, and sufficient time.

In any data collection project there is a fundamental trade-off between quality, speed, and price

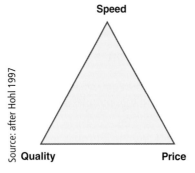

Figure 8.19 Relationship between quality, speed, and price in data collection.

Source: after Hohl 1997

Speed

Quality          Price

(Figure 8.19). Collecting high-quality data quickly is possible, but it is also very expensive. If price is a key consideration, then lower-quality data can be collected over a longer period.

Data collection projects can be carried out intensively over a short period or less extensively over a longer period. A key decision facing managers of such projects is whether to pursue a strategy of incremental or very rapid collection. Incremental data collection involves breaking the data collection project into small manageable subprojects. This allows data collection to be undertaken with lower annual resource and funding levels (although total project resource requirements may be larger). It is a good approach for inexperienced organizations that are embarking on their first data collection project because they can learn and adapt as the project proceeds. At the same time, these longer-term projects run the risk of employee turnover and burnout, as well as changing data, technology, and organizational priorities.

Whichever approach is preferred, a pilot project carried out on part of the study area and a selection of the data types can prove to be invaluable. A pilot project can identify problems in workflow, database design, personnel, and equipment. A pilot database can also be used to test equipment and to develop procedures for quality assurance. Many projects require a test database to carry out hardware and software acceptance tests, as well as to facilitate software customization. It is essential that project managers are prepared to discard all the data obtained during a pilot data collection project, so that the main phase can proceed unconstrained.

A further important decision is whether data collection should use in-house or external resources. It is now increasingly common to outsource geographic data collection to specialist companies that usually undertake the work in areas of the world with very low labor costs (e.g., India, Thailand, and Vietnam). Three factors influencing this decision are cost/schedule, quality, and long-term ramifications. Specialist external data collection agencies can often perform work faster, cheaper, and with higher quality than with in-house staff, but because of the need for real cash to pay external agencies, this may not be possible. In the short term, project costs, quality, and time are the main considerations, but over time dependency on external groups may become a problem.

Box 8.3 introduces the work of Carmen Reyes, a researcher with the National Autonomous University of Mexico with over 40 years of experience in GI projects and data collection.

## Biographical Box (8.3)

### Carmen Reyes

Carmen Reyes (Figure 8.20) is an expert and a researcher in GI science, cybercartography, geocybernetics, and geomatics; she has a BSc in Mathematics from the National Autonomous University of Mexico (UNAM) and a master's in Mathematics from the Metropolitan Autonomous University (UAM). She obtained her PhD in geographic information systems from Simon Fraser University, in Canada. During the past 40 years she has worked in over 80 geomatics/GI projects for local and international institutions, in both public and private sectors.

Dr. Reyes succeeded in her efforts to introduce geomatics and GI science in the Mexican scientific realm as founder in 1999 and, for a decade, general director of the J. L. Tamayo Center for Research in Geography and Geomatics (CentroGeo) under the (Mexican) National Science and Technology Council. She was presented with the Samuel Gill Gamble Award for Cartography by the Government of Canada through the PanAmerican

Institute of Geography and History (PAIGH) of the Organization of American States (OAS). Dr. Reyes is member of national and international organizations, networks, and academic committees. She is in the Board of Directors of the Global Spatial Network (GSN) and member of the global advisory council of the Open Geospatial Consortium (OGC). Currently she is a senior researcher at CentroGeo, teaches graduate courses, is senior supervisor of Master and PhD students, is the chief editor of the *Journal on Geocybernetics*, and manages GI science projects.

On geocybernetics and knowledge-based GI systems, Carmen says:

The term "geocybernetics" encompasses several avenues of research that explicitly incorporate the science of cybernetics, general systems theory, modeling, and complexity theory as theoretical building blocks. Currently, together

▶

with a research group at CentroGeo I am conducting empirical and theoretical work in geo-cybernetics that include: (1) cybercartography, (2) complex solutions in geomatics, (3) collective mental maps, (4) the geomatics prototype, (5) the Strabo technique and (6) the Reyes method.

In the tradition of science, experimentation is a key and invaluable resource, for which data collection and the processing and application of information to sustain or reject hypotheses is a common approach. What makes this approach different is its main thesis— the assumption that conversations between scientific and societal actors should be based on cognitive knowledge frameworks. Through the process, new knowledge frameworks emerge out of the fusion of one or more knowledge domains; i.e., a transdisciplinary process evolves through which cognitive bridges are built not only within geomatics/GI science but also with other knowledge domains, such as public policy, landscape ecology, and criminology.

This knowledge-based approach to GI science and geomatics has been very effective for the interaction between science and society and has resulted in novel scientific findings and

Courtesy: Carmen Reyes

**Figure 8.20** Carmen Reyes, GI Scientist.

outcomes. It can be stated that the main driving force in the processes to design and implement solutions from a geomatics/GI science perspective is the K in knowledge rather than the I in information or the D in data.

## Questions for Further Study

1. Evaluate the suitability of free geographic data for your home region or country for use in a GI project of your choice.
2. What are the advantages of batch vectorization over heads-up digitizing?
3. What quality assurance steps would you build into a data collection project designed to construct a database of land parcels for tax assessment?
4. Why do so many geographic data formats exist? Which ones are most suitable for selling vector data?

## Further Reading

Hohl, P. (ed.). 1997. *GIS Data Conversion: Strategies, Techniques and Management*. Santa Fe, NM: OnWord Press.

Kerski, J. J. and Clark, J. 2012. *The GIS Guide to Public Domain Data*. Redlands, CA: ESRI Press.

Lillesand, T. M., Kiefer, R. W., and Chipman, R. W. 2008. *Remote Sensing and Image Interpretation* (6th ed.). Hoboken, NJ: Wiley.

Weng, Q. 2012. *An introduction to contemporary remote sensing*. New York: McGraw-Hill.

# 9 Creating and Maintaining Geographic Databases

All large operational geographic information (GI) systems are built on the foundation of a geographic database. After the people who manage and operate GI systems, the database is arguably the most important part because of the costs of collection and maintenance and because the database forms the basis of all query, analysis, and decision-making activities. Today, virtually all large GI system implementations store data in a database management system (DBMS), a specialist piece of software designed to handle multiuser access to an integrated set of data. Extending standard DBMS to store geographic data raises several interesting challenges. Databases need to be designed with great care and should be structured and indexed to provide efficient query and transaction performance. A comprehensive security and transactional access model is necessary to ensure that multiple users can access the database at the same time. Ongoing maintenance is also an essential, but very resource-intensive, activity.

## LEARNING OBJECTIVES

After studying this chapter you will be able to:

- Understand the role of database management systems in GI systems.

- Recognize Structured (or Standard) Query Language (SQL) statements.

- Understand the key geographic database data types and functions.

- Be familiar with the stages of geographic database design.

- Understand the key techniques for structuring geographic information, specifically creating topology and indexing.

- Understand the issues associated with multiuser editing and versioning.

## 9.1 Introduction

A database can be thought of as an integrated set of data on a particular subject. Geographic databases are simply databases containing geographic data for a particular area and subject. It is quite common to encounter the term *spatial* in the database world. As discussed in Section 1.1.2, spatial refers to data about any type of space at both geographic and nongeographic scales. A geographic database is a critical part of an operational GI system both because of the cost of creation and maintenance and because of the impact of a geographic database on all analysis, modeling, and decision-making activities. Databases can be physically stored in files or by using specialist software programs

called a DBMS. Today, most large organizations use a combination of files and DBMS for storing data assets.

**A database is an integrated set of data on a particular subject.**

The database approach to storing geographic data offers a number of advantages over traditional file-based datasets.

- Collecting all data at a single location reduces redundancy.

- Maintenance costs decrease because of better organization and reduced data duplication.

- Applications become data independent so that multiple applications can use the same data and can evolve separately over time.

- User knowledge can be transferred between applications more easily because the database remains constant.

- Data sharing is facilitated, and a corporate view of data can be provided to all managers and users.

- Security and standards for data and data access can be established and enforced.

- DBMSs are better able to manage large numbers of concurrent users working with vast amounts of data.

Using databases when compared to files also has some disadvantages.

- The cost of acquiring and maintaining DBMS software can be quite high.

- A DBMS adds complexity to the problem of managing data, especially in small projects.

- Single-user performance will often be better for files, especially for more complex data types and structures where specialist indexes and access algorithms can be implemented.

In recent years geographic databases have become increasingly large and complex (see Table 1.2). For example, a U.S. National Image Mosaic will be over 25 terabytes (TB) in size, a Landsat satellite global image mosaic at 15-m resolution is 6.5 TB, and the Ordnance Survey of Great Britain has approximately 450 million vector features in its MasterMap database covering all of Britain. This chapter describes how to create and maintain geographic databases and presents the concepts, tools, and techniques that are available to manage geographic data in databases. Several other chapters provide additional information that is relevant to this discussion. In particular, the nature of geographic data and how to represent them in GI systems were described in Chapters 2, 3, and 4, and data modeling and data collection were discussed in Chapters 7 and 8, respectively. Later chapters introduce the tools and techniques that are available to query, model, and analyze geographic databases (Chapters 13, 14, and 15). Finally, Chapters 16 through 18 discuss the important management issues associated with creating and maintaining geographic databases.

## 9.2 Database Management Systems

**A DBMS is a software application designed to organize the efficient and effective storage and access of data.**

Small, simple databases that are used by a small number of people can be stored on computer hard disks in standard files. However, large, more complex databases with many tens, hundreds, or thousands of users require specialist database management system (DBMS) software to ensure data integrity and longevity. A DBMS is a software application designed to organize efficient and effective data storage and access. To carry out this function, DBMSs provide a number of important capabilities. These are introduced briefly here and are discussed further in this and other chapters.

DBMSs are successful because they are able to provide the following:

- *A data model.* As discussed in Chapter 7, a data model is a mechanism used to represent real-world objects digitally in a computer system. All DBMSs include standard general-purpose core data models suitable for representing several object types (e.g., integer and floating-point numbers, dates, and text). Several DBMSs now also support geographic (spatial) object types.

- *A data load capability.* DBMSs provide tools to load data into databases. Simple tools are available to load standard supported data types (e.g., character, number, and date) in well-structured formats.

- *Indexes.* An index is a data structure that is used to speed up searching. All databases include tools to index standard database data types.

- *A query language.* One of the major advantages of DBMS is that they support a standard data query/manipulation language called SQL (Structured/Standard Query Language).

- *Security.* A key characteristic of DBMSs is that they provide controlled access to data. This includes the ability to restrict user access to just part of a database. For example, a casual GI system user might have read-only access to only part of a database, but a specialist user might have read and write (create, update, and delete) access to the entire database.

- *Controlled update.* Updates to databases are controlled through a transaction manager responsible for managing multiuser access and ensuring that updates affecting more than one part of the database are coordinated.

- *Backup and recovery.* It is important that the valuable data in a database are protected from system failure and incorrect (accidental or deliberate) update. Software utilities are provided to back up all or part of a database and to recover the database in the event of a problem.

- *Database administration tools.* The task of setting up the structure of a database (the schema), creating and maintaining indexes, tuning to improve performance, backing up and recovering, and allocating user access rights is performed by a database administrator (DBA). A specialized collection of tools and a user interface are provided for this purpose.

- *Applications.* Modern DBMSs are equipped with standard, general-purpose tools for creating, using, and maintaining databases. These include applications for designing databases (CASE [computer-aided software engineering] tools) and for building user interfaces for data access and presentations (e.g., forms and reports).

- *Application programming interfaces (APIs).* Although most DBMSs have good general-purpose applications for standard use, most large, specialist applications will require further customization using a commercial off-the-shelf programming language and a DBMS programmable API.

This list of DBMS capabilities is very attractive to GI system users, and so, not surprisingly, virtually all large GI databases are managed using DBMS technology. Indeed, most GI system software vendors include DBMS software within their GI system software products or provide an interface that supports very close coupling to a DBMS. For further discussion of this subject, see Chapter 7.

**Today, virtually all large GI systems use DBMS technology for data management.**

## 9.2.1 Types of DBMSs

DBMSs can be classified according to the way they organize, store, and manipulate data. Three main types of DBMSs have been used in GI systems: relational (RDBMS), object (ODBMS), and object-relational (ORDBMS).

A relational database comprises a set of tables, each a two-dimensional list (or array) of records containing attributes about the objects under study. This apparently simple structure has proven to be remarkably flexible and useful in a wide range of application areas, such that historically over 95% of the data in DBMSs have been stored in RDBMSs. Most of the world's current DBMSs are built on a foundation of core relational concepts.

Object database management systems (ODBMSs) were initially designed to address several of the weaknesses of RDBMSs. These include the inability to store complete objects directly in the database (both object state and behavior; see Box 7.2 for an overview of objects and object technology). Because RDBMSs were focused primarily on business applications such as banking, human resource management, and stock control and

inventory, they were never designed to deal with rich data types, such as geographic objects, sound, and video. A further difficulty is the poor performance of RDBMSs for many types of geographic query. These problems are compounded by the difficulty of extending RDBMSs to support geographic data types and processing functions, which obviously limits their adoption for geographic applications. ODBMSs can store objects persistently (semipermanently on disk or other media) and provide object-oriented query tools. A number of commercial ODBMSs have been developed, including GemStone/S Object Server from GemStone Systems, Inc., Objectivity/DB from Objectivity, Inc., and Versant Object Database from Versant Object Technology Corp.

Despite the technical elegance of ODBMSs, they have not proven to be as commercially successful as some people initially predicted. This is largely because of the massive installed base of RDBMSs and the fact that RDBMS vendors have now added many of the important ODBMS capabilities to their standard RDBMS software systems to create hybrid object-relational DBMSs (ORDBMSs). An ORDBMS can be thought of as an RDBMS engine with some additional capabilities for dealing with objects. They can handle both the data describing what an object is (object attributes such as color, size, and age) and the behavior that determines what an object does (object methods or functions such as drawing instructions, query interfaces, and interpolation algorithms), and these can be managed and stored together as an integrated whole. Examples of ORDBMS software include IBM's DB2 and Informix, Microsoft's SQL Server, and Oracle Corp.'s Oracle DBMS. Because ORDBMSs and the underlying relational model are so important in GI systems, these topics are discussed at length in Section 9.3.

A number of ORDBMSs have now been extended to support geographic object types and functions through the addition of seven key capabilities (these are introduced here and discussed further later in this chapter):

- Query parser—the engines used to interpret queries by splitting them up and decoding them have been extended to deal with geographic types and functions.

- Query optimizer—software query optimizers have been enhanced so that they are able to handle geographic queries efficiently. Consider a query to find all potential users of a new brand of premier wine to be marketed to wealthy households from a network of retail stores. The objective is to select all households within 3 km of a store that have an income greater than $110,000. This could be carried out in two ways:

1. Select all households with an income greater than $110,000; from this selected set, select all households within 3 km of a store.

2. Select all households within 3 km of a store; from this selected set select all households with an income greater than $110,000.

Selecting households with an income greater than $110,000 is an attribute query that can be performed very quickly. Selecting households within 3 km of a store is a geometric query that takes much longer. Executing the attribute query first (Option 1 above) will result in fewer geometry query tests for store proximity, and therefore the whole query will be completed much more quickly.

- Query language—query languages have been improved to handle geographic types (e.g., points and areas) and functions (e.g., select areas that touch each other).

- Indexing services—standard unidimensional DBMS data index services have been extended to support multidimensional (i.e., x, y, z coordinates) geographic data types.

- Storage management—the large volumes of geographic records with different sizes (especially geometric and topological relationships) have been accommodated through specialized storage structures.

- Transaction services—standard DBMSs are designed to handle short (subsecond) transactions, and these have been extended to deal with the long transactions common in many geographic applications.

- Replication—services for replicating databases have been extended to deal with geographic types and with problems of reconciling changes made by distributed users.

## 9.2.2 Geographic DBMS Extensions

A number of the major commercial DBMS vendors have released spatial database extensions to their standard ORDBMS products. IBM offers two solutions—DB2 Spatial Extender and Informix Spatial; Microsoft has released spatial capabilities in the core of SQLServer; and Oracle has spatial capability in the core of Oracle DBMS and a Spatial option that adds more advanced features (see Box 9.1). The open-source DBMS PostgreSQL has also been extended with spatial types and functions (PostGIS).

**ORDBMSs provide core support for geographic data types and functions.**

Although these systems differ in technology, scope, and features, they all provide basic capabilities to store, manage, and query geographic objects. This

---

**Technical Box (9.1)**

### Oracle Spatial

Oracle Spatial is an extension to the Oracle DBMS that provides the foundation for the management of spatial (including geographic) data inside an Oracle database. The standard types and functions in Oracle (CHAR, DATE, or INTEGER, etc.) are extended with geographic equivalents. Oracle Spatial supports three basic geometric forms:

- Points: Points can represent locations such as buildings, fire hydrants, utility poles, oil rigs, boxcars, or roaming vehicles.

- Polylines: Polylines can represent things like roads, rivers, utility lines, or fault lines.

- Polygons and complex polygons with holes: Polygons can represent things like outlines of cities, districts, floodplains, or oil and gas fields. A polygon with a hole might geographically represent a parcel of land surrounding a patch of wetland.

These simple feature types can be aggregated to richer types using topology and linear referencing capabilities (see Section 7.2.3.3). In addition, Oracle

Spatial can store and manage georaster (image) data. Oracle Spatial extends the Oracle DBMS query engine to support geographic queries. There is a set of spatial operators to perform area-of-interest and spatial-join queries; length, area, and distance calculations; buffer and union queries; and administrative tasks. The Oracle Spatial SQL used to create a table and populate it with a single record is shown in the following script. (The characters after the dash on each line are comments that describe the operations. The discussion of SQL syntax in Section 9.4 will help decode this program.)

```
-- Create a table for routes (highways).
CREATE TABLE lrs_routes (
 route_id   NUMBER PRIMARY KEY,
 route_name   VARCHAR2(32),
 route_geometry   MDSYS.SDO_GEOMETRY);
-- Populate table with just one route
   for this example.
INSERT INTO lrs_routes VALUES(
 1,
 'Route1',
```

▶

```
MDSYS.SDO_GEOMETRY(
  3002, -- line string, 3 dimensions:
    X,Y,M
  NULL,
  NULL,
  MDSYS.SDO_ELEM_INFO_ARRAY(1,2,1), -- one
    line string, straight segments
  MDSYS.SDO_ORDINATE_ARRAY(
  2,2,0, -- Starting point - Exit1; 0
    is measure from start.
  2,4,2, -- Exit2; 2 is measure from
    start.
  8,4,8, -- Exit3; 8 is measure from
    start.
  12,4,12, -- Exit4; 12 is measure from
    start.
  12,10,NULL, -- Not an exit; measure
    will be automatically calculated &
    filled.
  8,10,22, -- Exit5; 22 is measure from
    start.
```

```
  5,14,27) -- Ending point (Exit6); 27
    is measure from start.
  )
);
```

Geographic data in Oracle Spatial can be indexed using R-tree and quadtree indexing methods (these terms are defined in Section 9.7.2). There are also capabilities for managing projections and coordinate systems, as well as long transactions (see discussion in Section 9.9.1). Finally, there are also some tools for elementary spatial data analysis (Chapters 13 and 14). Oracle Spatial can be used with all major GI software products, and developers can create specific-purpose applications that embed SQL commands for manipulating and querying data. Oracle has generated considerable interest among larger IT-focused organizations. IBM has approached this market in a similar way with its Spatial Extender for the DB2 DBMS and Spatial for Informix. Most recently Microsoft has added comparable spatial capabilities to its SQLServer DBMS. There are also open-source alternatives such as PostGIS.

is achieved by implementing the seven key database extensions described in the previous section. It is important to realize, however, that none of these is a complete GI system in itself. The focus of these extensions is data storage, retrieval, and management, and they lack the advanced capabilities for geographic editing, mapping, and analysis. Consequently, they must be used in conjunction with a conventional GI system except in the case of the simplest query-focused applications. Figure 9.1 shows how GI systems and DBMS software can work together in a generalized way and some of the tasks best carried out by each system.

## 9.3 Storing Data in DBMS Tables

The lowest level of user interaction with a geographic database is usually the object class (also sometimes called a layer or feature class), which is an organized collection of data on a particular theme (e.g., all pipes in a water network, all soil polygons in a river basin, or all elevation values in a terrain surface). Object classes are stored in standard database tables. A table is a two-dimensional array of rows and columns. Each object class is stored as a single database table in a DBMS. Table rows contain objects (instances of object classes, e.g., data for a single pipe), and the columns contain object properties, or attributes as they are frequently called (Figure 9.2: see also Figure 9.10 as a conceptual example). The data stored at individual row, column intersections are usually referred to as values. Geographic database tables are distinguished from nongeographic tables by the presence of a geometry column (often called the shape column). To save space and improve performance, the actual coordinate values may be stored in a highly compressed binary form.

Figure 9.1 The roles of GI systems and DBMS.

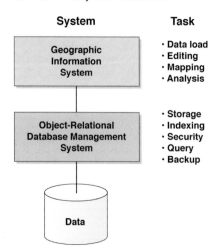

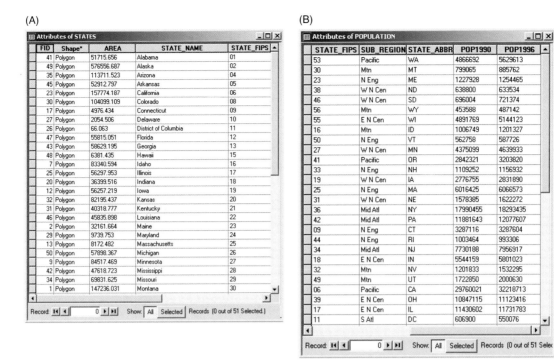

(A)

| FID | Shape* | AREA | STATE_NAME | STATE_FIPS |
|---|---|---|---|---|
| 41 | Polygon | 51715.656 | Alabama | 01 |
| 49 | Polygon | 576556.687 | Alaska | 02 |
| 35 | Polygon | 113711.523 | Arizona | 04 |
| 45 | Polygon | 52912.797 | Arkansas | 05 |
| 23 | Polygon | 157774.187 | California | 06 |
| 30 | Polygon | 104099.109 | Colorado | 08 |
| 17 | Polygon | 4976.434 | Connecticut | 09 |
| 27 | Polygon | 2054.506 | Delaware | 10 |
| 26 | Polygon | 66.063 | District of Columbia | 11 |
| 47 | Polygon | 55815.051 | Florida | 12 |
| 43 | Polygon | 58629.195 | Georgia | 13 |
| 48 | Polygon | 6381.435 | Hawaii | 15 |
| 7 | Polygon | 83340.594 | Idaho | 16 |
| 25 | Polygon | 56297.953 | Illinois | 17 |
| 20 | Polygon | 36399.516 | Indiana | 18 |
| 12 | Polygon | 56257.219 | Iowa | 19 |
| 32 | Polygon | 82195.437 | Kansas | 20 |
| 31 | Polygon | 40318.777 | Kentucky | 21 |
| 46 | Polygon | 45835.898 | Louisiana | 22 |
| 2 | Polygon | 32161.664 | Maine | 23 |
| 29 | Polygon | 9739.753 | Maryland | 24 |
| 13 | Polygon | 8172.482 | Massachusetts | 25 |
| 50 | Polygon | 57898.367 | Michigan | 26 |
| 9 | Polygon | 84517.469 | Minnesota | 27 |
| 42 | Polygon | 47618.723 | Mississippi | 28 |
| 34 | Polygon | 69831.625 | Missouri | 29 |
| 1 | Polygon | 147236.031 | Montana | 30 |

Record: 0  Show: All Selected  Records (0 out of 51 Selected.)

(B)

| STATE_FIPS | SUB_REGION | STATE_ABBR | POP1990 | POP1996 |
|---|---|---|---|---|
| 53 | Pacific | WA | 4866692 | 5629613 |
| 30 | Mtn | MT | 799065 | 885762 |
| 23 | N Eng | ME | 1227928 | 1254465 |
| 38 | W N Cen | ND | 638800 | 633534 |
| 46 | W N Cen | SD | 696004 | 721374 |
| 56 | Mtn | WY | 453588 | 487142 |
| 55 | E N Cen | WI | 4891769 | 5144123 |
| 16 | Mtn | ID | 1006749 | 1201327 |
| 50 | N Eng | VT | 562758 | 587726 |
| 27 | W N Cen | MN | 4375099 | 4639933 |
| 41 | Pacific | OR | 2842321 | 3203820 |
| 33 | N Eng | NH | 1109252 | 1156932 |
| 19 | W N Cen | IA | 2776755 | 2831890 |
| 25 | N Eng | MA | 6016425 | 6066573 |
| 31 | W N Cen | NE | 1578385 | 1622272 |
| 36 | Mid Atl | NY | 17990455 | 18293435 |
| 42 | Mid Atl | PA | 11881643 | 12077607 |
| 09 | N Eng | CT | 3287116 | 3287604 |
| 44 | N Eng | RI | 1003464 | 993306 |
| 34 | Mid Atl | NJ | 7730188 | 7956917 |
| 18 | E N Cen | IN | 5544159 | 5801023 |
| 32 | Mtn | NV | 1201833 | 1532295 |
| 49 | Mtn | UT | 1722850 | 2000630 |
| 06 | Pacific | CA | 29760021 | 32218713 |
| 39 | E N Cen | OH | 10847115 | 11123416 |
| 17 | E N Cen | IL | 11430602 | 11731783 |
| 11 | S Atl | DC | 606900 | 550076 |

Record: 0  Show: All Selected  Records (0 out of 51 Selec

(C)

| FID | Shape* | AREA | STATE_NAME | STATE_FIPS | SUB_REGION | STATE_ABBR | POP1990 | POP1996 |
|---|---|---|---|---|---|---|---|---|
| 0 | Polygon | 67286.875 | Washington | 53 | Pacific | WA | 4866692 | 5629613 |
| 1 | Polygon | 147236.031 | Montana | 30 | Mtn | MT | 799065 | 885762 |
| 2 | Polygon | 32161.664 | Maine | 23 | N Eng | ME | 1227928 | 1254465 |
| 3 | Polygon | 70810.156 | North Dakota | 38 | W N Cen | ND | 638800 | 633534 |
| 4 | Polygon | 77193.625 | South Dakota | 46 | W N Cen | SD | 696004 | 721374 |
| 5 | Polygon | 97799.492 | Wyoming | 56 | Mtn | WY | 453588 | 487142 |
| 6 | Polygon | 56088.066 | Wisconsin | 55 | E N Cen | WI | 4891769 | 5144123 |
| 7 | Polygon | 83340.594 | Idaho | 16 | Mtn | ID | 1006749 | 1201327 |
| 8 | Polygon | 9603.218 | Vermont | 50 | N Eng | VT | 562758 | 587726 |
| 9 | Polygon | 84517.469 | Minnesota | 27 | W N Cen | MN | 4375099 | 4639933 |
| 10 | Polygon | 97070.750 | Oregon | 41 | Pacific | OR | 2842321 | 3203820 |
| 11 | Polygon | 9259.514 | New Hampshire | 33 | N Eng | NH | 1109252 | 1156932 |
| 12 | Polygon | 56257.219 | Iowa | 19 | W N Cen | IA | 2776755 | 2831890 |
| 13 | Polygon | 8172.482 | Massachusetts | 25 | N Eng | MA | 6016425 | 6066573 |
| 14 | Polygon | 77328.336 | Nebraska | 31 | W N Cen | NE | 1578385 | 1622272 |
| 15 | Polygon | 48560.578 | New York | 36 | Mid Atl | NY | 17990455 | 18293435 |
| 16 | Polygon | 45359.238 | Pennsylvania | 42 | Mid Atl | PA | 11881643 | 12077607 |
| 17 | Polygon | 4976.434 | Connecticut | 09 | N Eng | CT | 3287116 | 3287604 |
| 18 | Polygon | 1044.850 | Rhode Island | 44 | N Eng | RI | 1003464 | 993306 |
| 19 | Polygon | 7507.302 | New Jersey | 34 | Mid Atl | NJ | 7730188 | 7956917 |
| 20 | Polygon | 36399.516 | Indiana | 18 | E N Cen | IN | 5544159 | 5801023 |
| 21 | Polygon | 110667.297 | Nevada | 32 | Mtn | NV | 1201833 | 1532295 |
| 22 | Polygon | 84870.187 | Utah | 49 | Mtn | UT | 1722850 | 2000630 |
| 23 | Polygon | 157774.187 | California | 06 | Pacific | CA | 29760021 | 32218713 |
| 24 | Polygon | 41192.863 | Ohio | 39 | E N Cen | OH | 10847115 | 11123416 |
| 25 | Polygon | 56297.953 | Illinois | 17 | E N Cen | IL | 11430602 | 11731783 |
| 26 | Polygon | 66.063 | District of Columbia | 11 | S Atl | DC | 606900 | 550076 |
| 27 | Polygon | 2054.506 | Delaware | 10 | S Atl | DE | 666168 | 724890 |

Record: 0  Show: All Selected  Records (0 out of 51 Selected.)  Options ▾

**Figure 9.2** Parts of GI systems database tables for U.S. states: (A) STATES table; (B) POPULATION table; (C) joined table—COMBINED STATES and POPULATION.

**Relational databases are made up of tables. Each geographic class (layer) is stored as a table.**

Tables are joined using common row/column values or keys, as they are known in the database world. Figure 9.2 shows parts of tables containing data about U.S. states. The STATES table (Figure 9.2A) contains the geometry (in the SHAPE field) and some basic attributes, an important one being a unique STATE FIPS (STATE_FIPS [Federal Information Processing Standard] code) identifier. The POPULATION table (Figure 9.2B) was created entirely independently, but also has a unique identifier column called STATE_FIPS. Using standard database tools, the two tables can be joined based on the common STATE_FIPS identifier column (the key) to create a third table, COMBINED STATES and POPULATION (Figure 9.2C). Following the join, these can be treated as a single table for all GI systems operations such as query, display, and analysis. There is additional discussion of relational joins and their generalization to spatial joins in Section 13.2.2.

**Database tables can be joined to create new views of the database.**

In a groundbreaking description of the relational model that underlies the vast majority of the world's databases, in 1970 Ted Codd of IBM defined a series of rules for the efficient and effective design of database table structures. The heart of Codd's idea was that the best relational databases are made up of simple, stable tables that follow five principles:

1. There is only one value in each cell at the intersection of a row and column.
2. All values in a column are about the same subject.
3. Each row is unique.
4. There is no significance to the sequence of columns.
5. There is no significance to the sequence of rows.

Figure 9.3A shows a database table of land parcels for tax assessment that contradicts several of Codd's principles. Codd suggests a series of transformations called normal forms that successively improve the simplicity and stability and reduce the redundancy of database tables (thus reducing the risk of editing conflicts) by splitting them into subtables that are rejoined at query time. Unfortunately, joining large tables is computationally expensive and can result in complex database designs that are difficult to maintain. For this reason, nonnormalized table designs are often used in GI systems.

Figure 9.3B is a cleansed version of 9.3A that has been entered into a GI system DBMS: There is now only one value in each cell (Date and AssessedValue are now separate columns); missing values have been added; an OBJECTID (unique system identifier) column has been added; and the potential confusion between Dave Widseler and D Widseler has been resolved. Figure 9.3C shows the same data after some normalization to make it suitable for use in a tax assessment application. The database now consists of three tables that can be joined using common keys. Figure 9.3C `Attributes of Tab10_3b` can be joined to Figure 9.3C `Attributes of Tab10_3a` using the common ZoningCode column, and Figure 9.3C `Attributes of Tab10_3c` can be joined using OwnersName to create Figure 9.3D. It is now possible to execute SQL queries against these joined tables as discussed in the next section.

---

**Figure 9.3** Tax assessment database: (A) raw data; (B) cleaned data in a GI systems DBMS (*continued*)

(A)

**Attributes of Tab10_1**

| OBJECTID* | ParcelNumb | OwnerNam | OwnerAddress | PostalCode | ZoningCode | ZoningType | DateAssessed | AssessedValue |
|---|---|---|---|---|---|---|---|---|
| 1 | 673-100 | Jeff Peters | 10 Railway Cuttings | 114390 | 2 | Residential | 2002 | 220000 |
| 2 | 673-101 | Joel Campbell | 1115 Center Place | 114390 | 2 | Residential | 2003 | 545500 |
| 3 | 674-100 | Dave Widseler | | 114391 | 3 | Commercial | 99 | 249000 |
| 4 | 674-100 | | 452 Diamond Plaza | 114391 | 3 | Commercial | 2000 | 275500 |
| 5 | 674-100 | D Widseler | 452 Diamond Plaza | 114391 | 3 | Commercial | 2001 | 290000 |
| 6 | 670-231 | Sam Camarata | 19 Big Bend Bld | 114391 | 2 | Residential | 2004 | 450575 |
| 7 | 674-112 | Chris Capelli | Hastings Barracks | 114392 | 2 | Residential | 2004 | 350000 |
| 8 | 674-113 | Sheila Sullivan | 10034 Endin Mansions | 114391 | 2 | Residential | 02 | 1005425 |

Record: 0  Show: All Selected  Records (0 out of 8 Selected.)  Options ▾

(B)

**Attributes of Tab10_2**

| OBJECTID* | ParcelNumb | OwnerNam | OwnerAddress | PostalCode | ZoningCode | ZoningType | DateAssessed | AssessedValue |
|---|---|---|---|---|---|---|---|---|
| 1 | 673-100 | Jeff Peters | 10 Railway Cuttings | 114390 | 2 | Residential | 2002 | 220000 |
| 2 | 673-101 | Joel Campbell | 1115 Center Place | 114390 | 2 | Residential | 2003 | 545500 |
| 3 | 674-100 | Dave Widseler | 452 Diamond Plaza | 114391 | 3 | Commercial | 1999 | 249000 |
| 4 | 674-100 | Dave Widseler | 452 Diamond Plaza | 114391 | 3 | Commercial | 2000 | 275500 |
| 5 | 674-100 | Dave Widseler | 452 Diamond Plaza | 114391 | 3 | Commercial | 2001 | 290000 |
| 6 | 670-231 | Sam Camarata | 19 Big Bend Bld | 114391 | 2 | Residential | 2004 | 450575 |
| 7 | 674-112 | Chris Capelli | Hastings Barracks | 114392 | 2 | Residential | 2004 | 350000 |
| 8 | 674-113 | Sheila Sullivan | 10034 Endin Mansions | 114391 | 2 | Residential | 2002 | 1005425 |

Record: 0  Show: All Selected  Records (0 out of 8 Selected.)  Options ▾

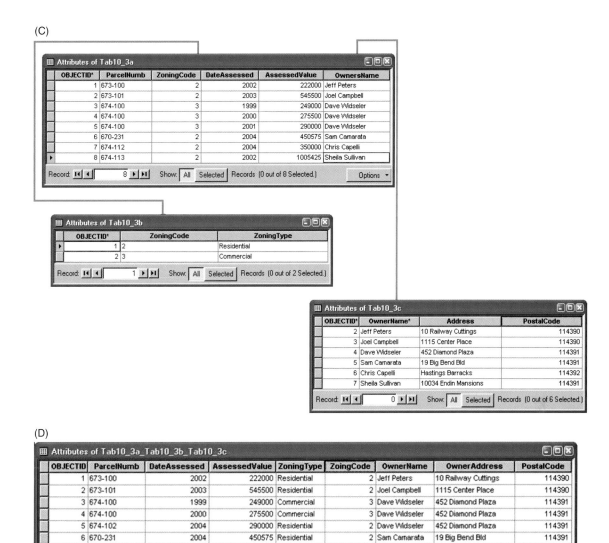

(C)

(D)

**Figure 9.3** (*continued*) (C) data partially normalized into three subtables; and (D) joined table.

## 9.4 SQL

The standard database query language adopted by virtually all mainstream databases is SQL (Structured or Standard Query Language: ISO Standard ISO/IEC 9075). There are many good background books and system implementation manuals on SQL, and so only brief details will be presented here. SQL may be used directly via an interactive command line interface; it may be compiled in a general-purpose programming language (e.g., C/C++/C#, Java, or Visual Basic); or it may be embedded in a graphical user interface (GUI). SQL is a set-based, rather than a procedural (e.g., Visual Basic) or object-oriented (e.g., Java or C#) programming language designed to retrieve sets (row and column combinations) of data from tables.

There are three key types of SQL statements: DDL (Data Definition Language), DML (Data Manipulation Language), and DCL (Data Control Language). A major revision of SQL in 2004 defined spatial types and functions as part of a multimedia extension called SQL/MM. A further revision in 2011 added temporal database queries.

The data in the database shown in Figure 9.3C may be queried to find parcels where the AssessedValue is greater than $300,000 and the ZoningType is Residential. This is an apparently simple query, but it requires three table joins to execute it. The SQL statements as implemented in the Microsoft Access DBMS are as follows:

```
SELECT Tab10_3a.ParcelNumb, Tab10_3c.Address,
    Tab10_3a.AssessedValue
```

```
FROM (Tab10_3b INNER JOIN Tab10_3a ON
    Tab10_3b.ZoningCode =
    Tab10_3a.ZoningCode) INNER JOIN Tab10_3c
    ON Tab10_3a.OwnersName =
    Tab10_3c.OwnerName
WHERE (((Tab10_3a.AssessedValue) >300000) AND
    ((Tab10_3b.ZoningType) ="Residential"));
```

The **SELECT** statement defines the columns to be displayed (the syntax is TableName.ColumnName). The **FROM** statement is used to identify and join the three tables (**INNER JOIN** is a type of join that signifies that only matching records in the two tables will be considered). The **WHERE** clause is used to select the rows from the columns using the constraints `(((Tab10_3a.AssessedValue) >300000) AND ((Tab10_3b.ZoningType) ="Residential"))`. The result of this query is shown in Figure 9.4. This triplet of **SELECT, FROM, WHERE** is the staple of SQL queries.

**SQL is the standard database query language. Today it has geographic capabilities.**

In SQL, Data Definition Language statements are used to create, alter, and delete relational database structures. The **CREATE TABLE** command is used to define a table, the attributes it will contain, and the primary key (the column used to identify records uniquely). For example, the SQL statement to create a table to store data about Countries, with two columns (name and shape (geometry)), is as follows:

```
CREATE TABLE Countries (
name          VARCHAR(200) NOT NULL PRIMARY
   KEY,
shape          POLYGON NOT NULL
CONSTRAINT  spatial reference
CHECK          (SpatialReference(shape)  = 14)
)
```

This SQL statement defines several table parameters. The name column is of type **VARCHAR** (variable character) and can store values up to 200 characters. Name cannot be null (**NOT NULL**); that is, it must have a value, and it is defined as the **PRIMARY KEY**, which

**Figure 9.4** Results of a SQL query against the tables in Figure 9.3C (see text for query and further explanation).

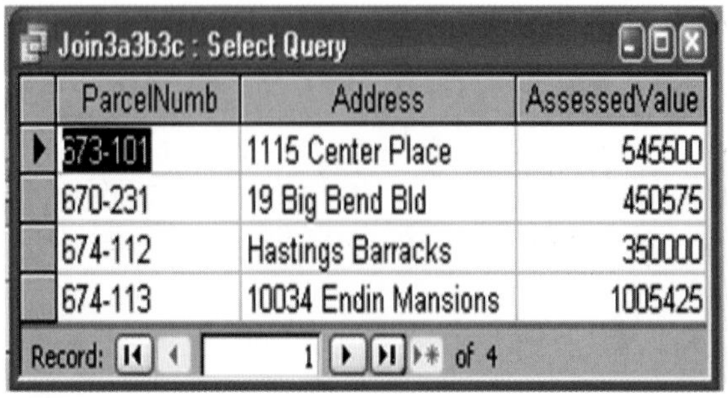
Join3a3b3c : Select Query

| ParcelNumb | Address | AssessedValue |
|---|---|---|
| 673-101 | 1115 Center Place | 545500 |
| 670-231 | 19 Big Bend Bld | 450575 |
| 674-112 | Hastings Barracks | 350000 |
| 674-113 | 10034 Endin Mansions | 1005425 |

Record: 1 of 4

means that its entries must be unique. The shape column is of type **POLYGON**, and it is defined as **NOT NULL**. It has an additional spatial reference constraint (projection), meaning that a spatial reference is enforced for all shapes (Type 14—this will vary by system, but could be Universal Transverse Mercator (UTM)—see Section 4.8.2).

Data can be inserted into this table using the SQL **INSERT** command:

```
INSERT INTO Countries
(Name, Shape) VALUES ('Kenya', Polygon
   ('((x y, x y, x y, x y) ,2))
```

Actual coordinates would need to be substituted for the *x*, *y* values. Several additions of this type would result in a table like the following:

| Name | Shape |
|---|---|
| Kenya | Polygon geometry |
| South Africa | Polygon geometry |
| Egypt | Polygon geometry |

Data Manipulation Language statements are used to retrieve and manipulate data. Objects with a size greater than 11,000 can be retrieved from the countries table using a **SELECT** statement:

```
SELECT Countries.Name,
FROM Countries
WHERE Area(Countries.Shape) > 11000
```

In this system, the area is computed automatically from the shape field using a DBMS function and does not need to be stored.

Data Control Language statements handle authorization access. The two main DCL keywords are **GRANT** and **REVOKE**, which authorize and rescind access privileges, respectively.

## 9.5 Geographic Database Types and Functions

Several attempts have been made to define a superset of geographic data types that can represent and process geographic data in databases. Unfortunately, space does not permit a review of them all here. This discussion will instead focus on the practical aspects of this problem and will be based on the widely accepted ISO (International Organization for Standards) and Open Geospatial Consortium (OGC) standards. The GI systems community working under the auspices of ISO and OGC has defined the core geographic types and functions to be used in a DBMS and accessed using the SQL query language (see Section 9.4 for a

**Figure 9.5** Geometry class hierarchy.

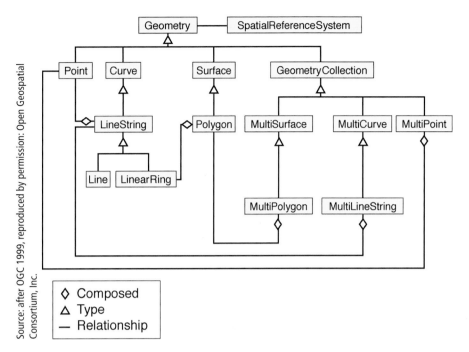

◇ Composed
△ Type
— Relationship

discussion of SQL). The geometry types are shown in Figure 9.5. In this hierarchy, the Geometry class is the root class. It has an associated spatial reference (coordinate system and projection, for example, Lambert Azimuthal Equal Area). The Point, Curve, Surface, and GeometryCollection classes are all subtypes of Geometry. The other classes (boxes) and relationships (lines) show how geometries of one type are aggregated from others (e.g., a LineString is a collection of Points).

For further explanation of how to interpret this object model diagram, see the discussion in Section 7.3.

According to this ISO/OGC standard, there are nine methods for testing spatial relationships between these geometric objects. Each method takes as input two geometries (collections of one or more geometric objects) and evaluates whether or not the relationship is true. Two examples of possible relations for all point, line, and area combinations are shown in Figure 9.6.

---

**Figure 9.6** Examples of possible relations for two geographic database operators: (A) Contains; and (B) Touches operators.

(A) **Contains**

Does the base geometry contain the comparison geometry?

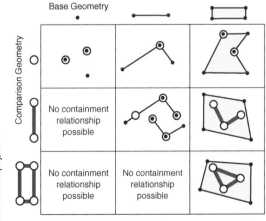

For the base geometry to contain the comparison geometry, it must be a superset of that geometry.

A geometry cannot contain another geometry of higher dimension.

(B) **Touches**

Does the base geometry touch the comparison geometry?

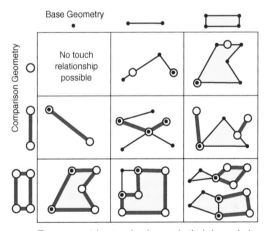

Two geometries touch when only their boundaries intersect.

In the case of the point–point Contain combination (upper left square in Figure 9.6A), two comparison geometry points (big circles) are contained in the set of base geometry points (small circles). In other words, the base geometry is a superset of the comparison geometry. In the case of the line–area Touches combination (middle right square in Figure 9.6B), the two lines touch the area because they intersect the area boundary. The full set of Boolean operators to test the spatial relationships between geometries is

- Equals—are the geometries the same?
- Disjoint—do the geometries share a common point?
- Intersects—do the geometries intersect?
- Touches—do the geometries intersect at their boundaries?
- Crosses—do the geometries overlap (can be geometries of different dimensions, for example, lines and polygons)?
- Within—is one geometry within another?
- Contains—does one geometry completely contain another?
- Overlaps—do the geometries overlap (must be geometries of the same dimension)?
- Relate—are there intersections between the interior, boundary, or exterior of the geometries?

Seven methods support spatial analysis on these geometries. Four examples of these methods are shown in Figure 9.7.

- Distance—determines the shortest distance between any two points in two geometries (Section 13.3.1).
- Buffer—returns a geometry that represents all the points whose distance from the geometry is less than or equal to a user-defined distance (Section 13.3.2).
- ConvexHull—returns a geometry representing the convex hull of a geometry (a convex hull is the smallest polygon that can enclose another geometry without any concave areas: think of stretching an elastic band around the polygon).
- Intersection—returns a geometry that contains just the points common to both input geometries.
- Union—returns a geometry that contains all the points in both input geometries.
- Difference—returns a geometry containing the points that are different between the two geometries.
- SymDifference—returns a geometry containing the points that are in either of the input geometries, but not both.

**Figure 9.7** Examples of spatial analysis methods on geometries: (A) Buffer; (B) Convex Hull (*continued*)

(A) **Buffer**

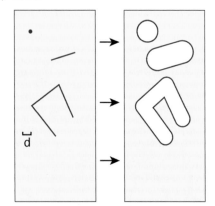

Given a geometry and a buffer distance, the buffer operator returns a polygon that covers all points whose distance from the geometry is less than or equal to the buffer distance.

(B) **Convex Hull**

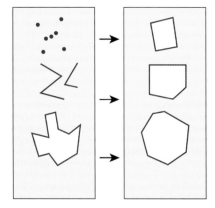

Given an input geometry, the convex hull operator returns a geometry that represents all points that are within all lines between all points in the input geometry.

A convex hull is the smallest polygon that wraps another geometry without any concave areas.

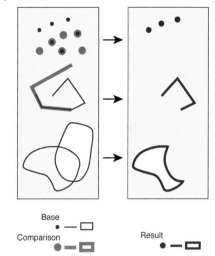

Source: after Zeiler and Murphy, 2010

The intersect operator compares a base geometry (the object from which the operator is called) with another geometry of the same dimension and returns a geometry that contains the points that are in both the base geometry and the comparison geometry.

(D) **Difference**

Base
Comparison
Result

The difference operator returns a geometry that contains points that are in the base geometry and subtracts points that are in the comparison geometry.

**Figure 9.7** (*continued*) (C) Intersection; (D) Difference.

## 9.6 Geographic Database Design

This section is concerned with the technical aspects of logical and physical geographic database design. Chapter 7 provides an overview of these subjects, and Chapters 16 to 18 discuss the organizational, strategic, and business issues associated with designing and maintaining a geographic database.

### 9.6.1 The Database Design Process

All GI system and DBMS software packages have their own core data model that defines the object types and relationships that can be used in an application (Figure 9.8). A DBMS package will define and implement a model for data types and access functions,

such as those implemented in SQL and discussed in Section 9.4. DBMSs are capable of dealing with simple geographic features and types (e.g., points, lines, areas, and sometimes also rasters) and relationships. A GI system can build on top of these simple feature types to create more advanced types and relationships (e.g., TINs, topologies, and feature-linked annotation geographic relationships; see Chapter 7 for a definition of these terms). The GI system types can be combined with domain data models that define specific object classes and relationships for specialist domains (e.g., water utilities, city parcel maps, and census geographies). Last, individual projects create specific physical instances of data models that are populated with data (objects for the specified object classes). For example, a city planning department may build a database of sewer lines that uses a water/wastewater (sewer) domain data model template that is built on top of core GI systems and DBMS models. Figure 7.2 and Section 7.1.2 show three increasingly abstract stages in data modeling: conceptual, logical, and physical. The result of a data-modeling exercise is a physical database design. This design will include specification of all data types and relationships, as well as the actual database configuration required to store them.

Database design involves the creation of conceptual, logical, and physical models in the six practical steps that are shown in Figure 9.9.

**Database design involves three key stages: conceptual, logical, and physical.**

**Figure 9.8** Four levels of data model available for use in GI systems projects.

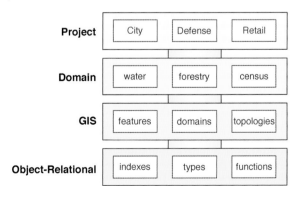

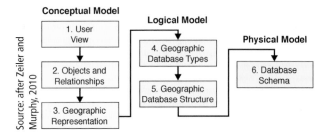

Figure 9.9 Stages in database design.

### 9.6.1.1 Conceptual Model

***Model the User's View***   This involves tasks such as identifying organizational functions (e.g., controlling forestry resources, finding vacant land for new buildings, and maintaining highways), determining the data required to support these functions, and organizing the data into groups to facilitate data management. This information can be represented in many ways; for example, a report with accompanying tables is often used.

***Define Objects and Their Relationships***   Once the functions of an organization have been defined, the object types (classes) and functions can be specified. The relationships between object types must also be described. This process usually benefits from the rigor of using object models and diagrams to describe a set of object classes and the relationships between them.

***Select Geographic Representation***   Choosing the types of geographic representation (discrete object or continuous field; see Section 3.5) will have profound implications for the way a database is used, and so it is a critical database design task. It is, of course, possible to change between representation types, but this is computationally expensive and results in loss of information.

### 9.6.1.2 Logical Model

***Match to Geographic Database Types***   This stage involves matching the object types to be studied to specific data types (point, line, area, georaster, etc.) supported by the GI systems that will be used to create and maintain the database. Because the data model of the GI system is usually independent of the actual storage mechanism (i.e., it could be implemented in Oracle, PostGIS, Microsoft Access, or a proprietary file system), this activity is defined as a logical modeling task.

***Organize Geographic Database Structure***   This stage includes tasks such as defining topological associations, specifying rules and relationships, and assigning coordinate systems.

### 9.6.1.3 Physical Model

***Define Database Schema***   The final stage is definition of the actual physical database schema that will hold the database data values. This is usually created using the DBMS software's Data Definition Language. The most popular of these is SQL with geographic extensions (see Section 9.4), although some nonstandard variants also exist in older DBMSs.

## 9.7 Structuring Geographic Information

Once data have been captured in a geographic database according to a schema defined in a geographic data model, it is often desirable to perform some structuring and organization in order to support efficient query, analysis, and mapping. There are two main structuring techniques relevant to geographic databases: topology creation and indexing.

### 9.7.1 Topology Creation

The subject of topology was covered in Section 7.2.3.2 from a conceptual data modeling perspective and is revisited here in the context of databases where the discussion focuses on the two main approaches to structuring and storing topology in a GI system DBMS.

Topology can be created for vector datasets using either batch or interactive techniques. Batch topology builders are required to handle CAD, survey, simple feature, and other unstructured vector data imported from nontopological systems. Creating topology is usually an iterative process because it is seldom possible to resolve all data problems during the first pass, and manual editing is required to make corrections. Some typical problems that may arise are shown in Figures 8.11, 8.12, and 8.13 and are discussed in Section 8.3.2.2. Interactive topology creation is performed dynamically at the time objects are added to a database using GI system editing software. For example, when adding water pipes using interactive vectorization tools (see Section 8.3.2.1), before each object is committed to the database topological connectivity can be checked to see if the object is valid (that is, it conforms to some preestablished rules for database object connectivity).

Two database-oriented approaches have emerged in recent years for storing and managing topology: normalized and physical. The normalized model focuses on the storage of an arc–node data structure. It is said to be normalized because each object is decomposed into individual topological primitives for storage in a database and then subsequent reassembly when a query is posed. For example, polygon objects are assembled at query time by joining tables containing the line segment geometries and topological primitives that define topological relationships (see Section 7.2.3.2 for a conceptual description of this process). In the physical model, topological primitives are not stored in the database, and the entire geometry is stored together for each object. Topological relationships are then computed on-the-fly whenever they are required by client applications.

Figure 9.10 is a simple example of a set of database tables that store a dataset according to the normalized topology model. The dataset (sketch in top left corner) comprises three feature classes (Parcels, Buildings, and Walls) and is implemented in three tables. In this example the three feature class tables have only one column (ID) and one row (a single instance of each feature in a feature class). The Nodes, Edges, and Faces tables store the points, polylines, and polygons for the dataset and some of the topology (for each Edge the From–To connectivity and the Faces on the Left–Right in the direction of digitizing). Three other tables (Parcel X Face, Wall X Edge, and Building X Face) store the cross-references for assembling Parcels, Buildings, and Walls from the topological primitives.

The normalized approach offers a number of advantages to GI systems users. Many people find it comforting to see topological primitives actually stored in the database. This model has many similarities to the arc–node conceptual topology model (see Section 7.2.3.2), and so it is familiar to many users and easy to understand. The geometry is only stored once, thus minimizing database size and avoiding "double digitizing" slivers (Section 8.3.2.2). Finally, the normalized approach easily lends itself to access via a SQL API. Unfortunately, there are three main disadvantages

associated with the normalized approach to database topology: query performance, integrity checking, and update performance/complexity.

Query performance suffers because queries to retrieve features from the database (the most common type of query) must combine data from multiple tables. For example, to fetch the geometry of Parcel P1, a query must combine data from four tables (Parcels, Parcel X Face, Faces, and Edges) using complex geometry/topology logic. The more tables that have to be visited, and especially the more that have to be joined, the longer it will take to process a query.

The standard referential integrity rules in DBMS are very simple and have no provision for complex topological relationships of the type defined here. There are many pitfalls associated with implementing topological structuring using DBMS techniques such as stored procedures (program code stored in a database), and in practice systems have resorted to implementing the business logic to manage things like topology external to the DBMS in a middle-tier application server or a set or server-side program code.

Updates are similarly problematic because changes to a single feature will have cascading effects on many tables. This raises attendant performance (especially scalability, that is, large numbers of concurrent queries) and integrity issues. Moreover, it is uncertain how multiuser updates will be handled that entail long transactions with design alternatives (see Sections 9.9 and 9.9.1 for coverage of these two important topics). For comparative purposes the same dataset used in the normalized model (Figure 9.10) is implemented using the physical model in Figure 9.11. In the physical model the three feature classes (Parcels, Buildings, and Walls) contain the same IDs, but differ significantly

**Figure 9.10** Normalized database topology model.

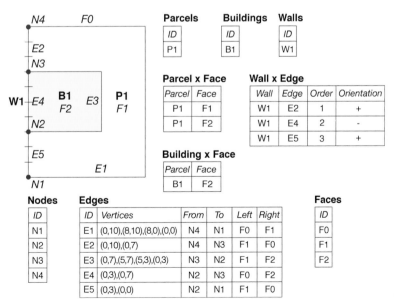

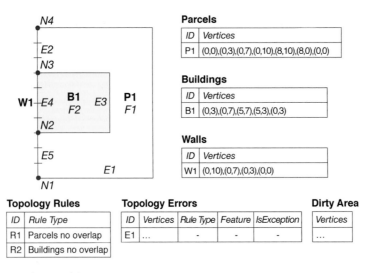

**Parcels**

| ID | Vertices |
|----|----------|
| P1 | (0,0),(0,3),(0,7),(0,10),(8,10),(8,0),(0,0) |

**Buildings**

| ID | Vertices |
|----|----------|
| B1 | (0,3),(0,7),(5,7),(5,3),(0,3) |

**Walls**

| ID | Vertices |
|----|----------|
| W1 | (0,10),(0,7),(0,3),(0,0) |

**Topology Rules**

| ID | Rule Type |
|----|-----------|
| R1 | Parcels no overlap |
| R2 | Buildings no overlap |

**Topology Errors**

| ID | Vertices | Rule Type | Feature | IsException |
|----|----------|-----------|---------|-------------|
| E1 | ... | - | - | - |

**Dirty Area**

| Vertices |
|----------|
| ... |

**Figure 9.11** Physical database topology model.

in that they also contain the geometry for each feature. The only other things required to be stored in the database are the specific set of topology rules that have been applied to the dataset (e.g., parcels should not overlap each other, and buildings should not overlap with each other), together with information about known errors (sometimes users defer topology cleanup and commit data with known errors to a database) and areas that have been edited but not yet validated (had their topology (re)built).

The physical model requires that an external client or middle-tier application server is responsible for validating the topological integrity of datasets. Topologically correct features are then stored in the database using a structure that is much simpler than the normalized model. When compared to the normalized model, the physical model offers two main advantages of simplicity and performance. Because all the geometry for each feature is stored in the same table column/row value, and there is no need to store topological primitives and cross-references, this is a very simple model. The biggest advantage and the reason for the appeal of this approach is query performance. Most DBMS queries do not need access to the topology primitives of features, and so the overhead of visiting and joining multiple tables is unnecessary. Even when topology is required, it is faster to retrieve feature geometries and recompute topology outside the database than to retrieve geometry and topology from the database.

In summary, the normalized and physical topology models have both advantages and disadvantages. The normalized model is implemented in Oracle Spatial and can be accessed via a SQL API, making it easily available to a wide variety of users. The physical model is implemented in ArcGIS and offers fast update and query performance for high-end GI system applications. Esri has also implemented a long transaction and versioning model based on the physical database topology model.

## 9.7.2 Indexing

Geographic databases tend to be very large and geographic queries computationally expensive. As a result, geographic queries, such as finding all the customers (points) within a store trade area (polygon), can take a very long time (perhaps 10 to 100 seconds or more for a 50 million customer database). The point has already been made in Section 7.2.3.2 that topological structuring can help speed up certain types of queries such as adjacency and network connectivity. A second way to speed up queries is to index a database and use the index to find data records (database table rows) more quickly. A database index is logically similar to a book index; both are special organizations that speed up searching by allowing random instead of sequential access. A database index is, conceptually speaking, an ordered list derived from the data in a table. Using an index to find data reduces the number of computational tests that have to be performed to locate a given set of records. In DBMS jargon, indexes avoid expensive full-table scans (reading every row in a table) by creating an index and storing it as a table column.

**A database index is a special representation of information about objects that improves searching.**

Figure 9.12 presents a simple example of the standard DBMS one-dimensional B-tree (balanced tree) index that is found in most major commercial DBMSs. Without an index, a search of the original data in this table to guarantee finding any given value would involve 16 tests/read operations (one for each data point). The B-tree index orders the data

**B-Tree Indexed Data**

| Original Data | Level 1 | Level 2 | Level 3 |
|---|---|---|---|
| 1 | | | 1 |
| 13 | | 22 | 13 |
| 69 | | | 14 |
| 52 | | | 22 |
| 25 | | 36 | 25 |
| 26 | | | 26 |
| 71 | | | 31 |
| 36 | 36 | | 36 |
| 22 | | | 52 |
| 72 | | 68 | 53 |
| 67 | | | 67 |
| 68 | | | 68 |
| 14 | | | 69 |
| 70 | | | 70 |
| 31 | | | 71 |
| 53 | | | 72 |

Figure 9.12 An example of a B-tree index.

and splits the ordered list into buckets of a given size (in this example it is four and then two), and the upper value for the bucket is stored (it is not necessary to store the uppermost value). To find a specific value, such as 72, using the index involves a maximum of six tests: one at Level 1 (less than or greater than 36), one at Level 2 (less than or greater than 68), and a sequential read of four records at Level 3. The number of levels and buckets for each level can be optimized for each type of data. Typically, the larger the dataset, the more effective indexes are in retrieval performance.

Unfortunately, creating and maintaining indexes can be quite time consuming; this is especially an issue when the data are very frequently updated. Because indexes can occupy considerable amounts of disk space storage, requirements can be very demanding for large datasets. As a consequence,

many different types of index have been developed to try to alleviate these problems for both geographic and nongeographic data. Some indexes exploit specific characteristics of data to deliver optimum query performance, others are fast to update, and still others are robust across widely different data types.

The standard indexes in DBMSs, such as the B-tree, are one dimensional and are very poor at indexing geographic objects. Many types of geographic indexing techniques have been developed, some of which are experimental and have been highly optimized for specific types of geographic data. Research shows that even a basic spatial index will yield very significant improvements in spatial data access and that further refinements often yield only marginal improvements at the costs of simplicity, as well as speed of generation and update. Three main methods of general practical importance have emerged in GI systems: grid indexes, quadtrees, and R-trees.

### 9.7.2.1 Grid Index
A grid index can be thought of as a regular mesh placed over a layer of geographic objects. Figure 9.13 shows a layer that has three features indexed using two grid levels. The highest (coarsest) grid (Index 1) splits the layer into four equal-sized cells. Cell A includes parts of Features 1, 2, and 3; Cell B includes a part of Feature 3; and Cell C has part of Feature 1. There are no features on Cell D. The same process is repeated for the second-level index (Index 2). A query to locate an object searches the indexed list first to find the object and then retrieves the object geometry or attributes for further analysis (e.g., tests for overlap, adjacency, or containment with other objects on the same or another layer). These two tests are often referred to as primary and secondary filters. Secondary filtering, which involves geometric processing, is much more computationally expensive. The performance of an index is clearly related to the relationship between grid and

Figure 9.13 A multilevel grid geographic database index.

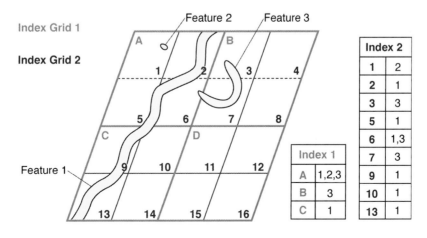

object sizes, and on object density. If the grid size is too large relative to the size of object, too many objects will be retrieved by the primary filter, and therefore a lot of expensive secondary processing will be needed. If the grid size is too small, any large objects will be spread across many grid cells, which is inefficient for draw queries (queries made to the database for the purpose of displaying objects on a screen). For data layers that have a highly variable object density (for example, administrative areas tend to be smaller and more numerous in urban areas in order to equalize population counts), multiple levels can be used to optimize performance. Experience suggests that three grid levels are normally sufficient for good all-round performance. Grid indexes are one of the simplest and most robust indexing methods. They are fast to create and update and can handle a wide range of types and densities of data. For this reason they have been quite widely used in commercial GI systems.

**Grid indexes are easy to create, can deal with a wide range of object types, and offer good performance.**

### 9.7.2.2 Quadtree Indexes

Quadtree is a generic name for several kinds of indexes that are built by recursive division of space into quadrants. In many respects, quadtrees are a special type of grid index. The difference here is that in quadtrees space is always recursively split into four quadrants based on data density. Quadtrees are data structures used for both indexing and compressing geographic database layers, although the discussion here will relate only to indexing. The many types of quadtrees can be classified according to the types of data that are indexed (points, lines, areas, surfaces, or rasters), the algorithm that is used to decompose (divide) the layer being indexed, and whether fixed or variable resolution decomposition is used.

In a point quadtree, space is divided successively into four rectangles based on the location of the points (Figure 9.14). The root of the tree corresponds to the region as a whole. The rectangular region is divided into four usually irregular parts based on the (x,y) coordinates of the first point. Successive points subdivide each new subregion into quadrants until all the points are indexed.

Region quadtrees are commonly used to index lines, areas, and rasters. The quadtree index is created by successively dividing a layer into quadrants. If a quadrant cell is not completely filled by an object, then it is subdivided again. Figure 9.15 is a quadtree of a woodland (red) and water (white) layer. Once a layer has been decomposed in this way, a linear index can be created using the search order shown in Figure 9.16. By reducing two-dimensional

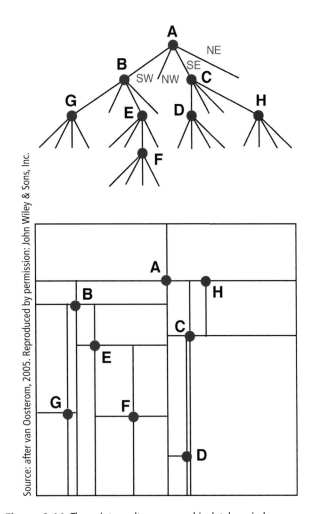

**Figure 9.14** The point quadtree geographic database index.

geographic data to a single linear dimension, a standard B-tree can be used to find data quickly.

**Quadtrees have found favor in GI systems because of their applicability to many types of data (both raster and vector), their ease of implementation, and their relatively good improvements in search performance.**

### 9.7.2.3 R-tree Indexes

R-trees group objects using a rectangular approximation of their location called a minimum bounding rectangle (MBR) or minimum enclosing rectangle (see Box 9.2). Groups of point, line, or area objects are indexed based on their MBR. Objects are added to the index by choosing the MBR that would require the least expansion to accommodate each new object. If the object causes the MBR to be expanded beyond some preset parameter, then the MBR is split into two new MBRs. This may also cause the parent MBR to

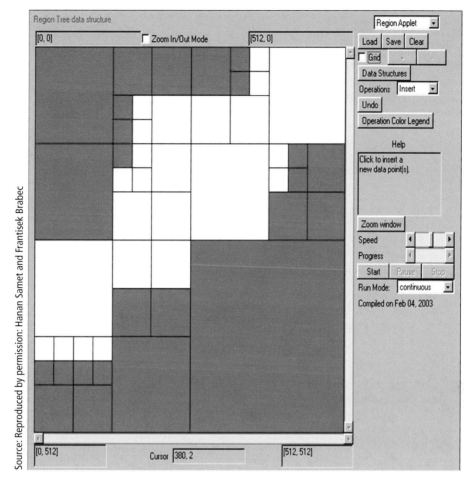

**Figure 9.15** The region quadtree geographic database index.

become too large, resulting in this also being split. The R-tree shown in Figure 9.17 has two levels. The lowest level contains three "leaf nodes"; the highest has one node with pointers to the MBR of the leaf nodes. The MBR is used to reduce the number of objects that need to be examined in order to satisfy a query.

**R-trees are popular methods of indexing geographic data because of their flexibility and excellent performance.**

R-trees are suitable for a range of geographic object types and can be used for multidimensional data. They offer good query performance, but the speed of update is not as fast as for grids and quadtrees. The spatial extensions to the IBM Informix DBMS and Oracle Spatial both use R-tree indexes.

**Figure 9.16** Linear quadtree search order.

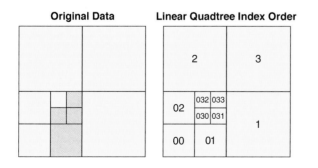

**Figure 9.17** The R-tree geographic database index.

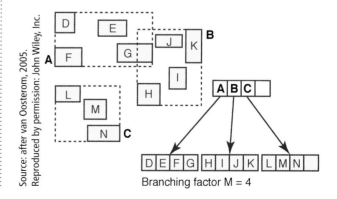

## Minimum Bounding Rectangle

Minimum bounding rectangles (MBRs) are very useful structures that are widely implemented in GI systems. An MBR essentially defines the smallest box whose sides are parallel to the axes of the coordinate system that encloses a set of one or more geographic objects. It is defined by the two coordinates at the bottom left (minimum $x$, minimum $y$) and the top right (maximum $x$, maximum $y$), as is shown in Figure 9.18.

MBRs can be used to generalize a set of data by replacing the geometry of each of the objects in the box with two pairs of coordinates defining the box. A second use is for fast searching. For example, all the area objects in a database layer that are within a given study area can be found by performing an area-on-area contains test (see Figure 9.18) for each object and the study area boundary. If the area objects have complex boundaries (as is normally the case in GI systems), this can be a very time-consuming task. A quicker approach is to split the task into two parts. First, screen out all the objects that are definitely in and definitely out by comparing their MBRs. Because very few coordinate comparisons are required, this is very fast. Then use the full geometry outline of the

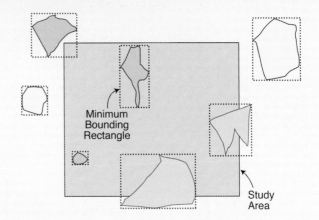

Figure 9.18 Area in area test using MBR. An MBR can be used to determine objects definitely in the study area (green) because of no overlap, definitely out (yellow), or possibly in (blue). Objects possibly in can then be analyzed further using their exact geometries. Note the red object that is actually completely outside, although the MBR suggests it may be partially within the study area.

remaining area objects to determine containment. This is computationally expensive for areas with complex geometries.

## 9.8 Editing and Data Maintenance

Editing is the process of making changes to a geographic database by adding new objects or changing existing objects as part of data load or database update and maintenance operations. A database update is any change to the geometry and/or attribute of one or more objects, a change to an object relationship, or any change to the database schema. A general-purpose geographic database will require many tools for tasks such as geometry and attribute editing, database maintenance (e.g., system administration and tuning), creating and updating indexes and topology, importing and exporting data, and georeferencing objects.

Contemporary GI systems come equipped with an extensive array of tools for creating and editing geographic object geometries and attributes. These tools form workflow tasks that are exposed within the framework of a WYSIWYG (what you see is what you get) editing environment. The objects are displayed using map symbology, usually in a

projected coordinate system space and frequently on top of "background" layers such as aerial photographs or satellite images, or street centerline files. Object coordinates can be digitized into a geographic database using many methods, including freehand digitizing on a digitizing table, on-screen heads-up vector digitizing by copying existing raster or vector sources, on-screen semiautomatic line following, automated feature recognition, and reading survey measurements from an instrument (e.g., GPS or Total Station) file (see Sections 4.9 and 8.2.2.1). The end result is always a layer of $(x,y)$ coordinates with optional $z$ and $m$ (height and attribute) values. Similar tools also exist for loading and editing raster data.

Data entered into the editor must be stored persistently in a file system or database, and access to the database must be carefully managed to ensure continued security and quality. The mechanism for managing edits to a file or database is called a transaction. There are many challenging issues associated with implementing multiuser access to geographic data stored in a DBMS, as discussed in the next section.

# 9.9 Multiuser Editing of Continuous Databases

For many years, one of the most challenging problems in GI data management was how to allow multiple users to edit the same continuous geographic database at the same time. In GI applications the objects of interest (geographic features) do not exist in isolation and are usually closely related to surrounding objects. For example, a tax parcel will share a common boundary with an adjacent parcel, and changes in one will directly affect the other; similarly, connected road segments in a highway network need to be edited together to ensure continued connectivity. It is straightforward to provide multiple users with concurrent read and query access to a continuous shared database, but more difficult to deal with conflicts and avoid potential database corruption when multiple users want write (update) access. However, solutions to both of these problems have been implemented in mainstream GI systems and DBMS. These solutions extend standard DBMS transaction models and provide a multiuser framework called versioning.

## 9.9.1 Transactions

A group of edits to a database, such as the addition of three new land parcels and changes to the attributes of a sewer line, is referred to as a "transaction." In order to protect the integrity of databases, transactions are atomic; that is, transactions are either completely committed to the database or they are rolled back (not committed at all). Many of the world's GI and non-GI databases are multiuser and transactional; that is, they have multiple users performing edit/update operations at the same time. For most types of databases, transactions take a very short (subsecond) time. For example, in the case of a banking system, a transfer from a savings account to a checking account takes perhaps 0.001 second. It is important that the transaction is coordinated between the accounts and that it is atomic; otherwise one account might be debited and the other not credited. Multiuser access to banking and similar systems is handled simply by locking (preventing access to) affected database records (table rows) during the course of the transaction. Any attempt to write to the same record is simply postponed until the record lock is removed after the transaction is completed. Because banking transactions, like many other transactions, take only a very short amount of time, users never even notice whether a transaction is deferred.

**A transaction is a group of changes that are made to a database as a coherent group. All the changes that form part of a transaction are either committed, or the database is rolled back to its initial state.**

Although some geographic transactions have a short duration (short transactions), many extend to hours, weeks, and months and are called long transactions. Consider, for example, the amount of time necessary to capture all the land parcels in a city subdivision (housing estate). This might take a few hours for an efficient operator working on a small subdivision, but an inexperienced operator working on a large subdivision might take days or weeks. This may cause three multiuser update problems. First, locking the whole or even part of a database to other updates for this length of time during a long transaction is unacceptable in many types of applications, especially those involving frequent maintenance changes (e.g., utilities and land administration). Second, if a system failure occurs during the editing, work may be lost unless there is a procedure for storing updates in the database. Also, unless the data are stored in the database, they are not easily accessible to others who would like to use them.

## 9.9.2 Versioning

Short transactions use what is called a pessimistic locking concurrency strategy. That is, it is assumed that conflicts will occur in a multiuser database with concurrent users and that the only way to avoid database corruption is to lock out all but one user during an update operation. The term *pessimistic* is used because this is a very conservative strategy, assuming that update conflicts will occur and that they must be avoided at all costs. An alternative to pessimistic locking is optimistic versioning, which allows multiple users to update a database at the same time. Optimistic versioning is based on the assumption that conflicts are very unlikely to occur, but if they do occur then software can be used to resolve them.

**The two strategies for providing multiuser access to geographic databases are pessimistic locking and optimistic versioning.**

Versioning sets out to solve the long transaction and pessimistic locking concurrency problem that was described earlier in this chapter. It also addresses a second key requirement peculiar to geographic databases—the need to support alternative representations of the same objects in the database. In some important geographic applications, it is a requirement to allow designers to create and maintain multiple object designs. For example, when designing a new housing subdivision, the water department manager may ask two designers to lay out alternative designs for a water system. The two designers would work concurrently to add objects to the same database layers, snapping to the same objects. At some point, they may wish to compare designs and perhaps create a third design based on parts of their two designs. While this design process is taking place, operational maintenance editors could be changing the same objects with which they are working. For example, maintenance updates resulting from new service connections or repairs to broken pipes will change the database and may affect the objects used in the new designs.

Figure 9.19 compares linear short transactions and branching long transactions as implemented in a versioned database. Within a versioned database, the different database versions are logical copies of their parents (base tables); that is, only the modifications (additions and deletions) are stored in the database (in version change tables). A query against a database version combines data from the base table with data in the version change tables. The process of creating two versions based on the same parent version is called branching. In Figure 9.19B, Version 4 is a branch from Version 2. Conversely, the process of combining two versions into one version is called merging. Figure 9.19B also illustrates the merging of Versions 6 and 7 into Version 8. A version can be updated at any time with any changes made in another version. Version reconciliation can be seen between Versions 3 and 5. Because the edits contained within Version 5 were reconciled with 3, only the edits in Versions 6 and 7 are considered when merging to create Version 8.

---

**Figure 9.19** Database transactions: (A) linear short transactions; (B) branching version tree.

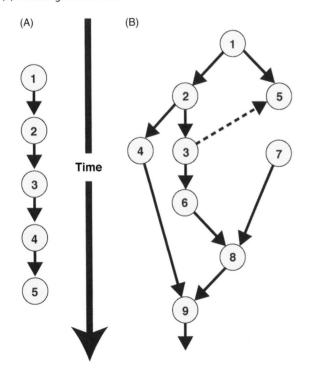

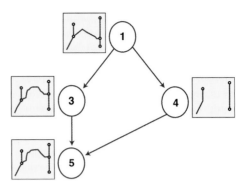

**Figure 9.20** Version reconciliation. For Version 5 the user chooses via the GUI the geometry edit made in Version 3 instead of 1 or 4.

There are no database restrictions or locks placed on the operations performed on each version in the database. The versioning database schema isolates changes made during the edit process. With optimistic versioning, it is possible for conflicting edits to be made within two separate versions, although normal working practice will ensure that the vast majority of edits made will not result in any conflicts (Figure 9.20). In the event that conflicts are detected, the data management software will handle them either automatically or interactively. If interactive conflict resolution is chosen, the user is directed to each feature that is in conflict and must decide how to reconcile the conflict. The GUI will provide information about the conflict and display the objects in their various states. For example, if the geometry of an object has been edited in the two versions, the user can display the geometry as it was in any of its previous states and then select the geometry to be added to the new database state.

## 9.10 Conclusion

Database management systems are now a vital part of large modern operational GI systems as well as related interdisciplinary fields such as computer science and engineering (see Box 9.3). They bring with them standardized approaches for storing and, more important, accessing and manipulating geographic data using the SQL query language. GI systems provide the necessary tools to load, edit, query, analyze, and display geographic data. DBMSs require a database administrator (DBA) to control database structure and security and to tune the database to achieve maximum performance. Innovative work in the GI system field has extended standard DBMS to store and manage geographic data and has led to the development of long transactions and versioning that have application across several fields.

## Oliver Günther, Computer Scientist

Oliver Günther (Figure 9.21) received a master's degree in industrial engineering from Karlsruhe Institute of Technology and M.S. and Ph.D. degrees in computer science from the University of California at Berkeley. After some time as junior faculty at UC Santa Barbara and its National Center for Geographic Information and Analysis (NCGIA), he set up a research group on environmental information systems at FAW Ulm (Germany). From 1993 until 2011 he held the chair of information systems at Humboldt-Universität zu Berlin. From 2005–2011 he also served as Dean of Humboldt's School of Business and Economics. In January 2012 he was appointed President of the University of Potsdam. Oliver's Ph.D. thesis, "Efficient Structures in Geometric Data Management" (Springer-Verlag, 1988), provided some theoretical foundations for spatial data management with a particular focus on index structures. His ACM Computing Surveys article on multidimensional access methods (1998, with Volker Gäde) remains one of the most-read references on the subject (Figure 9.22). His textbook, *Environmental Information Systems* appeared at about the same time

Figure 9.21  Oliver Günther, Computer Scientist.

(Springer-Verlag, 1998). In the late 1990s he contributed to the design of the first German Environmental Data Catalogue (UDK). His more recent work focuses on privacy and security issues. Oliver Günther has

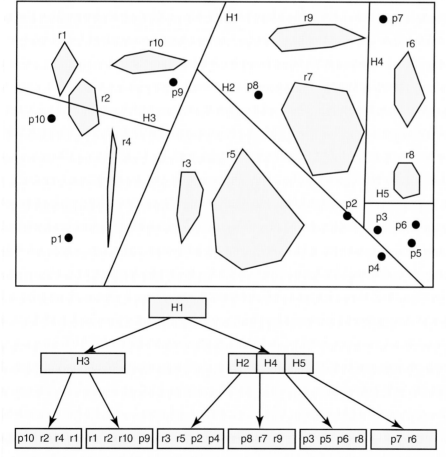

Figure 9.22  Cell tree data indexing structure. In the cell tree, each geometric object O is represented by a set of possibly overlapping convex cells whose union equals O. Cell tree nodes are split if they do not fit on one disk page anymore. The algorithm then looks for a splitting hyperplane that balances the entries approximately evenly among the two resulting subspaces and intersects a minimum number of cells. ▶

held visiting faculty positions at the European School of Management and Technology in Berlin, the École Nationale Supérieure des Télécommunications in Paris, at UC Berkeley, and the University of Cape Town. He has served as a consultant and board member to various government agencies and high-tech companies, including Poptel AG, Germany's first voice-over-IP company. In 2011–2012 he served as President of the German Informatics Society. As a high school student, Oliver was a two-time winner of Germany's National Mathematics Competition.

Looking back at the evolution of geographic and environmental information systems, Oliver remarks:

> When I first thought about GI systems and environmental applications, I differentiated between four distinct phases: data capture, data aggregation, data storage, and data analysis. Many of the problems we faced in the 1980s have now been solved. Data capture has changed fundamentally due to the ubiquitous availability of all kinds of sensors.

Data aggregation and also data analysis have profited from the excellent work of our data mining colleagues. Algorithms for routing and map generalization have also improved greatly, which led to the high-performance navigation systems which we got used to very quickly. Data storage has become more efficient, partly due to the smart data structures we developed back then. Now the ability to load databases into main memory will change the big picture one more time. Metadata management, finally, has become a crucial component of any major information system, taking care of search and access control. However, these technical advances must be balanced with rising concerns about privacy and security. Overall, GI systems and environmental applications have turned from a niche application to an essential part of our information infrastructure. It was nice to be able to contribute somewhat to these developments.

## Questions for Further Study

1. Identify a geographic database with multiple layers and draw a diagram showing the tables and the relationships between them. Which are the primary keys, and which keys are used to join tables? Does the database have a good relational design?

2. What are the advantages and disadvantages of storing geographic data in a DBMS?

3. Is SQL a good language for querying geographic databases?

4. Why are there multiple methods of indexing geographic databases?

## Further Reading

Date, C. J. 2005. *Database in Depth: Relational Theory for Practitioners.* Sebastopol, CA: O'Reilly Media, Inc.

Hoel, E., Menon, S., and Morehouse, S. 2003. Building a robust relational implementation of topology. In Hadzilacos, T., Manolopoulos, Y., Roddick, J. F., and Theodoridis, Y. (eds.). *Advances in Spatial and Temporal Databases. Proceedings of 8th International Symposium, SSTD 2003 Lecture Notes in Computer Science,* Vol. 2750.

OGC. 1999. *OpenGIS Simple Features Specification for SQL, Revision 1.1.* Available at www.opengis.org.

Samet, H. 2006. *The Foundations of Multidimensional and Metric Data Structures.* San Francisco: Morgan-Kaufmann.

Shekar, S. and Chawla, S. 2003. *Spatial Databases: A Tour.* Upper Saddle River, NJ: Prentice Hall.

van Oosterom, P. 2005. Spatial access methods. In Longley, P. A., Goodchild, M. F., Maguire, D. J., and Rhind, D. W. (eds.) *Geographic Information Systems: Principles, Techniques, Applications, and Management* (abridged ed.). Hoboken, NJ: Wiley. 385–400.

Yeung, A. K. W. and Hall, G. B. 2007. *Spatial Database Systems: Design, Implementation and Project Management* (GeoJournal Library). Dordrecht: Springer.

Zeiler, M. and Murphy, J. 2010. *Modeling Our World: The ESRI Guide to Geodatabase Concepts* (2nd ed.). Redlands, CA: ESRI Press.

# 10 The GeoWeb

Traditionally, the only practical way to solve a problem using a geographic information (GI) system was to assemble all the necessary parts in one place, on the user's desktop or on a convenient server. But today all the parts—the data and the software—can be accessed remotely. The GeoWeb is a vision for the future, already substantially realized, in which all these parts are able to operate together, in effect turning the Internet into a massive GI system that is accessible anywhere, at any time, from any device. The GeoWeb vision also includes real-time feeds of data from sensors and the ability to integrate these with other data. The Cloud provides a versatile, powerful, economical, and largely transparent host for online GI systems and services. This chapter introduces the basic concepts of the GeoWeb, describes current capabilities, and looks to a future in which GI systems and services are increasingly mobile and available everywhere.

## LEARNING OBJECTIVES

After studying this chapter you will be familiar with:

- How the parts of GI systems can be distributed instead of centralized.
- Geoportals, and the standards and protocols that allow remotely stored data to be discovered and accessed.
- The technologies that support real-time acquisition and distribution of geographic information.
- The service-oriented architectures and mashups that combine interoperable GI services from different Web sites.
- The capabilities of mobile devices, including mobile phones and wearable computers.
- The concepts of augmented and virtual reality.

## 10.1 Introduction

Early computers were extremely expensive, forcing organizations like universities to provide computing services centrally, from a single site, and to require users to come to the computing center to access its services. As the cost of computers fell, from millions of dollars in the 1960s, to hundreds of thousands in the late 1970s, to less than a thousand, it became possible for departments, small groups, and finally individuals to own computers and to install them on their desktops. In the past five years desktop and mobile phone technologies have converged, allowing many advanced services to be offered through a smartphone powered by a central processor with

the power and capacity of the super-computers of 20 years ago. Computers are being embedded in familiar devices such as car and entertainment systems, suggesting a future in which computing will be everywhere and to an increasing extent invisible to the user. To get there, however, will require extensive research, both on the technical issues and on the social implications. Box 10.1 describes the work of Yu Liu, a leading Chinese researcher in GI science, whose recent research illustrates the potential of the GeoWeb for gaining an improved understanding of human geography.

In Figure 1.12 we identified the six component parts of a GI system as its hardware, software, data, users, procedures, and network. This chapter describes how the network, to which almost all

### Yu Liu, leading GeoWeb researcher.

Yu Liu (Figure 10.1) is professor of GI science and director of the Geosoft Lab at the Institute of Remote Sensing and Geographical Information Systems, Peking University. His research interests span a wide spectrum of GI science topics, including digital gazetteers, geographic information retrieval, spatial relations, and uncertainty. Each of these topics can be applied toward extracting geographical information from textual documents, especially Web pages. In a recent study, he developed a method that uses the number of toponym cooccurrences retrieved from Web pages to conduct spatial analyses, demonstrating that Web pages can serve as a low-cost data source for geographical applications (Figure 10.2).

In the coming Big Data era, Yu also recognizes the high potential of geospatial Big Data (e.g., volunteered geographical information) for both GI systems and GI science. The broad spread of information and communication technologies has provided a wide range of individual-level and spatiotemporally tagged data sources (Figure 10.3). It thus becomes possible to investigate human mobility patterns and social networks from social media data, mobile phone data, taxi trajectory data, and so on. Recently the work of his group has focused on data mining and knowledge discovery from big data. He believes that such research will provide an innovative perspective on traditional geographical

Courtesy: Yu Liu

**Figure 10.1** Yu Liu, Peking University.

topics, including spatial interaction and distance-decay effects. Revealing the underlying geographical impacts of observed patterns leads to a better understanding of our socioeconomic environments.

Yu received a BS in geography (1994) and a PhD in software engineering (2003), both from Peking University. He led a team to develop a GI system platform 10 years ago, and although the project ultimately failed, this experience provided him with special insight into GI science research. Yu also coauthored a GIS textbook that is well known in China and participated in translating the second edition of this book to Chinese.

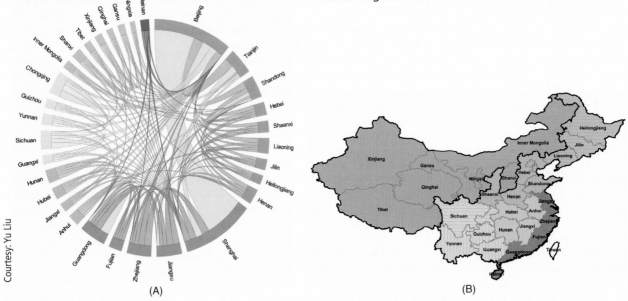

Courtesy: Yu Liu

(A)  (B)

**Figure 10.2** Relatedness of provinces of China, measured using toponym cooccurrences in Web documents. (A) Chord graph depicting the relatedness between provinces. For each city, only the five top relations are depicted. (B) Regionalization based on relatedness between provinces. Two-region clustering is delineated by heavy lines. Six-region clusters are represented by different colors, and the 14 regions are distinguished using different fill patterns. The clustering result identifies several well-known geographical zones, such as northeastern China (including Heilongjiang, Jilin, and Liaoning) and southwestern China (including Sichuan, Chongqing, Guizhou, Yunnan, and Guangxi). Note that Taiwan is blank due to data absence, and the South China Sea Islands are not shown for simplicity.

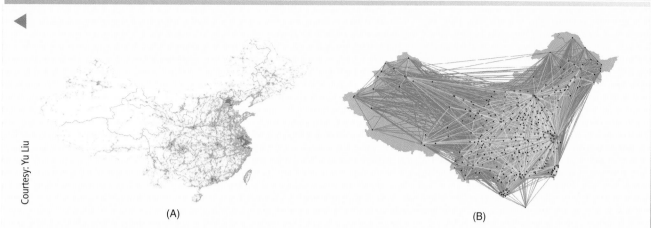

(A)

(B)

**Figure 10.3** Spatial distribution of check-in records collected from a social media Web site over the course of one year. The dataset covers 521,000 users and 370 cities in China. (A) Density map of all check-in points, clearly depicting the distributions of cities and transportation networks in China. (B) From the users' movements, we can obtain the interactions between the 370 cities. The red lines indicate stronger connections. Note that the South China Sea Islands are not shown for simplicity.

computers are now connected, has enabled a new vision of *online* GI systems, in which the component parts no longer need to be colocated. New technologies are making it possible for a GI project to be conducted not only on the desktop but also anywhere the user chooses to be, using data located anywhere on the network, and using services provided by remote sites on the network.

### Online GI systems allow the six component parts to be at different locations.

Although some of this chapter's discussion will be concerned with online services that are distributed across several independently managed sites, with all that implies about the need for standards and interoperability, much of the discussion will be about a more integrated architecture known as the Cloud. Cloud computing, as offered by companies such as Amazon or Microsoft, provides a massive online computing resource that is effectively transparent to the user, meaning that very little effort is needed to migrate from the user's own desktop to the Cloud: Software developed for the desktop is expected to operate equally easily in the Cloud. The advantages are obvious: Many users, say within one organization, are able to access the application simultaneously; data and software updates no longer need to be distributed to individual computers, but can be hosted in the Cloud; workloads can be distributed across the Cloud's resources much more cost-effectively, lowering costs to all users; and the responsibility for maintaining the Cloud no longer falls to the user's organization. Note that although this chapter refers frequently to "the Cloud," there are in reality many Clouds, some owned by organizations and restricted

to members, and others offered commercially for general purposes.

In Chapter 6 we discussed some aspects of this new concept of online GI systems and services. Many of those services are now available through smartphones, including wayfinding and mapping, and searches that are geographically targeted ("find the nearest coffee shop to my current location"). In Section 6.6.6 we termed these "handheld GI systems and services." Statistical agencies such as the U.S. Bureau of the Census or the U.S. Department of Agriculture are now using handheld devices widely for field data collection and routinely equip their field crews with such devices, allowing them to record data in the field and to upload the results wirelessly or back in the office; the field crews of utility companies use similar systems to record the locations and status of transformers, poles, and switches. Figures 10.4 and 10.5 show

**Figure 10.4** A soldier using a simple GIS in the field to collect data. The device on the pole is a GPS antenna, used to georeference data as they are collected.

**Figure 10.5** Recording data in the field, an increasingly common task in many occupations. In this case the device is used to record the location and other details of a delivery.

examples of such activities. Vendors are also offering various forms of online GI systems (Sections 6.6.2 and 6.6.3) that begin to offer services comparable to the analytic power of the traditional desktop.

Certain concepts need to be clarified at the outset. First, there are four distinct locations of significance to distributed and online GI systems and services:

- The location of the user and of the interface through which the user submits requests and obtains information, denoted by $U$ (e.g., the hand-held device in Figure 10.4).

- The location of the data being accessed by the user, denoted by $D$. Traditionally, data had to be moved to the user's computer before being used, but today's technology allows data to be accessed directly from data warehouses and archives, or indirectly through portals.

- The location where the data are processed, denoted by $P$. In Section 1.5.2.3 we introduced the concept of a GI service, a processing capability accessed at a remote site rather than provided locally by the user's desktop GI system. A Cloud will include many servers, perhaps distributed over a wide area.

- The area that is the focus of the GI project, or the *subject* location, denoted by $S$. All GI projects necessarily study some area, obtain data as a representation of the area, and apply GI system processes to those data.

In traditional GI projects three of these locations—$U$, $D$, and $P$—are the same because both the data and processing occur at the user's desktop. The subject location could be anywhere in the world, depending on the project. But in distributed and online GI projects there is no longer any need for $D$ and $P$ to be the same as $U$. Moreover, it is possible for the user to be located in the subject area $S$ and be able to see, touch, feel,

and even smell it, rather than being in a distant office. The GI interface might be held in the user's hand, or stuffed in a backpack, or mounted in a vehicle.

**In distributed and online GI projects, the user location and the subject location can be the same.**

Critical to distributed and online GI projects are the standards and specifications that make it possible for devices, data, and processes to operate together, or *interoperate*. Some of these are universal, such as ASCII, the standard for the coding of characters (Box 3.1), and XML, the extensible markup language that is used by many services on the Web. Others are specific to GI, and many of these standards have been developed through the Open Geospatial Consortium (OGC, www.opengeospatial.org), an organization set up to promote openness and interoperability in the world of GI. Among the many successes of OGC over the past decade are the simple feature specification, which standardizes many of the terms associated with primitive elements of GIS databases (polygons, polylines, points, etc.; see also Chapter 7); Geography Markup Language (GML), a version of XML that handles geographic features and enables open-format communication of geographic data; and specifications for Web services (Web Map Service, WMS; Web Feature Service, WFS; and Web Coverage Service, WCS) that allow a user's software to request data automatically from remote servers.

Distributed and online GI systems and services reinforce the notion that today's computer is not simply the device on the desk, with its hard drive, processor, and peripherals, but something more extended. The slogan "The Network Is the Computer" provided an early vision for at least one company—Sun Microsystems (www.sun.com)—and propelled many major developments in computing. The term *cyberinfrastructure* describes a contemporary approach to the conduct of science, relying on high-speed networks, massive processors, and distributed networks of sensors and data archives (www.cise.nsf.gov/sci/reports/toc.cfm). By integrating the world's computers it is possible to provide the kinds of massive computing power that are needed by projects such as SETI (the Search for Extra-Terrestrial Intelligence, www.seti.org), which processes terabytes of data per day in the search for anomalies that might indicate life elsewhere in the universe, and makes use of computer power wherever it can find it on the Internet (Figure 10.6). *Grid* computing is a generic term for such a fully integrated worldwide network of computers and data, and as we have already seen, the Cloud represents the convergence of a number of separate trends into a new vision for online computing.

**Figure 10.6** The power of this home computer would be wasted at night when its owner is sleeping, so instead its power has been "harvested" by a remote server and used to process signals from radio telescopes as part of a search for extraterrestrial intelligence.

In the early days of GI systems, all the data were derived from paper maps and described features on the Earth's surface that were largely static and unchanging. But an increasing amount of real-time data is becoming available, thanks in part to GPS, and is being fed over the Internet to users. It is now routine, for example, to keep track of the progress of an arriving flight before going to the airport (Figure 10.7), or to make available real-time data on

traffic congestion that are obtained by monitoring the speed of vehicles using GPS (see, for example, www.inrix.com). Real-time data are available from surveillance cameras, networks of environmental sensors, and an increasing range of small, cheap devices. RFID (radio frequency identification) allows the tracking of objects that have been implanted or tagged with small sensors and is now widely used in retailing, livestock management, and building construction. Many mobile phones carry RFID tags, as do new passports, items purchased from major retailers, vehicles using automatic highway toll gates, and even some oddly motivated individuals (Figure 10.8). A QR (quick response) code on an object (Figure 10.9) allows a sensor such as a mobile phone to immediately link to an Internet site where additional information about the object can be found. All of this raises the prospect that at some point in the future, it will be possible to know where *everything* is, or at least everything that is important in some specific area of human activity.

In Chapter 4 we introduced the concept of a *mashup*, which describes linking Web sites to create new services that none of the component sites can provide alone and has special significance when the linking is done through geographic location, when it is akin to overlay (see Section 13.2.4). Mashup, and the linking of services in general, is a key concept of the GeoWeb. Another is *service-oriented architecture* (SOA), the

**Figure 10.7** Tracking a flight: the real-time location of Alaska Airlines Flight 2551 from Portland to Santa Barbara, together with a base map and a depiction of current weather (light blue shows snow).

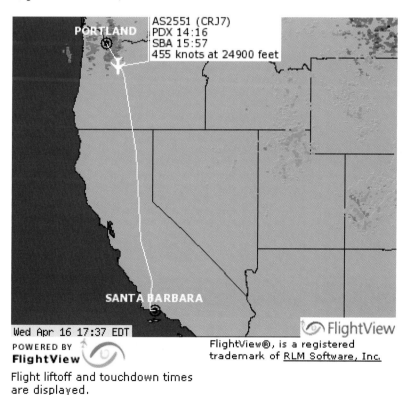

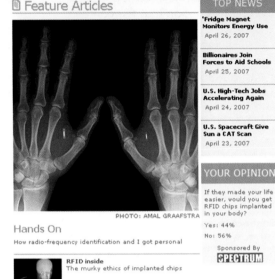

**Figure 10.8** RFID (radio frequency identification) uses small tags to identify and track various kinds of objects.

notion that any complex computer application can be decomposed into component parts, and that each of these parts can be provided by services that are distributed over the Internet. For example, suppose one is interested in developing an application to display current news stories, using a map interface that allows the user to zoom to any part of the world and to read about current events there. The service might combine a real-time feed of news stories, including text and images, from Web site A; sending the feed to a second site, B, that specializes in finding place-names in text and converting them to latitude and longitude; sending the results to a third site, C, that specializes in generating cartographic displays; and finally connecting the user with this third site. The locations of each of the three sites could be anywhere in the world that is connected to the Internet. SOA requires standards to ensure that the various services are interoperable and mechanisms for searching out services that meet specific requirements. A more elaborate example of SOA is given in Section 10.4.1.

**Figure 10.9** A QR code.

## 10.2 Distributing the Data

Since its popularization in the early 1990s, the Internet has had a tremendous and far-reaching impact on the accessibility of GI and on the ability of GI users to share datasets. As we saw in Chapter 8, a large and increasing number of Web sites offer GI for free, for sale, or for temporary use and also provide services that allow users to search for datasets that satisfy certain requirements.

In the past few years many private citizens have become involved in the distributed creation and dissemination of geographic information, in a process known as *volunteered geographic information* (VGI; see also Section 1.5.6). The notion that the content of the Web is increasingly created by its users has been termed Web 2.0 (introduced in Section 1.5.5), and today the Web is littered with thousands of sites that support the creation of VGI. Section 4.2 cited one prominent example, Wikimapia, in effect a place-name index created entirely by volunteers and with rich descriptive information. Another example is Open Street Map (OSM), one of many efforts around the world to enlist volunteers in the creation of open, free digital maps that is especially important in areas where such maps are not freely available (Table 1.4). VGI is becoming an important source of geographic information, particularly information that is difficult to obtain from any other source. On the other hand, its contributors have no authority, VGI sites do not generally conduct formal quality testing, and there is no equivalent of the trust that people place in traditional mapping sources (Chapter 17).

In effect, in a period of little more than two decades we have gone from a situation in which geographic data were available only in the form of printed maps from map libraries and retailers, to one in which petabytes (Table 1.2) of information are available for download and use at electronic speed (about 1.5 million CDs would be required to store 1 petabyte). For example, the NASA-sponsored EOSDIS (Earth Observing System Data and Information System; earthdata. nasa.gov) archives and distributes the geographic data from the EOS series of satellites, acquiring new data at over a terabyte per day, with an accumulated total of more than 1 petabyte at this site alone.

**Some GIS archives contain petabytes of data.**

The vision of the GeoWeb goes well beyond the ability to access and retrieve remotely located data, however, because it includes the concepts of *search*, *discovery*, and *assessment*. In the world of distributed GI systems, how do users search for data, discover their existence at remote sites, and assess their fitness for use? Four concepts are important in this respect: *object-level metadata*, *geolibraries*, *geoportals*, and *collection-level metadata*.

## 10.2.1 Object-Level Metadata

Strictly defined, metadata are data about data, and *object-level* metadata (OLM) describe the contents of a single dataset. We need information about data for many purposes, and OLM try to satisfy them all. First, we need OLM to automate the process of search and discovery over archives. In that sense OLM are similar to a library's catalog, which organizes the library's contents by author, title, and subject, and makes it easy for a user to find a book. But OLM are potentially much more powerful because a computer is more versatile than the traditional catalog in its potential for re-sorting items by a large number of properties, going well beyond author, title, and subject, and including geographic location. Second, we need OLM to determine whether a dataset, once discovered, will satisfy the user's requirements—in other words, to assess the fitness of a dataset for a given use. Does it have sufficient spatial resolution and acceptable quality? Such metadata may include comments provided by others who tried to use the data, or contact information for such previous users (users often comment that the most useful item of metadata is the phone number of the person who last tried to use the data). Third, OLM must provide the information needed to handle the dataset effectively. This may include technical specifications of format, or the names of software packages that are compatible with the data, along with information about the dataset's location and its volume. Finally, OLM may provide useful information on the dataset's contents. In the case of remotely sensed images, this may include the percentage of cloud obscuring the scene, or whether the scene contains particularly useful instances of specific phenomena, such as hurricanes.

**Object-level metadata are formal descriptions of datasets that satisfy many different requirements.**

OLM generalize and abstract the contents of datasets, and we would therefore expect the datasets to be smaller in volume than the data they describe. In reality, however, it is easy for the complete description of a dataset to generate a greater volume of information than the actual contents. OLM are also expensive to generate because they represent a level of understanding of the data that is difficult to assemble and require a high level of professional expertise. Generation of OLM for a geographic dataset can easily take much longer than it takes to catalog a book, particularly if it is necessary to deal with technical issues such as the precise geographic coverage of the dataset, its projection and datum details (Chapter 4), and other properties that may not be easily accessible. Thus the cost of OLM generation, and the incentives that motivate people to provide OLM, are important issues.

For metadata to be useful, it is essential that they follow widely accepted standards. If two users are to be able to share a dataset, they must both understand the rules used to create its OLM, so that the custodian of the dataset can first create the description and so that the potential user can understand it.

The most widely used standard for OLM is the U.S. Federal Geographic Data Committee's Content Standards for Digital Geospatial Metadata, or CSDGM, first published in 1993 and now the basis for many other standards worldwide. Box 10.2 lists some of its major features. As a *content standard*, CSDGM describes the items that should be in an OLM archive, but does not prescribe exactly how they should be formatted or structured. This allows developers to implement the standard in ways that suit their own software environments, but guarantees that one implementation will be understandable to another—in other words, that the implementations will be *interoperable*. For example, Esri's ArcGIS provides several formats for OLM, including one using the widely recognized XML standard and another using Esri's own format.

CSDGM was devised as a system for describing geographic datasets, and most of its elements make sense only for data that are accurately georeferenced and represent the spatial variation of phenomena over the Earth's surface. As such, its designers did not attempt to place CSGDM within any wider framework. But in the past decade a number of more broadly based efforts have also been directed at the metadata problem and at the extension of traditional library cataloging in ways that make sense in the evolving world of digital technology.

One of the best known of these is the Dublin Core (see Box 10.3), which is the outcome of an effort to find the minimum set of properties needed to support search and discovery for datasets in general, not only geographic datasets. Dublin Core treats both space and time as instances of a single property, "coverage," and unlike CSGDM does not lay down how such specific properties as spatial resolution, accuracy, projection, or datum should be described.

The principle of establishing a minimum set of properties is sharply distinct from the design of CSGDM, which was oriented more toward the capture of all knowable and potentially important properties of geographic datasets. Of direct relevance here is the problem of cost, and specifically the cost of capturing a full CSGDM metadata record. Although many organizations have wanted to make their data more widely available and have been driven to create OLM for their datasets, the cost of determining the full set of CSDGM elements is often highly discouraging. There is interest therefore in a concept of *light metadata*, a limited set of properties that is both comparatively cheap to capture and still useful to support search and discovery. Dublin Core represents this approach and thus sits at the opposite end of a spectrum from CSGDM. Every organization must somehow determine where its needs lie on this spectrum, which ranges from light and cheap to heavy and expensive.

**Light, or stripped-down, OLM provide a short but useful description of a dataset that is cheaper to create.**

## Technical Box (10.2)

### Major Features of the U.S. Federal Geographic Data Committee's Content Standards for Digital Geospatial Metadata

1. Identification Information—basic information about the dataset.

2. Data Quality Information—a general assessment of the quality of the dataset.

3. Spatial Data Organization Information—the mechanism used to represent spatial information in the dataset.

4. Spatial Reference Information—the description of the reference frame for, and the means to encode, coordinates in the dataset.

5. Entity and Attribute Information—details about the information content of the dataset, including the entity types, their attributes, and the domains from which attribute values may be assigned.

6. Distribution Information—information about the distributor of and options for obtaining the dataset.

7. Metadata Reference Information—information on the currentness of the metadata information and the responsible party.

8. Citation Information—the recommended reference to be used for the dataset.

9. Time Period Information—information about the date and time of an event.

10. Contact Information—identity of, and means to communicate with, person(s) and organization(s) associated with the dataset.

## The 15 Basic Elements of the Dublin Core Metadata Standard

1. TITLE. The name given to the resource by the CREATOR or PUBLISHER.

2. AUTHOR or CREATOR. The person(s) or organization(s) primarily responsible for the intellectual content of the resource.

3. SUBJECT or KEYWORDS. The topic of the resource, or keywords, phrases, or classification descriptors that describe the subject or content of the resource.

4. DESCRIPTION. A textual description of the content of the resource, including abstracts in the case of document-like objects or content description in the case of visual resources.

5. PUBLISHER. The entity responsible for making the resource available in its present form, such as a publisher, a university department, or a corporate entity.

6. OTHER CONTRIBUTORS. Person(s) or organization(s) in addition to those specified in the CREATOR element who have made significant intellectual contributions to the resource, but whose contribution is secondary to the individuals or entities specified in the CREATOR element.

7. DATE. The date the resource was made available in its present form.

8. RESOURCE TYPE. The category of the resource, such as home page, novel, poem, working paper, technical report, essay, or dictionary.

9. FORMAT. The data representation of the resource, such as text/html, ASCII, Postscript file, executable application, or JPEG image.

10. RESOURCE IDENTIFIER. String or number used to uniquely identify the resource.

11. SOURCE. The work, either print or electronic, from which this resource is delivered, if applicable.

12. LANGUAGE. Language(s) of the intellectual content of the resource.

13. RELATION. Relationship to other resources.

14. COVERAGE. The spatial locations and temporal durations characteristics of the resource.

15. RIGHTS MANAGEMENT. The content of this element is intended to be a link (a URL or other suitable URI as appropriate) to a copyright notice, a rights-management statement, or perhaps a server that would provide such information in a dynamic way.

## 10.2.2 Geolibraries and Geoportals

The use of digital technology to support search and discovery opens up many options that were not available in the earlier world of library catalogs and bookshelves. Books must be placed in a library on a permanent basis, and there is no possibility of reordering their sequence—but in a digital catalog it is possible to reorder the sequence of holdings in a collection almost instantaneously. So although a library's shelves are traditionally sorted by subject, it would be possible to re-sort them digitally by author name or title, or by any property in the OLM catalog. Similarly, the traditional card catalog allowed only three properties to be sorted—author, title, and subject—and discouraged sorting by multiple subjects. But the digital catalog can support any number of subjects.

Of particular relevance to GI users is the possibility of sorting a collection by the coverage properties: location and time. Both the spatial and temporal dimensions are continuous, so it is impossible to capture them in a single property analogous to author that can then be sorted numerically or alphabetically. In a digital system this is not a serious problem, and it is straightforward to capture the coverage of a dataset and to allow the user to search for datasets that cover an area or time of interest defined by the user. Moreover, the properties of location and time are not limited to geographic datasets because many types of information are associated with specific areas on the Earth's surface or with specific time periods. Searching based on location or time would enable users to find information about any place on the Earth's surface, or any time period—and to find reports, photographs, or even pieces of music, as long as they possessed geographical and temporal *footprints*.

The term *geolibrary* has been coined to describe digital libraries that can be searched for information about any user-defined geographic location. A U.S. National Research Council report (see Further Reading) describes the concept and its implementation, and many instances of geolibraries can be found on the Web.

**A geolibrary can be searched for information about a specific geographic location.**

Although geolibraries are very useful sources of data, it is often difficult for the user to predict *which* geolibrary is most likely to contain a given dataset, or the closest approximation to it. Rather than have to resort to trial and error, it would be much better if it were possible to search a single site for *all* datasets. This is the goal of a *geoportal*, which can be defined as a single point of entry to a distributed collection of geolibraries, with a single catalog, just as many networks of libraries provide a single *union catalog*

to all their holdings. There is a similarity here to the Web search engines such as Google, which provide a single point of entry for search over a large proportion of the entire Web. However, conventional search engines are not well designed for finding geographic datasets.

The Geospatial One-Stop, now incorporated into data.gov and accessible at geo.data.gov, is a good example of a geoportal, and in many ways it represents the state of the art in searching for geographic data. It was initiated by the U.S. government as a single point of entry to the holdings of government agencies, but it also catalogs many other geolibraries (though its coverage is essentially limited to the United States). A user visiting geo. data.gov is presented with several ways of specifying requirements (Figure 10.10) and is provided with a list of datasets that appear to satisfy them, together with links to the geolibraries that contain the data. Mechanisms are provided for handling datasets that have usage restrictions and may require licensing or the payment of fees. The site receives tens of thousands of visitors per day and

Figure 10.10 The U.S. government's geoportal geo.data.gov. The user is able to search through all datasets registered with the portal, selecting by geographic location and many other categories.

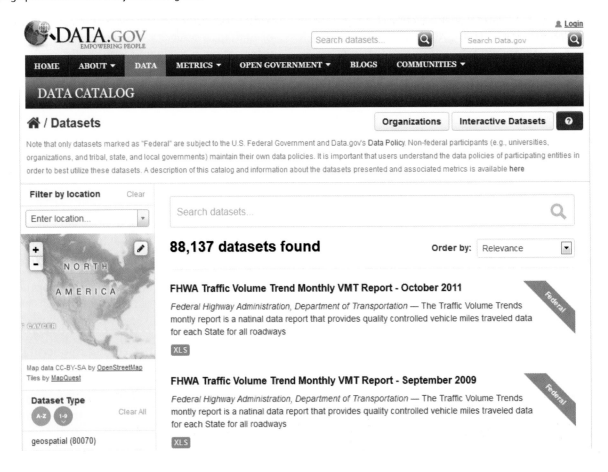

catalogs the contents of over a thousand geolibraries. It also supports automatic harvesting of catalog contents from geolibraries that are willing to contribute in this way.

Contemporary geoportals support several forms of use. The most traditional is the download, when an entire dataset is simply transferred to the user's hard drive and then incorporated in some form of analysis. But it is also possible to rely on many geoportals for simple online functions such as display, in which case the user needs no more than a Web browser. Finally, it may be possible to use a dataset "live" if the geoportal supports the appropriate standards. In this case the user's GI system behaves as if the geoportal's data were local, but sends requests over the Internet rather than to the local hard drive when specific data are needed in response to an operation. In this mode there is no need to download an entire dataset, and standards take care of such potential interoperability problems as differences of coordinate system or geographic coverage. Live data standards include the Open Geospatial Consortium's WMS, WFS, and WCS.

## 10.3 The Mobile User

The computer has become so much a part of our lives that for many people it is difficult to imagine life without it. We increasingly need computers to shop, to communicate with friends, to obtain the latest news, and to entertain ourselves. In the early days, the only place one could use computers was in a computing center, within a few meters of the central processor. Computing had extended to the office by the 1970s and to the home by the 1990s. The portable computers of the 1980s opened the possibility of computing in the garden, at the beach, or "on the road" in airports and airplanes. Wireless communication services such as WiFi (the wireless access technology based on the 802.11 family of standards) now allow broadband communication to the Internet from "hotspots" in many hotels, restaurants, airports, office buildings, and private homes (Figure 10.11). The range of mobile computing devices is also multiplying rapidly, from the relatively cumbersome but powerful laptop weighing several kilograms to the tablet, and the mobile

**Figure 10.11** Map of WiFi (802.11) wireless broadband "hotspots" within 1 mile of the White House (1600 Pennsylvania Avenue, Washington, DC) and using the T-Mobile Internet provider.

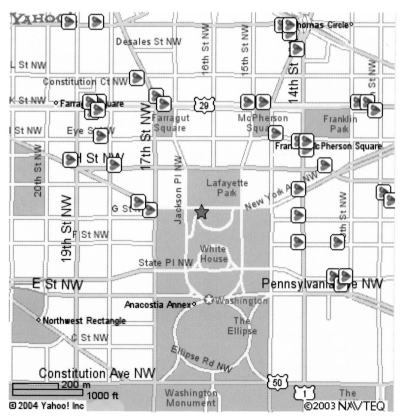

Courtesy: Krzysztof Janowicz

**Figure 10.12** Google's Glass, an example of a wearable, hands-free device, modeled here by Krzysztof Janowicz, University of California, Santa Barbara.

phone weighing a hundred or so grams. In many ways the ultimate endpoint of this progression is the *wearable* computer, a device that is worn on the body or fully embedded in the user's clothing, goes everywhere, and provides ubiquitous computing service. Such devices are already obtainable, including Google's Glass (Figure 10.12), which is being promoted as a hands-free phone, a camera, and a simple computing device.

> **Computing has moved from the computer center and is now close to, even inside, the human body. The ultimate form of mobile computing is the wearable computer.**

What possibilities do such systems create for GI? Of greatest interest is the case in which the user is situated in the subject location; that is, *U* is contained within *S*, and the system is used to analyze immediate surroundings. A convenient way to think about the possibilities is by comparing *virtual* reality with *augmented* reality.

## 10.3.1 Virtual Reality and Augmented Reality

One of the great strengths of GI systems is the window they provide on the world. A researcher in an office in Nairobi, Kenya, might use a GI system to obtain and analyze data on the effects of salinization in the Murray Basin of Australia, combining images from satellites with base topographic data, data on roads, soils, and population distributions, all obtained from different sources via the Internet. In doing so, the researcher would build a comprehensive picture of part of the world he or she might never have visited, all through the medium of digital geographic data and the GI system, and might learn almost as much through this window as from actually being there. The expense and time of traveling to Australia would be avoided, and the analysis could proceed almost instantaneously. Some aspects of the study area would be missing, of course—aspects of culture, for example, that can best be experienced by meeting with the local people (see the discussion of representation in Chapter 3).

Research environments such as this are termed *virtual realities* because they replace what humans normally gather through their senses—sight, sound, touch, smell, and taste—by presenting information from a database. In most GI system applications, such as those discussed in Section 12.4.3, only one of the senses, sight, is used to create this virtual reality, or VR. In principle it is possible to record sounds and store them in a GI database as attributes of features, but in practice very little use is made of any sensory channel other than vision. Moreover, in most GI system applications the view presented to the user is the view from *above*, even though our experience with looking at the world from this perspective is limited (for most of us, it is restricted to times when we requested a window seat in an airplane). GI systems have been criticized for what has been termed their *God's-eye*, or a *privileged* view by some writers, on the basis that it distances the researcher from the real conditions experienced by people on the ground.

> **Virtual environments can place the user in distant locations.**

More elaborate VR systems are capable of *immersing* the user by presenting the contents of a database in a three-dimensional environment using special eyeglasses or by projecting information onto walls surrounding the user and effectively *transporting* the user into the environment represented in the database. But even the standard personal computer is capable of creating remarkably close approximations to the physical appearance of the geographic landscape, and services such as Google Earth and Microsoft's Bing Maps have pushed the limits dramatically in recent years. Such *geobrowsers* or *virtual globes* create three-dimensional renderings of the Earth and rely on the powerful graphics capabilities now available in the standard office or home computer to support real-time movement—a "magic carpet ride" that even a child of 10 can learn to experience. Recent versions allow a smooth transition from the "God's-eye view" from above to street views assembled from hundreds of millions of

**Figure 10.13** View along East 64th Street in Manhattan generated in Google Maps using street-level photographs.

photographs (Figure 10.13), and the 3-D extension of Microsoft's Bing Maps presents photorealistic exteriors of buildings (Figure 10.14).

Roger Downs, professor of geography at Pennsylvania State University, uses Johannes Vermeer's famous painting popularly known as *The Geographer* (Figure 10.15) to make an important point about virtual realities. On the table in front of the figure is a map, taken to represent the geographer's window on a part of the world that happens to be of interest. But the subject figure is shown looking out the window, at the real world, perhaps because he needs the information he derives from his senses to understand the world as shown on the map. The idea of combining information from a database with information derived directly through the senses is termed *augmented reality*, or AR. In terms of the locations of computing discussed earlier, AR is clearly of most value when the location of the user $U$ is contained within the subject area $S$, allowing the user to augment what can be seen directly with information retrieved about the same area from a database. This might include historic information, or predictions about the future, or information that is for some other reason invisible to the user.

AR can be used to augment or replace the work of senses that are impaired or absent. A team led by the late Reginald Golledge, professor of geography at the University of California, Santa Barbara, experimented with AR as a means of helping visually impaired people to perform the simple task of navigation. The system, which was worn by the user (Figure 10.16), included a differential GPS for accurate positioning, a GI database that included very detailed information on the immediate environment, a compass to determine the position of the user's head, and a pair of earphones. The user gave the system information to identify the desired destination, and the system then generated sufficient information to replace the

**Figure 10.15** Johannes Vermeer's painting of 1669.

**Figure 10.14** Simulated view along East 64th Street in Manhattan generated in Microsoft's Bing Maps 3D, with three-dimensional buildings and exterior textures.

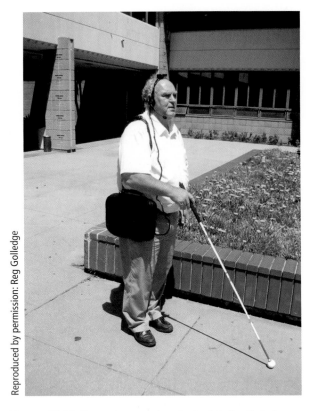

**Figure 10.16** The system worn here by the late Reg Golledge (a leader of the development team) uses a GI system and GPS to augment the senses of a visually impaired person navigating through a complex space such as a university campus.

**Figure 10.17** An augmented reality system developed at Columbia University by Professor Steve Feiner and his group.

normal role of sight, either through verbal instructions, or through the generation of stereo sounds that appeared to come from the appropriate direction.

### Augmented reality combines information from the database with information from the senses.

Steven Feiner, professor of computer science at Columbia University, has demonstrated another form of AR that superimposes historic images and other information directly on the user's field of view. For example, in Figures 10.17 and 10.18 a user wearing a head-mounted device coupled to a wearable computer is seeing both the Columbia University main library building and an image generated from a database showing the building that occupied the library's position prior to the university's move to this site in 1896: the Bloomingdale Insane Asylum.

## 10.3.2 Location-Based Services

A location-based service (LBS) is an information service provided by a device that knows where it is and is capable of modifying the information it provides based on that knowledge. Traditional computing devices, such as desktops or laptops, have no way of

knowing where they are, and their functions are in no way changed when they are moved. But increasingly the essential information about a device's location is available, from a GPS or RFID tag incorporated in the device or from the device's IP address (Section 4.4), and is being used for a wide range of purposes.

### A location-based service is provided by a computing device that knows where it is.

The simplest and most obvious form of locationally enabled device is the GPS receiver, and any device that includes a GPS capability is capable in principle of providing LBS. The most ubiquitous LBS-capable devices are the modern mobile phone and tablet. A variety of methods exist for determining device locations, including GPS embedded in the device itself, measurements made by the mobile phone of signals received from towers, and measurements made by the towers of signals received from the device. As long as the device is on, the operator of the mobile network is able to pinpoint its location at least to the accuracy represented by the size of the cell, on the order of 10 km, and frequently much more accurately, even to 10 m.

One of the strongest motives driving this process is emergency response. A large proportion of emergency calls come from mobile phones, and although the location of each landline phone is likely to be recorded in a database available to the emergency

**Figure 10.18** The user's field of view of Feiner's AR system, showing the Columbia University main library and the insane asylum that occupied the site in the 1800s.

Columbia University by Professor Steve Feiner and his group, © 1999, T. Höllerer, S. Feiner, & J. Pavlik Computer Graphics & User Interfaces Lab, Columbia University

responder, in a significant proportion of cases the user of a mobile phone is unable to report his or her current location to sufficient accuracy to enable effective response. Several well-publicized cases have drawn attention to the problem. The magazine *Popular Science*, for example, reported a case of a woman who lost control of her car in Florida in February 2001, skidding into a canal. Although she called 911 (the emergency number standard in the United States and Canada), she was unable to report her location accurately and died before the car was found.

### Emergency services provide one of the strongest motivations for LBS.

There are many other examples of LBS that take advantage of locationally enabled mobile phones. A Yellow Pages service responds to a user who requests information on businesses that are close to his or her current location (Where is the nearest pizza restaurant? Where is the nearest hospital? Where is the nearest WiFi hotspot?) by sending a request that includes location to a suitable Web server. The response might consist of an ordered list presented on the mobile phone screen, or a simple map centered on the user's current location (Figure 10.19). A trip planner gives the user the ability to find an optimum driving route from the current location to some defined destination. Similar services are now being provided by public transport operators, and in some cases these services make use of GPS transponders on buses and trains to provide information on actual, as distinct from scheduled, arrival and departure times. Some social networking sites allow a mobile phone user to display a map showing the current locations of nearby friends (provided their mobile phones are active and they are also registered for this service). Geocaching (see Figure 12.6) is a type of orienteering in which contestants navigate their way to sites using GPS and conventional navigational means.

**Figure 10.19** A map displayed on a mobile phone and centered on the user's current location.

**Figure 10.20** Credit card in use.

Direct determination of location, using GPS or measurement to or from towers, is only one basis on which a computing device might know its location, however. Other forms of LBS are provided by fixed devices and rely on the determination of location when the device was installed. For example, many point-of-sale systems that are used by retailers record the location of the sale, combining it with other information about the buyer obtained by accessing credit-card or store-affinity-card records (Figure 10.20). In exchange for the convenience or financial inducement of a card, the user effectively surrenders some degree of location privacy to the company whenever the card is used. One benefit is that it is possible for the company to analyze transactions, looking for patterns of purchase that are outside the card user's normal buying habits, perhaps because they occur in locations that the user does not normally frequent, or at anomalous times. Many of us will have experienced the embarrassment of having a credit card transaction refused in an unfamiliar city because the techniques used by the company have flagged the transaction as an indicator that the card might have been stolen. In principle, a store-affinity card gives the company access to information about buying habits and locations, in return for a modest discount.

**Location is revealed every time a credit, debit, or store-affinity card is used.**

## 10.3.3 Issues in Mobile GIS

Using GI systems in the field or "on the road" is very different from using them in the office. First, the location of the user is important and is directly relevant to the application. It makes good sense to center maps on the user's location, to provide the capability for maps that show the view from the user's location rather than from above, and to offer maps that are oriented to the user's direction of travel, rather than north. Second, the field environment may make certain kinds of interaction impractical or less desirable. In a moving vehicle,

for example, it would be dangerous to present the driver with visual displays, unless perhaps these are directly superimposed on the field of view (on the windshield). Instead, such systems often provide instructions through computer-generated speech and use speech recognition to receive instructions. With wearable devices that provide output on minute screens built into the user's eyeglasses, there is no prospect of conventional point-and-click with a mouse, so again voice communication may be more appropriate. On the other hand, many environments are noisy, creating problems for voice recognition.

One of the most important limitations to mobility remains the battery, and although great strides have been made in recent years, battery (or other wireless energy) technology has not advanced as rapidly as other components of mobile systems, such as processors and storage devices. Batteries typically account for the majority of a mobile system's weight, and they limit operating time severely.

**The battery remains the major limitation to LBS and to mobile computing in general.**

Although broadband wireless communication is possible using WiFi, connectivity remains a major issue. Wireless communication techniques tend to be

- Limited in spatial coverage. WiFi hotspots are limited to a single building or perhaps a hundred meters from the router; mobile phone–based techniques have wider coverage but lower bandwidth (communication speeds), and only the comparatively slow satellite-based systems approach global coverage.

- Noisy. Mobile phone and WiFi communications tend to "break up" at the edges of coverage areas, or as devices move between cells, leading to errors in communication.

- Limited in temporal coverage. A moving device is likely to lose signal from time to time, whereas some devices, particularly recorders installed in commercial vehicles and those used for surveys, are unable to upload data until within range of a home system or fixed beacon.

- Insecure. Wireless communications often lack adequate security to prevent unwanted intrusion. It is comparatively easy, for example, for someone to tap into a wireless session and obtain sensitive information.

- Progress is being made on all these fronts, but it will be some years before it becomes possible to use GI systems and services as efficiently, effectively, and safely anywhere in the field as it currently is in the office.

## 10.4 Distributing the Software: GI Services

This final section addresses distributed processing, the notion that the actual operations of a GI system might be provided from remote sites, rather than by the user's own computer. Despite the move to component-based software (Section 6.3.3), it is still true that almost all the operations performed on a user's data are executed in the same computer, the one on the user's desk. Each copy of a popular GI system includes the same functions, which are replicated in every installation of that system around the world. When new versions are released, they must be copied and installed at each site, and because not all copies are replaced, there is no guarantee that the system used by one user is identical to the system used by another, even though the vendor and product name are the same.

A GI service is defined as a program executed at a remote site that performs some specific task. The execution of the program is initiated remotely by the user, who may have supplied data, or may rely on data provided by the service, or both. A simple example of a GI service is that provided by wayfinding sites such as MapQuest (www.mapquest.com) or Yell.com. The user's current location and desired destination are provided to the service (the current location may be entered by the user or provided automatically from GPS), but the data representing the travel network are provided by the GI service. The results are obtained by solving for the shortest path between the origin and the destination (often a compromise between minimizing distance and minimizing expected travel time), a function that exists in many GI systems but in this case is provided remotely by the GI service. Finally, the results are returned to the user in the form of driving instructions, a map, or both.

**A GI service replaces a local function with one provided remotely by a server.**

In principle, any GI system function could be provided in this way, based on server software (Section 6.6.3). In practice, however, certain functions tend to have attracted more attention than others. One obvious problem is commercial: How would a GI service pay for itself, would it charge for each transaction, and how would this compare to the normal sources of income for GI system vendors based on software sales? Some services are offered free and generate their revenue by sales of advertising space or by offering the service as an add-on to some other service. MapQuest and Yell are good examples, generating much of their revenue through direct advertising and through embedding their service in other Web sites, such as travel services. Other services, such as Esri's ArcGIS Online, are offered on a subscription basis where the user is charged based on actual use.

In general, the characteristics that make a function suitable for offering as a GI service appear to be

- Reliance on a database that must be updated frequently and is too expensive for the average user to acquire. Both geocoding (Box 4.3) and wayfinding services fall into this category, as do gazetteer services (Section 4.10).

- Reliance on GI system operations that are complex and can be performed better by a specialized service than by a generic system.

The number of available GI services and GeoWeb services is growing steadily, creating a need for directories and other mechanisms to help users find them and standards and protocols for interacting with them. Esri's ArcGIS Online (arcgis.com; Figure 10.21) provides such a directory, in addition to its role as a geoportal to distributed data. The next section describes the state of the art in exploiting distributed GI services.

**Figure 10.21** ArcGISOnline provides a directory of remote GIServices. In this case a search for services related to transportation networks has identified fourteen maps and services.

Courtesy: Esri

## 10.4.1 Service-Oriented Architecture

When managing recovery in the aftermath of disasters, it is essential that managers have a complete picture of the situation: a map showing the locations and status of incidents, rescue vehicles and hospitals, traffic conditions, weather, and many other relevant assets and factors. A report of the U.S. National Research Council discusses the use of GI technology in all aspects of emergency management (see Further Reading).

To illustrate the capabilities of current technology, particularly distributed GI services and the GeoWeb, the Loma Linda University Medical Center and Esri teamed up to develop the Advanced Emergency GIS (AEGIS). It employs mashups and a service-oriented architecture to gather a wide variety of relevant data and to present it in readily understood form to emergency managers and other users. Figure 10.22 shows a typical synoptic visualization during the series of wildfires that hit Southern California in late 2007. The various icons signal the availability of information on traffic conditions (real-time video feeds from cameras on the freeway network), traffic incidents, hospitals (e.g., the number of available beds in each hospital's emergency room), helicopters, the current footprint of each fire, and much other useful information, any of which can be displayed by clicking the icon.

Figure 10.23 shows the architecture of the system. Roughly ten servers maintained by the relevant agencies, such as the California Department of Transportation and the U.S. Geological Survey, provide real-time feeds of data. The server in the center polls each server at regular intervals, with a frequency depending on the rate of change of the respective server's data, which may range from seconds in the case of emergency vehicle locations to months in the case of base mapping. The format of data contributed by each server varies and in some cases requires the use of an additional service, such as geocoding, or special software to extract relevant information. The central server integrates all the feeds into a composite mashup and distributes the result to users such as the one symbolized on the right. Several standards are involved, including SOAP (Simple Object Access Protocol), RSS (originally Rich Site Summary), and HTTP (Hypertext Transfer Protocol).

It would be dishonest to suggest that this kind of application is easy to develop. Although standards exist, there are a large number of them, and each requires its own specialized approach. In this example the most problematic case is the incident report feed from the California Department of Transportation, which requires analyzing the text of Web pages posted by the agency using a specially developed program. Moreover, AEGIS relies on the correct operation of many servers and feeds and is clearly vulnerable to power and network outages. However, there is no doubt that in future it will be easier to develop complex SOA-based systems, given the obvious power and immediacy that is possible.

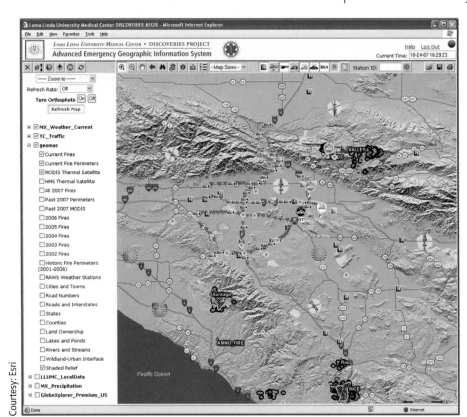

Courtesy: Esri

**Figure 10.22** The Advanced Emergency GIS (AEGIS) in operation during the wildfire outbreak in Southern California, late 2007.

**Figure 10.23** The service-oriented architecture of AEGIS.

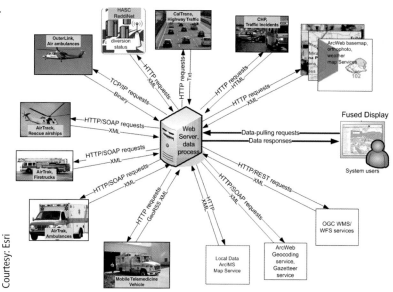

Courtesy: Esri

## 10.5 Prospects

Distributed, online GI services offer enormous benefits: They reduce duplication of effort, allow users to take advantage of remotely located data and services through simple devices, and provide ways of combining information gathered through the senses with information obtained from digital sources. Many issues continue to impede progress, however: complications resulting from the difficulties of interacting with devices in field settings; limitations placed on communication bandwidth and reliability; and limitations inherent in battery technology. Some older issues have largely been overcome, including problems of incompatible georeferencing systems (Chapter 4), but other incompatibilities continue to exist, including differences in the meaning of terms used to describe geographic features. Perhaps more problematic

than any of these at this time is the difficulty of imagining the full potential of distributed GI services. We are used to GI technology on the desktop and conscious that we have not yet fully exploited its potential. So it is hard to imagine what might be possible when the technology can be carried anywhere and its information combined with the window on the world provided by our senses.

More broadly, the GeoWeb offers a vision of a future in which geographic location is central to a wide range of applications of information technology, and one in which individuals are able to access detailed information about the dynamic state of transportation, weather, and a host of other socially important domains. We are just beginning to see the potential of this vision and to recognize that although its benefits are often compelling, it raises invasion of privacy and surveillance issues that society as a whole will have to address (Box 10.4).

---

### Biographical Box (10.4)

### Sarah Elwood and the social implications of the GeoWeb

Sarah Elwood (Figure 10.24) is a professor of geography at the University of Washington whose research bridges GI science, urban geography, and critical poverty studies. In the 1990s and 2000s, she and other critical scholars showed that GI systems and services are more than digital technologies for storing and representing spatial data. Rather, they are also an array of social and political practices for the production, regulation, and use of geo-

graphic information, and a way of knowing and making knowledge. This tradition of social critique of GI technologies emphasizes that geospatial data, geographic technologies, and geovisual representations are nonneutral. They tend to advance the interests of society's most powerful people and places, and as such, geographers and others have an important role to play in developing alternative applications of GI technologies that combat inequality.

Sarah argues in her work that these core critiques can help us better understand the societal implications of

▶

the GeoWeb. For instance, looking at the social and political relationships of data creation we see a new world of bottom-up, user-generated and user-modified data, produced and circulated online, transforming the nature of privacy. Personal information from the GeoWeb can be assembled into detailed time–space profiles of individuals without their knowledge or consent. The role of global corporations as GI actors makes it difficult for governments to regulate or even know about these problems. The GeoWeb's new forms of immediate, individualized, and photorealistic representation (e.g., Google Street View) shape what we know or assume about other people and places in ways that are different from the more abstracted and generalized visual language of conventional cartography. Her research shows that activists, disadvantaged social groups, even schoolchildren can use the GeoWeb to create new kinds of geovisual politics, creating multimedia visual and locative representations that rely on humor, parody, witness, and storytelling.

Courtesy: Sarah Elwood

**Figure 10.24** Sarah Elwood, University of Washington.

## Questions for Further Study

1. Design a location-based game based on GPS-enabled mobile phones.

2. Find a selection of geolibraries on the Web and identify their common characteristics. How do each of them allow the user to (a) specify locations of interest, (b) browse through datasets that have similar characteristics, and (c) examine the contents of data sets before acquiring them?

3. To what extent do citizens have a right to locational privacy? What laws and regulations control the use of locational data on individuals?

4. How many computers are there currently in your home, and how many items in your home have RFID tags? If you have a mobile phone, is it GPS-enabled? Do you have other GPS devices in the home?

## Further Reading

Fu, P. and Sun, J. 2011. *Web GIS: Principles and Applications*. Redlands, CA: Esri Press.

National Research Council. 1999. *Distributed Geolibraries: Spatial Information Resources*. Washington, DC: National Academy Press.

National Research Council. 2007. *Successful Response Starts with a Map: Improving Geospatial Support for Disaster Management*. Washington, DC: National Academy Press.

Scharl, A. and Tochtermann, K. (eds.). 2007. *The Geospatial Web: How Geobrowsers, Social Software and the Web 2.0 Are Shaping the Network Society*. Berlin: Springer.

Sui, D. Z., Elwood, S., and Goodchild, M. F. (eds.). 2013. *Crowdsourcing Geographic Knowledge: Volunteered Geographic Information (VGI) in Theory and Practice*. New York: Springer.

# 11 Cartography and Map Production

This chapter is the first of a series of chapters that examine geographic information (GI) system outputs; the next, Chapter 12, deals with the closely related but distinct subject of visualization. This chapter reviews the nature of cartography and the ways that users interact with GI systems in order to produce digital and hardcopy reference and thematic maps. Standard cartographic conventions and graphic symbology are discussed, as is the range of transformations used in map design. Map production is reviewed in the context of creating map series, as well as maps for specific applications. Some specialized types of mapping are introduced that are appropriate for particular applications areas.

## LEARNING OBJECTIVES

After studying this chapter, you will understand:

- The nature of maps and cartography.
- Key map-design principles.
- The choices that are available to compose maps.
- The many types of map symbology.
- Concepts of map-production flow lines.

## 11.1 Introduction

System outputs such as maps represent the pinnacle of many GI projects. Because the purpose of information systems is to produce results, this aspect of GI systems is vitally important to many managers, technicians, and scientists. Maps are the most important type of output from most GI systems because they are a very effective way of summarizing and communicating the results of GI system operations to a wide audience. The importance of map output is further highlighted by the fact that many consumers of geographic information only interact with GI systems through their use of map products.

For the purposes of organizing the discussion in this book, it is useful to distinguish between two types of GI system output: formal maps, created according to well-established cartographic conventions, that are used as a reference or communication product (e.g., a military mapping agency 1:250,000-scale topographic map, or a geological survey 1:50,000-scale paper map—see Box 11.3); and

transitory map and map-like visualizations, used simply to display, analyze, edit, and query geographic information (e.g., results of a database query to retrieve areas with poor health indicators viewed on a desktop computer, or routing information displayed on a mobile device). Both can exist in digital form on interactive display devices or in hardcopy form on paper and other media. In practice, this distinction is somewhat arbitrary, and there is considerable overlap, but at the core the motivations, tools, and techniques for map production and map visualization are quite different. The current chapter will focus on maps and the next on visualization.

Cartography is concerned with the art, science, and techniques of making maps or charts. Conventionally, the term *map* is used for terrestrial areas (Figure 11.1) and *chart* for marine areas (Figure 11.2), but they are both maps in the sense the word is used here. In statistical or analytical fields, statistical graphics provide the pictorial representation of data, but this does not form part of the discussion in this chapter. Note, however, that statistical graphics can

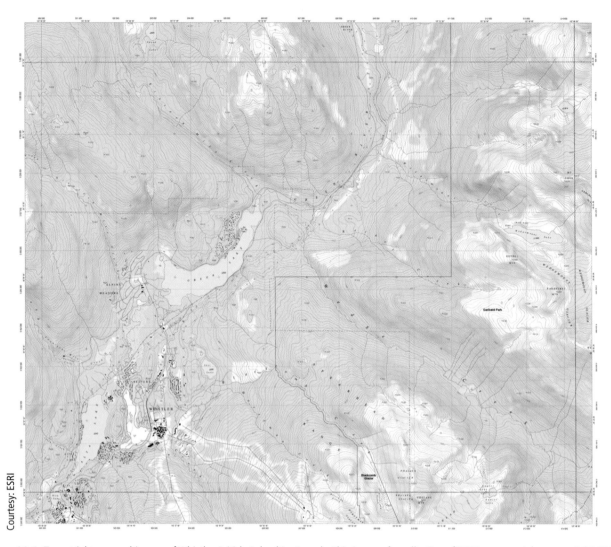

**Figure 11.1** Terrestrial topographic map of Whistler, British Columbia, Canada. This is one of a collection of 7016 commercial maps at 1:20,000 scale covering the province.

be used on maps (e.g., Figure 11.3A, B). Cartography dates back thousands of years to a time before paper, but the main visual display principles were developed during the paper era; thus many of today's digital cartographers still use the terminology, conventions, and techniques from the paper era. Box 11.1 illustrates the importance of maps in a historic military context.

### Maps are important communication and decision support tools.

Historically, the origins of many national mapping organizations can be traced to the need for mapping for "geographical campaigns" of infantry warfare, for colonial administration, and for defense. Today such organizations fulfill a far wider range of needs of many more user types (see Chapter 17). Although the military remains a heavy user of mapping, such territorial changes as arise from today's conflicts reflect a more

subtle interplay of economic, political, and historical considerations—though, of course, the threat or actual deployment of force remains a pivotal consideration. Today, GI system–based terrestrial mapping underpins a wide range of activities, such as the support of humanitarian relief efforts (Figure 11.3) and the partitioning of territory through negotiation rather than force (e.g., GI systems were used by senior decision makers in the Darfur Peace Agreement in 2006). The time frame over which events unfold is also much more rapid: It is inconceivable to think of politicians, managers, and officials being able to neglect geographic space for months, weeks, or even days, never mind years.

Paper maps remain in widespread use because of their transportability, reliability, ease of use, and the routine application of printing technology. They are also amenable to conveying straightforward

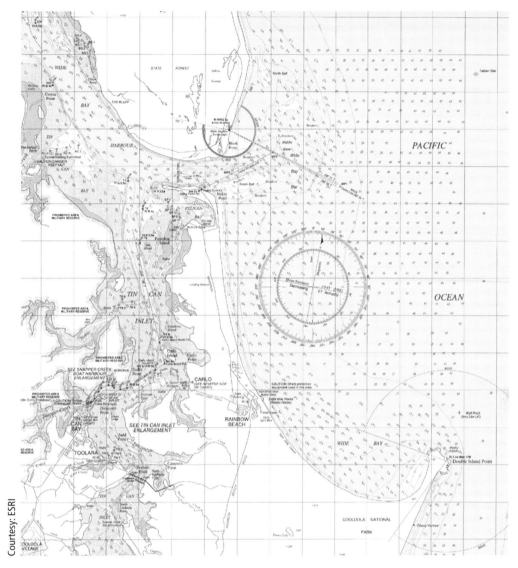

Courtesy: ESRI

**Figure 11.2** Marine chart of Great Sandy Strait (South), Queensland, Australia Boating Safety Chart. This chart conforms to international charting standards.

---

## Applications Box (11.1)

### Military Maps in History

"Roll up that map; it will not be wanted these ten years." British Prime Minister William Pitt the Younger made this remark after hearing of Napoleon's victory over Austria and Russia at the Battle of Austerlitz in 1805, when it became clear that the prospects of a British military campaign in Continental Europe had been thwarted for the foreseeable future. The quote illustrates the crucial historic role of mapping as a tool of decision support in warfare, in a world in which nation–states were far more insular than they are today. It also identifies two other defining characteristics of the use of geographic

information in 19th-century society. First, the principal, straightforward purpose of much terrestrial mapping was to further national interests by infantry warfare. Second, the time frame over which changes in geographic activity patterns unfolded was, by today's standards, incredibly slow—Pitt envisaged that no British citizen would revisit this territory for a quarter of a (then) average lifetime! In the 19th century, printed maps were available in many countries for local, regional, and national areas, and their uses extended to land ownership, tax assessment, and navigation, among other activities.

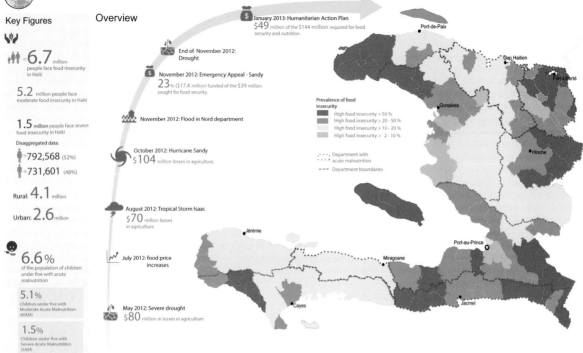

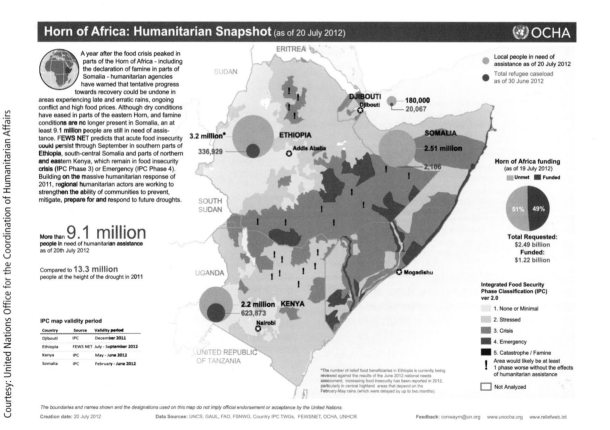

**Figure 11.3** Humanitarian relief maps: (A) Haiti food security and nutrition; (B) Horn of Africa Food Crisis.

messages and supporting decision making. Yet our increasingly detailed understanding of the natural environment and the accelerating complexity of society mean that the messages that mapping can convey are increasingly sophisticated and, for the reasons set out in Chapter 5, uncertain. Greater democracy and accountability, coupled with the pervasive use of maps on smartphones and increased spatial reasoning abilities that better education brings, indicate that more people than ever feel motivated and able to contribute to all kinds of spatial policy. This makes the decision-support role immeasurably more challenging, varied, and demanding of visual media. Today's mapping must be capable of communicating an extensive array of messages and emulating the widest range of what-if scenarios (see Section 15.1.1).

**Both paper and digital maps have an important role to play in many economic, environmental, and social activities.**

The visual medium of a given application must also be open to the widest community of users. Technology has led to the development of an enormous range of devices to bring mapping to the greatest range of users in the widest spectrum of decision environments. In-vehicle displays, smartphones, and wearable computers are all important in this regard (see Chapters 6 and 10 for examples). Most important of all, the development of the Web makes societal representations of space a real possibility for the first time. That is to say, it is now very easy to create cartographic products that represent everything from consumer preferences about how they will vote in an election to indicators of social deprivation, and ethnicity. For example, each month Google creates over 1 billion maps for users. The technology and applications of transient maps and other visualizations are explored more fully in the next chapter.

## 11.2 Maps and Cartography

There are many possible definitions of a map; here we use the term to describe digital or analog (softcopy or hardcopy) output from a GI system that shows geographic information using well-established cartographic conventions. A map is the final outcome of a series of GI processing steps (Figure 11.4) beginning with data collection, editing, and maintenance (Chapter 8), extending through data management (Chapter 9) and analysis (Chapters 13–15), and concluding with a map. Each of these activities successively transforms a database of geographic information until it is in the form appropriate to display on a given technology.

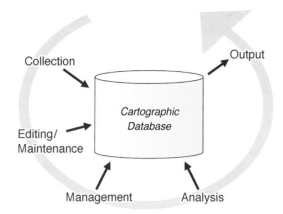

**Figure 11.4** GI system processing transformations needed to create a map.

Central to any GI system is the creation of a data model that defines the scope and capabilities of its operation (Chapter 7) and the management context in which it operates (Chapters 16–18). There are two basic types of map (Figure 11.5): reference maps, such as topographic maps from national mapping agencies (see also Figure 11.1), that convey general information; and thematic maps that depict specific geographic themes, such as population census statistics, soils, or climate zones (see Figures 11.14 and 11.15 for further examples).

**Maps are both storage and communication mechanisms.**

Maps fulfill two very useful functions, acting as both storage and communication mechanisms for geographic information. The old adage "A picture is worth a thousand words" connotes something of the efficiency of maps as a storage container. The modern equivalent of this saying is "A map is worth a million bytes." Before the advent of GI systems, the paper map was the database (see Section 3.7), but now a map can be considered merely one of many possible products generated from a digital geographic database. Maps are also a mechanism to communicate information to viewers. They can present the results of analyses (e.g., the optimum site suitable for locating a new store, or analysis of the impact of an oil spill). They can communicate spatial relationships between phenomena across the same map, or between maps of the same or different areas. As such they can assist in the identification of spatial order and differentiation. Effective decision support requires that the message of the map be readily interpretable in the mind of the decision maker. A major function of a map is not simply to marshal and transmit known information about the world, but also to create or reinforce a particular message.

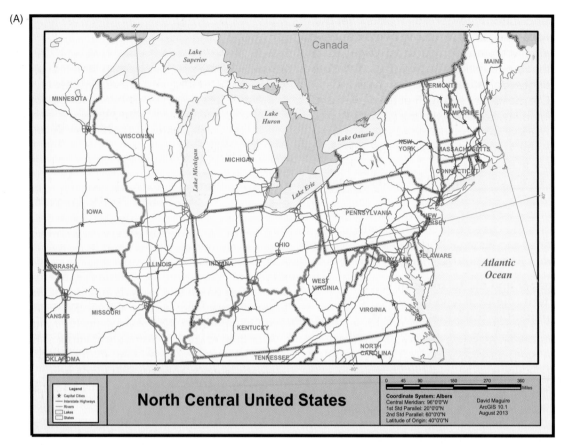

(A)

**North Central United States**

Coordinate System: Albers
Central Meridian: 96°0'0"W
1st Std Parallel: 20°0'0"N
2nd Std Parallel: 60°0'0"N
Latitude of Origin: 40°0'0"N

David Maguire
ArcGIS 10.1
August 2013

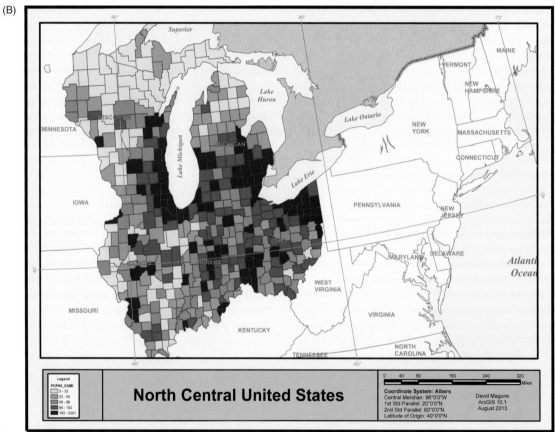

(B)

**North Central United States**

Coordinate System: Albers
Central Meridian: 96°0'0"W
1st Std Parallel: 20°0'0"N
2nd Std Parallel: 60°0'0"N
Latitude of Origin: 40°0'0"N

David Maguire
ArcGIS 10.1
August 2013

**Figure 11.5** Two maps of North Central United States: (A) a reference map; and (B) a thematic map showing population density.

Box 11.2 concerns the work of Ferjan Ormeling, who has written extensively on the importance of maps. Maps also have several limitations:

- Maps can be used to miscommunicate or obfuscate either accidentally or on purpose. For example, incorrect use of symbols can convey a misleading message to users by highlighting one type of feature at the expense of another (see Figure 11.14 for an example of different choropleth map classifications of the same data).

- Maps are a single realization of a spatial process. If we think for a moment about maps from a statistical perspective, then each map instance represents the outcome of a sampling trial and is therefore a single occurrence generated from all possible maps on the same subject for the same area (see Section 5.4.2). The significance of this is that other sample maps drawn from the same population would exhibit variations, and consequently, we need to be careful in drawing inferences from a single map sample. For example, a map of soil textures is derived by interpolating soil

### Ferjan Ormeling, Cartographer

Ferjan Ormeling (Figure 11.6) received his graduate degree in geography from Groningen University, the Netherlands; he paid for his studies by working for Wolters-Noordhoff Atlas Productions as a part-time assistant atlas-editor from 1961–1968. In Utrecht University from 1969 onward he cooperated in setting up a graduate program in cartography, from 1985 as a professor.

After an earlier focus on map perception research, he specialized in atlas cartography and toponymy. He helped produce two editions of the national atlas of the Netherlands, in 1977 and 1991. He edited two cartographic manual series, *Basic Cartography* (4 volumes with Roger Anson) and *Cartography, visualization of geospatial data* (with Menno-Jan Kraak), both translated widely. Ferjan has been vice chair of the United Nations Group of Experts on Geographical Names (UNGEGN, 2007–2017) and as such produced an online toponymy Web course in 2011. Since 2000 he has been a member of the Explokart research group on the history of Dutch cartography (now located at the University of Amsterdam) and has published books on the development of Dutch colonial cartography (of the area of present Indonesia) and on the history of Dutch school atlases.

Ferjan believes that without names maps are no use, but if names are used, they should be standardized (Figure 11.7). As can be seen in Figure 11.7, national standardization is not enough, however, and conversion systems between different writing systems should be standardized as well, making sure that every name uniquely refers to one specific topographic object only (this principle is called univocity). Even then, the standardization work should be sustained as name versions for specific topographic objects may quickly be superseded, so updating is required regularly.

Ferjan is concerned with the post-1990 mass resurgence of exonyms in Central and Eastern Europe, which he considers as detrimental to cartographic communication,

Courtesy: Ferjan Ormeling

**Figure 11.6** Ferjan Ormeling, Cartographer.

although he has supported the right of minority language groups (like the Welsh, Frisians, Basques, etc.) to have their own geographical name versions also rendered on topographic maps; currently this has also been recognized as a crucial aspect of one's cultural heritage. Through the EU EuroGeoNames project he has helped to shape the geographical names model in INSPIRE, Europe's exchange format for geospatial data files.

The inclusion on maps of the proper geographical names renders these maps suitable for use as decision support tools. Ferjan's other concern is that decision makers need to be sufficiently trained to interpret map names. A course on map analysis for decision makers is long overdue!

▶

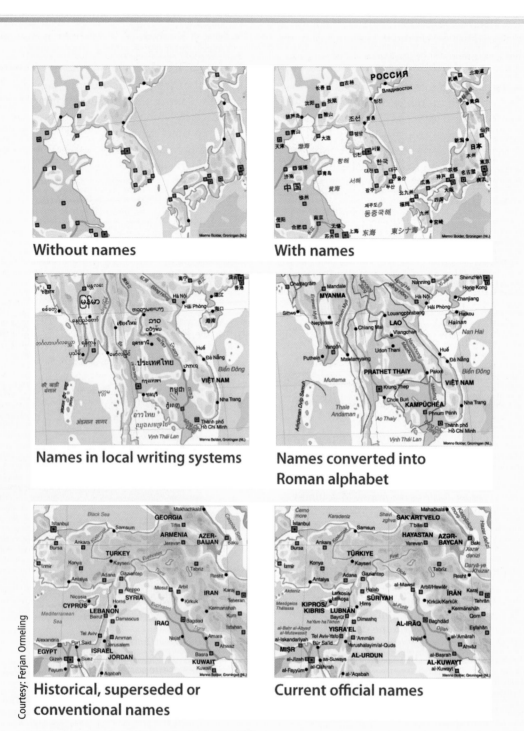

Figure 11.7 The importance of standardized, up-to-date names on maps.

**Without names**

**With names**

**Names in local writing systems**

**Names converted into Roman alphabet**

**Historical, superseded or conventional names**

**Current official names**

Courtesy: Ferjan Ormeling

sample texture measurements. Repeated sampling of soils will show natural variation in the texture measurements.

- Maps are often created using complex rules, symbology, and conventions and can be difficult to understand and interpret for the untrained viewer.

This is particularly the case, for example, in multivariate statistical thematic mapping where the idiosyncrasies of classification schemes and color symbology can be challenging to comprehend.

**Uncertainty pertains to maps just as it does to other geographic information.**

## 11.2.1 Maps and Media

Without question, GI systems and science have fundamentally changed cartography and the way we create, use, and think about maps (see Box 11.3). The digital cartography of GI systems frees mapmakers from many of the constraints inherent in traditional paper mapping (see also Section 3.7).

- The paper map is of fixed scale. Generalization procedures (Section 3.8) can be invoked in order to maintain clarity during map creation. This detail

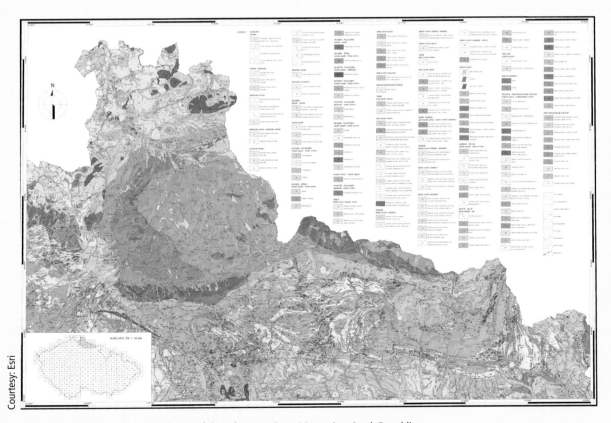

is not recoverable, except by reference back to the data from which the map was compiled. The zoom facility of GI systems can allow mapping to be viewed at a range of scales and permit detail to be filtered out as appropriate at a given scale.

- The paper map is of fixed extent, and adjoining map sheets must be used if a single map sheet does not cover the entire area of interest. (An unwritten law of paper map usage is that the most important map features always lie at the intersection of four paper map sheets!) GI systems, by contrast, can provide a seamless medium for viewing space, and users are able to pan across wide swaths of geography.

- Most paper maps present a static view of the world, whereas conventional paper maps and charts are not adept at portraying dynamics. GI representations are able to achieve this through animation.

- The paper map is flat and hence limited in the number of perspectives that it can offer on three-dimensional data. 3-D visualization is much more effective within GI systems, which can support interactive pan and zoom operations (see Figure 8.15 for a 3-D view example).

- Paper maps provide a view of the world as essentially complete. GI system-based mapping allows the supplementation of base-map material with further data. Data layers can be turned on and off to examine data combinations.

- Paper maps provide a single, map-producer-centric, view of the world. GI system users are able to create their own, user-centric map images in an interactive way. Side-by-side map comparison is also possible in GI systems.

**GI systems offer flexible ways to produce many types of maps.**

## 11.3 Principles of Map Design

Map design is a creative process during which the cartographer, or mapmaker, tries to convey the message of the map's objective. The primary goals in map design are to share information, highlight patterns and processes, and illustrate results. A secondary objective is to create a pleasing and interesting picture, but this must not be at the expense of fidelity to reality and meeting the primary goals. Map design is quite a complex procedure requiring the simultaneous optimization of many variables and harmonization of multiple methods. Cartographers must be prepared to compromise and balance choices. It is difficult to define exactly what constitutes a good design, but the general consensus is that a good design is

one that looks good, is simple and elegant, and most important, leads to a map that is fit for its purpose.

Robinson et al. identify seven controls on the map-design process:

- *Purpose.* The purpose for which a map is being made will determine what is to be mapped and how the information is to be portrayed. Reference maps are multipurpose, whereas thematic maps tend to be for a single purpose (Figure 11.5). With the digital technology of GI systems, it is easier to create maps, and many more are digital and interactive. As a consequence, today's maps are increasingly created for a specific purpose.

- *Reality.* The phenomena being mapped will usually impose some constraints on map design. For example, the orientation of the country—whether it be predominantly southwest–northeast (Japan) or north–south (Chile)—can determine layout in no small part.

- *Available data.* The specific characteristics of data (e.g., raster or vector, continuous or discrete, or point, line, or area) will affect the design. There are many different ways to symbolize map data of all types, as discussed in Section 11.3.2.

- *Map scale.* Scale is an apparently simple concept, but it has many ramifications for mapping (see Box 2.3 for further discussion). It will control the quality of data that can appear in a map frame, the size of symbols, the overlap of symbols, and much more. Although one of the early promises of digital cartography and GI systems was scale-free databases that could be used to create multiple maps at different scales, this has never been realized because of technical complexities.

- *Audience.* Different audiences want different types of information on a map and expect to see information presented in different ways. Usually, executives (and small children!) are interested in summary information that can be assimilated quickly, whereas advanced users often want to see more information. Similarly, those with restricted eyesight find it easier to read bigger symbols.

- *Conditions of use.* The environment in which a map is to be used will impose significant constraints. Maps for outside use in poor or very bright light will need to be designed in ways different to maps for use indoors where the light levels are less extreme.

- *Technical limits.* The display medium, be it digital or hardcopy, will impact the design process in several ways. For example, maps to be viewed in an Internet browser or on a smartphone, where resolution and bandwidth are limited, should be simpler and based on fewer data than equivalents to be displayed on a desktop PC monitor.

## 11.3.1 Map Composition

Map composition is the cartographic design process of creating a map comprising several closely interrelated elements (Figure 11.9):

- *Map body.* The principal focus of the map is the main map body, or in the case of comparative maps there will be two or more map bodies. It should be given the necessary space and use symbols appropriate to its significance.

- *Inset/overview map.* Inset and overview maps may be used to show, respectively, an area of the main map body in more detail (at a larger scale) and the general location or context of the main body.

- *Title.* One or more map titles are used to identify the map and to inform the reader about its content.

- *Legend.* This lists the items represented on the map and how they are symbolized. Many different layout designs are available, and a considerable body of information is available about legend design.

- *Scale.* The map scale provides an indication of the size of objects and the distances between them. A paper map scale is a ratio, where one unit on the map represents some multiple of that value in the real world. The scale can be symbolized numerically (1:1,000), graphically (a scalebar), or texturally ("one inch equals 10,000 inches"). The scale is a representative fraction (see Section 3.7), and so a 1:1,000 scale is larger (finer) than 1:100,000. A small- (coarse-) scale map displays a larger area than a large- (fine-) scale map, but with less detail. See also Box 2.3 for more discussion of the many meanings of scale.

- *Direction indicator.* The direction and orientation of a map can be conveyed in one of several ways, including grids, graticules, and directional symbols (usually north arrows). A grid is a network of parallel and perpendicular lines superimposed on a map (see Figure 11.1). A graticule is a network of longitude and latitude lines on a map that relates points on a map to their true location on the Earth (see Figure 11.5).

- *Map metadata.* Map compositions can contain many other types of information, including the map projection, date of creation, data sources, and authorship (see also Section 10.2).

A key requirement for a good map is that all map elements are composed into a layout that has good

**Figure 11.9** The principal components of a map composition layout.

Courtesy: Esri

visual balance. On large-scale maps, such as 1:10,000 national mapping agency topographic series, all the contextual items (everything in the preceding list except the map body) usually appear as marginal notations (or marginalia). In the case of map series or atlases (see Section 11.4), some of the common information may be in a separate document. On small- or medium-scale maps this information usually appears within the map border (Figure 11.9).

## 11.3.2 Map Symbolization

The data to be displayed on a map must be classified and represented using graphic symbols that conform to well-defined and accepted conventions. The choice of symbolization is critical to the usefulness of any map. Unfortunately, the seven controls on the design process listed in Section 11.3.1 also conspire to mean that there is not a single universal symbology model applicable everywhere, but rather one for each combination of factors. Again, we see that cartographic design is a compromise reached by simultaneously optimizing several factors.

Good mapping requires that spatial objects and their attributes can be readily interpreted in applications. In Chapter 2, attributes were classified as being measured on the nominal, ordinal, interval, ratio, or cyclic scales (see Box 2.1). Also, in Chapter 3 points, lines, and areas were described as types of discrete objects (see Section 3.5.1), and surfaces were discussed as a type of continuous field (see Section 3.5.2). We have already seen how attribute measures that we think of as continuous are actually discretized to levels of precision imposed by measurement or design (see Chapter 5). The representation of spatial objects is similarly imposed: cities might be captured as points, areas, mixtures of points, lines, and areas (as in a street map: Figure 6.12) or 2.5-D and 3-D objects (see Section 12.4.2), depending on the base scale of a representation and the importance of city objects to the application. Measurement scales and spatial object types are thus one set of conventions that are used to abstract reality. The choice of output format, digital or paper, may entail reclassification or transformation of attribute measures.

The process of mapping attributes frequently entails further problems of classification because many spatial attributes are inherently uncertain (Chapter 5). For example, in order to create a map of occupational type, individuals' occupations will be classified first into socioeconomic groups (e.g., "factory worker") and perhaps then into supergroups, such as "blue collar" (see Figure 1.17). At every stage in the aggregation process, we inevitably do injustice to many individuals who perform a mix of white- and blue-collar, intermediate, and skilled functions, by lumping them into a single group. In practice, the validity and usefulness of an occupational or geodemographic classification will have become established over repeated applications, and the task of mapping is to convey thematic variation in as efficient a way as possible.

### 11.3.2.1 Attribute Representation and Transformation

Humans are good at interpreting visual data—much more so than interpreting numbers, for example—but conventions are still necessary to convey the message that the mapmaker wants the data to impart. Many of these conventions relate to the use of symbols (such as the way highway shields denote route numbers on many U.S. medium- and fine-scale maps) and colors (blue for rivers, green for forested areas, etc.) and have been developed over a period of hundreds of years. Mapping of different themes (such as vegetation cover, surface geology, and socioeconomic characteristics of human populations) has a more recent history. Here too, however, mapping conventions have developed, and sometimes they are specific to particular applications.

Attribute mapping entails the use of graphic symbols, which (in two dimensions) may be referenced by points (e.g., historic monuments and telecoms antennae), lines (e.g., roads and water pipes), or areas (e.g., forests and urban areas). Basic point, line, and area symbols are modified in different ways in order to communicate different types of information. The ways in which these modifications take place adhere to cognitive principles and the accumulated experience of application implementations. The nature of these modifications was first explored by Jacques Bertin in 1967 and was extended to the typology illustrated in Figure 11.10 by Alan MacEachren. The size and orientation of point and line symbols are varied principally to distinguish between the values of ordinal and interval/ratio data using graduated symbols (such as the divided pie symbols shown in Figure 11.11). Figure 11.12 illustrates how orientation and color can be used to depict the properties of locations, such as ocean current strength and direction.

Hue refers to the use of color, principally to discriminate between nominal categories, as in agricultural or urban land-use maps (Figure 11.13). Different hues may be combined with different textures or shapes if there are a large number of categories, in order to avoid difficulties of interpretation. The shape of map symbols can be used to communicate information about either a spatial attribute (e.g., a viewpoint or the start of a walking trail), or its spatial location (e.g., the location of a road or boundary of a particular type: Figure 11.13), or spatial relationships (e.g., the relationship between subsurface topography and ocean currents). Arrangement, texture, and focus refer

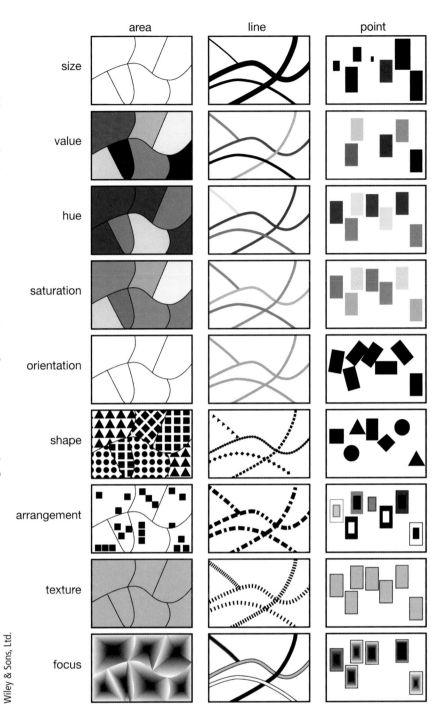

Source: MacEachren 1994; from *Visualization in Geographical Information Systems*, Hearnshaw H. M. and Unwin D. J. (eds.). Reproduced by permission of John Wiley & Sons, Ltd.

area   line   point

size

value

hue

saturation

orientation

shape

arrangement

texture

focus

**Figure 11.10** Bertin's graphic primitives, extended from seven to ten variables (the variable location is not depicted).

to within- and between-symbol properties that are used to signify pattern. A final graphic variable in the typologies of MacEachren and Bertin is location (not shown in Figure 11.10), which refers to the practice of offsetting the true coordinates of objects in order to improve map intelligibility, or changes in map projection. We discuss this in more detail later in this section. Some of the common ways in which these graphic variables are used to visualize spatial object types and attributes are shown in Table 11.1.

The selection of appropriate graphic variables to depict spatial locations and distributions presents one set of problems in mapping. A related task is how best to position symbols on the map to optimize map interpretability. The representation of nominal data by graphic symbols and icons is apparently trivial, although in practice automating placement presents some challenging analytical problems. Most GI system software packages include generic algorithms for positioning labels and symbols in relation to geographic objects.

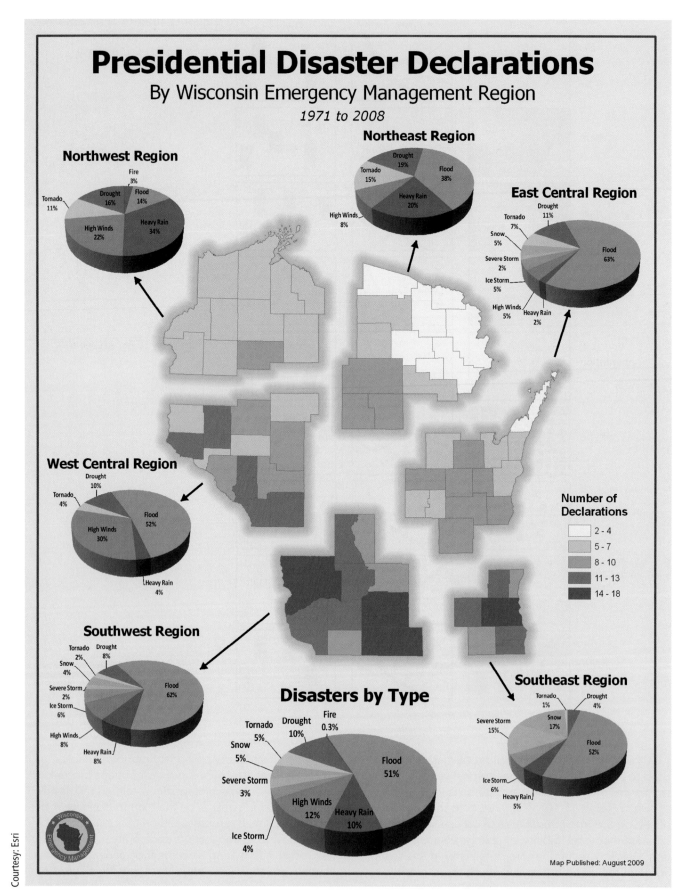

**Figure 11.11** Wisconsin Presidential Disaster Declarations 1971–2008.

**Figure 11.12** Tauranga Harbor Tidal Movements, Bay of Plenty, NZ. The arrows indicate speed (color) and direction (orientation).

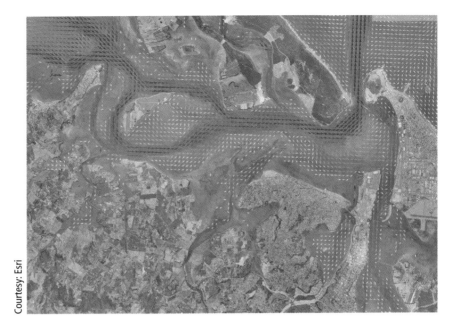

Courtesy: Esri

**Figure 11.13** Use of hue (color) to discriminate between Bethlehem, West Bank, Israel urban land-use categories and use of symbols to communicate location and other attribute information.

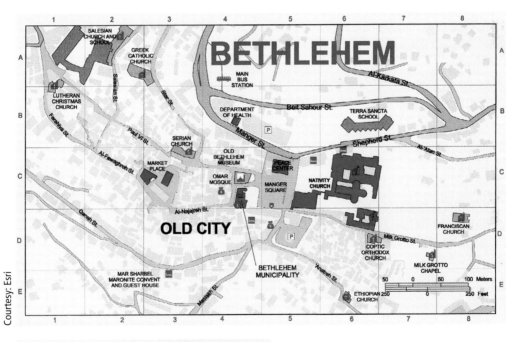

Courtesy: Esri

**Table 11.1** Common methods of mapping spatial object types and attribute data, with examples.

| Spatial object type | Attribute type | | |
|---|---|---|---|
| | **Nominal** | **Ordinal** | **Interval/Ratio** |
| **Point (0-D)** | Symbol map (each category a different class of symbol—color, shape, orientation), and/or use of lettering: e.g., presence/absence of building type (Figure 11.13) | Hierarchy of symbols or lettering (color and size): e.g., small/medium/large depots, city size (Figure 11.7) | Graduated symbols (color and size): e.g., disease incidence (Figures 11.3, 11.16) |
| **Line (1-D)** | Network connectivity map (color, shape, orientation): e.g., presence/absence of connection (Figure 6.1) | Graduated line symbology (color and size): e.g., road classifications (Figure 6.12) | Flow map with width or color lines proportional to flows (color and size): e.g., traffic flows (Figure 11.24) |
| **Area (2-D)** | Unique category map (color, shape, orientation, pattern): e.g., soil types or geology (Figure 11.8) | Graduated color or shading map: e.g., timber yield low/medium/high, name concentration (Figure 1.7) | Continuous hue/shading, e.g., dot-density or choropleth map: e.g., percentage of retired population (Figure 11.14) |
| **Surface (2.5-D)** | One color per category (color, shape, orientation, pattern), e.g., relief classes: mountain/valley, children's activity plots (Figure 3.1) | Ordered color map, e.g., areas of gentle/steep/very steep slopes, land use risk (Figure 7.14) | Contour map (e.g., isobars/isohyets: e.g., topography contours, diabetes contours (Figure 5.2) |

Point labels are positioned to avoid overlap by creating a window, or mask (often invisible to the user), around text or symbols. Linear features, such as rivers, roads, and contours, are often labeled by placing the text using a spline function to give a smooth, even distribution, and distinguishing by use of color. Area labels are assigned to central points (see Figure 11.9), using geometric algorithms similar to those used to calculate geometric centroids (Section 14.2.1). These generic algorithms are frequently customized to accommodate common conventions and rules for particular classes of application, such as topographic (Figure 11.1), utility, transportation, and seismic maps. Generic and customized algorithms also include color conventions for map symbolization and lettering.

Ordinal attribute data are assigned to point, line, and area objects in the same rule-based manner, with the ordinal property of the data accommodated through use of a hierarchy of graphic variables (symbol and lettering sizes, types, colors, intensities, etc.). As a general rule, the typical user is unable to differentiate between more than seven (plus or minus two) ordinal categories, and this provides an upper limit on the normal extent of the hierarchy.

A wide range of conventions is used to visualize interval- and ratio-scale attribute data. Proportional circles and bar charts are often used to assign interval- or ratio-scale data to point locations (Figures 11.11 and 11.16). Variable line width (with increments that correspond to the precision of the interval measure) is a standard convention for representing continuous variation in flow diagrams.

A variety of ways can be used to ascribe interval- or ratio-scale attribute data to areal entities that are predefined. In practice, however, none is unproblematic. The standard method of depicting areal data is in zones (Figure 11.5B). However, as was discussed in Box 2.5, the choropleth map brings the dubious visual implication of within-zone uniformity of attribute value. Moreover, conventional choropleth mapping also allows any large (but possibly uninteresting) areas to dominate the map visually. A variant on the conventional choropleth map is the dot-density map, which uses points as a more aesthetically pleasing means of representing the relative density of zonally averaged data—but not as a means of depicting the precise locations of point events. Proportional circles provide one way around this problem; here the circle is scaled in proportion to the size of the quality being mapped, and the circle can be centered on any convenient point within a zone. However, there is a tension between using circles that are of sufficient size to convey the variability in the data and the problems of overlapping circles on busy areas of maps that have large numbers of symbols. Circle location also entails the same kind of positioning problem as that of name and symbol placement outlined earlier.

If the richness of the map presentation is to be equivalent to that of the representation from which it was derived, the intensity of color or shading should directly mirror the intensity or magnitude of attributes. The human eye is adept at discerning continuous variations in color and shading, and Waldo Tobler has advanced the view that continuous scales present the best means of representing geographic variation. There is no natural ordering implied by use of different colors, and the common convention is to represent continuous variation on the red-green-blue (RGB) spectrum. In a similar fashion, difference in the hue, lightness, and saturation (HLS) of shading is used in color maps to represent continuous variation. International standards on intensity and shading have been formalized.

At least four basic classification schemes have been developed to divide interval and ratio data into categories (Figure 11.14):

1. *Natural (Jenks) breaks.* Here classes are defined according to apparently natural groupings of data values. The breaks may be imposed on the basis of break points that are known to be relevant to a particular application, such as fractions and multiples of mean income levels, or rainfall thresholds known to support different thresholds of vegetation ("arid," "semiarid," "temperate," etc.). This is "top-down" or deductive assignment of breaks. Inductive ("bottom-up") classification of data values may be carried out by using GI software to look for relatively large jumps in data values.

2. *Quantile breaks.* Here each of a predetermined number of classes contains an equal number of observations. Quartile (four-category) classifications are widely used in statistical analysis, whereas quintile (five-category; Figure 11.14) classifications are well suited to the spatial display of uniformly distributed data. Yet because the numeric size of each class is rigidly imposed, the result can be misleading. The placement of the boundaries may assign almost identical attributes to adjacent classes, or features with quite widely different values in the same class. The resulting visual distortion can be minimized by increasing the number of classes—assuming the user can assimilate the extra detail that this creates.

3. *Equal-interval breaks.* These are best applied if the data ranges are familiar to the user of the map, such as temperature bands or shaded-relief contour maps.

4. *Standard deviation.* Such classifications show the distance of an observation from the mean. The GI system calculates the mean value and then generates class breaks in standard deviation measures above and below it. This would be an appropriate choice for a map of crimes showing variations above and below the mean for the whole map. Use of a two-color ramp helps to emphasize values above and below the mean (Figure 11.14).

**Classification procedures are used in map production in order to ease user interpretation.**

The selection of classification is very much the outcome of choice, convenience, and the experience of the cartographer. The automation of mapping in GI systems has made it possible to evaluate different possible classifications. Looking at distributions allows us to see if the distribution is strongly skewed—which might justify using unequal class intervals in a particular application. A study of poverty, for example, could quite happily class millionaires along with all those earning over $50,000, as they would be equally irrelevant to the outcome.

### 11.3.2.2 Multivariate Mapping
Multivariate maps show two or more variables for comparative purposes. For example, Figure 11.3B shows funding and food security in the Horn of Africa. Many maps are a compilation of composite maps based on a range of constituent indicators. Climate maps, for example, are compiled from direct measures such as amount and distribution of precipitation, diurnal temperature variation, humidity, and hours of sunshine, plus indirect measures such as vegetation coverage and type. It is unlikely there will ever be a perfect correspondence between each of these components, and historically it has been the role of the cartographer to integrate disparate data. In some instances, data may be averaged over zones that are devoid of any strong meaning, whereas the different components that make up a composite index may have been measured at a range of scales. Mapping can mask scale and aggregation problems where composite indicators are created at large scales using components that were only intended for use at small scales.

Figure 11.15 illustrates how multivariate data can be displayed on a shaded area map. In this soil texture map of Mexico, three variables are displayed simultaneously, as indicated in the legend, using color variations to display combinations of percent sand (base), silt (right), and clay (left). For another example of multivariate mapping see Box 11.4.

# Mobile Homes Density

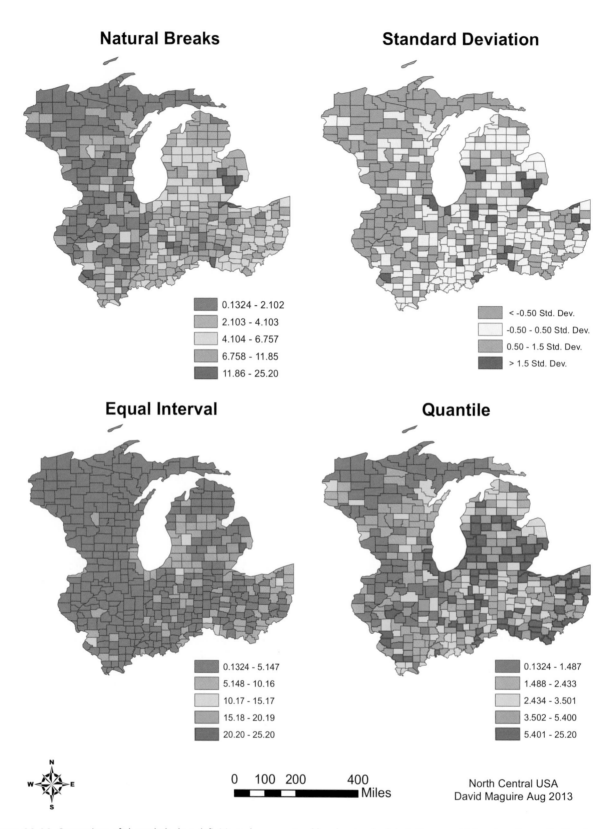

**Natural Breaks**

0.1324 - 2.102
2.103 - 4.103
4.104 - 6.757
6.758 - 11.85
11.86 - 25.20

**Standard Deviation**

< -0.50 Std. Dev.
-0.50 - 0.50 Std. Dev.
0.50 - 1.5 Std. Dev.
> 1.5 Std. Dev.

**Equal Interval**

0.1324 - 5.147
5.148 - 10.16
10.17 - 15.17
15.18 - 20.19
20.20 - 25.20

**Quantile**

0.1324 - 1.487
1.488 - 2.433
2.434 - 3.501
3.502 - 5.400
5.401 - 25.20

0  100  200  400
Miles

North Central USA
David Maguire Aug 2013

**Figure 11.14** Comparison of choropleth class definition schemes: natural breaks, standard deviation, equal interval, and quantile. The data are Mobile Homes Density for North Central USA, 2004.

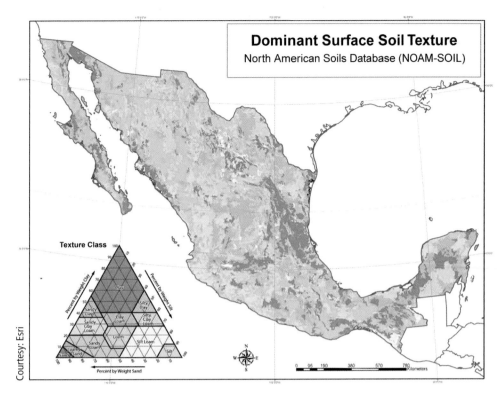

**Figure 11.15** Multivariate data map of dominant surface soil texture.

---

## Applications Box (11.4)

### Multivariate Mapping of Groundwater Analyte over Time

An increasing number of GI system users need to represent time-dependent (temporal) data on maps in a meaningful yet concise manner. The ability to analyze data for both spatial and temporal patterns in a single presentation provides a powerful incentive to develop tools for communication to map users and decision makers. Although there are many different methods to display scientific data to help discern either spatial or temporal trends, few visualization software packages allow for a single graphical presentation within a geographic context. Chemical concentration data for many different groundwater monitoring wells are widely available for one or more chemical or analyte samples over time.

Figure 11.16 shows how this complex temporal data can be represented on a map so that the movement and change in concentration of the analyte can be observed, both horizontally and vertically, and the potential for the analyte reaching the water table after many years can be assessed.

This map is innovative because it uses a clock diagram code for temporal and spatial visualization. The clock diagrams are analogous to the rose diagrams often used to depict wind direction in meteorological maps or strike direction in geologic applications. The clock diagram method consists of three steps: (1) sample location data, including depth, are compiled along with attribute data such as concentration and time; (2) all data are read into the software application and clock diagram graphics are created; and (3) clock diagram graphics are placed on the map at the location of the well where samples were collected. Although this method of representing temporal data through the use of clock diagrams is not entirely original, the use of multiple clock diagrams as symbols to emphasize movement of sampled data over time is a new concept. The clock diagram code and method have proven to be an effective means of identifying temporal and spatial trends when analyzing groundwater data.

▶

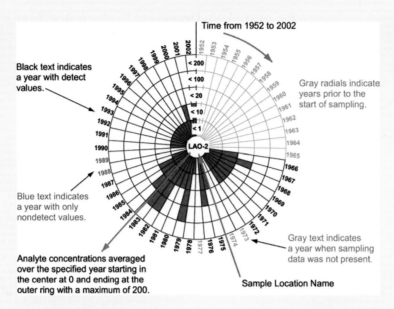

Time from 1952 to 2002

Black text indicates a year with detect values.

Gray radials indicate years prior to the start of sampling.

Blue text indicates a year with only nondetect values.

Analyte concentrations averaged over the specified year starting in the center at 0 and ending at the outer ring with a maximum of 200.

Gray text indicates a year when sampling data was not present.

Sample Location Name

< 200
< 100
< 20
< 10
< 1

LAO-2

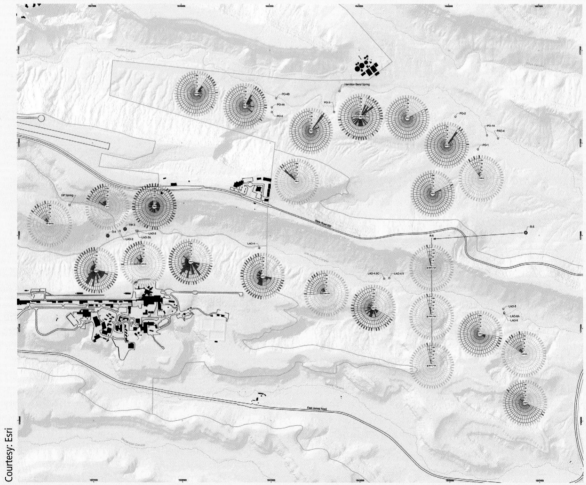

Courtesy: Esri

**Figure 11.16** Representing temporal data as clock diagrams on maps.

## 11.4 Map Series

The discussion thus far has focused on the general principles of map design that apply equally to a single map or a collection of maps for the same area or a common theme being considered over multiple areas. The real power of digital cartography using GI systems is revealed when changes need to be made to maps or when collections of similar maps need to be created. Editing or copying a map composition to create similar maps of any combination of areas and data themes is relatively straightforward with a GI system.

Many organizations use GI systems to create collections or series of topographic or thematic maps. Examples include a topographic map series to cover a state or country (e.g., USGS 1:24,000 quad sheets of the United States or Eurogeographics 1:250,000 maps of Europe), an atlas of reference maps (e.g., the *National Geographic Eighth Edition Atlas of the World*), a series of geology maps for a country (see Box 11.3), land parcel maps within a jurisdiction (Figure 11.17), and utility asset maps that might be bound into a map book. Map series by definition share a number of common elements (for example, projection, general layout, and symbology), and a

number of techniques have been developed to automate the map-series production process.

**GI systems make it much easier to create collections of maps with common characteristics from map templates.**

Figure 11.4 shows the main GI system subcomponents and workflows used to create maps. In principle, this same process can be extended to build production flow lines that result in collections of maps such as an atlas, map series, or map book. Such a GI system requires considerable financial and human-resource investments, to both create and maintain it, and many operatives will use it over a long period of time. In an ideal system, an organization would have a single "scale-free," continuous database, which is kept up to date by editing staff and is capable of feeding multiple flow lines at several scales, each with different content and symbology. Unfortunately, this vision remains elusive because of several significant scientific and technical issues. Nevertheless, considerable progress has been made in recent years toward this goal.

The heart of map production through GI systems is a geographic database covering the area and data

**Figure 11.17** A page from the Town of Brookline, Massachusetts, Tax Assessor's map book. Each map book page uses a common template (as shown in the legend), but has a different map extent: (A) page 26. (*continued*)

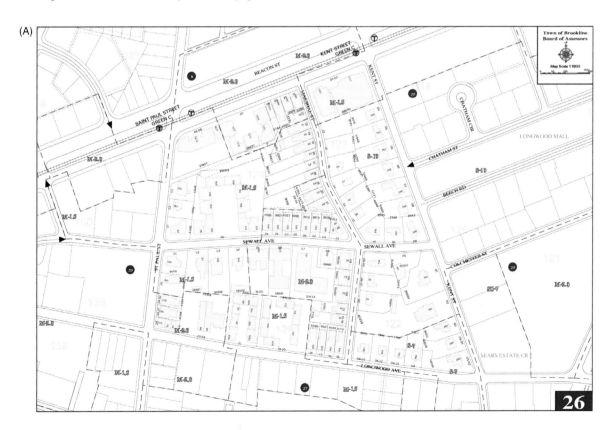

# ATLAS MAP LEGEND

(B)

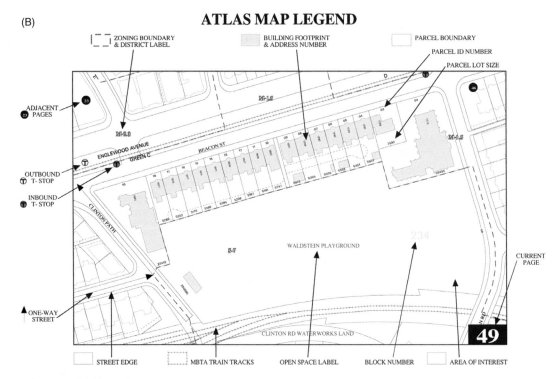

**Figure 11.17** (*continued*) (B) legend. (Courtesy: Esri)

layers of interest (see Chapter 9 for background discussion about the basic principles of creating and maintaining geographic databases). This database is built using a data model that represents interesting aspects of the real world in the GI system (see Chapter 7 for background discussion of GI data models). Such a base cartographic data model is often referred to as a digital landscape model (DLM) because its role is to represent the landscape in the GI database as a collection of features that are independent of any map-product representation. Figure 11.18 is an extended version of Figure 11.4 that shows the key components and flows in a map-production GI system.

The DLM continuous geographic database is usually stored in a database management system (DBMS)

that is managed by the GI system. The database is created and maintained by editors that typically use sophisticated desktop software applications that must be carefully tailored to enforce strict data integrity rules (e.g., geometric connectivity, attribute domains, and topology—see Section 9.7.1) to maintain database quality.

In an ideal world, it will be possible to create multiple different map products from this DLM database. Example products that might be generated by a civilian or military national mapping organization include a 1:10,000-scale topographic map in digital format (e.g., Adobe Portable Document Format (pdf)), a 1:50,000 topographic map sheet in paper format, a 1:100,000 paper map of major streets/highways, a 1:250,000 map suitable for inclusion in a digital or printed atlas, and a 1:250,000 digital database in GML (Geography Markup Language) digital format. In practice, however, for reasons of efficiency and because cartographic database generalization is still not fully automated, a series of intermediate data model layers are employed (see Section 3.8 for a discussion of generalization). The base, or fine-scale, DLM is generalized by both automated and manual means to create coarser-scale DLMs (Figure 11.18 shows two additional medium- and coarse-scale DLMs). From each of these DLMs one or more cartographic data models (digital cartographic models [DCMs]) can be created that derive cartographic representations from real-world features.

**Figure 11.18** Key components and information flows in a GI map-production system.

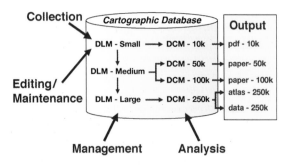

For example, for a medium-scale DCM representation a road centerline network can be simplified to remove complex junctions and information about overpasses and bridges because they are not appropriate to maps at this scale.

Once the necessary DCMs have been built, the process of map creation can proceed. Each individual map will be created in the manner described in Section 11.3, but because many similar maps will be required for each series or collection, some additional work is necessary to automate the map production process. Many similar maps can be created efficiently from a common map template that includes any material common to all maps (for example, inset/overview maps, titles, legends, scales, direction indicators, and map metadata; see Section 11.3.1 and Figure 11.9). Once a template has been created, it is simple to specify the individual map-sheet content (e.g., map data from DCM data layers, specific title, and metadata information) for each separate sheet. The geographic extent of each map sheet is typically maintained in a separate database layer. This could be a regular grid (see inset to Figure 11.8), but commonly there is some overlap in sheets to accommodate irregular-shaped areas and to ensure that important features are not split at sheet edges. An automated batch process can then "stamp out" each sheet from the database as and when required. Finally, if the sheets are to form a map book or atlas, it is convenient to generate an index of names (for example, places or streets) that shows on which sheet they are located.

In Box 11.5, Danny Dorling reflects on the current state of cartography and describes some of his work.

## Danny Dorling, Cartographer

Danny Dorling is the Halford Mackinder Professor of Geography at the University of Oxford (Figure 11.19). With a group of colleagues he helped create the Web site www.worldmapper.org which shows who has most and least in the world. He has published with others more than 25 books on issues related to social inequalities and several hundred journal papers. His cartographic work includes three co-authored texts: *Identity in Britain: A Cradle-to-Grave Atlas, The Atlas of the Real World: Mapping the Way We Live*, and *Bankrupt Britain: An Atlas of Social Change*. His most recent book is *Population 10 Billion*. He is currently President of the Society of Cartographers.

Over many years Danny has used cartograms (discussed in greater detail in Chapter 12) to highlight geographic distributions by distorting area in proportion to a geographic variable. Figure 11.20 shows four examples of continuous area cartograms of the British population.

Writing more widely about cartography in general in a note in *The Cartographic Journal* recently, Danny suggested that:

> Cartography is about choices. There is no single right way to depict the world that we are a part of. How we choose to depict it will alter how we see it and treat others, the land, energy, air travel, see the seas of the Mediterranean, the lands of the middle east, whether we see a place as being in the middle, or as being east of somewhere we consider now more central, and whether we worry more about volcanoes or

**Figure 11.19** Danny Dorling.

Courtesy: Danny Dorling

the excesses of leopard print clothing, let alone what happens when you first built a city with no streets.

Cartography is both art and science, but also a part of the humanities and of popular culture, of new technology and of ancient history. "Making maps or charts" is as old as we are and changes as fast as we change. More new maps and charts will have been made in our lifetimes than over the course of the lives of every human who has ever lived. With new technology it may also be possible for our children to be able to say the same thing again, and consign us to part

▶

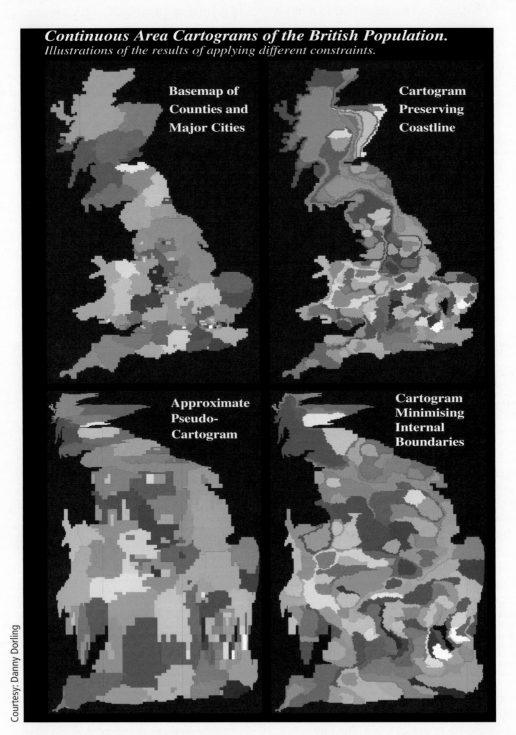

**Continuous Area Cartograms of the British Population.**
*Illustrations of the results of applying different constraints.*

Basemap of Counties and Major Cities

Cartogram Preserving Coastline

Approximate Pseudo-Cartogram

Cartogram Minimising Internal Boundaries

Courtesy: Danny Dorling

**Figure 11.20** Continuous Area Cartograms of the British Population.

of that strange and obscure history, no longer being at the forefront that we write so much about.

Always, what matters most is whether the sense that is being made of the world does itself make sense. If it does not then we need new maps, more maps, and better maps.

Dorling, D. (2013) Cartography: making sense of our worlds: *The Cartographic Journal*, 50, 2, 152–154.

## 11.5 Applications

Obviously, a huge number and wide range of applications use maps extensively. The goal here is to highlight a few examples that raise some interesting cartographic issues.

Shaded area maps of continuous internal/ratio scale data are very widely used throughout the world for displaying socioeconomic data, such as population censuses. When combined with charts and other graphics they can synthesize very large amounts of data allowing patterns to be observed and processes inferred. Figure 11.21 is a classic example of a population census choropleth map showing population density data for the United States.

The relative importance of representing space and attributes will vary within and between different applications, as will the ability to broker improved measures of spatial distributions through integration of ancillary sources. These tensions are not new. In a topographic map, for example, the width of a single-carriageway road or electric utility cable may be exaggerated considerably (Figure 11.22). This is done in order to enhance the legibility of features that are central to general-purpose topographic mapping. In some instances, these prevailing conventions will have evolved over long periods of time, while in others the newfound capabilities of GI systems entail a distinct break from the past. As a general rule, where accuracy and precision of georeferencing are important, the standard conventions of topographic mapping will be applied (e.g., Figure 11.1).

Utility applications use GI systems that have come to be known as Automated Mapping and

**Figure 11.21** U.S. Census Bureau population census map product from the 2010 Census and 2011 Population Estimates program.

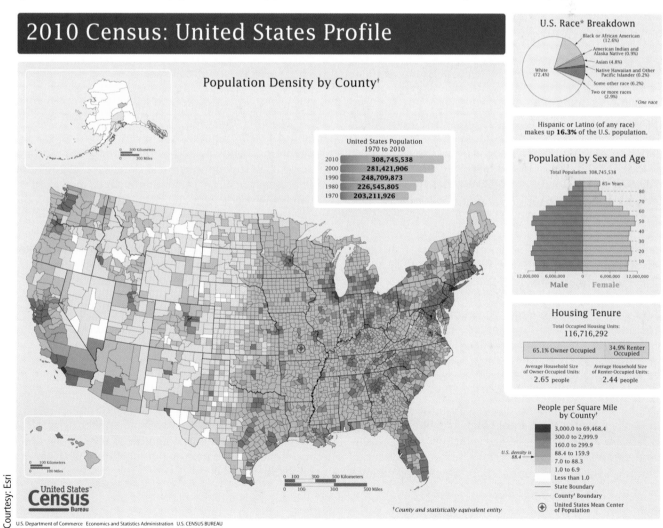

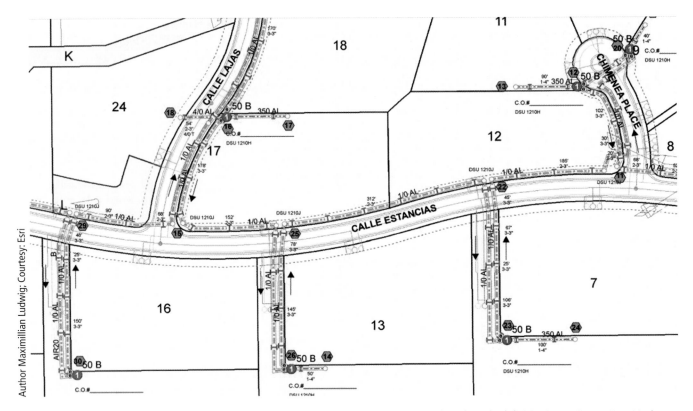

**Figure 11.22** El Paso Electric Company map showing the design for an electric underground residential subdivision in Las Cruces, New Mexico.

Facilities Management (AM/FM) systems. The prime objectives of such systems are asset management and the production of maps that can be used in work orders that drive asset-maintenance projects (Figure 11.22). Some of the maps produced from AM/FM and other types of GI system use conventional cartographic representations, but others prefer schematic representations for the purposes of speed, simplicity, and clarity. Figure 11.23 is a schematic map showing the approach to an airport (Stockholm Arlanda) that an airline pilot would use in a cockpit map book.

Transportation applications use a procedure known as linear referencing (Section 4.5) to visualize point (such as street furniture road signs) and linear (such as parking restrictions, speed limits, and road-surface quality measures) objects or events. In linear referencing, two-dimensional geography is collapsed into one-dimensional linear space. Data are collected as linear measures from the start of a route (a path through a network). The linear measures are added at display time, usually dynamically, and segment the route into smaller sections (hence the term *dynamic segmentation*, which is sometimes

used to describe this type of model). Figure 11.24 is a map of Montpelier, Vermont, that shows the results of several highway analysis studies (average daily traffic, roadway and bridge conditions, and frequent crash locations). The data were collected as linear events using several methods, including driving the roads with an in-vehicle sensor. For more information on linear referencing, see Sections 4.5 and 7.2.3.3.

We began this chapter with a discussion of the military uses of mapping, and such applications also have special cartographic conventions—as in the operational overlay maps used to communicate battle plans (Figure 11.25). On these maps, friendly, enemy, and neutral forces are shown in blue, red, and green, respectively. The location, size, and capabilities of military units are depicted with special multivariate symbols that have operational and tactical significance. Other subjects of significance, such as minefields, impassable vegetation, and direction of movement, also have special symbols. Animations of such maps can be used to show the progression of a battle, including future what-if scenarios.

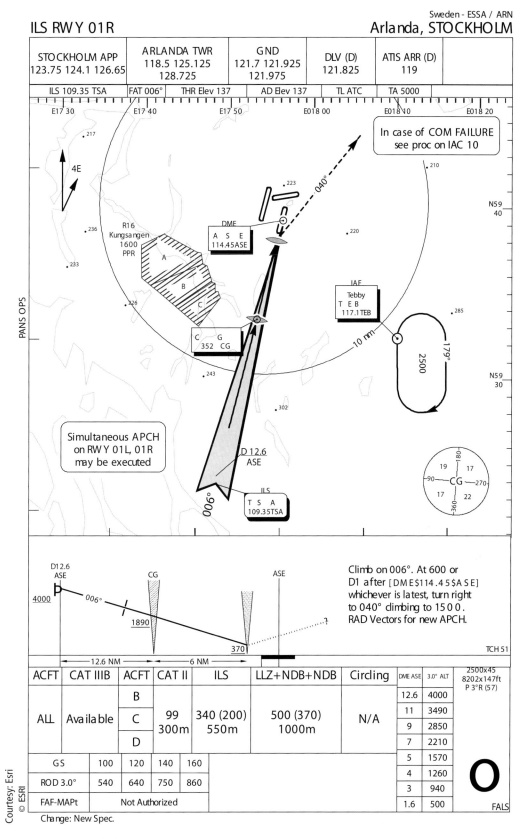

**Figure 11.23** Aeronautical Approach Chart.

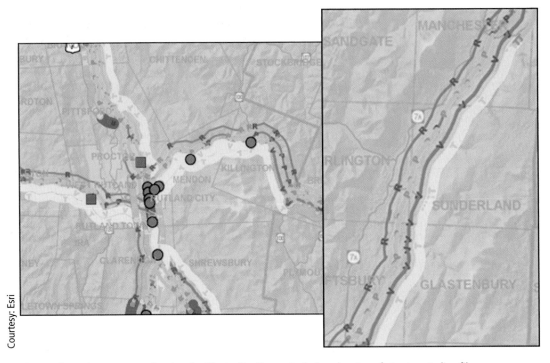

Courtesy: Esri

**Figure 11.24** Linear-referencing transportation data for Montpelier, Vermont, displayed on top of street centerline files.

**Figure 11.25** Military map showing overlay graphics on top of base-map data. Blue rectangles are friendly forces units, red diamonds are enemy force units, and black symbols are military operations—maneuvers, obstacles, phase lines, and unit boundaries. These "tactical graphics" depict the battle space and the tactical environment for the operational units.

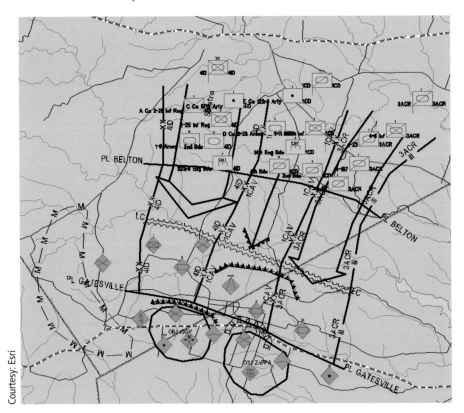

Courtesy: Esri

## 11.6 Conclusion

Cartography is both an art and a science. The modern cartographer must be very familiar with the application of computer technology. The very nature of cartography and mapmaking has changed profoundly over the past few decades and will never be the same again. Nevertheless, there remains a need to understand the nature and representational characteristics of what goes into maps if they are to provide robust and defensible aids to decision making, as well as tactical and operational support tools. In cartography, there are few hard-and-fast rules to drive map composition, but a good map is often obvious once complete. Modern advances in GI system-based cartography make it easier than ever to create large numbers of maps very quickly using automated techniques once databases and map templates have been built. The creation of databases and map templates continues to be an advanced task requiring the services of trained professionals. The type of data that is used on maps is also changing—today's maps often reuse and recycle different datasets, obtained over the Internet, that are rich in detail but may be unsystematic in collection and incompatible in terms of scale. This all underpins the importance of metadata (Section 10.2) to evaluate datasets in terms of scale, aggregation, and representativeness prior to mapping. Collectively, these changes are driving the development of new applications founded on the emerging advances in scientific visualization that will be discussed in the next chapter.

## Questions for Further Study

1. Identify the criteria that you would use to design a military mapping system for use during the Afghanistan War. How might the system subsequently be adapted for use in humanitarian relief?

2. Why are there so many classification schemes for area-based choropleth maps? What criteria would you use to select the best scheme?

3. Using Table 11.1 as a guide, create an equivalent table that identifies example maps for all the map types.

4. Using a GI system of your choice, create a map of your local area containing all seven elements listed in Section 11.3.1. What makes this a good map?

## Further Reading

Brewer, C. A. 2005. *Designing Better Maps: A Guide for GIS Users*. Redlands, CA: Esri Press.

Kraak, M.-J. and Ormerling, F. 2009. *Cartography: Visualization of Spatial Data*. (3rd ed.). Harlow, UK: Pearson Education.

Robinson, A. H., Morrison, J. L., Muehrcke, P. C., Kimerling, A. J., and Guptill, S.C. 1995. *Elements of Cartography* (6th ed.). Hoboken, NJ: Wiley.

Slocum, T. A., McMaster, R. B., Kessler, F. C., and Howard, H. H. 2009. *Thematic Cartography and Geographic Visualization*. (3rd ed.). Upper Saddle River, NJ: Prentice Hall.

Tufte, E. R. 2001. *The Visual Display of Quantitative Information*. Cheshire, CT: Graphics Press.

# 12 Geovisualization

This chapter builds on the cartographic design principles set out in Chapter 11 to describe a range of novel ways in which information can be presented visually to the user. Using techniques of geovisualization, geographic information (GI) systems provide a far richer and more flexible medium for portraying attribute distributions than paper mapping. First, through techniques of spatial query, geovisualization allows users to explore, synthesize, present (communicate), and analyze the meaning of a representation (see Chapter 2). Second, it facilitates map transformation using techniques such as cartograms and dasymetric mapping. Third, geovisualization allows the user to interact with the real world from a distance, through interaction with and even immersion in digital environments. Together, these functions provide new insights into geographic data, broaden the user base of GI systems, and have implications for the supply and use of volunteered geographic information (VGI).

## LEARNING OBJECTIVES

After studying this chapter you will understand:

- How GI systems facilitate visual communication.
- The ways in which good user interfaces can help to resolve spatial queries.
- Some of the ways in which geographic representations may be transformed.
- How 3-D geovisualization and augmented reality can improve our understanding of the world.

## 12.1 Introduction: Uses, Users, Messages, and Media

Effective decision support through GI systems requires that computer-held representations (as discussed in Chapters 2 and 5) are readily interpretable in the minds of the users who need them to make decisions. Because representations are necessarily partial and selective, they are unlikely to be accepted as objective. Critiques of GI systems (Section 1.5.5) have suggested that even widely available digital maps and virtual Earths (as provided by Google, Apple, and Microsoft) present a "privileged" or "God's-eye" view of the world because the selectivity inherent in representation may be used to reinforce

the message or objectives of the interest groups that created them. This issue predates digital GI: it is evident in the early motivations for creating maps to support military operations (Box 11.1), and the subject of geopolitics is rife with examples of the use of mapping to reinforce political propaganda. Today, Web mapping and digital spatial data infrastructures can open up politically sensitive issues and, in the era of VGI, allow them to be openly contested (see Box 12.1).

Today's Big Data and other sources are much more numerous, varied, voluminous, frequently updated, and complicated than was the case even a decade ago. Geovisualization techniques offer the prospect of rendering them intelligible to users and avoiding information overload, by weeding out

## The Legacy of Conflict, "Tag Wars," and Cohesive Communities

OpenStreetMap (www.openstreetmap.org and Section 18.6) is an ongoing collaborative volunteered geographic information (VGI) project that is seeking to create a free digital map base of the entire world. Online maps are created using global positional system (GPS) traces that are recorded, uploaded, and annotated by volunteers, along with aerial photography and other free sources. Most volunteers have detailed local knowledge of the areas for which they capture data, and all volunteers are empowered to edit the database as well as add to it, in order to generate coverage and to correct for errors. Local knowledge is, of course, situated in direct experience (Chapter 3), and open-source maps may be heavily contested if, as in the case of Cyprus, received wisdom is heavily conditioned by cultural affiliation.

Cyprus has been divided since 1974, when Turkish forces invaded the north part of the island. The south is the Greek-dominated Republic of Cyprus (which is internationally recognized), and the north is the Turkish Republic of Northern Cyprus (which is not). After 1974, places in the north part of the island were renamed in Turkish, and road signs were made consistent with them. Some volunteers believe that Turkish names should be used, with Greek names relegated to the metadata; this stance is pragmatic not least because OpenStreetMap is used as a navigational aid. Yet other volunteers, mindful perhaps of the island's unhappy recent history and the illegal status of the north, have persisted in replacing Turkish names with their pre-1974 Greek counterparts. The ensuing "tag war", marked by the successive editing of Turkish and Greek names, can be viewed at www.openstreetmap.org/browse/node/276379679/history, and some of the changes to the map are illustrated in Figure 12.1. Geovisualization enables dynamic mapping of the tensions that surround this disputed territory and that underpinned Greek-Cypriot rejection of the United Nations' Annan Plan to reunite the island in 2004.

Source: © OpenStreetMap contributors

**Figure 12.1** Greek and Turkish edits to OpenStreetMap in Cyprus.

distracting or extraneous detail. They also make it possible to identify and assess outliers—apparently rogue observations that stand out from an overall pattern or trend—that are not revealed by other Big Data analytics. At the other end of the data volume spectrum, historical GI system applications are concerned with assembling digital shards of evidence that may be scattered across space and time or with providing wider spatial context to painstaking local archive studies. In such cases, geovisualization provides the context and framework for building representations, sometimes around sparse records. In these various instances, geovisualization is different from cartography and map production (see Chapter 11) in that it typically uses interactive computing for data exploration, it entails the creation of multiple (including 3-D) representations of spatial data, or it allows the animated representation of changes over time.

Geovisualization may entail interrogation of multiple representations of large and complex datasets, through a series of interactions that pursue different lines of inquiry, with the objective of developing cumulative findings. Mapping makes it possible to communicate the meaning of a spatial representation of the real world to users. Historically, the paper map was the only available interface between the mapmaker and the user: it was permanent, contained a fixed array of attributes, and was of predetermined and invariant scale. Typically, national mapping agencies would provide metadata (Section 10.2) that demonstrated that the map was fit for the general purpose for which a map was designed, although this did not invariably make it easy for users to ascertain whether it was safe to use for specific applications. Moreover, it was not possible to differentiate between the needs of different users. These attributes severely limit the usefulness of the paper map in today's applications environment, which can entail applications of seemingly unfathomable complexity. Indeed it is often now necessary to visualize data that are very detailed, continuously updated, or have been assembled from diverse sites across the Internet. These developments are reinvigorating the debate over the meaning of maps and the potential of mapping.

Like paper mapping, digital visualization is open to negligent or malevolent use. However, in general terms, geovisualization makes the selective nature of representation more transparent and open to scrutiny. An interactive computer environment makes it easier to evaluate data quality and to present alternative constructions and representations of geographic information.

In technical terms, geovisualization builds on the established tenets of map production and display. It entails the creation and use of visual representations

to facilitate thinking, understanding, and knowledge construction about human and physical environments, at geographic scales from the architectural to the global. It is also a research-led field that integrates a wide range of approaches from scientific computing, cartography (Chapter 11), image analysis (Section 8.2.1), information visualization, and exploratory spatial data analysis (ESDA: Chapters 13 and 14). The core motivation for this activity is to develop theories, methods, and tools for visual exploration, analysis, synthesis, and presentation of spatial data.

**Geovisualization is used to explore, analyze, synthesize, and present spatial data.**

As such, today's geovisualization is much more than conventional map design. It has developed into an applied area of activity that leverages geographic data resources (including Big Data) to meet a very wide range of scientific and social needs. It is also a research field that is developing new visual methods and tools to facilitate better user interaction with such data. In this chapter, we discuss how this is achieved through *query* and *transformation* of geographic data and also user *immersion* within them. Query and transformation are discussed in a substantially different context in Chapters 13 and 14, where they are encountered as types of spatial analysis.

## 12.2 Geovisualization, Spatial Query, and User Interaction

### 12.2.1 Overview

Fundamental to effective geovisualization is an understanding of how human cognition shapes the usage of GI systems, how people think about space and time, and how spatial environments might be better represented using computers and digital data. The conventions of map production, presented in Section 11.3, are central to the use of mapping as a decision support tool. Many of these conventions are common to paper and digital mapping, although GI systems allow far greater flexibility and customization of map design. Geovisualization develops and extends these concepts in a number of new and innovative ways in pursuit of the following objectives:

1. *Exploration*: for example, to establish whether and to what extent the general message of a dataset is sensitive to inclusion or exclusion of particular data elements.

2. *Synthesis*: to present the range, complexity, and detail of one or more datasets in ways that can be readily assimilated by users. Good geovisualization

should enable the user to differentiate the "wood from the trees" and may be particularly important in highlighting the time dimension.

3. *Presentation:* to communicate the overall message of a representation in a readily intelligible manner and to enable the user to understand the likely overall quality of the representation.

4. *Analysis:* to provide a medium to support a range of methods and techniques of spatial analysis (see Chapters 13 and 14).

These objectives cut across the full range of different applications tasks and are pursued by users of varying expertise that desire different degrees of interaction with their data.

**Geovisualization allows users to explore, synthesize, present, and analyze their data more thoroughly.**

These motivations may be considered in relation to the conceptual model of uncertainty that was presented in Figure 5.1, which is presented with additions in Figure 12.2. This extended conceptual model encourages us to think of geographic analysis

not as an endpoint, but rather as the start of an iterative process of feedbacks and what-if scenario testing. It is thus appropriate to think of geovisualization as imposing a further filter (U4 in Figure 12.2) both on the results of analysis and on the conception and representation of geographic phenomena. The most straightforward way in which reformulation and evaluation of a representation of the real world can take place is through posing *spatial queries* to ask generic spatial and temporal questions such as:

- Where is . . .?
- What is at location . . .?
- What is the spatial relation between . . .?
- What is similar to . . .?
- Where has . . . occurred?
- What has changed since . . .?
- Is there a general spatial pattern, and what are the anomalies?

These questions are articulated through graphical user interfaces (GUIs) that remain based upon windows, icons, menus, and pointers—the so-called

**Figure 12.2** Filters U1–U4: conception, measurement, analysis, and visualization. Geographic analysis is not an end point, but rather the start of an iterative process of feedbacks and "what-if?" scenario testing. (See also Figure 5.1)

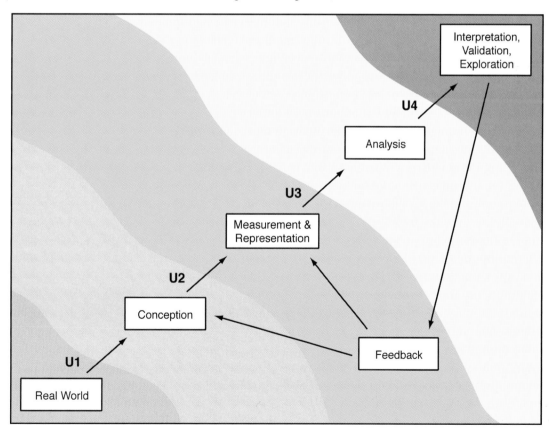

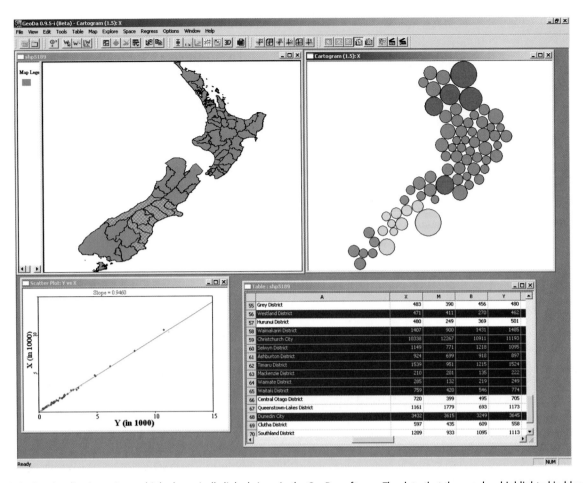

**Figure 12.3** Geovisualization using multiple dynamically linked views in the GeoDa software. The data that the user has highlighted in blue in the table are identified in the maps using yellow hatching or coloring, whereas the red circles denote outliers in the analysis. The scatterplot summarizes the relationship between two variables under consideration.

WIMP interface illustrated in Figure 12.3. Although eclipsed by different design principles in computer gaming, the familiar actions of pointing, clicking, and dragging windows and icons remain the most common ways of interrogating a geographic database and summarizing results in map and tabular form. The wide use of multitouch screen and touchpad technology of smartphones and tablets also makes it possible to manipulate data and maps through gestures. Zooming is one such interactive gesture, accomplished by emulating a pinching motion on the screen of the hardware device. High-end GI systems use multiple displays and projections of maps (see Section 4.8) along with aspatial summaries such as bar charts and scatterplots. Together these make it possible to build up a picture of the spatial and other properties of a representation. The user is able to link these representations by "brushing" data in one of the multiple representations and viewing the same observations in the other views. This facilitates learning about a representation in a data-led way.

### Spatial querying is performed through pointing, clicking, and dragging windows and icons.

Although most computer interfaces can run software that enables spatial query, the greatest volume of spatial queries is made through smartphones and other handheld devices, many of which use integral GPS receivers or phone networks to render them location aware. These devices have increased the challenges to geovisualization because of the much more restricted screen size. They have also led to a vast range of new spatially enabled applications, mainly for use by nonexpert users.

The implications for geovisualization are profound and far reaching in that a new, larger, and much more diverse user base is seeking to use smartphones to solve an ever wider range of location-specific problems, on the fly and in real time. Figure 12.4 shows the user interface of a locationally aware smartphone used to identify travel options (in terms of travel cost, travel time, and

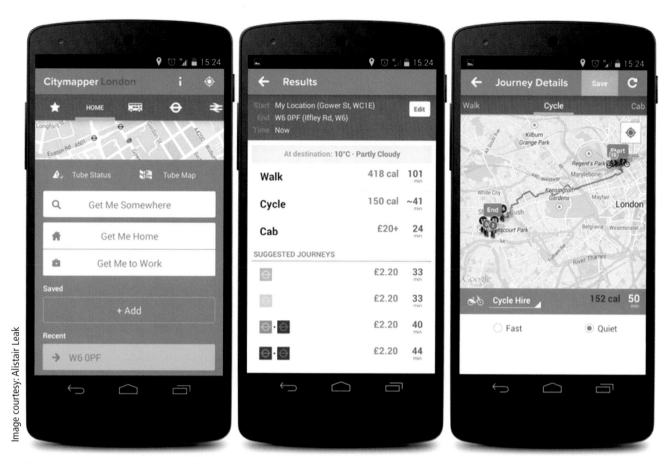

**Figure 12.4** The locationally aware Citymapper mobile phone application for London, presenting travel options between two points, taking into account some aspects of travel conditions.

calorific usage) between two locations in London. Such services can also accommodate real-time spatial information feeds such as prevailing weather, current network disruptions, or levels of traffic congestion and may allow users to pose a range of additional spatial queries.

Applications such as these are currently developed for the general market and are not tailored to personal circumstances (e.g., visual acuity), prior experience, decision-making strategies, and basic cognitive abilities. Looking to the future, it is likely that interfaces will be tailored to serve group or individual needs: for example it is possible to envisage that a navigation service offering landmark-based wayfinding instructions might use only particular types of landmarks, or that the number and type of landmarks might be changed according to the user's familiarity with particular physical settings. This area of current research is termed *cognitive engineering,* and in the future it is likely that cognitively engineered applications will be more closely attuned to the user's ability to interpret spatiotemporal information, in order to meet specific user requirements. In practice, the software used to accomplish this may come from a wide variety of proprietary and open sources.

The advent of VGI and Web 2.0 GI systems is increasing the range of sources that may be used to create, serve, and exchange GI; such sources are increasingly being used to devise niche maps for specialist applications. Figure 12.5 illustrates this with regard to the creation of maps for use in orienteering, created using public-domain software and OpenStreetMap data. A closely related application is geocaching, which is an outdoor treasure-hunting game in which participants use a GPS receiver or other navigational techniques to hide and seek caches, or containers, as shown in Figure 12.6.

**Cognitive engineering matches device functions to user requirements.**

## 12.2.2 Spatial Query Online and the Geoweb

The spatial query functions illustrated in Figure 12.4 are also central to many Internet GI systems applications, and the different computer and network architectures set out in Chapter 10 can each be used

(A)

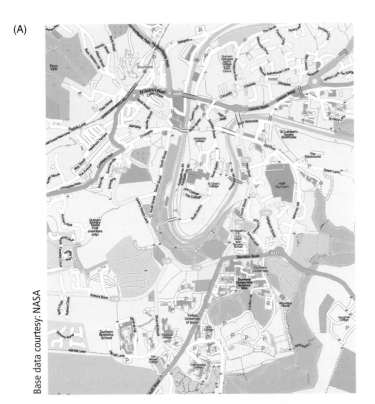

Base data courtesy: NASA

(B)

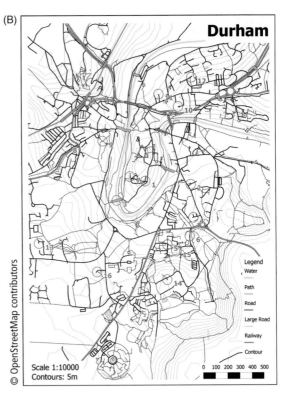

© OpenStreetMap contributors

(C)

Images courtesy: Ollie O'Brien

**Figure 12.5** (A) The general-purpose map layer base of OpenStreetMap for Durham, UK; (B) a specialist orienteering map created using the OpenStreetMap base along with the contour data derived from Shuttle Radar Topography Mission data. These sources were integrated using the MapWindow open-source GIS (www.mapwindow.org); and (C) a competitor registering at an orienteering post.

**Figure 12.6** Geocaching is a global treasure-hunting game where participants locate containers, called geocaches, hidden outdoors and then share their experiences online.

to query remote datasets. For many users, spatial query is the end objective of a GI systems application, as for example with queries about the taxable value of properties, customer-care applications about service availability, or Internet-site queries about real-time traffic conditions. In other applications, spatial query is a precursor to more advanced spatial analysis. Spatial query may appear routine, but may entail complex operations, particularly when the datasets are continuously updated (refreshed)

in real time—as, for example, when optimal traffic routes are updated in the light of traffic conditions or road closures.

The innovation of Web 2.0 has enabled bi-directional collaboration between Web sites (see Section 1.5.1) and has important implications for geovisualization, including the development of consolidator sites that take real-time feeds from third-party applications. Box 12.2 describes one such application, MapBox.

## Interactive Mapping using MapBox

Tilemill (www.mapbox.com/tilemill/) is an open-source software tool that allows users to create interactive maps with pan and zoom functionality. Users can create map tiles that are assembled into a single portable database (MBTiles—www.mapbox.com/developers/mbtiles/) that can either be hosted by Mapbox or on the user's own server.

MBTiles are an adaptation of SQLite, a self-contained, transactional structured query language (SQL) database engine (see Section 9.4). The Mapbox business model is innovative in that it is based upon user subscription for Web mapping hosting services, with all software developed as open-source products.

Applications using this infrastructure include Eric Fischer's maps of Twitter social media data to derive smartphone device usage (www.mapbox.com/labs/twitter-gnip/brands/#5/38.000/-95.000), and of a map separating locals from tourists (www.mapbox.com/labs/twitter-gnip/locals/#5/38.000/-95.000; see also Figure 17.7B). Figure 12.7 presents a MapBox hosted map of the currency of OpenStreetMap data for Rio de Janeiro (www.mapbox.com/osm-data-report/).

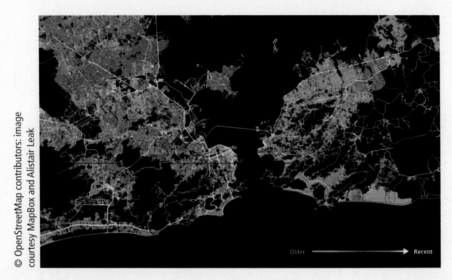

© OpenStreetMap contributors: image courtesy MapBox and Alistair Leak

Older ⟶ Recent

**Figure 12.7** A MapBox-hosted map of the currency of OpenStreetMap data for Rio de Janeiro (red colors denote old data, with more recent uploads and amendments in blue).

## 12.3 Geovisualization and Transformation

### 12.3.1 Overview

Chapter 11 illustrated circumstances in which it was appropriate to adjust the *classification intervals* of maps, in order to highlight the salient characteristics of spatial distributions. Changing the *scale* at which attributes are measured can help us to represent spatial phenomena, as with the scaling relations that may be used to characterize a fractal coastline (Box 2.7). Other components of a map can be adjusted in order to present or highlight spatial information: for example, the work of computer scientist Ross Maciejewski, illustrated in Box 12.3, illustrates how the linear features of maps can be used in novel ways to depict spatial distributions.

We have previously discussed the inherent uncertainties in representing geographic data (Chapter 5) and some of the consequences of the absence of natural units for most geographic analysis. In addition we have seen (Box 2.5) how representing *count* data using choropleth maps is unwise because areal extent of mapped units is not taken into account. Standard map production and display functions in GI systems allow us to standardize according to numerical base categories, yet large but unimportant (in attribute terms) zones then achieve greater visual dominance than is warranted.

These related problems can be addressed by using geovisualization to transform the shape and extent of areal units. Our use of the term *transformation* here is in the cartographic sense of taking one vector space (the real, or observable, world) and transforming it into another (the geovisualization). We

## Ross Maciejewski, Computer Scientist

Ross Maciejewski (Figure 12.8) completed his PhD at Purdue University in Computer Engineering in 2009 under the direction of David S. Ebert. His doctoral work in syndromic surveillance and spatial analysis brought him into contact with members of the geography community, notably visual analytics expert Alan MacEachren at Penn State University.

Following the completion of his PhD, Ross served as a visiting faculty member at Purdue as a member of the Department of Homeland Security's Center of Excellence focusing on visual analytics (VACCINE) exploring Coast Guard Search and Rescue Cases, criminal incident reports, and a variety of other spatiotemporal data. His work at Purdue's VACCINE Center was honored by the U.S. Coast Guard with a Meritorious Team Commendation as part of his work on the Port Resilience for Operational/Tactical Enforcement to Combat Terrorism (PROTECT) Team.

In 2011, Ross joined Arizona State University's School of Computing, Informatics, and Decision System Engineering, where he continues to develop visual analytics methods. His bristle maps are an innovative multivariate mapping technique in which geographically locatable statistical measures are assigned color, orientation, and length offsets along a geographical network. The image shown in Figure 12.9 presents statistical estimates of the density of vandalism complaints in Lafayette, Indiana. The linear extent of each bristle from the underlying street network allows the user to gain an impression of the severity of vandalism crimes at that point on the network. Bristle length and color are used to encode an estimate of the number of crimes, whereas orientation and color are used to differentiate between daytime and nighttime occurrences (red for night, blue for day, opposite sides of a street for day/night). This technique provides a neat depiction of the geotemporal pattern of crime in the area.

Speaking of the prospects of geovisualization, Ross says: "This is an exciting time for students to get involved in spatial data analysis. Currently, new technology is allowing us to collect spatial data at unprecedented rates, and students can learn how to mine these data for information that can in turn help us to shape our world. There are great opportunities to develop visual analytics for business intelligence, urban planning, and operational research that exploit the ever growing volume of spatial data that is becoming available in real time."

Ross's own research applications in geographic visualization and visual analytics focus on public health, social media use, criminal incident reporting, and dietary analysis. He is a regular participant in research competitions concerned with visual analytics. You can check out the current work of the Visual Analytics and Data Exploration Research Lab at vader.lab.asu.edu.

Courtesy: Ross Maciejewski

**Figure 12.8** Ross Maciejewski, computer scientist.

Courtesy: Ross Maciejewski

**Figure 12.9** Bristle map of vandalism complaints in Lafayette, Indiana.

**Table 12.1** Some examples of coordinate and cartographic transformations of spatial phenomena of different dimensionality. (Based on D. Martin. 1996. *Geographic Information Systems: Socioeconomic Applications* (2nd ed.). London: Routledge: 65)

| Dimension | 0 | 1 | 2 | 3 |
|---|---|---|---|---|
| Conception of spatial phenomenon | Population distribution | Coastline | Agricultural field | Land surface |
| *Coordinate transformation of real world arising in GI systems representation/measurement* | | | | |
| Measurement | Imposed areal aggregation | Sequence of digitized points | Digitized polygon boundary | Arrangement of spot heights |
| *Cartographic transformation to aid interpretation of representation* | | | | |
| Visualization | Cartogram or dasymetric map | Generalized line | Integral/natural area | Digital elevation model |

consider cartographic transformation here, rather than in Chapter 11 because this is likely to be the outcome of an interactive series of map transformations.

**GI systems provide a flexible medium for the cartographic transformation of maps.**

Some examples of the ways in which real-world phenomena of different dimensions (Box 2.2) may or may not be transformed are shown in Table 12.1. These illustrate how the representation filter of Figure 12.2 may have the effect of transforming some objects, such as population distributions, but not others, such as agricultural fields. Further transformation may occur in order to present the information to the user in the most intelligible and parsimonious way. Thus, cartographic transformation through the U3 filter in Figure 12.2 may result from the imposition of artificial units, such as census tracts for population, or selective abstraction of data, as in the generalization of cartographic lines (Section 3.8). The standard conventions of map production and display are not always sufficient to make the user aware of the transformations that have taken place between conception and measurement—as in choropleth mapping, for example, where mapped attributes are presented using the proportions of the zones to which they pertain.

Geovisualization techniques make it possible to manipulate the shape and form of mapped boundaries using GI systems, and in many circumstances it may be appropriate to do so. Indeed, where a standard mapping projection obscures the message of the attribute distribution, map transformation becomes necessary. There is nothing untoward in doing so. One of the messages of Chapter 4 was that all conventional mapping of geographically extensive areas entails transformation, in order to represent the curved surface of the Earth on a flat screen or a sheet of paper. Similarly, in Section 2.8, we described how

generalization procedures may be applied to line data in order to maintain their structure and character in small-scale maps. There is nothing sacrosanct about popular or conventional representations of the world, although the widely used transformations and projections do confer advantages in terms of wide user recognition and hence interpretability.

## 12.3.2 Cartograms

Cartograms are maps that lack planimetric correctness and distort area or distance in the interests of some specific objective. The usual objective is to reveal patterns that might not be readily apparent from a conventional map or, more generally, to promote legibility. Thus, the integrity of the spatial object, in terms of areal extent, location, contiguity, geometry, or topology, is made subservient to an emphasis on attribute values or particular aspects of spatial relations. One of the best-known cartograms (strictly speaking a *linear cartogram*) is the London Underground map, devised in 1933 by Harry Beck to fulfill the specific purpose of helping travelers to navigate across the network. The central-area cartogram that is widely used today is a descendant of Beck's 1933 map and provides a widely recognized representation of connectivity in London, using conventions that are well suited to the attributes of spacing, configuration, scale, and linkage of the London Underground system. The attributes of public transit systems differ between cities, as do the cultural conventions of transit users, and thus it is unsurprising that cartograms pertaining to transit systems elsewhere in the world appear quite different.

**Cartograms are map transformations that distort area or distance in the interests of some specific objective.**

A central tenet of Chapter 2 was that all representations are necessarily selective abstractions of reality.

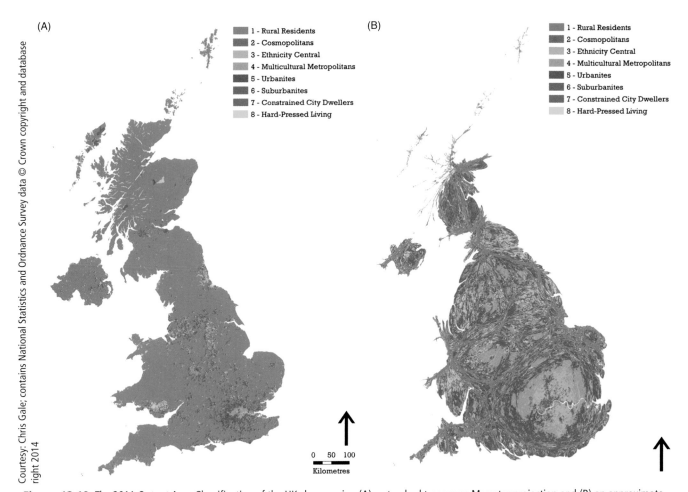

Legend (A):
1 - Rural Residents
2 - Cosmopolitans
3 - Ethnicity Central
4 - Multicultural Metropolitans
5 - Urbanites
6 - Suburbanites
7 - Constrained City Dwellers
8 - Hard-Pressed Living

Legend (B):
1 - Rural Residents
2 - Cosmopolitans
3 - Ethnicity Central
4 - Multicultural Metropolitans
5 - Urbanites
6 - Suburbanites
7 - Constrained City Dwellers
8 - Hard-Pressed Living

0   50   100
Kilometres

**Figure 12.10** The 2011 Output Area Classification of the UK shown using (A) a standard transverse Mercator projection and (B) an approximate area cartogram.

Cartograms depict transformed, hence artificial, realities, using particular exaggerations that are deliberately chosen. Figure 12.10A presents a transverse Mercator map of a *geodemographic classification* based upon the 2011 UK Census, in which colors are used to denote the lifestyle characteristics of neighborhoods. The most rural group, "Rural Residents," occupy the least densely settled parts of the country and so visually dominate the map, whereas the geographies of the other seven Super Groups are far less clear. Figure 12.10B presents an approximate area cartogram of the same data, in which the area of each census zone is made proportional to its population size, and the areas are stretched in order to maintain contiguity wherever possible. This makes the densely populated areas of London and the other major urban conurbations appear much larger, whereas sparsely populated areas, particularly in North Wales and Scotland, are diminished in extent. Every individual in the population is thus accorded approximately equal weight in Figure 12.10B. The overall shape and compass orientation of the country are kept roughly correct—although the real-world pattern of zone shape, contiguity, and topology is to some extent compromised. When attributes are mapped, the variations within cities are revealed, whereas the variations in the countryside are reduced in size so as not to dominate the image and divert the eye from the circumstances of the majority of the population. Cartograms like these have been widely used by Danny Dorling (Box 11.5) to depict the considerable diversity in social structure and wealth that occurs in many densely populated city areas.

Although cartograms usually require human judgment and design, automated routines (such as that in ArcGIS used to produce Figure 12.10B) are also available. Throughout this book, one of the recurrent themes has been the value of GI systems as a medium for data sharing, and this is most easily achieved if pooled data share common coordinate systems, or can be reprojected for comparison. Yet this convenience, as well as wide recognition of the mapped data, sometimes needs to be evaluated against the value of cartometric transformations that accommodate the way that humans think about or experience space, as with the measurement of distance as travel time or travel cost. As such, the visual transformations inherent in cartograms can provide improved means of envisioning spatial structure, and geovisualization offers an interactive and dynamic medium for experimentation with different transformations.

## 12.3.3 Remodeling Spatial Distributions as Dasymetric Maps

The cartogram offers a radical means of transforming space and hence achieving spatial balance, but sacrifices the familiar spatial framework valued by most users. It forces the user to make a stark choice between assigning each mapped element equal weight and being able to relate spatial objects to real locations on the Earth's surface. The interactive environment of geovisualization and scalable cached datasets provides a means by which the data-integrative power of GI systems can be used to remodel spatial distributions and hence assign spatial attributes to meaningful, recognizable spatial objects. One example of the way in which ancillary sources of information may be used to improve representation of a spatial distribution is known as *dasymetric mapping*. Here, the intersection of two datasets is used to suggest more precise estimates of a spatial distribution.

Figure 12.11A presents a census tract geography for which small-area population totals are known, whereas Figure 12.11B shows the spatial distribution of built structures in an urban area (which might be obtained from a cadaster or very-fine-resolution satellite imagery, for example). A reasonable assumption (in the absence of evidence of mixed land use or very different residential structures such as high-rise apartments and widely spaced bungalows) is that all the built structures house resident populations at uniform density. Figure 12.11C shows how this assumption, plus an overlay of the areal extent of built structures, allows population figures to be allocated to smaller areas than census tracts and allows calculation of indicators of residential density. The practical usefulness of dasymetric mapping is illustrated in Figure 12.12.

Figure 12.12 shows the location of an elongated census block in the City of Bristol, UK, which is characterized by both a high unemployment *rate* (Figure 12.12A) and a low absolute *incidence* of unemployment (Figure 12.12B). The resolution to this seeming paradox is that the tract houses few people, an unusually large proportion of whom are unemployed (see Section 5.4.3 for discussion of related issues of ecological fallacy). However, this hypothesis alone would not help us much in deciding whether or not the zone (or part of it) should be included in an inner-city workfare program, for example. Use of GI systems to overlay fine-resolution aerial photography (Figure 12.12C) reveals the tract to be largely empty of population apart from a small extension to a large housing estate. It would thus appear sensible to assign this zone the same policy status as the zone to its west.

**Dasymetric mapping uses the intersection of two datasets (or layers in the same dataset) to obtain more precise estimates of a spatial distribution.**

Dasymetric mapping and related techniques present a window on reality that looks more convincing than conventional choropleth mapping. However, it is important to remain aware that the visualization of reality is only as good as the assumptions that are used to create it. The information about population concentration used in Figure 12.12, for example, is defined only on the basis of common sense and is likely to be error prone (some built forms may be offices and shops, for example), and this will inevitably result in inaccurate visual

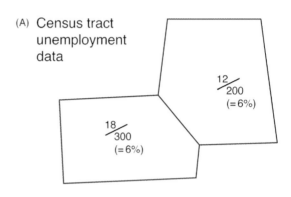

(A) Census tract unemployment data

$\frac{12}{200}$ (=6%)

$\frac{18}{300}$ (=6%)

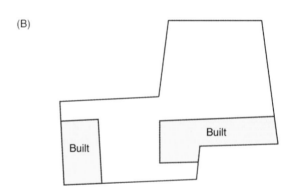

(B)

Built

Built

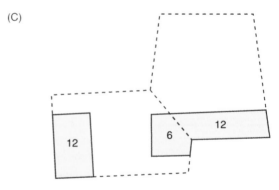

(C)

12

6    12

**Figure 12.11** Modeling a spatial distribution in an urban area using dasymetric mapping: (A) zonal distribution of census population; (B) distribution of built structures; and (C) overlay of (A) and (B) to obtain a more contained measure of population distribution.

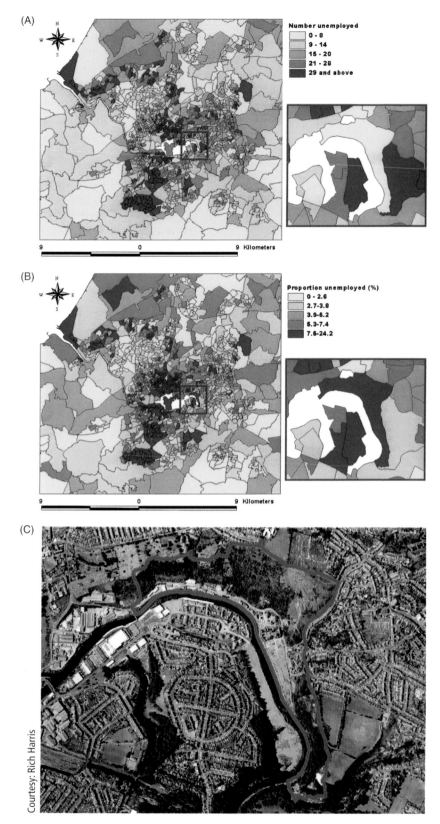

**Figure 12.12** Dasymetric mapping in practice, in Bristol, UK: (A) numbers employed viewed using lowest available census geography (block); (B) proportions unemployed using the same geography; and (C) orthorectified aerial photograph of part of the area of interest.

representation (see also Figure 12.20 below). Inference of land use from land cover, in particular, is an uncertain and error-prone process (see Section 5.3.2), and there is a developing literature on best practice for the classification of land use employing land cover information and classifying different (e.g., domestic versus nondomestic) land uses.

## 12.4 Participation, Interaction, Augmentation, and Dynamic Representation

### 12.4.1 Public Participation and Participatory GI systems (PPGIS)

Geovisualization can be used to facilitate the active involvement of many different groups in the discussion and management of change through planning, and may act as a bulwark against officialdom or big business. Good human–computer interaction is key to this process, and the use of geovisualization to foster public participation in (the use of) GI systems is often termed PPGIS. (The same acronym is also used to denote "Participatory GI Systems.") The focus of PPGIS is on how people perceive, manipulate, and interact with representations of the real world as manifested in GI systems (see Box 12.4). Its other concerns include the way people evaluate options through multicriteria decision making (Section 15.4), and the social issues of how GI systems usage remains concentrated within networks of established interests (Section 1.5.5). A related theme is the use of GI systems to create multiple representations— capturing and maintaining the different perspectives of stakeholders rather than framing debate in the terms of a single prevailing authoritative view.

---

## Applications Box 12.4

### User Interaction by Indigenous Tribespeople

The Extreme Citizen Science (ExCiteS) group is an interdisciplinary initiative, based at University College London that brings together expertise in anthropology, computer science, and geography, among other disciplines. The group is involved in several participatory mapping projects with indigenous forest communities, most of which are nonliterate, have never used information and communications technology (ICT) devices, and have little or no experience with reading maps, let alone creating them. Allowing these people to engage in spatial data collection and map making therefore requires the development of bespoke software as well as communication and training protocols.

One ExCiteS project, set in the African Republic of the Congo (also known as Congo-Brazzaville), aims to engage indigenous people in monitoring the ecological and social impacts of rainforest logging, using a smartphone application. Initial design work focused upon the task of measurement and representation (Figure 12.13),

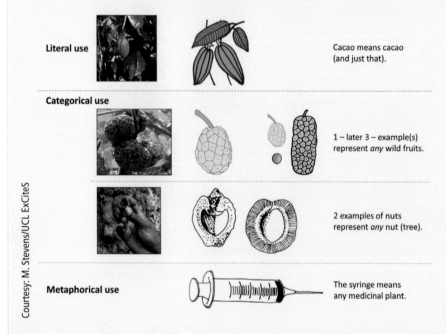

Literal use
Cacao means cacao (and just that).

Categorical use
1 – later 3 – example(s) represent *any* wild fruits.

2 examples of nuts represent *any* nut (tree).

Metaphorical use
The syringe means any medicinal plant.

Courtesy: M. Stevens/UCL ExCiteS

Figure 12.13 User interface design principles for native Congolese tribespeople.

▶

using pictorial icons to represent literal objects (e.g., cacao trees), categories of objects (e.g., examples of the category of native fruits), and metaphors (e.g., the syringe as a metaphor for medicinal plants). Other design work devised a user interface that allows for easy navigation between object classes (arranged in decision trees) and for linking instances to geographic locations and augmenting them with appropriate media (e.g., photography and voice recording).

Fieldwork was used to familiarize users with the devices and their interfaces (Figure 12.14A) so that they might be used to link attributes to locations throughout the study area (Figure 12.14B). Users were asked to comment on interface and icon designs, and their feedback was used to iteratively improve the software. As a result of this participatory design exercise it was possible to establish a set of icons that is well understood by the participants and that represent attributes of the environment that they considered relevant.

Check out this and related projects at the ExCiteS website (www.ucl.ac.uk/excites).

Images courtesy: J. Lewis/UCL ExCiteS

**Figure 12.14** (A) Preparatory work for (B) linkage of attributes to locations in the Congolese study area.

Geovisualization has a range of uses in PPGIS, including

- Making the growing complexity of land-use planning, resource use, and community development intelligible to communities and different government departments.
- Radically transforming the planning profession through use of new tools for community design and decision making.
- Unlocking the potential of the many Open Data (Section 17.4) sources that are becoming available at the local level.
- Helping communities shift land-use decisions from regulatory processes to performance-based strategies, and making the community decision-making process more proactive and less reactive.
- Improving community education about local environmental and social issues.

- Improving the feed of information between public and government in emergency planning and management.

In each of these applications, geovisualization has a strong cognitive component. Users need to feel equipped and empowered to interrogate representations in order to reveal otherwise-hidden information. This requires dynamic and interactive *software* environments and the *people* skills that are key to extracting meaning from a representation. Geovisualization allows people to use software to manipulate and represent data in multiple ways, in order to create what-if scenarios or to pose questions that prompt the discovery of useful relations or patterns. This is a core remit of PPGIS, where the geovisualization environment is used to support a process of knowledge construction and acquisition that is guided by the user's knowledge of the application. PPGIS research entails usability evaluations of structured tasks, using a mixture of computer-based usability evaluation techniques and traditional qualitative research methods, in order to identify cognitive activities and user problems associated with GI systems applications.

## 12.4.2 User Interaction and Representation in 2.5-D and 3-D

As indicated in Section 11.3.2, extruded 2.5-D representations may be used to reveal aspects of data that are not readily observeable in two dimensions. This is illustrated in Figure 12.15. The map shows London's strongly monocentric structure and the heavy specialization in office facilities both in the historic City of London and London's Docklands. The use of the extruded 2.5-D representation highlights urban density and areas of mixed land use, which are relevant to issues of sustainability and local travel. Metropolitan centers such as Croydon (south of the city) are dwarfed by the main center, suggesting they are struggling to compete. Another trend is the emergence of monofunctional and car-dependent "edge city" developments around Heathrow Airport.

The third dimension can also be used to represent built form, and in recent years, online 3-D models have become available for geographically extensive parts of many cities throughout the world. The virtual globe offerings from Microsoft and Google have arisen following the wide availability of very-fine-resolution height data from airborne instruments, such as Light Detection and Ranging (LiDAR; see Section 8.2.2.2). Augmentation of these models is increasingly straightforward, given the wide availability of free software such as Google Sketchup that enables even the nonspecialist to create 3-D representations of individual buildings and site layouts or to allow users to walk through the interiors of buildings. These representations provide the general user with valuable context to augment online two-dimensional maps and street views, and they have been used for general applications as diverse as city marketing and online social networking. Research applications can benefit from general comparisons with 3-D built form, as with geographer Paul Torrens's work relating the built form of the city to its WiFi geography (Figure 12.16). "Off-the-shelf" 3-D representations can provide very useful contextual information, although it is important to check the provenance of the data that were used to create them, especially if the motivation for their creation was more directly aligned with the generation of advertising revenues than provision of spatial data infrastructure (Section 18.6). Detailed 3-D city models have been used to contextualize the effects of general processes (in the ways suggested in Section 1.3), as for example with modeling the concentration and circulation of pollutants in cities (Figure 12.17).

In many respects, such applications crystallize both what is most exciting and what is most frustrating about GI science. They are typically assembled from diverse sources by stakeholders in local and central government, in utilities, in transportation, and

**Figure 12.15** A 3-D representation of office and retail land use in London. The height of the bars identifies density of office and retail floorspace on 500-meter grid squares.

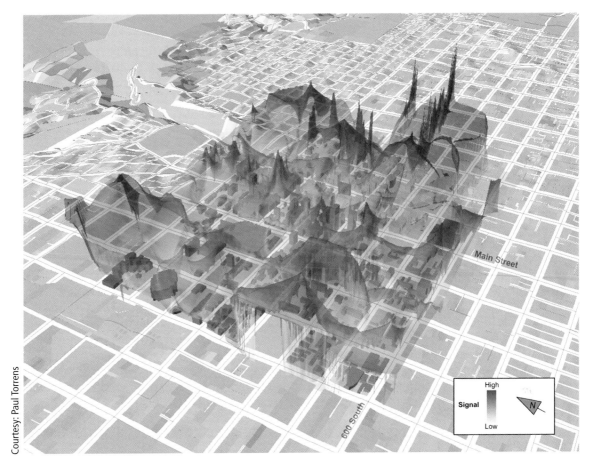

Courtesy: Paul Torrens

**Figure 12.16** A dense network of WiFi infrastructure viewed above the built environment of Salt Lake City, Utah. The pattern of WiFi transmissions suggests no centralized spatial organization, yet does mirror urban built form. WiFi activity is most prominent in the city's traditional commercial core, yet also reaches out to interstitial and peripheral parts of the city. (Areal extent of the WiFi cloud on the ground is ~ 3 km$^2$: "high" and "low" refers to the strength of WiFi signals, measured in decibels relative to 1 mW.)

Courtesy: Steve Evans: Ordnance Survey data © Crown Copyright; NOx data courtesy of Environmental Research Group, Kings College London and the Greater London Authority

**Figure 12.17** Visualizing nitrogen oxide (NOx) pollution in the Virtual London model. This pollutant, often nicknamed "urban smog," is largely derived from vehicle emissions: red identifies higher levels, whereas blue represents lower levels.

increasingly from members of the public. The very detailed yet disparate nature of data holdings suitable for inclusion in 3-D models makes networked GI systems the ideal medium for assembling and integrating data and for communicating and sharing the results. Yet it is crucial to ensure that each of the data sources used is suitable for the task, particularly if multiscale process models are to be applied to them.

It is important to consider the development of fine-scale city models in the context of the seamless scale-free mapping and street-view systems that allow users to view raster and vector information, including features such as buildings, trees, and automobiles, alongside synthetic and photorealistic global and local displays. In global visualization systems, datasets are projected onto a global TIN-based data structure (see Section 7.2.3.4). Rapid interactive rotation and panning of the globe is enabled by caching data in an efficient in-memory data structure. Multiple levels of detail, implemented as nested TINs and reduced-resolution datasets, allow fast zooming in and out. When the observer is close to the surface of the globe, these systems allow the display angle to be tilted to provide a perspective view of the Earth's surface. The user is able to roam smoothly over large volumes of global geographic information. This enables a better understanding of the distribution and abundance of globally distributed phenomena (e.g., ocean currents, atmospheric temperatures, and shipping lanes), as well as detailed examination of local features in the natural and built environment (e.g., optimum location of cell phone towers, or the impact of tree felling on the viewscape of tourist areas).

## 12.4.3 Handheld Computing and Augmented Reality

Immersion of desk- and studio-based users in 3-D virtual reality (see also Section 10.3.1 for a discussion of virtual and augmented reality) presents one way of promoting better remote interaction with the real world, using high-power computing and very-high-bandwidth computer networks. The same digital infrastructure is also being used to foster improved and more direct interaction with the real world by the development of a range of handheld, in-vehicle, and wearable computer devices. These are discussed, along with some of the geovisualization conventions they entail, in Sections 6.6 and 10.3. Figure 12.18 illustrates how semi-immersive systems are used in field computing, taking the example of a geography field course in the UK Lake District.

Figure 12.18A illustrates how students previously visualized past landscapes, by aligning a computer-generated viewshed with the real world. This approach exemplifies many of the shortcomings of traditional mapping: perfect alignment is possible from only a single point on the landscape, only a single prerendered coverage can be viewed, and there is no way of interrogating the database that was used to create the augmented representation. Use of the equipment shown in Figure 12.18B enables the user to establish the locations and directions of viewpoints, and loaded images are automatically displayed as the user approaches any of the numbered points. The handheld device makes it possible to annotate the view and to display a range of surface and subsurface characteristics (Figure 12.18C). As discussed in Section 10.3, the current generation of augmented-reality devices remains unable to render viewsheds on the fly, for reasons of computer and bandwidth capacity. Nonetheless, progress continues to be made in this direction, for example, through the integration of digital compasses into mobile devices.

Still other media and software are used to immerse users in the artificial worlds of *virtual reality*. Although not mainstream GI system applications, such tools can be used to help users better understand geographic real-world patterns and processes. The advent of immersive and semi-immersive systems has important implications for participation by a broad user base because they allow

- Users to access virtual environments and select different views of phenomena.
- Incremental changes in these perspectives to permit real-time *fly-throughs*.
- Repositioning or rearrangement of the objects that make up virtual scenes.
- Users to be represented graphically as *avatars*—digital representations of themselves.
- Engagement with avatars connected at different remote locations, in a networked virtual world.
- The development, using avatars, of new kinds of representation and modeling.
- The linkage of networked virtual worlds with *virtual reality* (VR) systems.

The developments that have fostered the emergence of 3-D visualization in GI systems have also encouraged the development of online virtual worlds (e.g., Second Life: www.secondlife.com) in which online users (represented as avatars) engage, interact, and collaborate in a peer-to-peer Web 2.0 environment.

There exists the prospect of infusing the collaborative environment of Second Life and other online spaces into virtual Earth applications (see Section 10.3.1) and creating an occupied Digital Earth. It is also likely that the kinds of advanced models discussed in Chapter 15 (e.g., agent-based models: Section 15.2) will also be assimilated into these virtual worlds in the near future.

**Figure 12.18** (A) Aligning a viewshed transparency to augment a view of the landscape; (B) use of computer devices to achieve the same effect for (C) a number of field sites and different surface characteristics, with the facility to annotate landscape views.

## 12.4.4 Visualizing Geotemporal Dynamics

Early GI systems were poorly adapted to representing temporal change, although improvements in computer processing power and software enhancements now make animation of representations routine and fairly straightforward. The wider use of geocomputational methods is also increasing the scope for reducing Big Data to mappable phenomena, using a range of visual techniques. It is not possible to summarize the wide range of work in this short overview, and we focus instead upon three instances of mapping geographic phenomena that become apparent at temporal scales ranging from the intergenerational to the diurnal.

The first (Figure 12.19) comes from the work of geographer Jens Kandt, who has mapped some of the data from Walter Bodmer's (Box 1.4) "People of the British Isles" project. The project took genetic samples from carefully selected individuals whose families had remained in the same localities in the UK for at least three generations. Clustering the important markers on the human genome is a significant Big Data project, and Figure 12.19 presents one of Kandt's attempts to develop a "genetic map" of the long-settled residents of the UK, using distance decay from sample locations. The message of this map is that it is still possible to trace the effects of successive waves of invaders of the British Isles upon contemporary population structure.

The dynamics of change in the built environment of cities is captured in socioeconomic classifications of social structure, which are often updated and mapped every ten years when the results of censuses or other major population surveys are released. Figure 12.20A shows part of London for one such classification, the 2011 UK Output Area Classification, depicted as a choropleth map. Such representations make it possible to depict the kaleidoscope of change between censuses, within the constraint of the administrative geographies for

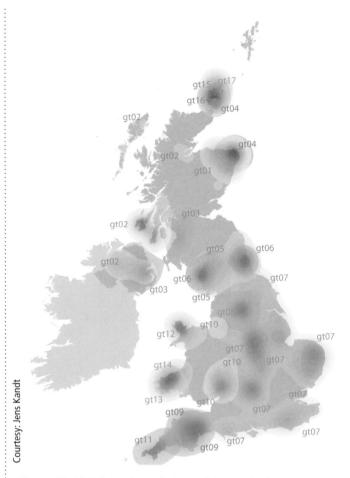

**Figure 12.19** A "genetic map" of part of the British Isles. The map uses kernel density estimation to impute values around the biological samples used to create Figure 1.13B.

(A)          (B)

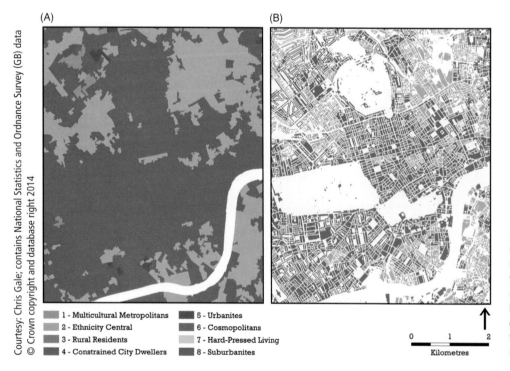

1 - Multicultural Metropolitans
2 - Ethnicity Central
3 - Rural Residents
4 - Constrained City Dwellers
5 - Urbanites
6 - Cosmopolitans
7 - Hard-Pressed Living
8 - Suburbanites

0    1    2
Kilometres

**Figure 12.20** Two representations of the 2011 Output Area Classification for the UK in Central London: (A) a standard choropleth map based upon Census Output Area geography and (B) using building footprints from Ordnance Survey (GB) Open Data.

which census and other data are released. Geographer Chris Gale has remodeled this distribution as a kind of dasymetric map (Section 12.3.3), which gives a better impression of changes in residential structure with respect to the built environment. Figure 12.20B uses Open Data made available through Great Britain's Ordnance Survey, and provides a means of mapping socioeconomic distributions that are not visually dominated by large vacant areas such as parks, although this Open Data source does not yet make it possible to discriminate between residential and other land uses.

At the other end of the temporal spectrum, Figure 12.21 presents computer scientist Muhammad Adnan's analysis of geotagged Tweets, selected for different times of day and segmented according to the ethnic and linguistic group of their originator. Such mapping makes it possible to map the flows of different groups within London according to time of day, week, or season. A related map of

**Figure 12.21** Geotemporal snapshots of the pattern of Twitter usage in Greater London by individuals classified as being of Polish or Sikh origin.

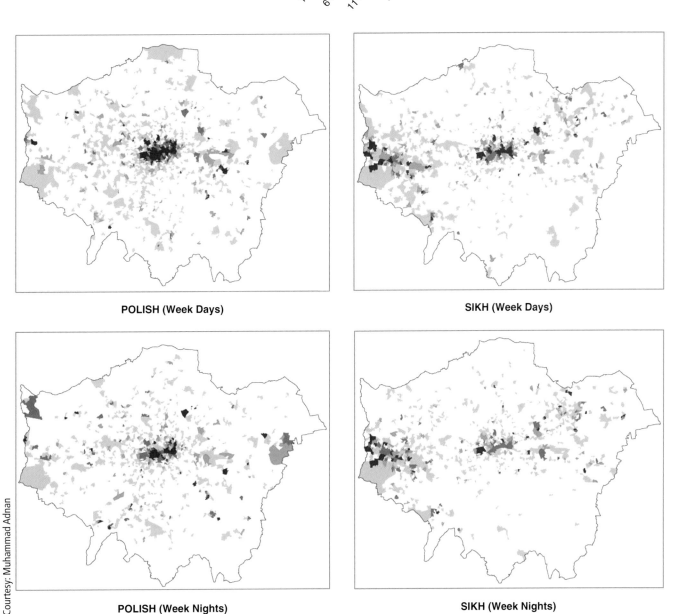

Number of Tweets

0    1 - 5    6 - 10    11 - 20    >20

POLISH (Week Days)          SIKH (Week Days)

POLISH (Week Nights)         SIKH (Week Nights)

Courtesy: Antonio Lima, map tiles by Stamen Design, map data © OpenStreetMap contributors

**Figure 12.22** The use of mobile telephone data to characterize daily movement patterns. Hue is used to identify time of day for each individual and travel mode.

activity patterns is provided in Figure 12.22. Here, computer scientists Antonio Lima and Mirco Musolesi have mapped the movements of three mobile phone users over time, using circles of different colors and hues. Hue is used to identify time of day, allowing the user to visualize the correlation between these movement patterns and any synchronization. Such visualizations may be used in attempts to predict future locations or activity patterns of the users.

## 12.5 Consolidation

Geovisualization can make a powerful contribution to decision making and can be used to simulate changes to reality. Yet, although it may be governed by scientific principles, the limitations of human cognition mean that user interpretation inevitably presents a further selective filter on the reality that is being represented. Mapping is about seeing the detail as well as the big picture, yet the wealth of detail that is available from today's Big Data can

sometimes create information overload and threaten to overwhelm the message of the map. Geovisualization seeks to foster effective user interaction and participation in the use of GI systems as decision support tools.

At its worst, greater sophistication of display may simply confer information overload to the user; or it may oversimplify and fail to communicate the inherent uncertainty in the spatial data that are being represented. But at its best, geovisualization empowers users to make informed and balanced assessments of geographic representations and thence to create better descriptions, explanations, predictions, and understanding. These benefits need not accrue only to specialists: geovisualization and PPGIS extend the community of engaged users and encourage it to visualize geography, to exchange spatially literate views, and to participate in effective decision making. Yet although the media and software may be changing at breathtaking pace, the old adage that "seeing is believing" only holds if visualization and user interaction are founded upon representations that are fit for purpose.

## Questions for Further Study

1. Figure 12.23 is a cartogram redrawn from a newspaper feature on the costs of air travel from London in 1992. Using current advertisements in the press and on the Internet, create a similar cartogram of travel costs, in local currency, from the nearest international air hub to your place of study.

2. How can Web-based multimedia GI systems be used to improve community participation in decision making?

3. Produce two (computer-generated) maps of the Israeli West Bank security fence barrier to illustrate opposing views of its effects. For example, the first might illustrate how it helps to preserve an Israeli "Lebensraum," whereas the second might emphasize its negative effects on Palestinian communities. In a separate short annotative commentary, describe the structure and character of the fence at a range of spatial scales.

4. Review the common *sources* of uncertainty in geographic representation and the ways in which they can be *manifested* through geovisualization.

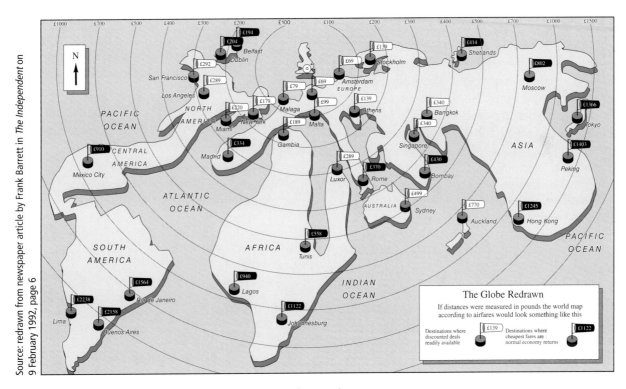

**Figure 12.23** The globe redrawn in terms of 1992 travel costs from London.

## Further Reading

Dorling, D. 2012. *The Visualization of Spatial Social Structure*. Chichester, UK: John Wiley & Sons.

Dykes, J., MacEachren, A. M., and Kraak, M.-J. (eds.) 2005. *Exploring Geovisualization*. London: Elsevier Science.

Haklay, M. (ed.) 2010. *Interacting with Geospatial Technologies*. Chichester, UK: John Wiley & Sons.

Kraak, M. J. and Ormeling, F. J. 2011. *Cartography visualization of spatial data* (3rd ed.). New York: Guilford Press.

Tufte, E. R. 2001. *The Visual Display of Quantitative Information* (2nd ed.). Cheshire, CT: Graphics Press.

# 13 Spatial Data Analysis

## LEARNING OBJECTIVES

This chapter is the first in a set of three dealing with geographic analysis and modeling methods. The chapter begins with a review of the relevant terms and presents an outline of the major topics covered in the three chapters. Analysis and modeling are grounded in spatial concepts, allowing the investigator to examine the role those concepts play in understanding the world around us. This chapter also examines methods constructed around the concepts of location, distance, and area. Location provides a common key between datasets, allowing the investigator to discover relationships and correlations between properties of places. Distance defines the separation of places in geographic space and acts as an important variable in many of the processes that impact the geographic landscape. Areas define the neighborhood context of processes and events and also capture the important concept of scale.

After studying this chapter you will understand:

- Definitions of spatial data analysis and related terms and tests to determine whether a method is spatial.

- Techniques for detecting relationships between the various properties of places and for preparing data for such tests.

- Methods to examine distance effects, in the creation of clusters, hotspots on *heat maps*, and anomalies.

- The applications of convolution in GI systems, including density estimation and the characterization of neighborhoods.

## 13.1 Introduction: What Is Spatial Analysis?

Many terms are used to describe the techniques that are the focus of these chapters. Although spatial analysis and spatial data analysis are the ones preferred here for the techniques discussed in Chapters 13 and 14, it is also common to find them being described as data *mining* and data *analytics*. Data mining often implies large volumes of data, so is often encountered in discussions of Big Data and the general search for patterns without particular reference to hypotheses about what those patterns should look like. The term data *analytics* is increasingly used in business applications and again is popular in the context of Big Data. To all intents and purposes, however, the terms are

equivalent. The definition of spatial modeling is left to Chapter 15.

The techniques covered in these three chapters are generally termed *spatial* rather than *geographic* because they can be applied to data arrayed in any space, not only geographic space (see Section 1.1.2). Many of the methods might potentially be used in analysis of outer space by astronomers, or in analysis of brain scans by neuroscientists. So the term *spatial* is used consistently throughout these chapters. It is important to recognize at the outset that spatial analysis is *different* because the standard assumptions made in many statistical tests are not valid when dealing with spatial data, a topic addressed at length in Chapter 14. It is important, therefore, to keep a clear distinction between statistical analysis and spatial analysis.

Spatial analysis is in many ways the crux of geographic information science and systems (GISS) because it includes all the transformations, manipulations, and methods that can be applied to geographic data to add value to them, to support decisions, and to reveal patterns and anomalies that are not immediately obvious. In other words, spatial analysis is the process by which we turn raw data into useful information, in pursuit of scientific discovery, or more effective decision making. If a geographic information (GI) system is a method of communicating information about the Earth's surface from one person to another, then the transformations of spatial analysis are ways in which the sender tries to inform the receiver, by adding greater informative content and value, and by revealing things that the receiver might not otherwise see.

Some methods of spatial analysis were developed long before the advent of GI systems, when they were carried out by hand or by the use of measuring devices like the ruler. The term *analytical cartography* is sometimes used to refer to methods of analysis that can be applied to maps to make them more useful and informative, and spatial analysis using GI systems is in many ways its logical successor. But it is much more powerful because it covers not only the contents of maps but also any type of geographic data.

### Spatial analysis can reveal things that might otherwise be invisible—it can make what is implicit explicit.

In this chapter we will look first at some definitions and basic concepts of spatial analysis. Concepts such as location, distance, and area—the topics discussed in this chapter—have already been encountered at various points in this book, but here they serve to provide an organizing framework for the vast array of methods that fall under the heading of spatial analysis. More advanced concepts, as well as the methods used to elucidate them, are discussed in Chapter 15, along with the use of GI systems in design, when their power is directed not to understanding the world but to improving it according to specific goals and objectives. Chapter 16 addresses the use of GI systems to examine dynamic processes, primarily by simulation.

### Spatial analysis is the crux of GISS, the means of adding value to geographic data and of turning data into useful information.

Methods of spatial analysis can be very sophisticated, but they can also be very simple. A large body of methods of spatial analysis has been developed over the past century or so, and some methods are highly mathematical—so much so that it might sometimes seem that mathematical complexity is an indicator of the importance of a technique. But the human eye and brain are also very sophisticated processors of geographic data and excellent detectors of patterns and anomalies in maps and images. So the approach taken here is to regard spatial analysis as spread out along a continuum of sophistication, ranging from the simplest types that occur very quickly and intuitively when the eye and brain look at a map to the types that require complex software and sophisticated mathematical understanding. Spatial analysis is best seen as a *collaboration* between the computer and the human, in which both play vital roles.

### Effective spatial analysis requires an intelligent user, not just a powerful computer.

There is an unfortunate tendency in the GI systems community to regard the making of a map using a GI system as somehow less important than the performance of a mathematically sophisticated form of spatial analysis. According to this line of thought, *real* use of GI systems involves number crunching, and users who *just* use GI systems to make maps are not serious users. But every cartographer knows that the design of a map can be very sophisticated and that maps are excellent ways of conveying geographic information and knowledge by revealing patterns and processes to us (see Chapter 11). We agree; moreover, we believe that mapmaking is potentially just as important as any other application of GI systems.

Spatial analysis may be defined in many possible ways, but all definitions in one way or another express the basic idea that information on locations is essential—that analysis carried out without knowledge of locations is not spatial analysis. One fairly formal statement of this idea is the following:

- Spatial analysis is a set of methods whose results are not invariant under changes in the locations of the objects being analyzed.

The double negative in this statement follows convention in mathematics, but for our purposes we can remove it:

- Spatial analysis is a set of methods whose results change when the locations of the objects being analyzed change.

On this test the calculation of an average income for a group of people is not spatial analysis because it in no way depends on the locations of the people. But the calculation of the center of the U.S. population is spatial analysis because the results depend on knowing where all U.S. residents are located. A GI system is an ideal platform for spatial analysis because its data structures accommodate the storage of object locations, as will have become clear from Chapters 6 and 7.

The techniques discussed in these chapters are no more than the tip of the spatial analysis iceberg. Some GI systems are more sophisticated than others in the range of techniques they support, and others are strongly oriented toward certain domains of application. For a more advanced review including many techniques not mentioned here, and for a detailed analysis of the techniques supported by each package, the reader is encouraged to examine the book *Geospatial Analysis: A Comprehensive Guide to Principles, Techniques and Software Tools* by De Smith, Goodchild, and Longley, which is available online at www.spatialanalysisonline.com.

## 13.1.1 Examples

Spatial analysis can be used to further the aims of science, by revealing patterns that were not previously recognized and that hint at undiscovered generalities and laws. Patterns in the occurrence of a disease may hint at the mechanisms that cause the disease, and some of the most famous examples of spatial analysis are of this nature, including the work of Dr. John Snow in unraveling the causes of cholera (see Box 13.1).

---

### Biographical Box (13.1)

### Dr. John Snow and the Causes of Cholera

In the 1850s cholera was poorly understood, and massive outbreaks were a common occurrence in major industrial cities. (Even today cholera remains a significant health hazard in many parts of the world, despite progress in understanding its causes and advances in treatment.) An outbreak in London in 1854 in the Soho district was typical of the time, and the deaths it caused are mapped in Figure 13.1. The map was made by Dr. John Snow (Figure 13.2), who had conceived the hypothesis that cholera was transmitted through the drinking of polluted water rather than through the air, as was commonly believed. He noticed that the outbreak appeared to be centered on a public drinking water pump in Broad Street (Figure 13.3)—and if his hypothesis was correct, the pattern shown on the map would reflect the locations

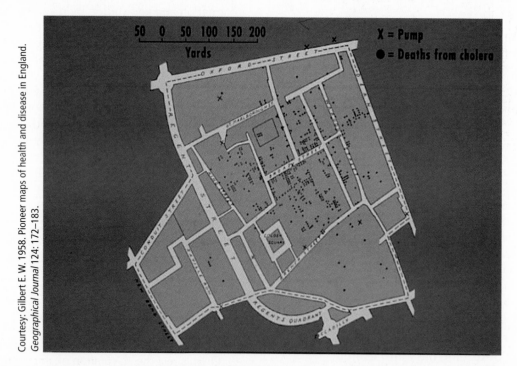

Courtesy: Gilbert E. W. 1958. Pioneer maps of health and disease in England. *Geographical Journal* 124: 172–183.

**Figure 13.1** A redrafting of the map made by Dr. John Snow in 1854, showing the deaths that occurred in an outbreak of cholera in the Soho district of London. The existence of a public water pump in the center of the outbreak (the cross in Broad Street) convinced Snow that drinking water was the probable cause of the outbreak. Stronger evidence in support of this hypothesis was obtained when the water supply was cut off, and the outbreak subsided.

▶

of people who drank the pump's water. There appeared to be anomalies, in the sense that deaths had occurred in households that were located closer to other sources of water, but he was able to confirm that these households also drew their water from the Broad Street pump. Snow had the handle of the pump removed, and the outbreak subsided, providing direct causal evidence in favor of his hypothesis. The full story is much more complicated than this largely apocryphal version, of course; much more information is available at www.jsi.com.

Today, Snow is widely regarded as the father of modern epidemiology.

Figure 13.3 A modern replica of the pump that led Snow to the inference that drinking water transmitted cholera, located in what is now Broadwick Street in Soho, London.

© TopFoto/The Image Works

Figure 13.2 Dr. John Snow.

It is interesting to speculate on what would have happened if early epidemiologists like Snow had had access to a modern GI system. The rules governing research today would not have allowed Snow to remove the pump handle, except after lengthy review, because the removal constituted an experiment on human subjects. To get approval, he would have had to have shown persuasive evidence in favor of his hypothesis, and it is doubtful that the map would have been sufficient because several other hypotheses might have explained the pattern equally well. First, it is conceivable that the population of Soho was inherently at risk of cholera, perhaps by being comparatively elderly or because of poor housing conditions. The map would have been more convincing if it had *normalized* the data by showing the *rate* of incidence relative to the population at risk. For example, if cholera was highest among the elderly, the map could have shown the number of cases in each small area of Soho as a proportion of the population over 50 in each area. Second, it is still conceivable that the hypothesis of transmission through the air between carriers could have produced the same observed pattern, if the first carrier happened to live in the center of the outbreak near the pump, though requiring a coincidence makes this hypothesis less attractive. Snow could have eliminated this alternative if he had been able to produce a sequence of maps, showing the locations of cases as the outbreak developed. Both of these options involve simple spatial analysis of the kind that is readily available today in GI systems. He might also have cited the scientific principle known as *Occam's Razor*, which favors accepting the simplest hypothesis when more than one is available.

## GI systems provide tools that are far more powerful than the map in suggesting causes of disease.

Today the causal mechanisms of diseases like cholera, which results in short, concentrated outbreaks, have long since been worked out. Much more problematic are the causal mechanisms of diseases that are rare and not so sharply concentrated in space and time. The now-classic work of Stan Openshaw at the University of Leeds, using one of his Geographical Analysis Machines, illustrates the kinds of applications that make good use of the power of GI systems in a more contemporary context.

Figure 13.4 shows an application of one of Openshaw's techniques to a comparatively rare but devastating disease whose causal mechanisms remain largely a mystery—childhood leukemia. The study area is northern England, the region from the Mersey to the Tyne Rivers. The analysis begins with two datasets: one of the locations of cases of the disease and the other of the numbers of people at risk in standard census reporting zones. Openshaw's technique then generates a large number of circles, of random sizes, and places them randomly over the map. The computer generates and places the circles and then analyzes their contents by dividing the number of cases found in the circle by the size of the population at risk within the circle. If the ratio is anomalously high, the circle is drawn. After a large number of circles have been generated, and a small proportion have been drawn, a pattern emerges. Two large concentrations, or clusters of cases, are evident in the figure. The one on the left is located around Sellafield, the location of the British Nuclear Fuels processing plant and a site of various kinds of past leaks of radioactive material. The other, in the upper right, is in the Tyneside region, and Openshaw and his colleagues discuss possible local causes.

Both of these examples are instances of the use of spatial analysis for scientific discovery and decision making, specifically for public health.

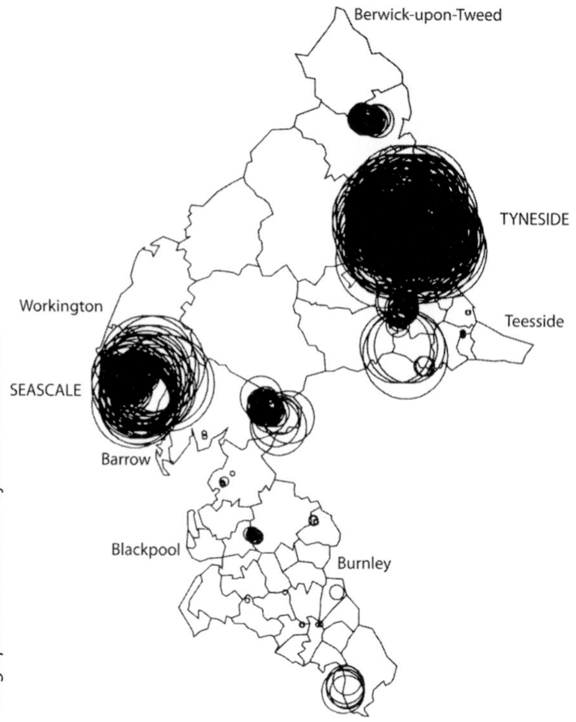

Source: Openshaw S., Charlton M., Wymer C., and Craft A. 1987. A Mark I geographical analysis machine for the automated analysis of point datasets. *International Journal of Geographical Information Systems* 1: 335–358.

**Figure 13.4** The map made by Openshaw and colleagues by applying their Geographical Analysis Machine to the incidence of childhood leukemia in northern England. A very large number of circles of random sizes is randomly placed on the map, and a circle is drawn if the number of cases it encloses substantially exceeds the number expected in that area given the size of its population at risk.

Box 13.2 is devoted to Sara McLafferty, who teaches and researches the use of GI systems in public health.

Sometimes spatial analysis is used *inductively*, to examine empirical evidence in the search for patterns that might support new theories or general principles, in this case with regard to disease causation (we suggested earlier that the term *data mining* is often used in this context). Inductive reasoning is often identified as characteristic of Big Data, as we noted in Section 1.4. Other uses of spatial analysis are *deductive*, focusing on the testing of known theories or principles against data. (Snow already had a theory of how cholera was transmitted and used the map as powerful confirmation to convince others.) Induction and deduction were also discussed in Section 2.7 in the context of the nature of spatial data. A third type of application is *normative*, using spatial analysis to develop or prescribe new or better designs, for the locations of new retail stores, or new roads, or a new manufacturing plant. Examples of this type appear in Section 14.4.

## Sara McLafferty

Sara McLafferty (Figure 13.5) is Professor of Geography and Geographic Information Science at the University of Illinois at Urbana-Champaign. Throughout her career she has been fascinated by the question: How do the places where people work and interact affect their health and well-being? Where we live influences our exposure to environmental pollution; our access to health services, well-paying jobs, parks, and recreational spaces; our exposure to crime and violence; and our ability to have supportive social interactions. Two decades ago, Sara and her students began using GI systems and spatial analysis methods to better understand the connections between place environments and health. An early study looked at the spatial clustering of breast-cancer cases in the town of West Islip in Long Island, NY, and analyzed whether spatial clusters were in close proximity to certain environmental hazards. Community members had an important role in the project, raising concerns about environmental factors underlying the high incidence of breast cancer that Sara's team investigated using GI systems. As the field of GI systems and health began to take off during the 1990s, Sara and Ellen Cromley decided to write a book describing the dynamic new field of GI systems and health. Published in 2002, with a second edition

Courtesy: Sara McLafferty

**Figure 13.5** Sara McLafferty.

in 2011, their book, *GIS and Public Health*, presents a foundation for the field and discusses applications of GI systems in exploring the determinants of health and planning health interventions in fields like infectious disease control, environmental health, and access to health services. Public-health researchers and planners from across the globe have adopted the book.

## 13.2 Analysis Based on Location

The concept of location—identifying *where* something exists or happens—is central to GISS, and the ability to compare different properties of the same place, and as a result to discover relationships and correlations and perhaps even explanations, is often presented as the field's greatest advantage. Take Figure 13.6 as an example. It shows the age-adjusted rate of death in the United States due to cancers of the throat and lung among adult males in a 20-year period 1950–1969, compiled by county. Maps such as this immediately prompt us to ask: Do I know of other properties of these counties that might explain their rates? Within a fraction of a second the mind is busy identifying counties, recalling general knowledge that might suggest cause, and checking other counties to see if they confirm suspicions. All of this depends, of course, on having a basis of knowledge (Section 1.2) in one's mind that is sufficient to identify particular counties and their general characteristics. One might note, for example, that Chicago, Detroit, Cleveland, and several other major cities are *hotspots* where cancer rates are significantly high, but that

other major cities such as Minneapolis are not—and ask why? It is important to note that rather than the *rate* of cancer as a proportion of some estimate of the total population at risk, the map highlights counties where the rate is *statistically significant* according to standard statistical tests. To reach this threshold, a county needs both a high rate and a large population because rates in counties with small populations can be extreme just by chance. But in some counties, notably Silver Bow County in Montana, the home of Butte and a massive copper-mining operation, the rate in that period was so high as to be significant even within a comparatively small population.

**GI systems allow us to look for explanations among the different properties of a location.**

The map shows many intriguing patterns, but the pattern that is of greatest significance to the history of cancer research concerns counties distributed around the Pacific, Gulf, and Atlantic coasts. These were counties involved in ship construction during World War II, when large amounts of asbestos were used for insulation. In the period 5 to 25 years later, the consequences of

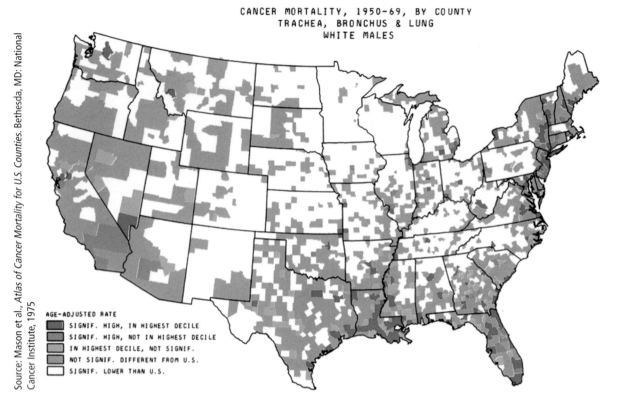

Source: Mason et al., *Atlas of Cancer Mortality for U.S. Counties*. Bethesda, MD: National Cancer Institute, 1975

CANCER MORTALITY, 1950-69, BY COUNTY
TRACHEA, BRONCHUS & LUNG
WHITE MALES

AGE-ADJUSTED RATE

- SIGNIF. HIGH, IN HIGHEST DECILE
- SIGNIF. HIGH, NOT IN HIGHEST DECILE
- IN HIGHEST DECILE, NOT SIGNIF.
- NOT SIGNIF. DIFFERENT FROM U.S.
- SIGNIF. LOWER THAN U.S.

**Figure 13.6** Age-adjusted rates of mortality due to cancers of the trachea, bronchus, and lung, among white males between 1950 and 1969, by county.

breathing asbestos fibers into the respiratory system are plain to see in the high rates of cancers in the counties containing Mobile, Alabama; Norfolk, Virginia; Jacksonville, Florida; and many other port cities.

It may seem that comparing the properties of places should be straightforward in a technology that insists on giving locations to all the information it stores. The next subsection discusses examples of such cases. In other cases, however, comparison can be quite difficult and complex. The data models adopted in GI systems and described in Chapter 8 are designed primarily to achieve efficiency in representation and to emphasize storing *information about one property for all places* over *information about all properties for one place*. We sometimes term this *horizontal* rather than *vertical* integration of data. For example, it is traditional to store all the elevation data for a given county together, perhaps as a digital elevation model, and all the soil data for the same county together, perhaps as a set of topologically related soil polygons. These are very efficient approaches, but they are not designed for a point-by-point comparison of elevation and soil type, or for answering questions such as "Are certain soil types more likely to be found at high elevations in this county?" Subsequent subsections discuss some of the GI system techniques designed specifically for situations such as this.

## 13.2.1 Analysis of Attribute Tables

In the example shown in Figure 13.6, it is quite likely that the kinds of factors responsible for high rates of cancer are already available in the attribute table of the counties, along with the cancer rates. In such cases our interest is in comparing the contents of two columns of the table, looking for possible relationships or correlations—are there counties for which cancer rates and the values of potentially causative variables are both high, or both low? Figure 13.7 shows a suitable example, where the interest lies in a possible relationship between two columns of a county attribute table. In this case the investigator suspects a pattern in the relationship between average value of house and percent black, variables that are collected and disseminated by the U.S. Bureau of the Census as county attributes. One way to examine this suspicion is to plot one variable against the other as a *scatterplot*. In Figure 13.7 median house value is shown on the vertical or *y*-axis, and percent black on the horizontal or *x*-axis.

In a formal statistical sense, these scatterplots allow us to examine in detail the *dependence* of one variable on one or more *independent* variables. For example, we might hypothesize that the median value of houses in a county is correlated with a number of variables such

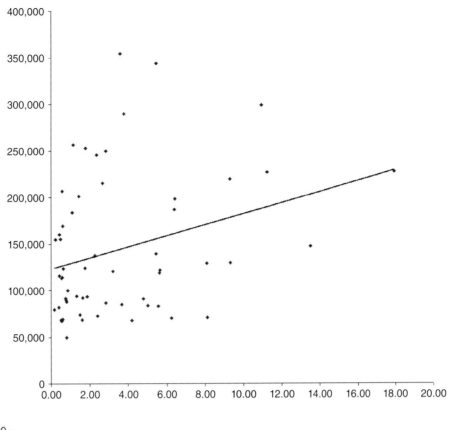

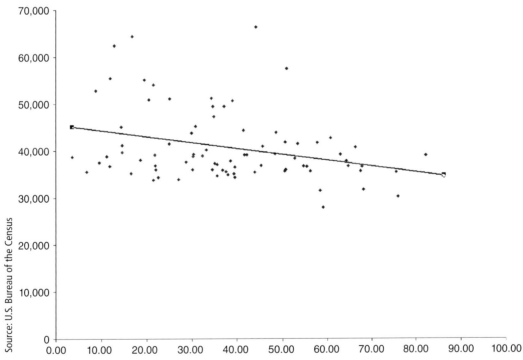

**Figure 13.7** Scatterplots of median house value (*y*-axis) versus percent black (*x*-axis) for U.S. counties in 1990, with linear regressions: (A) California; and (B) Mississippi.

as percent black, average county income, percent 65 and over, average county air pollution levels, and so forth, all stored in the attribute table associated with the properties. Formally this may be written as:

$$Y = f(X_1, X_2, X_3, \ldots, X_k)$$

where $Y$ is the dependent variable and $X_1$ through $X_k$ are all the possible independent variables that might affect housing value or be correlated with it. It is important to note that it is the independent variables that together predict the dependent variable, and that any hypothesized causal relationship is one way—that is, that median house value is *responsive* to average income, percent 65 and over, and so forth, and not vice versa. For this reason the dependent variable is termed the *response* variable, and the independent variables are termed *predictor* variables in some statistics textbooks.

In our case there is only a single predictor variable, and it is clear from the scatterplots that the relationship is far from perfect—that many other unidentified factors contribute to the determination of median housing value. In general this will always be true in the social and environmental sciences because it will never be possible to capture all the factors responsible for a given outcome. So we modify the model by adding an error term $\varepsilon$ to represent that unknown variation.

*Regression* analysis is the term commonly used to describe this kind of investigation, and it focuses on finding the simplest relationship indicated by the data. That simplest relationship is linear, implying that a unit increase in percent black always corresponds to a constant corresponding increase or decrease in the dependent variable. Linear relationships plot as straight lines on a scatterplot and have the equation:

$$Y = b_0 + b_1 X_1 + b_2 X_2 + b_3 X_3 + \ldots + b_k X_k + \varepsilon$$

though in the case of Figure 13.7 there is only one independent variable $X_1$. $b_1$ through $b_k$ are termed regression *parameters* and measure the direction and strength of the influence of the independent variables $X_1$ through $X_k$ on $Y$. $b_0$ is termed the *constant* or *intercept* term.

Figure 13.7A shows the data for the counties of California, and Figure 13.7B for the counties of Mississippi. Notice the difference—counties with a high percentage of blacks have *lower* housing values in Mississippi but *higher* housing values in California. In Figure 13.7 *best fit* lines have been drawn through the scatters of points. The gradient of this line is calculated as the $b_1$ parameter of the regression; it is positive in Figure 13.7A and negative in Figure 13.7B, indicating positive and negative trends, respectively. The value where the regression line intersects the y-axis identifies the (hypothetical) median housing value when percent black is zero and gives us the intercept value $b_0$. The more general multiple regression case works by extension of this principle, and each of the $b$ parameters gauges the marginal effects of its respective $X$ variable.

There are two lessons to be learned here. First, the relationship we have uncovered varies across space, from state to state, being an example of the general issue termed *spatial heterogeneity* that we discussed briefly in Section 2.2. Traditionally, science has concerned itself with finding patterns and relationships that exist everywhere—with *general* principles and laws that we described in Section 1.3 as *nomothetic*. In that sense what we have uncovered here is something different, a general principle (housing values are correlated with percent black) that varies in its specifics from state to state (a positive relationship in California, a negative one in Mississippi). Recently, geographers have developed a set of techniques that recognize such heterogeneity explicitly and focus not so much on what is true everywhere, but on how things vary across the geographic world. A prominent example is *geographically weighted regression* (GWR), a technique originally developed by Stewart Fotheringham, Martin Charlton, and Chris Brunsdon at the University of Newcastle-upon-Tyne. Rather than look for a single regression line, it examines how the slope and intercept vary across space in ways that may be related to other factors. Box 13.3 is devoted to Tomoki Nakaya, who among other contributions to spatial analysis has developed some of the widely used software for GWR.

The second lesson to be learned concerns the use of counties to unveil this relationship. Counties in the United States are particularly awkward units of observation because they vary enormously in size and population and mask variations in space that are often dramatic. The list of 58 counties of California, for example, includes one (Alpine) with a population of about 1,000 and another (Los Angeles) with a population of about 10 million. In Virginia individual cities are often their own counties and may be tiny in comparison with the size of San Bernardino County, California, the largest county in the continental United States. The lesson to be learned here is similar to that for English counties in Section 5.4.3: The results of this analysis depend intimately on the units of analysis chosen. Thus if we had repeated the analysis with other units, such as watersheds, our results might have been entirely different. California reverses the Mississippi trend in this case because the wealthy urban counties of California (San Francisco, Los Angeles, Alameda, etc.) are also the counties where most blacks live, whereas in Mississippi it is the rural counties with low housing values that house most blacks. But this pattern

## Tomoki Nakaya

Tomoki Nakaya is Professor of Geography at Ritsumeikan University in Japan and codirector of the Institute of Disaster Mitigation for Urban Cultural Heritage. He obtained his PhD from Tokyo Metropolitan University in 1997. His research specializes in spatial statistics and mathematical modeling in human geography, and he played a principal role in developing the software for fitting geographically weighted regression (GWR4 is available from the National Centre for Geocomputation in Ireland).

One of his current research themes is the integration of spatial or space–time statistics and geovisualization, for better understanding of such phenomena as crime, health, and disaster hazards. A good example of such visualization based on space–time statistical analysis is found in the study with the local police department for epidemiological analysis on snatching crime in the city of Kyoto. Figure 13.8 shows the results of one of his analyses, in this case the space–time density of bag-snatching crimes in Kyoto. An outbreak of crimes in one area is followed by increased policing and citizen vigilance, displacing the outbreak to a new area. As policing and vigilance increases in the second area and relaxes in the first, crime returns to the first area—and the pattern repeats itself. Visualization of data in space and time can thus lead to

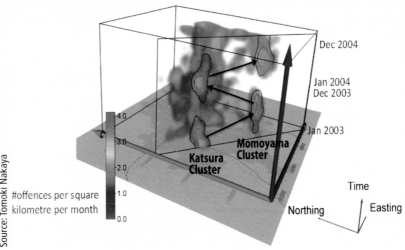

Source: Tomoki Nakaya

**Figure 13.8** Alternating occurrence of bag-snatching clusters in a pair of areas in Kyoto.

new insights into criminal behavior, as well as providing a good storytelling tool.

Nakaya published the first book on applications of GI systems in spatial epidemiology in Japan in 2004, and since then has been involved in numerous related projects, such as GI systems-based research into cancer-registry data, focusing on social inequalities in the cancer burden at the local level. With his colleague Keiji Yano he also has led the Virtual Kyoto project, which collects and stores detailed 2-D and 3-D geographical data about the city of Kyoto in different historical periods.

---

might not hold at all if we were to use watersheds rather than counties as the units of analysis. This issue is known in general as the *modifiable areal unit problem* and is also discussed in Section 5.4.3.

Earlier we described spatial analysis as a set of techniques whose results depend on the locations of the objects of study. But note that in this case the locations of the counties do not affect the results, and latitude and longitude nowhere play a role. Locations are useful for making maps of the inputs, but this analysis hardly rates the term *spatial*. In fact, many types of statistical analysis can be applied to the contents of attribute tables, without ever needing to know the locations or geometric shapes of features. On the other hand space can enter into models such as the ones discussed in this section if the researcher feels that some of the influences on Y come from the values of variables in nearby areas rather than from variables measured in the same area. For example, we might believe that housing

values in county *i* are influenced by conditions both in county *i* and also in neighboring counties. This kind of model is explicitly spatial, embodying the concept of influence *at a distance*, and calls for a special kind of regression termed *spatial regression* (a technique strongly related to GWR) in which values of independent variables in nearby areas are added to the model. A full discussion of these methods is outside the scope of this chapter, but can be found in several of the books listed in the Further Reading section.

### 13.2.2 Spatial Joins

In the previous section the variables needed for the analysis were all present in the same attribute table. Often, however, it is necessary to perform some basic operations first to bring the relevant variables together; recall the earlier discussion of how GI systems favor horizontal over vertical integration. Suppose, for example, that we wish to conduct a nationwide analysis of the average income of major cities and have a GI database containing cities as points, together with some relevant attributes of those cities: the dependent variable average income, plus some independent variables such as percent with college degrees and percent retired people. But other variables that are potentially relevant to the analysis are available only at the state level, including the state's percent unemployed. In Section 9.3 we described a *relational join* or simply a *join*—a fundamental operation in databases that is used to combine the contents of two tables using a common key. In this case the key is "state," a variable that exists in each city record to indicate the state containing the city, and in each state record as an identifier. The result of the join will be that each city record now includes the attributes of the city's containing state, allowing us to add the state-level variables to the analysis. Figure 9.2 shows a simple example of a join. One of the most powerful features of a GI system is the ability to join tables in this way based on common geographic location. To return to the point made at the outset of this discussion of location, we have now achieved what GI systems have always promised—vertical integration, or the ability to link disparate information based on location.

But this example is simple because common location is represented here by a key indicating the state containing the city. This spatial relationship was explicitly coded in the city attribute table in this case, but in other cases it will need to be computed using the functions of a GI system. The next two subsections describe circumstances in which the performance of a spatial join is a much more complex operation.

## 13.2.3 The Point-in-Polygon Operation

Comparing the properties of points to those of containing areas is a common operation in a GI system. It occurs in many areas of social science, when an investigator is interested in the extent to which the behavior of an individual is determined by properties of the individual's neighborhood. It occurs when point-like events, such as instances of a disease, must be compared to properties of the surrounding environment. In other applications of GI systems it occurs when a company needs to determine the property on which an asset such as an oil well or a transformer lies. In its simplest form, the point-in-polygon operation determines whether a given point lies inside or outside a given polygon. In more elaborate forms there may be many polygons, and many points, and the task is to assign points to polygons. If the polygons overlap, it is possible that a given point lies in one, many, or no polygons, depending on its location. Figure 13.9 illustrates the task.

The point-in-polygon operation makes sense from both the discrete-object and the continuous-field

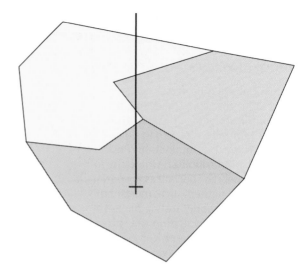

**Figure 13.9** The point-in-polygon problem, shown in the continuous-field case (the point must by definition lie in exactly one polygon, or outside the project area). In only one instance (the orange polygon) is there an odd number of intersections between the polygon boundary and a line drawn vertically upward from the point.

perspectives (see Section 3.5 for a discussion of these two perspectives). From a discrete-object perspective both points and polygons are objects, and the task is simply to determine enclosure. From a continuous-field perspective, polygons representing a variable such as land ownership cannot overlap because each polygon represents the land owned by one owner, and overlap would imply that a point is owned simultaneously by two owners. Similarly from a continuous-field perspective there can be no gaps between polygons. Consequently, the result of a point-in-polygon operation from a continuous-field perspective must assign each point to exactly one polygon.

**The point-in-polygon operation is used to determine whether a point lies inside or outside a polygon.**

Although the actual methods used by programmers to perform standard GI system operations are not normally addressed in this book, the standard approach or *algorithm* for the point-in-polygon operation is sufficiently simple and interesting to merit a short description. In essence, it consists of drawing a line from the point to infinity (see Figure 13.9), in this case parallel to the y-axis, and determining the number of intersections between the line and the polygon's boundary. If the number is odd, the point is inside the polygon, and if it is even, the point is outside. The algorithm must deal successfully with special cases—for example, if the point lies directly below a vertex (corner point) of the polygon. Some algorithms extend the task to include a third option, when the point lies exactly on the boundary. But others ignore this, on the grounds that it is never possible in practice to determine location with perfect accuracy, and so it is never possible to determine whether an infinitely small point lies on or off an infinitely thin boundary line.

## 13.2.4 Polygon Overlay

Polygon overlay is similar to the point-in-polygon operation in the sense that two sets of objects are involved, but in this case both are polygons. It again exists in two forms, depending on whether a continuous-field or discrete-object perspective is taken. The development of effective algorithms for polygon overlay was one of the most significant challenges of early GI systems, and the task remains one of the most complex and difficult to program.

> **The complexity of computing a polygon overlay was one of the greatest barriers to the development of vector GI systems.**

From the discrete-object perspective, the task is to determine whether two area objects overlap, to determine the area of overlap, and to define the area formed by the overlap as one or more new area objects (the overlay of two polygons can produce a large number of distinct area objects; see Figure 13.10). This operation is useful for determining answers to such queries as:

- How much of this proposed clear-cut lies in this riparian zone?
- How much of the projected catchment area of this proposed retail store lies in the catchment of this other existing store in the same chain?
- How much of this land parcel is affected by this easement?

**Figure 13.10** Polygon overlay, in the discrete-object case. Here the overlay of two polygons produces nine distinct polygons. One has the properties of both polygons, four have the properties of the yellow shaded polygon but not the blue (bounded) polygon, and four are outside the yellow polygon but inside the blue polygon.

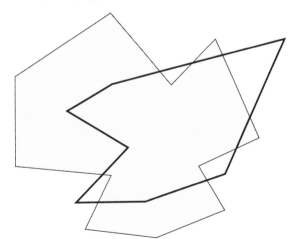

- What proportion of the land area of the United States lies in areas managed by the Bureau of Land Management?

From the continuous-field perspective the task is somewhat different. Figure 13.11 shows two datasets, both of which are representations of fields: one differentiates areas according to land ownership, and the other differentiates the same region according to land-cover class. In the terminology of Esri's ArcGIS, both datasets are instances of *area coverages*, or fields of nominal variables represented by nonoverlapping polygons. The methods discussed earlier in this chapter could be used to interrogate either dataset separately, but there are numerous queries that require simultaneous access to both datasets. For example:

- What is the land-cover class, and who is the owner of the point indicated by the user?
- What is the total area of land owned by X and with land cover class A?
- Where are the areas that lie on publicly owned land and have land-cover class B?

None of these queries can be answered by interrogating one of the datasets alone—the datasets must somehow be combined (vertically integrated) so that interrogation can be directed simultaneously at both of them.

The continuous-field version of polygon overlay does this by first computing a new dataset in which the region is partitioned into smaller areas that have uniform characteristics on both variables. Each area in the new dataset will have two sets of attributes—those obtained from one of the input datasets and those obtained from the other. In effect, then, we will have performed a spatial join by creating a new table that combines both sets of attributes, though in this case

**Figure 13.11** Polygon overlay in the continuous-field case. Here a dataset representing two types of land cover (A on the left, B on the right) is overlaid on a dataset representing three types of ownership (the two datasets have been offset for visual clarity). The result will be a single dataset in which every point is identified with one land cover type and one ownership type. It will have five polygons because land cover A intersects with two ownership types, and land cover B intersects with three.

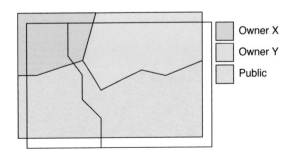

Owner X
Owner Y
Public

we will also have created a new set of features. All the boundaries will be retained, but they will be broken into shorter fragments by the intersections that occur between boundaries in one input dataset and boundaries in the other. Note the unusual characteristics of the new dataset shown in Figure 13.11. Unlike the two input datasets, where boundaries meet in junctions of three lines, the new map contains a new junction of four lines, formed by the new intersection discovered during the overlay process. Because the results of overlay are distinct in this way, it is almost always possible to discover whether a geographically referenced dataset was formed by overlaying two earlier datasets.

### Polygon overlay has different meanings from the continuous-field and discrete-object perspectives.

With a single dataset that combines both inputs, it is an easy matter to answer all the queries listed earlier through simple interrogation, or to look for relationships between the attributes. It is also easy to reverse the overlay process. If neighboring areas that share the same land-cover class are merged, for example, the result is the land-ownership map, and vice versa.

Polygon overlay is a computationally complex operation, and as noted earlier, much work has gone into developing algorithms that function efficiently for large datasets. One of the issues that must be tackled by a practically useful algorithm is known as the *spurious polygon* or *coastline weave* problem. It is almost

inevitable that there will be instances in any practical application where the same line on the ground occurs in both datasets. This happens, for example, when a coastal region is being analyzed because the coastline is almost certain to appear in every dataset of the region. Rivers and roads often form boundaries in many different datasets—a river may function both as a land-cover-class boundary and as a land-ownership boundary, for example. But although the same line is represented in both datasets, its representations will almost certainly not be the same. They may have been digitized from different maps, digitized using different numbers of points, subjected to different manipulations, obtained from entirely different sources (an air photograph and a topographic map, for example), and subjected to different measurement errors. When overlaid, the result is a series of small slivers. Paradoxically, the more care one takes in digitizing or processing, the worse the problem appears, as the result is simply more slivers, albeit smaller in size.

### In two vector datasets of the same area, there will almost certainly be instances where lines in each dataset represent the same feature on the ground.

Table 13.1 shows an example of the consequences of slivers and how a GI system can be rapidly overwhelmed if it fails to anticipate and deal with them adequately. Today, a GI system will offer various methods for dealing with the problem, the most common of which is the specification of a *tolerance*. If two

Table 13.1 Numbers of polygons resulting from an overlay of five datasets, illustrating the spurious polygon problem. The datasets come from the Canada Geographic Information System discussed in Section 1.5.1, and all are representations of continuous fields. Dataset 1 is a representation of a map of soil capability for agriculture, Datasets 2 through 4 are land-use maps of the same area at different times (the probability of finding the same real boundary in more than one such map is very high), and Dataset 5 is a map of land capability for recreation. The final three columns show the numbers of polygons in overlays of three, four, and five of the input datasets.

| Acres | 1 | 2 | 3 | 4 | 5 | 1+2+5 | 1+2+3+5 | 1+2+3+4+5 |
|---|---|---|---|---|---|---|---|---|
| 0–1 | 0 | 0 | 0 | 1 | 2 | 2,640 | 27,566 | 77,346 |
| 1–5 | 0 | 165 | 182 | 131 | 31 | 2,195 | 7,521 | 7,330 |
| 5–10 | 5 | 498 | 515 | 408 | 10 | 1,421 | 2,108 | 2,201 |
| 10–25 | 1 | 784 | 775 | 688 | 38 | 1,590 | 2,106 | 2,129 |
| 25–50 | 4 | 353 | 373 | 382 | 61 | 801 | 853 | 827 |
| 50–100 | 9 | 238 | 249 | 232 | 64 | 462 | 462 | 413 |
| 100–200 | 12 | 155 | 152 | 158 | 72 | 248 | 208 | 197 |
| 200–500 | 21 | 71 | 83 | 89 | 92 | 133 | 105 | 99 |
| 500–1,000 | 9 | 32 | 31 | 33 | 56 | 39 | 34 | 34 |
| 1,000–5,000 | 19 | 25 | 27 | 21 | 50 | 27 | 24 | 22 |
| >5,000 | 8 | 6 | 7 | 6 | 11 | 2 | 1 | 1 |
| Totals | 88 | 2,327 | 2,394 | 2,149 | 487 | 9,558 | 39,188 | 90,599 |

lines fall within this distance of each other, the GI system will treat them as a single line, and not create slivers (see also Section 8.3.2.2). The resulting overlay contains just one version of the line, not two. But at least one of the input lines has been moved, and if the tolerance is set too high, the movement can be substantial and can lead to problems later.

## 13.2.5 Raster Analysis

Many of the complications addressed in the previous subsections disappear if the data are structured in raster form and if the cells in each layer of data are geometrically identical. For example, suppose we are interested in the agricultural productivity of land and have data in the form of a raster, each 10 m by 10 m cell giving the average annual yield of corn in the cell. We might investigate the degree to which these yield values are predictable from other properties of each cell, using rasters of fertilizer quantity applied, depth to water table, percent organic matter, and so forth, each provided for the same set of 10 m by 10 m cells.

In such cases there is no need for complicated overlay operations because all the attributes are already available for the same set of spatial features, the cells of the raster. Overlay in raster is thus an altogether simpler operation, and this has often been cited as a good reason to adopt raster rather than vector structures. Attributes from different rasters can be readily combined for a variety of purposes, as long as the rasters consist of identically defined arrays of cells. The power of this raster-based approach is such that it deserves its own section; it is discussed in Section 15.2.4 under the heading "Cartographic Modeling and Map Algebra."

In other cases, however, the different variables needed for an analysis may not use identical rasters. For example, it may be necessary to compare data derived from the AVHRR (Advanced Very High Resolution Radiometer) sensor with a cell size of 1 km by 1 km with other data derived from the MODIS (Moderate Resolution Imaging Spectroradiometer) sensor with a cell size of 250 m by 250 m—and the two rasters will also likely be at different orientations and may even use different map projections to flatten the Earth (see Section 8.2.1 for a discussion of satellite remote sensing and Section 4.8 for a discussion of map projections). In such cases it is necessary to employ some form of *resampling* to transform each dataset to a common raster. Box 13.4 shows a simple example of the use of resampling to join point data. Resampling is a form of spatial interpolation, a technique discussed at length in Section 13.3.6.

---

## Application Box (13.4)

### Comparing Attributes When Points Do Not Coincide Spatially

GI system users often encounter situations in which attributes must be compared, but for different sets of features. Figure 13.12 shows a study area with two sets of points, one being the locations where levels of ambient sound were measured using recorders mounted on telephone poles, the other the locations of interviews conducted with local residents to determine attitudes to noise. We would like to know about the relationship between sound and attitudes, but the locations and numbers of cases (the *spatial supports*) are different. A simple expedient is to use spatial interpolation (Section 13.3.6), a form of intelligent guesswork that provides estimates of the values of continuous fields at locations where no measurements have been taken. In other words, it *resamples* each field at different points. Given such a method it would be possible to conduct the analysis in any of three ways:

- By interpolating the second dataset to the locations at which the first attribute was measured

- By the reverse—interpolating the first dataset to the locations at which the second attribute was measured

- By interpolating both datasets to a common geometric base, such as a raster

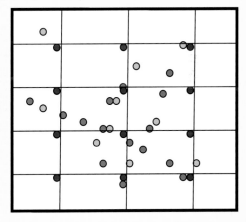

Figure 13.12 Coping with the comparison of two sets of attributes when the respective objects do not coincide. In this instance, attitudes regarding ambient noise have been obtained through a household survey of 15 residents (green dots) and are to be compared to ambient noise levels measured at 10 observation points (brown dots). The solution is to interpolate both sets of data to 12 comparison points (blue dots), using the methods discussed in Section 13.3.6.

▶

In the third case, note that it is possible to create a vast amount of data by using a sufficiently detailed raster. In essence, all these options involve manufacturing information, and the results will depend to some extent on the nature and suitability of the method used to do the spatial interpolation. We normally think of a relationship that is demonstrated with a large number of observations as stronger and more convincing than one based on a small number of observations. But the ability to manufacture data upsets this standard view. The implications of this power of GI systems to manufacture data are addressed in Section 14.5.

Figure 13.12 shows a possible solution using this third option. The number of grid points has been determined by the smaller number of cases—in this case, approximately 10 (12 are shown)—to minimize concerns about manufacturing information.

## 13.3 Analysis Based on Distance

The second fundamental spatial concept considered in this chapter is *distance*, the separation of places on the Earth's surface. The ability to calculate and manipulate distances underlies many forms of spatial analysis, some of the most important of which are reviewed in this section. All are based on the concept that the separation of features or events on the Earth's surface can tell us something useful, either about the mechanisms responsible for their presence or properties—in other words, to *explain* their patterns—or as input to decision-making processes. The first subsection looks at the measurement of distance and length in a GI system, as well as some of the issues involved. The second discusses the construction of buffers, together with their use in a wide range of applications. The identification of an anomalous concentration of points in space was the trigger that led to Dr. Snow's work on the causes of cholera transmission, and it is discussed in its general case as *cluster detection* in the third subsection. The concept of *spatial dependence* or *dependence at a distance*, first introduced in Chapter 2 as a fundamental property of geographic data, is given operational meaning in the fourth subsection. The fifth subsection addresses *density estimation*, based on the concept of averaging over defined distances, and the final subsection discusses the related distance-based operation of *spatial interpolation*.

### 13.3.1 Measuring Distance and Length

A *metric* is a rule for determining distance between points in a space. Several kinds of metrics are used in GI systems, depending on the application. The simplest is the rule for determining the shortest distance between two points in a flat plane, called the Pythagorean or straight-line metric. If the two points are defined by the coordinates $(x_1, y_1)$ and $(x_2, y_2)$, then the distance $D$ between them is the length of the hypotenuse of a right-angled triangle (Figure 13.13). Pythagoras's theorem tells us that the square of this length is equal to the sum of the squares of the lengths of the other two sides. So a simple formula results:

$$D = \sqrt{(x_2 - x_1)^2 + (y_2 - y_1)^2}$$

**A metric is a rule for determining the distance between points in space.**

The Pythagorean metric gives a simple and straightforward solution for a plane, if the coordinates $x$ and $y$ are comparable, as they are in any coordinate system based on a conformal projection, such as the Universal Transverse Mercator (UTM) or Web Mercator (see Chapter 4). But the metric will not work for latitude and longitude, reflecting a common source of problems in GI systems—the temptation to treat latitude and longitude as if they were equivalent to plane coordinates. This issue is discussed in detail in Section 4.8.1.

For points widely separated on the curved surface of the Earth the assumption of a flat plane leads to

---

**Figure 13.13** Pythagoras's Theorem and the straight-line distance between two points on a plane. The square of the length of the hypotenuse is equal to the sum of the squares of the lengths of the other two sides of the right-angled triangle.

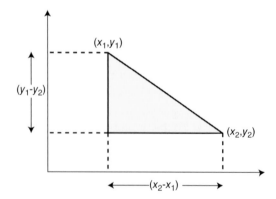

**Figure 13.14** The effects of the Earth's curvature on the measurement of distance, and the choice of shortest paths. The map shows the North Atlantic on the Mercator projection. The red line shows the track made by steering a constant course of 79 degrees from Los Angeles and is 9807 km long. The shortest path from Los Angeles to London is actually the black line, the trace of the great circle connecting them, with a length of roughly 8800 km. This is typically the route followed by aircraft flying from London to Los Angeles. Flying in the other direction, aircraft may sometimes follow other, longer tracks, such as the red line, if by doing so they can take advantage of jet stream winds.

significant distortion, and distance must be measured using the metric for a spherical Earth given in Section 4.6 and based on a great circle. For some purposes even this is not sufficiently accurate because of the non-spherical nature of the Earth, and even more complex procedures must be used to estimate distance that take nonsphericity into account. Figure 13.14 shows an example of the differences that the curved surface of the Earth makes when flying long distances.

In many applications the simple rules—the Pythagorean and great-circle equations—are not sufficiently accurate estimates of actual travel distance, and we are forced to resort to summing the actual lengths of travel routes. In GI systems this normally means summing the lengths of links in a network representation, and many forms of spatial analysis use this approach, along with Web services for obtaining driving directions. If a line is represented as a polyline, or a series of straight segments, then its length is simply the sum of the lengths of each segment, and each segment length can be calculated using the Pythagorean formula and the coordinates of its endpoints. But it is worth being aware of two problems with this simple approach.

First, a polyline is often only a rough version of the true object's geometry. A river, for example, never makes the sudden changes of direction of a polyline, and Figure 13.15 shows how smoothly curving streets

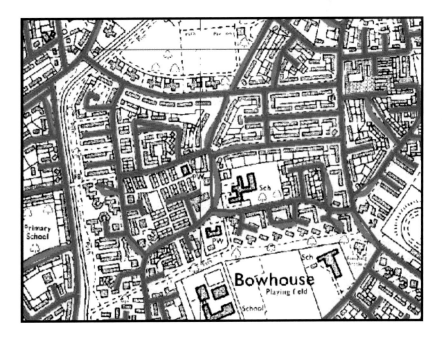

**Figure 13.15** The polyline representations of smooth curves tend to be shorter in length, as illustrated by this street map (note how curves are replaced by straight-line segments). But estimates of area tend not to show systematic bias because the effects of overshoots and undershoots tend to cancel out to some extent.

have to be approximated by the sharp corners of a polyline. Because there is a general tendency for polylines to shortcut corners, *the length of a polyline tends to be shorter than the length of the object it represents.* There are some exceptions, of course—surveyed boundaries are often truly straight between corner points, and streets are often truly straight between intersections. But in general the lengths of linear objects estimated in a GI system—and this includes the lengths of the perimeters of areas represented as polygons—are often substantially shorter than their counterparts on the ground (and note the discussion of fractals in Box 2.7). Note that this is not similarly true of area estimates because shortcutting corners tends to produce both underestimates and overestimates of area, and these tend to cancel out. So although estimates of line length tend to be systematically biased, estimates of area are not.

**A GI system will almost always underestimate the true length of a geographic line.**

Second, the length of a line in a two-dimensional representation will always be the length of the line's planar projection, not its true length in three dimensions, and the difference can be substantial if the line is steep (Figure 13.16). In most jurisdictions the area of a parcel of land is the area of its horizontal projection, not its true surface area. A system that stores the third dimension for every point is able to calculate both versions of length and area, but not a system that stores only the two horizontal dimensions.

## 13.3.2 Buffering

One of the most important GI system operations available to the user is the calculation of a *buffer* (see also Section 9.5). Given any set of objects, which may include points, lines, or areas, a buffer operation builds a new object or objects by identifying all areas that are within a certain specified distance of the original objects. Figure 13.17 shows instances of a point,

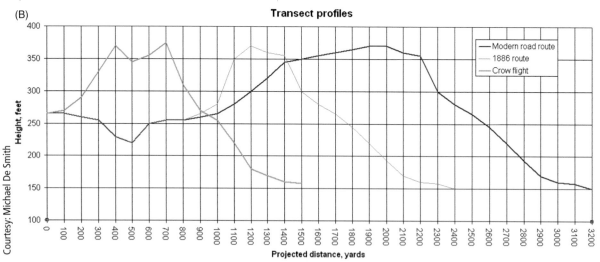

**Figure 13.16** The length of a path as traveled on the Earth's surface (red line) may be substantially longer than the length of its horizontal projection as evaluated in a two-dimensional GIS. (A) shows three paths across part of Dorset in the UK. The green path is the straight route, the red path is the modern road system, and the gray path represents the route followed by the road in 1886. (B) Shows the vertical profiles of all three routes, with elevation plotted against the distance traveled horizontally in each case. 1 ft = 0.3048 m, 1 yd = 0.9144 m.

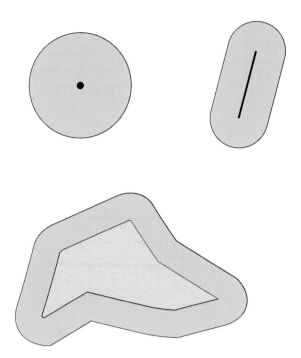

**Figure 13.17** Buffers (dilations) of constant width drawn around a point, a polyline, and a polygon.

a line, and an area, as well as the results of buffering. Buffers have many uses, and they are among the most popular of GI system functions:

- The owner of a land parcel has applied for planning permission to rebuild—the local planning authority could build a buffer around the parcel in order to identify all homeowners who live within the legally mandated distance for notification of proposed redevelopments.

- A logging company wishes to clear-cut an area but is required to avoid cutting in areas within 100 m of streams; the company could build buffers 100 m wide around all streams to identify these protected riparian areas.

- A retailer is considering developing a new store on a site, of a type that is able to draw consumers from up to 4 km away from its stores. The retailer could build a buffer around the site to identify the number of consumers living within 4 km of the site, in order to estimate the new store's potential sales.

- A new law is passed requiring people once convicted of sexual offenses involving young children to live more than ½ mile (0.8 km) from any school. Figure 13.18 shows the implications of such a law for an area of Los Angeles. The buffers are based on point locations; if the school grounds had been represented as polygons the buffered areas would be larger.

Buffering is possible in both raster and vector GI systems. In the raster case, the result is the classification of cells according to whether they lie inside or outside the buffer, whereas the result in the vector case is a new set of objects (Figure 13.17). But there is an additional possibility in the raster case that makes buffering more useful in some situations. Rather than buffer according to distance, we can ask a raster GI system to *spread* outward from one or more features at rates determined by *friction*, *travel speed*, or *cost* values stored in each cell. This form of analysis is discussed in Section 14.3.2.

**Buffering is one of the most useful transformations in a GI system and is possible in both raster and vector formats.**

**Figure 13.18** Buffers representing 1/2-mile exclusion zones around all schools in part of Los Angeles.

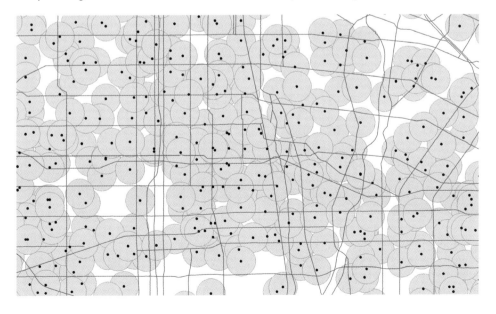

### 13.3.3 Cluster Detection

One of the questions most commonly asked about distributions of features, particularly point-like features, is whether they display a random pattern (see Figure 2.1), in the sense that all locations are equally likely to contain a point, or whether some locations are more likely than others—and particularly, whether the presence of one point makes other points either more or less likely in its immediate neighborhood. This leads to three possibilities:

- The pattern is *random* (points are located independently, and all locations are equally likely).
- The pattern is *clustered* (some locations are more likely than others, and the presence of one point may attract others to its vicinity).
- The pattern is *dispersed* (the presence of one point may make others less likely in its vicinity).

Establishing the existence of clusters is often of great interest because it may point to possible causal factors, as, for example, with the case of childhood leukemia studied by Stan Openshaw (Figure 13.4). Dispersed patterns are the typical result of competition for space, as each point establishes its own territory and excludes others. Thus such patterns are commonly found among organisms that exhibit territorial behavior, as well as among market towns in rural areas and among retail outlets.

> **Point patterns can be identified as clustered, dispersed, or random.**

It is helpful to distinguish two kinds of processes responsible for point patterns. *First-order* processes involve points being located independently, but may still result in clusters because of varying point density. For example, the drinking-water hypothesis investigated by Dr. John Snow and described in Box 13.1 led to a higher density of points around the pump because of greater access. Similarly, the density of organisms of a particular species may vary over an area because of varying suitability of habitat. *Second-order* processes involve interaction between points, leading to clusters when the interactions are attractive in nature and to dispersion when they are competitive or repulsive. In the cholera case, the contagion hypothesis rejected by Snow is a second-order process and results in clustering even in situations when all other density-controlling factors are perfectly uniform. Unfortunately, as argued in Section 13.1, Snow's evidence did not allow him to resolve with complete confidence between first-order and second-order processes, and in general it is not possible to determine whether a given clustered point pattern was created by varying density factors or by interactions. On the

other hand, dispersed patterns can only be created by second-order processes.

> **Clustering can be produced by two distinct mechanisms, identified as first-order and second-order.**

Many tests are available for clusters, and some excellent books have been written on the subject. Only one test will be discussed in this section, to illustrate the method. This is the $K$ function, and unlike many such statistics it provides an analysis of clustering and dispersion over a range of scales. In the interests of brevity the technical details will be omitted, but they can be found in the texts listed at the end of this chapter. They include procedures for dealing with the effects of the study area boundary, which is an important distorting factor for many pattern statistics.

$K(d)$ is defined as the expected number of points within a distance $d$ of an arbitrarily chosen point, divided by the density of points per unit area. When the pattern is random, this number is $\pi d^2$, so the normal practice is to plot the function:

$$\hat{L}(d) = \sqrt{K(d)/\pi}$$

because $\hat{L}(d)$ will equal $d$ for all $d$ in a random pattern, and a plot of $\hat{L}(d)$ against $d$ will be a straight line with a slope of 1. Clustering at certain distances is indicated by departures of $\hat{L}(d)$ above the line and dispersion by departures below the line.

Figures 13.19 and 13.20 show a simple example, used to discover how trees are spaced relative to

---

**Figure 13.19** Point pattern of individual tree locations. A, B, and C identify the individual trees analyzed in Figure 13.20.

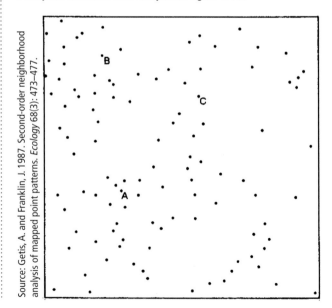

*Source: Getis, A. and Franklin, J. 1987. Second-order neighborhood analysis of mapped point patterns. Ecology 68(3): 473–477.*

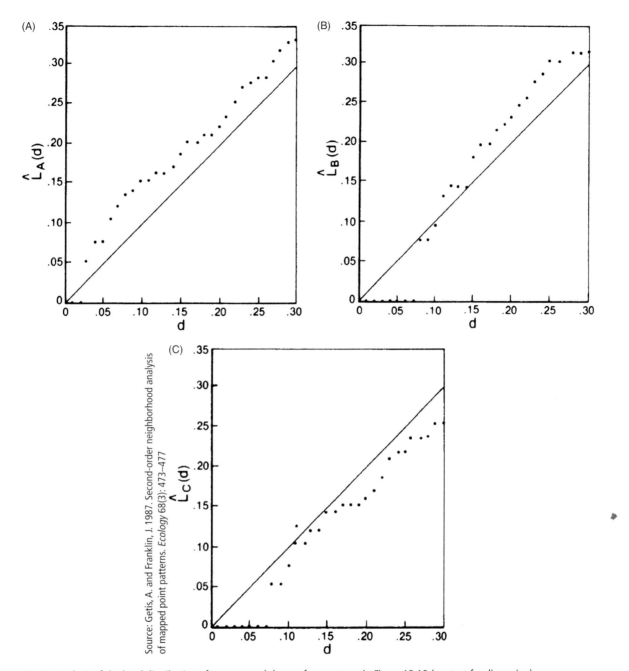

**Figure 13.20** Analysis of the local distribution of trees around three reference trees in Figure 13.18 (see text for discussion).

Source: Getis, A. and Franklin, J. 1987. Second-order neighborhood analysis of mapped point patterns. *Ecology* 68(3): 473–477

each other in a forest. The locations of trees are shown in Figure 13.19. In Figure 13.20A locations are analyzed in relation to Tree A. At very short distances there are fewer trees than would occur in a random pattern, but for most of the range up to a distance of 30% of the width (and height) of the study area there are *more* trees than would be expected, showing a degree of clustering. Tree B (Figure 13.20B) has no nearby neighbors, but shows a degree of clustering at longer distances. Tree C (Figure 13.20C) shows fewer trees than expected in a random pattern over most

distances, and it is evident in Figure 13.19 that C is in a comparatively widely spaced area of forest.

## 13.3.4 Dependence at a Distance

The concept of distance decay—that interactions and similarities decline over space in ways that are often systematic—was introduced in Section 2.6 as a fundamental property of geographic data. It is so fundamental in fact that it is often identified somewhat informally as Tobler's First Law of Geography, first discussed in Chapter 2. This subsection examines techniques for measuring spatial dependence effects—in other words, the ways in which the characteristics of one location are correlated with characteristics of other nearby locations. Underlying this are concepts similar to those examined in the previous subsection on cluster detection—the potential for *hotspots* of anomalously high (or low) values of some property.

One way to look for such effects is to think of the features as fixed and their attributes as displaying interesting or anomalous patterns. Are attribute values randomly distributed over the features, or do extreme values tend to cluster: high values surrounded by high values, and low values surrounded by low values? In such investigations, the processes that determined the locations of features, and were the major concern in the previous section, tend to be ignored and may have nothing to do with the processes that created the pattern of their attributes. For example, the concern might be with some attribute of counties—their average house value, or percent married—and hypotheses about the processes that led to counties having different values on these indicators. The processes that led to the locations and shapes of counties, on the other hand, which were political processes operating perhaps 100 years ago, would be of no interest. The usefulness, or otherwise, of different areal units was discussed in Chapter 5.

The Moran statistic described in Chapter 2 is designed precisely for this purpose, to indicate the general properties of the pattern of attributes. It distinguishes between positively autocorrelated patterns, in which high values tend to be surrounded by high values and low values by low values; random patterns, in which neighboring values are independent of each other; and dispersed patterns, in which high values tend to be surrounded by low and vice versa. Section 2.7 described various ways of defining the weights needed to calculate the Moran statistic and also showed how it is possible to use various measures of separation or distance as a basis for the weights. A common expedient, described in Box 2.6, is to use a simple binary indicator of whether or not two areas are adjacent as a surrogate for the distance between them.

**The Moran statistic looks for patterns among the attributes assigned to features.**

In recent years there has been much interest in going beyond these global measures of spatial dependence to identify dependences locally. Is it possible, for example, to identify individual hotspots, areas where high values are surrounded by high values, and coldspots where low values are surrounded by low? Is it possible to identify anomalies, where high values are surrounded by low or vice versa? Local versions of the Moran statistic are among this group, and along with several others now form a useful resource that is easily implemented in GI systems.

Figure 13.21 shows an example, using the Local Moran statistic to differentiate states according to their roles in the pattern of one variable, the median value of housing. The weights have been defined by adjacency, such that pairs of states that share a common boundary are given a weight of 1 and all other pairs a weight of zero.

## 13.3.5 Density Estimation

One of the strongest arguments for spatial analysis is that it is capable of addressing *context*, of asking to what extent events or properties at some location are related to or determined by the location's surroundings. Does living in a polluted neighborhood make a person more likely to suffer from diseases such as asthma; does living near a liquor store make binge drinking more likely among young people; can obesity be blamed in part on a lack of nearby parks or exercise facilities? Buffering is one approach, defining a neighborhood as contained within a circle of appropriate radius around the person's location. The liquor store question, for example, could be addressed by building buffers around each individual and through a point-in-polygon operation determining the number of liquor stores within a defined distance. But this would imply that any liquor store within the buffer was equally important regardless of its distance, and every store outside the buffer was irrelevant. Instead some kind of attenuating effect of distance, such as the options shown in Figure 2.7, would seem to be more appropriate and realistic.

The general term for this kind of technique is *convolution*. A weighting function is chosen, based on a suitable distance decay function, and applied to the nearby features. So the net impact $P$ of a set of liquor stores on a person at a given location $\mathbf{x}$ might be represented as the sum:

$$P(\mathbf{x}) = \sum_i w_i z_i$$

where $i$ denotes a liquor store, $z_i$ is a measure of the liquor store's relative size or importance, and $w_i$ is a weight that declines with distance according to a distance decay function. Functions such as $P$ defined in this way are often termed *potential* functions and have many uses in spatial analysis. They can be used to measure the potential impact of different population centers on a recreational facility, or the potential expenditures of the surrounding populations at a proposed retail facility. In all such cases the intent is to capture influence *at a distance*.

**Figure 13.21** The Local Moran statistic, applied using the GeoDa software (available via geodacenter.asu.edu) to describe local aspects of the pattern of housing value among U.S. states. In the map window, the states are colored according to median value, with the darker shades corresponding to more expensive housing. In the scatterplot window, the median value appears on the x-axis, whereas the y-axis is the weighted average of neighboring states. The three points colored yellow are instances where a state of below-average housing value is surrounded by states of above-average value. The windows are linked, and the three points are identified as Oregon, Arizona, and Pennsylvania. The global Moran statistic is +0.4011, indicating a general tendency for clustering of similar values.

In effect, $P$ measures the density of features in the neighborhood of **x**. The most obvious example is the estimation of population density, and that example is used in this discussion. But it could be equally well applied to the density of different kinds of diseases, or animals, or any other set of well-defined points. Consider a collection of point objects, such as those shown in Figure 13.22. The surface shown in the figure is an example of a *kernel function*, the central idea in density estimation. Any kernel function has an associated length measure, and in the case of the function shown, which is a Gaussian distribution, the length measure is a parameter of the distribution. We can generate Gaussian distributions with any value of this parameter, and they become flatter and wider as the value increases.

In density estimation, each point is replaced by its kernel function, and the various kernel functions are added to obtain an aggregate surface, a continuous field of density. If one thinks of each kernel as a pile of sand, then each pile has the same total weight of one unit. The total weight of all piles of sand is equal to the number of points, and the total weight of sand within a given area, such as the area shown in the figure, is an estimate of the total population in that area. Mathematically, if the population density is represented by a field $\rho(x,y)$, then the total population within area $A$ is the integral of the field function over that area, that is:

$$P = \int_A \rho dA$$

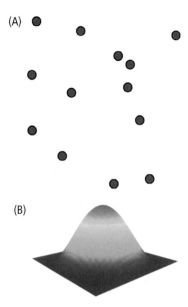

(A)

(B)

**Figure 13.22** (A) A collection of point objects, and (B) a kernel function. The kernel's shape depends on a distance parameter—increasing the value of the parameter results in a broader and lower kernel, and reducing it results in a narrower and sharper kernel. When each point is replaced by a kernel and the kernels are added, the result is a density surface whose smoothness depends on the value of the distance parameter.

A variety of kernel functions are used in density estimation, but the form shown in Figure 13.22B is perhaps the most common. This is the traditional bell curve or Gaussian distribution of statistics and is encountered elsewhere in this book in connection with errors in the measurement of position in two dimensions (Section 5.3.2.2). By adjusting the width of the bell, it is possible to produce a range of density surfaces of different amounts of smoothness. Figure 13.23 contrasts two density estimations from the same data, one using a comparatively narrow bell to produce a complex surface and the other using a broader bell to produce a smoother surface.

Both parts of Figure 13.23 are illustrations of the concept of a *heat map*. Such mappings are commonly employed in studies of public health and in crime mapping to identify areas where the density of cases is especially high. Note, however, that the patterns shown in such maps may be confounded by other factors. A hot spot on a heat map of disease, for example, indicates the presence of a high density of the disease. But a hot spot or local high on a map of density of crime may not be of much interest if it simply mirrors a high density of population—we expect more crime where there are more people and do not expect much crime in rural areas where there are few people, other things being equal, so crime density that is simply proportional to local population density will be nothing special. To echo the discussion of Snow's map earlier, we cannot tell from the map

(A)

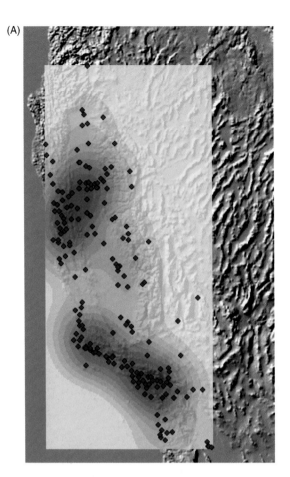

(B)

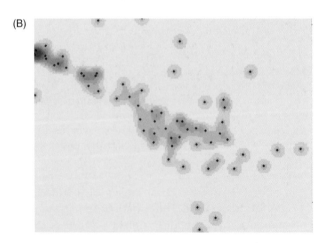

**Figure 13.23** Density estimation using two different distance parameters in the respective kernel functions. In (A), the surface shows the density of ozone-monitoring stations in California, using a kernel radius of 150 km. In (B), zoomed to an area of Southern California, a kernel radius of 16 km is too small for this dataset, as it leaves each kernel isolated from its neighbors.

whether the hot spot is due to a higher local rate of crime, or a constant rate superimposed on a higher density of people. When looking for hot spots, therefore, it is important to carefully select an appropriate *denominator* or *normalizing* variable and to show a rate rather than a simple density. A hot spot on a map of crimes *per capita* will be much more insightful than one on a map of crime density alone. Figure 13.6 resolved this issue by mapping the rate (actually the statistical significance of the rate) rather than the density of disease.

An important lesson to extract from this discussion concerns the importance of scale. Density is an abstraction, created by taking discrete objects and convolving their distribution with a kernel function. The result depends explicitly on the width of the kernel, which we recognize as a length measure. Change the measure and the resulting density surface changes, as Figure 13.23 shows. Density estimation is just one example of a common phenomenon in geographic data—the importance of scale in the definition of many data types and hence the importance of knowing what that scale is explicitly when dealing with geographic data.

> **Many geographic data types have a scale built into their definition; it is important therefore to know what that scale is as precisely as possible.**

## 13.3.6 Spatial Interpolation

Spatial interpolation is a pervasive operation in GI systems. Although it is often used explicitly in analysis, it is also used implicitly in various operations such as the preparation of a contour map display, where spatial interpolation is invoked without the user's direct involvement. Spatial interpolation is a process of intelligent guesswork in which the investigator (and the GI system) attempt to make a reasonable estimate of the value of a continuous field at places where the field has not actually been measured (spatial interpolation deals only with measured variables, that is, interval or ratio, not nominal or ordinal variables). All the methods use distance, based on the belief that the value at a location is more similar to the values measured at nearby sample points than to the values at distant sample points, a direct use of Tobler's First Law (discussed throughout Chapter 2). Spatial interpolation is an operation that makes sense only from the continuous-field perspective. The principles of spatial interpolation are discussed in Section 2.6; here the emphasis is on practical applications of the technique and on commonly used implementations of the principles.

Spatial interpolation finds applications in many areas:

- In estimating rainfall, temperature, and other attributes at places that are not weather stations, and where no direct measurements of these variables are available.

- In estimating the elevation of the surface between the measured locations of a digital elevation model (DEM).

- In *resampling* rasters (Section 13.2.5), the operation that must take place whenever raster data must be transformed to another grid.

- In contouring, when it is necessary to guess where to place contours between measured locations.

In all these instances spatial interpolation calls for intelligent guesswork, and the one principle that underlies all spatial interpolation is the Tobler Law (Section 2.6): "nearby things are more related than distant things." In other words, the best guess as to the value of a field at some point is the value measured at the closest observation points—the rainfall *here* is likely to be more similar to the rainfall recorded at the nearest weather stations than to the rainfall recorded at more distant weather stations. A corollary of this same principle is that in the absence of better information, it is reasonable to assume that any continuous field exhibits relatively smooth variation; fields tend to vary slowly and to exhibit strong positive spatial autocorrelation, a property of geographic data discussed in Section 2.6.

> **Spatial interpolation is the GISS version of intelligent guesswork.**

In this section three methods of spatial interpolation are discussed, all distance-based: Thiessen polygons; inverse-distance weighting (IDW), which is the simplest commonly used method; and Kriging, a popular statistical method that is grounded in the theory of regionalized variables and falls within the field of *geostatistics*. These methods are discussed at greater length in de Smith, Goodchild, and Longley's *Geospatial Analysis* reader and at www.spatialanalysisonline.com.

### 13.3.6.1 Thiessen Polygons

Thiessen polygons were suggested by Thiessen as a way of interpolating rainfall estimates from a few rain gauges to obtain estimates at other locations where rainfall had not been measured. The method is very simple: to estimate rainfall at any point, take the rainfall measured at the closest gauge. This leads to a map in which rainfall is constant within polygons surrounding each gauge and changes sharply as polygon boundaries are crossed. Although many users associate polygons defined in this way with Thiessen, they are also known as Voronoi and Dirichlet polygons. They have many other uses besides spatial interpolation:

- Thiessen polygons can be used to estimate the trade areas of each of a set of retail stores or shopping centers.

- They are used internally in a GI system as a means of speeding up certain geometric operations, such as search for a point's nearest neighbor.
- They are the basis of some of the more powerful methods for generalizing vector databases (Section 3.8).
- As a method of spatial interpolation they leave something to be desired, however, because the sharp change in interpolated values at polygon boundaries is often implausible.

Figure 13.24 shows a typical set of Thiessen polygons. If each pair of points that share a Thiessen polygon boundary is connected, the result is the network of irregular triangles. These are named after Delaunay and are frequently used as the basis for the triangles of a triangulated irregular network (TIN) representation of terrain (Section 7.2.3.4).

### 13.3.6.2 Inverse-Distance Weighting

IDW is the workhorse of spatial interpolation, the method that is most often used in GI analysis. It employs the Tobler Law by estimating unknown measurements as weighted averages over the known measurements at nearby points, giving the greatest weight to the nearest points.

More specifically, denote the point of interest as **x**, and the points where measurements were taken as $\mathbf{x}_i$, where $i$ runs from 1 to $n$, if there are $n$ data points. Denote the unknown value as $z(\mathbf{x})$ and the known measurements as $z_i$. Give each of these points a weight $w_i$, which will be evaluated based on the distance $d_i$ from $\mathbf{x}_i$ to **x**. Figure 13.25 explains this

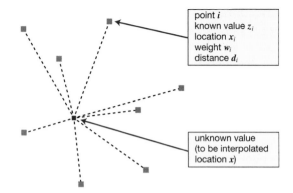

**Figure 13.25** Notation used in the equations defining spatial interpolation.

notation with a diagram. Then the weighted average computed at **x** is

$$z(\mathbf{x}) = \sum_i w_i z_i / \sum_i w_i$$

In other words, the interpolated value is an average over the observed values, weighted by the $w$'s. Notice the similarity between this equation and that used to define potential $P(\mathbf{x})$ in the previous subsection. The only difference is the presence here of a denominator, reflecting the nature of IDW as an averaging process rather than a summation.

There are various ways of defining the weights, but the option most often employed is to compute them as the inverse squares of distances—in other words (compare the options discussed in Section 2.5):

$$w_i = 1/d_i^2$$

This means that the weight given to a point drops by a factor of 4 when the distance to the point doubles (or by a factor of 9 when the distance triples). In addition, most software gives the user the option of ignoring altogether points that are further than some specified distance away, or of limiting the average to a specified number of nearest points, or of averaging over the closest points in each of a number of direction sectors. But if these values are not specified the software will assign default values to them.

> **IDW provides a simple way of guessing the values of a continuous field at locations where no measurement is available.**

IDW achieves the desired objective of creating a smooth surface whose value at any point is more like the values at nearby points than the values at distant points. If it is used to determine $z$ at a location where $z$ has already been measured, it will return the measured value because the weight assigned to a point at zero distance is infinite. For this reason IDW is described as an *exact* method of interpolation

**Figure 13.24** Thiessen polygons drawn around each station in part of the Southern California ozone-monitoring network. Note how the polygons, which enclose the area closest to each point, in theory extend off the map to infinity and so must be truncated by the system at the edge of the map.

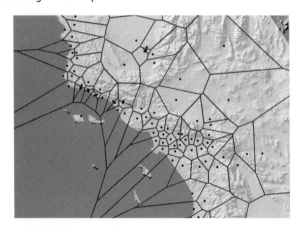

because its interpolated results honor the data points exactly. (An *approximate* method is allowed to deviate from the measured values in the interests of greater smoothness, a property that is often useful if deviations are interpreted as indicating possible errors of measurement, or local deviations that are to be separated from the general trend of the surface.)

But because IDW is an average it suffers from certain specific characteristics that are generally undesirable. A weighted average that uses weights that are never negative must always return a value that is between the limits of the measured values; no point on the interpolated surface can have an interpolated $z$ that is more than the largest measured $z$, or less than the smallest measured $z$. Imagine an elevation surface with some peaks and pits, but suppose that the peaks and pits have not actually been measured, but are merely indicated by the values of the measured points. Figure 13.26 shows a cross section of such a surface. Instead of interpolating peaks and pits as one might expect, IDW produces the kind of result shown in the figure—small pits where there should be peaks, and small peaks where there should be pits. This behavior is often obvious in GI system output that has been generated

using IDW. A related problem concerns extrapolation: if a trend is indicated by the data, as shown in Figure 13.26, IDW will inappropriately indicate a regression to the mean outside the area of the data points.

**IDW interpolation may produce counter-intuitive results in areas of peaks and pits and outside the area covered by the data points.**

In short, the results of IDW are not always what one would want. There are many better methods of spatial interpolation that address the problems that were just identified, but IDW's ease of programming and its conceptual simplicity make it among the most popular. Users should simply beware and take care to examine the results of interpolation to ensure that they make good sense.

### 13.3.6.3 Kriging
Of all the common methods of spatial interpolation, Kriging makes the most convincing claim to be grounded in good theoretical principles. The basic idea is to discover something about the general properties of the surface, as revealed by the measured values, and then to apply these properties in

**Figure 13.26** Potentially undesirable characteristics of IDW interpolation. Data points located at 20, 30, 40, 60, 70, and 80 have measured values of 2, 3, 4, 4, 3, and 2, respectively. The interpolated profile shows a pit between the two highest values and regression to the overall mean value of 3 outside the area covered by the data.

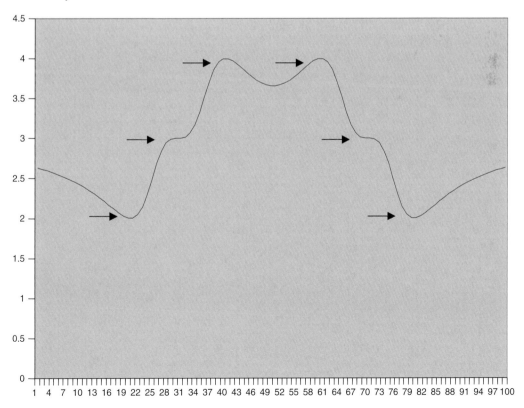

estimating the missing parts of the surface. Smoothness is the most important property (note the inherent conflict between this and the properties of fractals, Section 2.8), and it is operationalized in Kriging in a statistically meaningful way. There are many forms of Kriging, and the overview provided here is very brief. Further reading is identified at the end of the chapter.

**There are many forms of Kriging, but all are firmly grounded in theory.**

Suppose we take a point **x** as a reference and start comparing the values of the field there with the values at other locations at increasing distances from the reference point. If the field is smooth (if the Tobler Law is true, i.e., if there is positive spatial autocorrelation), the values nearby will not be very different—$z(\mathbf{x})$ will not be very different from $z(\mathbf{x}_i)$. To measure the amount, we take the difference and square it because the sign of the difference is not important: $(z(\mathbf{x}) - z(\mathbf{x}_i))^2$. We could do this with any pair of points in the area.

As distance increases, this measure will likely increase also, and in general a monotonic (consistent) increase in squared difference with distance is observed for most geographic fields. (Note that, as noted earlier, $z$ must be measured on a scale that is at least interval, though *indicator Kriging* has been developed to deal with the analysis of nominal fields.) In Figure 13.27, each point represents one pair of values drawn from the total set of data points at which measurements have been taken. The vertical axis represents one-half of the squared difference (one-half is taken for mathematical reasons), and the graph is known as the *semivariogram* (or

*variogram* for short—the difference of a factor of two is often overlooked in practice, though it is important mathematically). To express its contents in summary form, the distance axis is divided into a number of ranges or *bins*, as shown, and points within each range are averaged to define the heavy points shown in the figure. This semivariogram has been drawn without regard to the *directions* between points in a pair. As such, it is said to be an *isotropic* variogram. Sometimes there is sharp variation in the behavior in different directions, and *anisotropic* semivariograms are created for different ranges of direction (e.g., for pairs in each 90 degree sector).

**An anisotropic variogram asks how spatial dependence changes in different directions.**

Note how the points of this typical variogram show a steady increase in squared difference up to a certain limit and how that increase then slackens off and virtually ceases. Again, this pattern is widely observed for continuous fields, and it indicates that difference in value tends to increase up to a certain limit, but then to increase no further. In effect, there is a distance beyond which there are no more geographic surprises. This distance is known as the *range* and the value of difference at this distance as the *sill*.

Note also what happens at the other, lower end of the distance range. As distance shrinks, corresponding to pairs of points that are closer and closer together, the semivariance falls, but there is a suggestion that it never quite falls to zero, even at zero distance. In other words, if two points were sampled a vanishingly small distance apart, they would give different values. This is known as the *nugget* of the semivariogram. A nonzero nugget occurs when there is substantial error in the measuring instrument, such that measurements taken a very small distance apart would be different due to error, or when there is some other source of local noise that prevents the surface from being truly smooth. Accurate estimation of a nugget depends on whether there are pairs of data points sufficiently close together. In practice, the sample points may have been located at some time in the past, outside the user's control, or may have been spread out to capture the overall variation in the surface, so it is often difficult to make a good estimate of the nugget.

**The nugget can be interpreted as the variation among repeated measurements at the same point.**

To make estimates using Kriging, we need to reduce the semivariogram to a mathematical

---

**Figure 13.27** A semivariogram. Each cross represents a pair of points. The solid circles are obtained by averaging within the ranges or buckets of the distance axis. The solid line is the best fit to these five points, using one of a small number of standard mathematical functions.

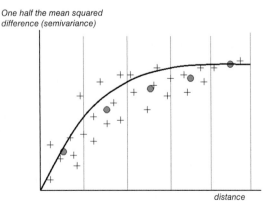

One half the mean squared difference (semivariance)

distance

function, so that semivariance can be evaluated at any distance, not just at the midpoints of buckets as shown in Figure 13.27. In practice, this means selecting one from a set of standard functional forms and fitting that form to the observed data points to get the best possible fit. This is shown in the figure. The user of a Kriging function in a GI system will have control over the selection of distance ranges and functional forms and whether a nugget is allowed.

Finally, the fitted semivariogram is used to estimate the values of the field at points of interest. As with IDW, the estimate is obtained as a weighted combination of neighboring values, but the estimate is designed to be the best possible given the evidence of the semivariogram. In general, nearby values are given greater weight, but unlike IDW direction is also important—a point can be *shielded* from influence if it lies behind another point because the latter's greater proximity suggests greater importance in determining the estimated value, whereas relative direction is unimportant in an IDW estimate. The process of maximizing the quality of the estimate is carried out mathematically, using the precise measures available in the semivariogram.

**Kriging responds both to the proximity of sample points and to their directions.**

Unlike IDW, Kriging has a solid theoretical foundation, but it also includes a number of options (e.g., the choice of the mathematical function for the semivariogram) that require attention from the user. In that sense it is definitely not a *black box* that can be executed blindly and automatically, but instead forces the user to become directly involved in the estimation process. For that reason GI software designers will likely continue to offer several different methods, depending on whether the user wants something that is quick, despite its obvious faults, or better but more demanding of the user.

### 13.3.6.4 A Final Word of Caution
Spatial interpolation and density estimation are in many ways logical twins: both begin with points and end with surfaces. Moreover, we already noted the similarity in the equations for density estimation and IDW. But conceptually the two approaches could not be more different because spatial interpolation seeks to estimate the missing parts of a continuous field from samples of the field taken at data points, whereas density estimation creates a continuous field from discrete objects. The values interpolated by spatial interpolation have the same measurement scale as the input values, but in

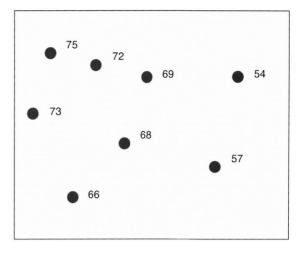

**Figure 13.28** A dataset with two possible interpretations: first, a continuous field of atmospheric temperature measured at eight irregularly spaced sample points, and second, eight discrete objects representing cities, with associated populations in thousands. Spatial interpolation makes sense only for the former and density estimation only for the latter.

density estimation the result is simply a count per unit area.

Figure 13.28 illustrates this difference. The dataset can be interpreted in two sharply different ways. In the first, it is interpreted as sample measurements from a continuous field, and in the second as a collection of discrete objects. In the discrete-object view there is nothing between the objects but empty space—no missing field to be filled in through spatial interpolation. It would make no sense at all to apply spatial interpolation to a collection of discrete objects and no sense at all to apply density estimation to samples of a field.

**Density estimation makes sense only from the discrete-object perspective and spatial interpolation only from the field perspective.**

### 13.4 Conclusion

This chapter has discussed some basic methods of spatial analysis based on two concepts: location and distance. Chapter 14 continues with techniques based on more advanced concepts, and Chapter 15 examines spatial modeling. Several general issues have been raised throughout the discussion: issues of scale and resolution and accuracy and uncertainty, which are discussed in greater detail in Chapter 5.

## Questions for Further Study

1. Did Dr. John Snow actually make his inference strictly from looking at his map? What information can you find on the Web on this issue (try www.jsi.com)?

2. You are given a map showing the home locations of the customers of an insurance agent and are asked to construct a map showing the agent's market area. Would spatial interpolation or density estimation be more appropriate, and why?

3. What is conditional simulation, and how does it differ from Kriging? Under what circumstances might it be useful?

4. What are the most important characteristics of the three methods of spatial interpolation discussed in this chapter? Using a test dataset of your own choosing, compute and describe the major features of the surfaces interpolated by each.

## Further Reading

De Smith, M. J., Goodchild, M. F., and Longley, P. A. 2009. *Geospatial Analysis: A Comprehensive Guide to Principles, Techniques and Software Tools* (3rd ed.). Winchelsea: Winchelsea Press. www.spatialanalysisonline.com.

Fotheringham, A. S., Brunsdon, C., and Charlton, M. 2002. *Geographically Weighted Regression: The Analysis of Spatially Varying Relationships.* Hoboken, NJ: Wiley.

McLafferty, S. and Cromley, E. 2011. *GIS and Public Health* (2nd ed.). New York: Guilford.

O'Sullivan, D. and Unwin, D. J. 2010. *Geographic Information Analysis* (2nd ed.). Hoboken, NJ: Wiley.

Silverman, B. W. 1986. *Density Estimation for Statistics and Data Analysis.* New York: Chapman and Hall.

# 14 Spatial Analysis and Inference

## LEARNING OBJECTIVES

The second of two chapters on spatial analysis focuses on five areas: analyses that address concepts of area and centrality, analyses of surfaces, analyses that are oriented to design, and statistical inference. The chapter begins with area-based techniques, including measures of area and shape. Techniques based on the concept of centrality seek to find central or representative points for geographic distributions. Surface techniques include measures of slope, aspect, and visibility along with the delineation of watersheds and channels. Design is concerned with choosing locations on the Earth's surface for various kinds of activities and with the positioning of points, lines, and areas and associated attributes to achieve defined objectives. Statistical inference concerns the extent to which it is possible to reason from limited samples or study areas to conclusions about larger populations or areas, but is often complicated by the nature of geographic data.

After studying this chapter you will understand:

- Methods for measuring properties of areas.
- Measures that can be used to capture the centrality of geographic phenomena.
- Techniques for analyzing surfaces and for determining their hydrologic properties.
- Techniques for the support of spatial decisions and the design of landscapes according to specific objectives.
- Methods for generalizing from samples, and the problems of applying methods of statistical inference to geographic data.

### 14.1 The Purpose of Area-Based Analyses

One of the ways in which humans simplify geography and address the Earth's infinite complexity is by ascribing characteristics to entire areas rather than to individual points. *Regional* geography relies heavily on this process to make parsimonious descriptions of the geographic world, a practice discussed earlier in Chapter 3 in the context of representation, and in Chapter 2 as one aspect of the nature of geographic data. At a technical level, this results in the creation not of points but of polygons (Section 7.2, and note that the term *polygon* is often used in geographic information science and systems [GISS] to refer to any area, whether or not its edges are straight.) In

Chapter 5 we discussed some of the issues of uncertainty that arise as a result of this process, when areas are not truly homogeneous or when their boundaries are not precisely known.

### 14.1.1 Measurement of Area

Many types of interrogation ask for measurements—we might want to know the total area of a parcel of land, or the distance between two points, or the length of a stretch of road—and in principle all these measurements are obtainable by simple calculations inside a geographic information (GI) system. Comparable measurements by hand from maps can be very tedious and error-prone. In fact, it was the ability of the computer to make accurate evaluations of area quickly that led the Canadian government to fund the development of the world's first GI system, the Canada Geographic Information System, in the mid-1960s (see the brief history of GI systems in Table 1.4), despite the primitive state and high costs of computing at that time. Evaluation of area by hand is a messy and soul-destroying business. The *dot-counting* method uses transparent sheets on which randomly located dots have been printed—an area on the map is estimated by counting the number of dots falling within it. In the *planimeter* method a mechanical device is used to trace the area's boundary, and the required measure accumulates on a dial on the machine.

**Humans have never devised good manual tools for making measurements from maps, especially measurements of area.**

By comparison, measurement of the area of a digitally represented polygon is trivial and totally reliable. The common procedure or algorithm calculates and sums the areas of a series of trapezia, formed by dropping perpendiculars to the x-axis, as shown in Figure 14.1. By making a simple change to the algorithm, it is also possible to use it to compute a polygon's centroid (see Section 14.2.1 for a discussion of centroids; note that the term *centroid* is often used in GISS whenever a polygon is collapsed to a point, whether or not the point is technically the mathematical centroid of the polygon). It is advisable to be cautious, however, to ensure that the coordinate system is appropriate. For example, if a polygon's vertices are coded in latitude and longitude, the result of computing area may be a measurement in "square degrees." But except at the

equator, the length of a degree of longitude varies with latitude and is always less than the length of a degree of latitude. On the other hand, measurement of area using projected coordinates, such as Universal Transverse Mercator (UTM) or U.S. State Plane (Sections 4.8.2 and 4.8.3), will give a result in square meters or square feet, respectively. Note, however, that neither of these projected coordinate systems has equal-area properties, so the result will be somewhat distorted. Studies that require accurate measures of area should use only equal-area projections.

## 14.1.2 Measurement of Shape

GI systems are also used to characterize the *shapes* of areas. In many countries the system of political representation is based on the concept of districts or constituencies, which are used to define who will vote for each place in the legislature (Box 14.1). In the United States and the UK, and in many other countries that derived their system of representation from the UK, one place is reserved in the legislature for each district. It is expected that districts will be compact in shape, and the manipulation of a district's shape to achieve certain overt or covert objectives is termed gerrymandering, after an early governor of Massachusetts, Elbridge Gerry (the shape of one of the state's districts was thought to resemble a salamander, with the implication that it had been manipulated to achieve a certain outcome in the voting). The construction of voting districts is an example of the principles of aggregation and zone design discussed in Section 5.2.1.

**Anomalous shape is the primary means of detecting gerrymanders of political districts.**

Geometric shape was the aspect that alerted Gerry's political opponents to the manipulation of districts, and today shape is measured whenever GI systems are used to aid in the drawing of political district boundaries, as must occur by law in the United States after every decennial census. An easy way to define shape is by comparing the perimeter length of an area to its area measure. Normally the square root of area is used to ensure that the numerator and denominator are both measured in the same units. A common measure of shape or compactness is

$$P/2\sqrt{\pi A}$$

where $P$ is the perimeter length and $A$ is the area. The factor of twice the square root of $\pi$ (3.54) ensures that the most compact shape, a circle, returns a shape of 1.0, and the most distended and contorted shapes return much higher values. Box 14.1 explores these ideas for a Congressional District in California.

**Figure 14.1** The algorithm for calculation of the area of a polygon given the coordinates of the polygon's vertices. The polygon consists of the three black arrows, plus the blue arrow forming the fourth side. Trapezia are dropped from each edge to the x-axis, and their areas are calculated as (difference in x) times (average of y). The trapezia for the first three edges, shown in yellow, orange, and blue, are summed. When the fourth trapezium is formed from the blue arrow, its area is negative because its start point has a larger x than its endpoint. When this area is subtracted from the total, the result is the correct area of the polygon.

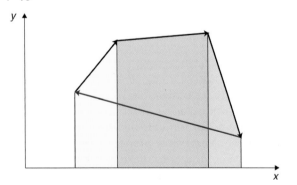

## Shape and Congressional Districts

One way to bias the outcome of an election in the United States is to draw (or *gerrymander*) district boundaries to include a majority of people likely to vote for a particular party. In Southern California the 23rd Congressional District had long been drawn to include the liberal-leaning coastal communities such as Santa Barbara and to exclude the more conservative agricultural areas of the interior. Its long, thin shape following the coastline had a shape index of 4.45. Representative Lois Capps, a Democrat, had held the seat through seven biennial elections.

Redistricting is required by law following each U.S. decennial census. To try to reduce the creation of "safe" seats by gerrymandering, a new process of districting was put in place in California that required districts to be drawn by a committee of citizens, rather than politicians. The result, after the 2010 census, was a new electoral map of California. Figure 14.2 shows how the old 23rd District was merged into a new 24th District that included both liberal-leaning and conservative-leaning areas; its much more compact shape has an index of 2.28. After a stiff fight, Lois Capps held the seat in the 2012 election with a majority of 53.8%.

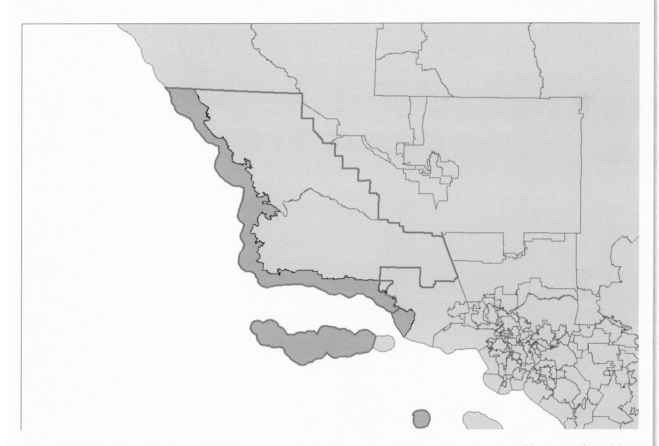

**Figure 14.2** Redistricting following the 2010 U.S. census, coupled with a new opposition to gerrymandering in California, transformed the old 23rd Congressional District (yellow) into a new, more compact 24th District (outlined in red).

## 14.2 Centrality

The topics of generalization and abstraction were discussed in Section 3.8 as ways of reducing the complexity of data. This section reviews a related topic, that of numerical summaries. If we want to describe the nature of summer weather in an area, we cite the *average* or *mean*, knowing that there is substantial variation around this value, but that it nevertheless gives a reasonable *expectation* about what the weather will be like on any given day. The mean (the more formal term) is one of a number of measures of *central tendency*, all of which attempt to create a summary description of a series of numbers in the form of a single number. Another is the *median*, the value such that one-half of the numbers are larger and one-half are smaller. Although the mean can be computed only for numbers measured on interval or ratio scales, the median can be computed for ordinal data. For nominal data the appropriate measure of central tendency is the *mode*, or the most common value. For definitions of nominal, ordinal, interval, and ratio see Box 2.1. Special methods must be used to measure central tendency for cyclic data; they are discussed in texts on directional data, for example, by Mardia and Jupp.

## 14.2.1 Centers

The spatial equivalent of the mean would be some kind of center, calculated to summarize the positions of a number of points. Early in U.S. history the Bureau of the Census adopted a practice of calculating a representative center for the U.S. population. As agricultural settlement advanced across the West in the Nineteenth Century, the repositioning of the center every 10 years captured the popular imagination. Today, the movement west has slowed and shifted more to the south (Figure 14.3) and by the next census may even have reversed.

### Centers are the two-dimensional equivalent of the mean.

The mean of a set of numbers has several properties. First, it is calculated by summing the numbers and dividing by the number of numbers. Second, if we take any value $d$ and sum the squares of the differences between the numbers and $d$, then when $d$ is set equal to the mean this sum is minimized (Figure 14.4). Third, the mean is the point about which the set of numbers would balance if we made a physical model such as the one shown in Figure 14.5 and suspended it.

**Figure 14.3** The mean center of the U.S. population, determined from the results of the 2010 census. Also shown is the median center, such that half of the population is to the north and half to the south, half to the east and half to the west.

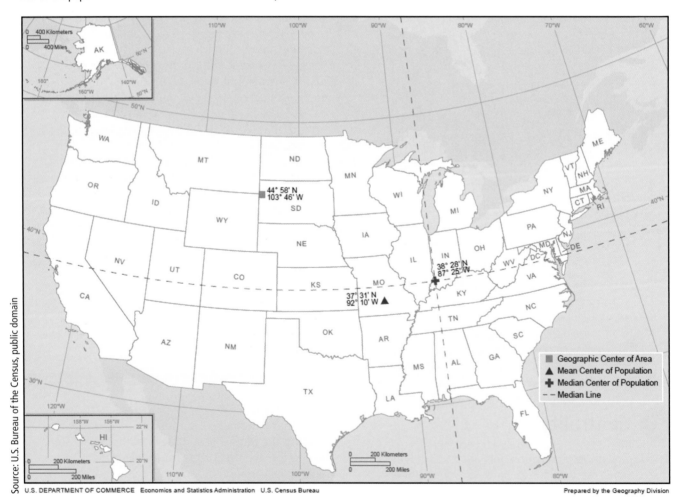

U.S. DEPARTMENT OF COMMERCE  Economics and Statistics Administration  U.S. Census Bureau

Prepared by the Geography Division

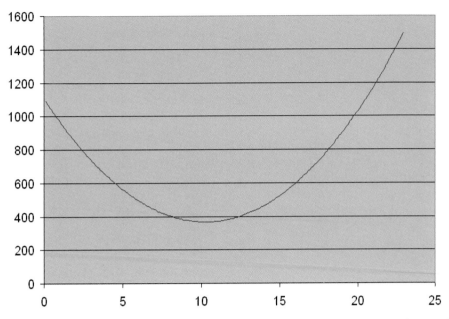

**Figure 14.4** Seven points are distributed along a line, at coordinates 1, 3, 5, 11, 12, 18, and 22. The curve shows the sum of distances squared from these points and how it is minimized at the mean [(1+3+5+11+12+18+22)/7 = 10.3].

These properties extend easily into two dimensions. Figure 14.6 shows a set of points on a flat plane, each one located at a point $(x_i, y_i)$ and with weight $w_i$. The centroid or mean center is found by taking the weighted average of the x and y coordinates:

$$\bar{x} = \sum_i w_i x_i \Big/ \sum_i w_i$$

$$\bar{y} = \sum_i w_i y_i \Big/ \sum_i w_i$$

It also is the point that minimizes the sum of squared distances, and it is the balance point.

Just like the mean, the centroid is a useful summary of a distribution of points. Although any single centroid may not be very interesting, a comparison of centroids for different sets of points or for different times can provide useful insights.

**The centroid is the most convenient way of summarizing the locations of a set of points.**

The property of minimizing functions of distance (the square of distance in the case of the centroid) makes centers useful for different reasons. Of particular interest is the location that minimizes the sum of distances, rather than the sum of squared distances because this could be the most effective location for any service that is intended to serve a dispersed population. The point that minimizes total straight-line distance is known as the point of minimum aggregate travel, or MAT, and is discussed in detail in Section 14.4, which is devoted to methods of design.

All the methods described in this and the following section are based on plane geometry and simple x,y coordinate systems. If the curvature of the Earth's surface is taken into account, the centroid of a set of points must be calculated in three dimensions and will always lie under the surface. More useful perhaps are versions of the MAT and centroid that minimize

**Figure 14.6** The centroid or mean center replicates the balance-point property in two dimensions—the point about which the two-dimensional pattern would balance if it were transferred to a weightless, rigid plane and suspended.

**Figure 14.5** The mean is also the balance point, the point about which the distribution would balance if it were modeled as a set of equal weights on a weightless, rigid rod.

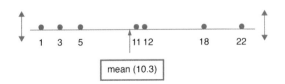

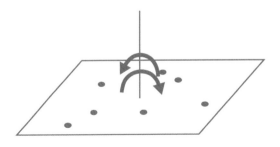

distance over the curved surface (the minimum total distance and the minimum total of squared distances, respectively, using great circle distances: see Sections 4.7 and 13.3.1).

## 14.2.2 Dispersion

Central tendency is the obvious choice if a set of measurements must be summarized in a single value, but what if there is the opportunity for a second summary value? Here the measure of choice for measurements with interval or ratio properties is the *standard deviation*, or the square root of the mean squared difference from the mean:

$$s = \sqrt{\sum_i (x_i - \bar{x})^2 / n}$$

where $n$ is the number of observations, $s$ is the standard deviation, $x_i$ refers to the $i$th observation, and $\bar{x}$ is the mean of the observations. In weighted form the equation becomes

$$s = \sqrt{\sum_i w_i (x_i - \bar{x})^2 / \sum_i w_i}$$

where $w_i$ is the weight given to the $i$th observation. The *variance*, or the square of the standard deviation (the mean squared difference from the mean), is often encountered, but it is not as convenient a measure for descriptive purposes. Standard deviation and variance are considered more appropriate measures of dispersion than the *range* (the difference between the highest and lowest numbers) because as averages they are less sensitive to the specific values of the extremes.

The standard deviation has also been encountered in Section 5.3.2.2 in a different guise, as the root mean squared error (RMSE), a measure of dispersion of observations about a true value. Just as in that instance, the Gaussian distribution provides a basis for generalizing about the contents of a sample of numbers, using the mean and standard deviation as the parameters of a simple bell curve. If data follow a Gaussian distribution, then approximately 68% of values lie within one standard deviation of the mean and approximately 5% of values lie outside two standard deviations.

These ideas convert very easily to the two-dimensional case. A simple measure of dispersion in two dimensions is the *mean distance from the centroid*. In some applications it may be desirable to give greater weight to more distant points. For example, if a school is being located, then students living at distant locations are comparatively disadvantaged. They can be given greater weight if each distance is squared, such that a student twice as far away receives four times the weight. This property is minimized by locating the school at the centroid.

**Mean distance from the centroid is a useful summary of dispersion.**

Measures of dispersion can be found in many areas of GI systems. The breadth of the kernel function of density estimation (Section 13.3.5) can be thought of as a measure of how broadly a pile of sand associated with each point is dispersed. RMSE is a measure of the dispersion inherent in positional errors (Section 5.3.2.2).

## 14.3 Analysis of Surfaces

Continuous fields of elevation provide the basis for many types of analysis in GI systems. More generally, any field formed by measurements of an interval or ratio variable, such as air temperature, rainfall, or soil pH, can also be conceptualized as a surface and analyzed using the same set of tools, though the results may make little sense in some cases. This section is devoted first to simple measurement of surface slope and aspect. It then introduces techniques for determining paths over surfaces, watersheds and channels, and finally intervisibility.

The various ways of representing continuous fields were first discussed in Section 3.5.2 and later in greater technical detail in Section 7.2. The techniques discussed in this section begin with a raster representation, termed a digital elevation model (DEM) in the case of terrain representation. In most cases the value recorded will be the elevation at the center of each raster cell, though in some cases it may be the mean elevation over the cell; it is always important to check a dataset's documentation on this issue before using the dataset.

**The digital elevation model is the most useful representation of terrain in a GI system.**

## 14.3.1 Slope and Aspect

Knowing the exact elevation of a point above sea level is important for some applications, including prediction of the effects of global warming and rising sea levels on coastal cities. For many applications, however, the value of a DEM lies in its ability to produce derivative measures through transformation, specifically measures of slope and aspect, both of which are also conceptualized as fields. Imagine taking a large sheet of plywood and laying it on the Earth's surface so that it touches at the point of interest. The magnitude of steepest tilt of the sheet defines the *slope* at that point, and the direction of steepest tilt defines the *aspect* (Box 14.2).

## Calculation of Slope Based on the Elevations of a Point and Its Eight Neighbors

There are many ways of estimating the slope in each cell of a DEM, depending on the equations that are used. Described here is the most common, though by no means the only, approach (refer to Figure 14.7 for point numbering):

$$b = (z_3 + 2z_6 + z_9 - z_1 - 2z_4 - z_7)/8D$$

$$c = (z_1 + 2z_2 + z_3 - z_7 - 2z_8 - z_9)/8D$$

where $b$ and $c$ are tan(slope) in the $x$ and $y$ directions, respectively, $D$ is the grid point spacing, and $z_i$ denotes elevation at the $i$th point, as shown in Figure 14.7. These equations give the four diagonal neighbors of Point 5 only half the weight of the other four neighbors in determining slope at Point 5.

Now we can calculate slope and aspect, as follows:

$$\tan(slope) = \sqrt{b^2 + c^2}$$

where *slope* is the angle of slope in the steepest direction.

$$\tan(aspect) = b/c$$

where *aspect* is the angle between the *y*-axis and the direction of steepest slope, measured clockwise.

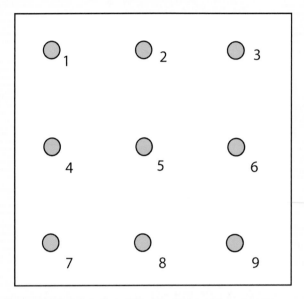

Figure 14.7 Calculation of the slope at Point 5 based on the elevation of it and its eight neighbors.

Because *aspect* varies from 0 to 360, an additional test is necessary that adds 180 to *aspect* if $c$ is positive.

This sounds straightforward, but it is complicated by a number of issues. First, what if the plywood fails to sit firmly on the surface, but instead pivots because the point of interest happens to be a peak or a ridge? In mathematical terms, we say that the surface at this point *lacks a well-defined tangent*, or that the surface at this point is *not differentiable*, meaning that it fails to obey the normal rules of continuous mathematical functions and differential calculus. The surface of the Earth has numerous instances of sharp breaks of slope, rocky outcrops, cliffs, canyons, and deep gullies that defy this simple mathematical approach to slope, and this is one of the issues that led Benoît Mandelbrot to develop his theory of fractals (see Section 2.8).

A simple and satisfactory alternative is to take the view that slope must be measured at a particular resolution. To measure slope at a 30-m resolution, for example, we evaluate elevation at points 30 m apart and compute slope by comparing them (equivalent in concept to using a plywood sheet 30 m across). The value this gives is specific to the 30-m spacing, and a different spacing (or different-sized sheet of plywood)

would have given a different result. In other words, *slope is a function of resolution* or scale, and it makes no sense to talk about slope without at the same time talking about a specific resolution or level of detail. This is convenient because slope is easily computed in this way from a DEM with the appropriate resolution.

**The spatial resolution used to calculate slope and aspect should always be specified.**

A second issue is the existence of several alternative *measures* of slope, and it is important to know which one is used in a particular software package and application. Slope can be measured as an *angle*, varying from 0 to 90 degrees as the surface ranges from horizontal to vertical. But it can also be measured as a percentage or ratio, defined as *rise over run*, and unfortunately there are two different ways of defining run. Figure 14.8 shows the two options, depending on whether run means the horizontal distance covered between two points, or the diagonal distance (the *adjacent* or the *hypotenuse* of the right-angled

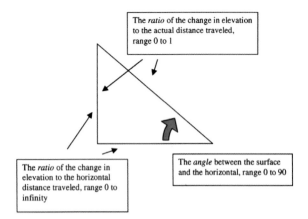

The *ratio* of the change in elevation to the actual distance traveled, range 0 to 1

The *ratio* of the change in elevation to the horizontal distance traveled, range 0 to infinity

The *angle* between the surface and the horizontal, range 0 to 90

**Figure 14.8** Three alternative definitions of slope. To avoid ambiguity we use the angle, which varies between 0 and 90 degrees.

triangle, respectively). In the first case (opposite over adjacent), slope as a ratio is equal to the tangent of the angle of slope and ranges from zero (horizontal) through 1 (45 degrees) to infinity (vertical). In the second case (opposite over hypotenuse), slope as a ratio is equal to the sine of the angle of slope and ranges from zero (horizontal) through 0.707 (45 degrees) to 1 (vertical). To avoid confusion we will use the term *slope* only to refer to the measurement in degrees and will call the other options tan(slope) and sin(slope), respectively.

When a GI system calculates slope and aspect from a DEM, it does so by estimating slope at each of the data points of the DEM, by comparing the elevation at that point to the elevations of surrounding points. But the number of surrounding points used in the calculation varies, as do the weights given to each of the surrounding points in the calculation. Box 14.2 shows this idea in practice, using one of the most common methods, which employs eight surrounding points and gives them different weights depending on how far away they are.

## 14.3.2 Modeling Travel on a Surface

One way to interpret a buffer drawn around a point (Figure 13.17) is that it represents the distance that could be traveled from the point in a given time, assuming constant travel speed. But if travel speed were not uniform and instead were represented by a continuous field of *friction*, then the buffer would be modified to expand more in some directions than in others. This function is often termed a *spread*, and like many techniques discussed in this section, it is an example of a function that is far easier to execute on a raster representation. The total friction associated with a route is calculated by summing the friction values of the cells along the route, and because there are many possible routes from an origin to a

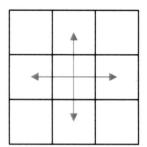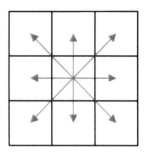

**Figure 14.9** The rook's-case (left) and queen's-case (right) move sets, defining the possible moves from one cell to another in solving the problem of optimum routing across a friction surface.

destination, it is necessary to select the route that minimizes total friction. But first one must select a *move set*, the set of possible moves that can be made between cells along the route. Figure 14.9 shows the two most commonly used move sets. When a diagonal move is made, it is necessary to multiply the associated friction by 1.414 ($\sqrt{2}$) to account for the move's greater length.

Figure 14.10 shows an example DEM, in this case with 30-m point spacing and covering an area roughly corresponding to Orange County, California. Archaeologists are often interested in the impact of terrain on travel because it may help to explain ancient settlement patterns and communication paths. If we assume that travelers will avoid steep slopes and high terrain, then a possible measure of friction would combine slope $s$ and elevation $e$ in an expression such as $s + e/100$ where $s$ is measured in degrees and $e$ in feet. Applying this to the Orange County DEM,

**Figure 14.10** A digital elevation model of an area of Southern California, including most of Orange County. High ground is shown in blue. The Pacific Ocean covers the lower left, and the red symbol is located at Santa Ana (John Wayne) Airport.

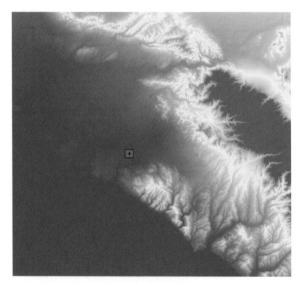

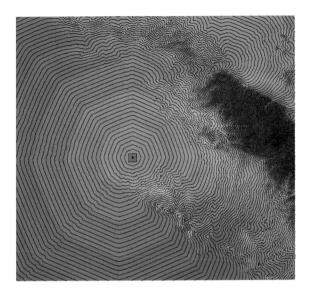

**Figure 14.11** Contours of equal travel cost from an origin. Friction or travel "cost" is determined by a combination of ground elevation and ground slope, reflecting the conditions faced by pre-Columbian populations in this part of California. Note the effect of the Santa Ana River gorge in providing easy access to the Inland Empire (upper right).

Figure 14.11 shows the effects of traveling away from a location near the current Santa Ana airport. Contours show points of equal total friction ("cost") from the start point. Notice how the gorge formed by the Santa Ana River provides an easy route across the high, steep ground and into what is now the Inland Empire.

### 14.3.3 Computing Watersheds and Channels

A DEM provides an easy basis for predicting how water will flow over a surface, and therefore of many useful hydrologic properties. Consider the DEM shown in Figure 14.12. We assume that water can flow from a cell to any of its eight neighboring cells, down the direction of steepest slope. Because only eight such directions are possible in this raster representation, we assume that water will flow to the lowest of the eight neighbors, provided at least one neighbor is lower. If no neighbor is lower, water is instead assumed to pond, perhaps forming a lake, until it rises high enough to overflow. Figure 14.12A shows the outflow directions predicted for each cell and outflow accumulation in the one cell that has no lower neighbor. Because this cell is at the edge of the area, we assume that it will spill over the boundary, forming the area's outflow.

A *watershed* is defined as the area upstream of a point—in other words, the area that drains through that point. Once we know the flow directions, it is

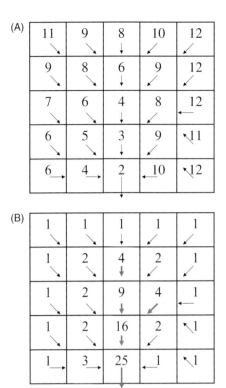

**Figure 14.12** Hydrologic analysis of a sample DEM: (A) the DEM and inferred flow directions using the queen's case move set; and (B) accumulated flows in each cell and eroded channels based on a threshold flow of 4 units.

easy to identify a point and the associated upstream area that forms that point's watershed. Note that any point on the map has an associated watershed and that watersheds therefore may overlap.

This solution represents what a hydrologist would term *overland* flow. When flow accumulates sufficiently, it begins to erode its bed and form a channel. If we could establish an appropriate threshold, expressed in terms of the number of upstream cells that drain through a given cell, then we could map a network of channels. Figure 14.12B shows an example using a threshold of four cells.

In reality some landscapes have closed depressions that fill with water to form lakes. Other landscapes, particularly those developed on soluble rocks such as limestone or gypsum, contain closed depressions that drain underground. But a DEM generated by any of the conventional means will also likely contain elevation errors, some of which will appear as artificial closed depressions. So a GI system will commonly include a routine to "fill" any such "closed depressions," allowing them to overflow and add to the general surface runoff. This filling step will need to be conducted before any useful hydrologic analysis of a landscape can be made.

Figure 14.13 shows the result of applying a filling step and then computing the drainage network for

**Figure 14.13** Analysis of the Orange County DEM predicts the river channels shown in red in this Google Earth mashup. The Santa Ana River appears to flow out of the gorge shown in the upper right, and then far to the west before emptying into the Pacific near Seal Beach. In reality, it turns south and empties near Newport Beach in the bottom center. See text for explanation.

the Orange County DEM, using a channel-forming threshold of 10,000 cells or 9 sq km. The analysis correctly predicts a large river flowing into the top right of the area shown, through the Santa Ana River gorge. However, rather than flowing south to reach the Pacific below the middle center of the figure as the real river does, the predicted channel flows several kilometers further west before turning south, entering the Pacific near Seal Beach. Interestingly, this is the historic course of the river, before several severe floods in the Nineteenth Century. The current concrete channel, clearly visible in the figure, is delimited by levées, which are not large enough to affect the DEM given its 30-m point spacing.

## 14.3.4 Computing Visibility

One of the most powerful forms of surface analysis centers on the ability to compute intervisibility: can Point A on the surface be seen from Point B? This often takes the form of determining a point's *viewshed*—the area of surface that can be seen by an observer located at the point whose eye is elevated some specified height above the surface. Viewsheds have been used to plan the locations of observation points, transmitters, and mobile-phone towers. They have been used to analyze the locations adopted for prehistoric burial mounds, to test the hypothesis that people sought to be buried at conspicuous points where their remains could dominate the landscape. An interesting analysis by GI scientist Anne Knowles

showed that a significant factor in the outcome of the Battle of Gettysburg, a turning point in the American Civil War, was the visibility of the battlefield achieved by the two commanders from their respective headquarters.

Figure 14.14 shows the viewshed of an observation point located on the low hills near Newport

**Figure 14.14** The area visible from a 200-ft tower located on the hills near Newport Beach, California.

Beach, again using the Orange County DEM. In this case the eye has been elevated 200 ft above the surface, enough to command a view of the flatter areas in the northwest, but still not enough to view areas to the north and east of the higher hills in the area.

## 14.4 Design

In this section we look at the analysis of spatial data not for the purpose of discovering anomalies, or testing hypotheses about process, as in previous sections, but with the objective of creating improved designs—in short, changing the world. These objectives might include designs that minimize travel distance, maximize someone's profit, or minimize the costs and environmental impacts of construction of some new development. Recently the term *geodesign* has become popular as a description of this kind of application; geodesign can be defined as "geography by design," or the application of GI systems to improving the world at geographic scales. Box 14.3 describes the work of Carl Steinitz, one of the leaders of the geodesign movement.

The three principles of retailing are often said to be *location*, *location*, and *location*, and over the years many GIS applications have been directed at applications that involve, in one way or another, the search for optimum designs. The methods for finding centers described in Section 14.2.1 were shown to have useful design-oriented properties, and the modeling of travel across a surface discussed in Section 14.3.2 can also be interpreted as a design problem, for routing power lines, highways, or tanks. This section includes discussion of a wider selection of these so-called *normative* methods, or methods developed for application to the solution of practical problems of design.

**Normative methods apply well-defined objectives to the design of systems.**

Design methods are often implemented as components of systems built to support decision making—so-called *spatial-decision support systems*, or SDSSs. Complex decisions are often contentious, with many *stakeholders* interested in the outcome and in arguing for one position or another. SDSSs are specially adapted GI systems that can be used during the decision-making process to provide instant feedback on the implications of various proposals and the

### Carl Steinitz, Geodesigner

Carl Steinitz (Figure 14.15) is the Alexander and Victoria Wiley Professor of Landscape Architecture and Planning, Emeritus, at Harvard University Graduate School of Design. In 1966, Steinitz received his PhD degree in City and Regional Planning, with a major in urban design, from the Massachusetts Institute of Technology (MIT). He also holds a Master of Architecture degree from MIT and a Bachelor of Architecture degree from Cornell University. In 1965 he began his affiliation with the Harvard Graduate School of Design as an initial research associate in the Laboratory for Computer Graphics and Spatial Analysis. He has been Professor of Landscape Architecture and Planning at the Graduate School of Design since 1973.

Professor Steinitz has devoted much of his academic and professional career to improving methods to analyze large land areas and make design decisions about conservation and development. His applied research and teaching focus on highly valued landscapes that are undergoing substantial pressures for change. Professor Steinitz has directed studies in as wide ranging locales as the Gunnison region of Colorado; the Monadnock

Courtesy: Carl Steinitz

**Figure 14.15** Carl Steinitz, Geodesigner.

region of New Hampshire; the region of Camp Pendleton, California; the Gartenreich Worlitz in Germany; the West Lake in Hangzhou, China; Coiba National Park in Panama; Cagliari, Italy; and the regions of Castilla La Mancha and Valencia in Spain. Figure 14.16 summarizes the process that has evolved through these projects.

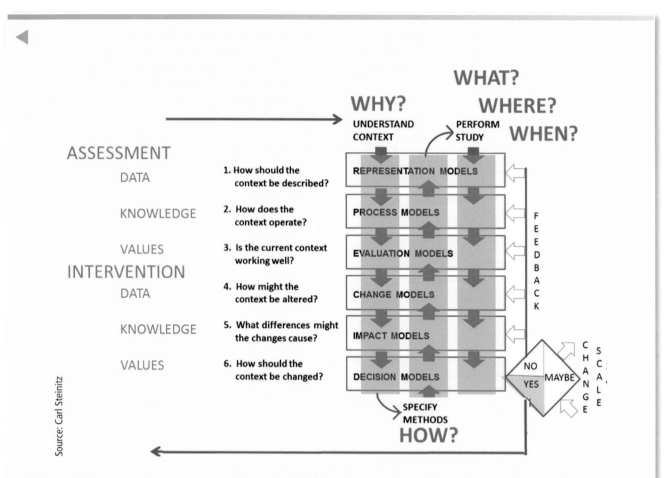

WHAT?

WHY?  WHERE?

UNDERSTAND    PERFORM    WHEN?
CONTEXT       STUDY

ASSESSMENT

DATA        1. How should the
               context be described?        REPRESENTATION MODELS

KNOWLEDGE    2. How does the
               context operate?             PROCESS MODELS        F
                                                                  E
VALUES       3. Is the current context                            E
               working well?               EVALUATION MODELS      D
INTERVENTION                                                      B
                                                                  A
DATA         4. How might the                                     C
               context be altered?          CHANGE MODELS         K

KNOWLEDGE    5. What differences might
               the changes cause?           IMPACT MODELS                    C      S
                                                                             H      C
VALUES       6. How should the                                      NO       A      A
               context be changed?          DECISION MODELS                  N   L  L
                                                              YES  MAYBE      G      E
                                                                             E

                                            SPECIFY
                                            METHODS
                                            HOW?

**Figure 14.16** A schematic model of the geodesign process, developed by Carl Steinitz to capture the various stages used in his practical applications of geodesign principles.

Source: Carl Steinitz

evaluation of what-if scenarios. SDSSs typically have special user interfaces that present only those functions relevant to the application.

The methods discussed in this section fall into several categories. The next section discusses methods for the optimum location of points and extends the method introduced earlier for the MAT. The second section discusses routing on a network and its manifestation in the *traveling-salesperson problem* (TSP). The methods also divide between those that are designed to locate points and routes on networks and those designed to locate points and routes in continuous space without respect to the existence of roads or other transportation links.

## 14.4.1 Point Location

The MAT problem is an instance of location in continuous space and finds the location that minimizes total distance with respect to a number of points. The analogous problem on a network would involve

finding that location on the network that minimizes total distance to a number of points, also located on the network, using routes that are constrained to the network. Figure 14.17 shows the contrast between continuous and network views, and Chapter 7 discusses data models for networks.

A very useful theorem first proved by Louis Hakimi reduces the complexity of many location problems on networks. Figure 14.17B shows a typical basis for network location. The links of the network come together in *nodes*. The weighted points are also located on the network and also form nodes. For example, the task might be to find the location that minimizes total distance to a distribution of customers, with all customers aggregated into these weighted points. The weights in this case would be counts of customers. The Hakimi theorem proves that for this problem of minimizing distance the only locations that have to be considered are the nodes; it is impossible for the optimum location to be anywhere else. It is easy to see why this should be so. Think of

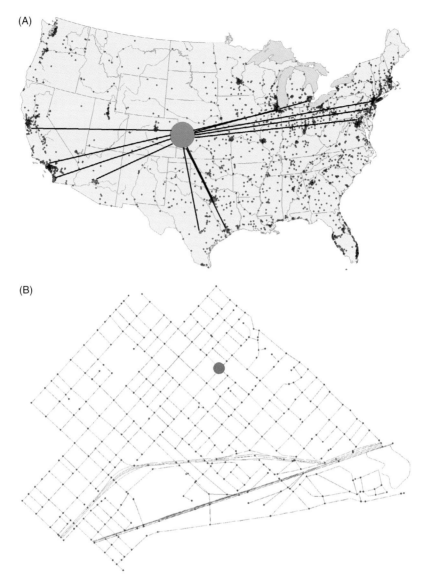

(A)

(B)

**Figure 14.17** Search for the best locations for a central facility to serve dispersed customers. In (A) the problem is solved in continuous space, with straight-line travel, for a warehouse to serve the 12 largest U.S. cities. In continuous space there is an infinite number of possible locations for the site. In (B) a similar problem is solved at the scale of a city neighborhood on a network, where Hakimi's theorem states that only junctions (nodes) in the network and places where there is weight need to be considered, making the problem much simpler, but where travel must follow the street network.

a trial point located in the middle of a link, away from any node. Think of moving it slightly in one direction along the link. This moves it toward some weights and away from others, but every unit of movement results in the same increase or decrease in total weighted distance. In other words, the total distance traveled to the location is a linear function of the location along the link. Because the function is linear, it cannot have a minimum mid-link, so the minimum must occur at a node.

> **Optimum location problems can be solved in either discrete or continuous space, depending largely on scale.**

The MAT problem on a network is known as the *1-median* problem, and the *p-median* problem seeks optimum locations for any number *p* of central facilities such that the sum of the distances between each weight and the *nearest* facility is minimized. A typical practical application of this problem is in the location of central public facilities, such as libraries, schools, or agency offices, when the objective is to locate for maximum total accessibility.

Many problems of this nature have been defined for different applications and implemented in GI systems. Whereas the median problems seek to minimize total distance, the *coverage* problems seek to minimize the *furthest* distance traveled, on the grounds that dealing with the worst case of accessibility is often more attractive than dealing with average accessibility. For example, it may make more sense to a city fire department to locate so that a response is possible to *every* property in less than five minutes, than to worry about minimizing the *average* response time. Coverage problems find applications in the location of emergency facilities, such as fire stations (Figure 14.18), where it is desirable that every possible emergency be covered within a fixed number of minutes of response time, or when the objective is to minimize the worst-case response time, to the furthest possible point.

All these problems are referred to as *location-allocation* problems because they involve two types of decisions: where to *locate* and how to *allocate* demand for service to the central facilities. A typical location-allocation problem might involve the selection of sites for supermarkets. In some cases the allocation of demand to sites is controlled by the designer, as it is in the case of school districts when students have no choice of schools. In other cases

allocation is a matter of choice, and good designs depend on the ability to predict how consumers will choose among the available options. Models that make such predictions are known as *spatial interaction models*, and their use is an important application of GI systems in market research.

**Location-allocation involves two types of decisions: where to locate and how to allocate demand for a service.**

## 14.4.2 Routing Problems

Point-location problems are concerned with the design of fixed locations. Another area of optimization is in routing and scheduling, or decisions about the optimum tracks followed by vehicles. A commonly encountered example is in the routing of delivery vehicles (Box 14.4). These examples show a base location, a depot that serves as the origin and final destination of delivery vehicles; and a series of stops that need to be made. There may be restrictions on the times at which stops must be made. For example, a vehicle delivering home appliances may be required to visit certain houses at certain times, when the residents are home. Vehicle routing and scheduling solutions are used by parcel delivery companies, school buses, on-demand public transport vehicles, and many other applications.

Underlying all routing problems is the concept of the *shortest path*—the path through the network between a defined origin and destination that minimizes distance or some other measure based on distance, such as travel time. Attributes associated with the network's links, such as length, travel speed, restrictions on travel direction, and level of congestion, are often taken into account. Many people are now familiar with the routine solution of the shortest path problem by Web sites such as Google Maps or MapQuest.com (Section 1.5.1), which solve many millions of such problems per day for travelers, and by similar methods to those used by in-vehicle navigation systems. They use standard algorithms developed decades ago, long before the advent of GI systems. The path that is strictly shortest is often not suitable because it involves too many turns or uses too many narrow streets, and algorithms will often be programmed to find longer routes that use faster highways, particularly freeways. Routes in Los Angeles, for example, can often be caricatured as (1) shortest route from origin to nearest freeway, (2) follow freeway network, and (3) shortest route from nearest freeway to destination, even though this route may be far from the shortest. The latest generation of route finders is also beginning to accommodate time of day, time

Figure 14.18 GI systems can be used to find locations for fire stations that result in better response times to emergencies.

© Shaun Lowe/Getty Images, Inc.

## Routing Service Technicians for Sears

Sears manages one of the largest home-appliance repair businesses in the world, with six distinct geographic regions that include 50 independent districts. More than 10,000 technicians throughout the United States complete approximately 11 million in-home service orders each year. Decisions on dividing the daily orders among the company's technician teams, and on routing each team optimally, used to be made by dispatchers who relied on their own intuition and personal knowledge. But human intuition can be misleading and is lost when dispatchers are sick or retire.

Several years ago Sears teamed with Esri (Redlands, CA) to build the Computer-Aided Routing System (CARS) and the Capacity Area Management System (CAMS). CAMS manages the planned capacity of available service

technicians assigned to geographic work areas, and CARS provides daily street-level geocoding and optimized routing for the mobile service technicians (Figure 14.19). The mobile Sears Smart Toolbox application provides service technicians with repair information for products, such as schematic diagrams. It also contains a GI system module for mobile mapping and routing, which gives in-vehicle navigation capabilities to assist in finding service locations and minimizing travel time. TeleAtlas ('s-Hertogenbosch, the Netherlands, now part of TomTom) provides the accurate street data that is critical for supporting geocoding and routing.

The system has had a major impact on Sears' business, saving the business tens of millions of dollars annually.

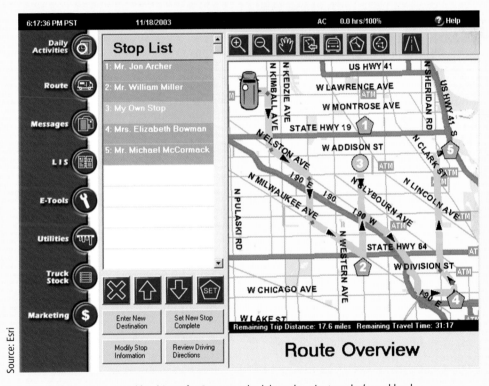

**Figure 14.19** Screenshot of the system used by drivers for Sears to schedule and navigate a day's workload.

of the week, and real-time information on traffic congestion.

The simplest routing problem with multiple destinations is the so-called traveling-salesperson problem, or TSP. In this problem there are a number

of places that must be visited in a tour from the depot, and the distances between pairs of places are known. The problem is to select the best tour from all possible orderings, in order to minimize the total distance traveled. In other words, the optimum is to

**Table 14.1** The number of possible tours in a traveling-salesperson problem.

| Number of places to visit | Number of possible tours |
|:---:|:---:|
| 3 | 1 |
| 4 | 3 |
| 5 | 12 |
| 6 | 60 |
| 7 | 360 |
| 8 | 2520 |
| 9 | 20160 |
| 10 | 181440 |

be selected out of the available tours. If there are $n$ places to be visited, including the depot, then there are $(n-1)!$ possible tours (the symbol ! indicates the product of the integers from 1 up to and including the number, known as the number's *factorial*). Because it is irrelevant, however, whether any given tour is conducted in a forward or backward direction, the effective number of options is $(n-1)!/2$. Unfortunately, this number grows very rapidly with the number of places to be visited, as Table 14.1 shows.

**A GI system can be very effective at solving routing problems because it is able to examine vast numbers of possible solutions quickly.**

The TSP is an instance of a problem that becomes quickly unsolvable for large $n$. Instead, designers adopt procedures known as *heuristics*, which are algorithms designed to work quickly and to come close to providing the best answer, although not guaranteeing that the best answer will be found. One not-very-good heuristic for the TSP is to proceed always to the closest unvisited destination (a so-called *greedy* approach) and finally to return to the start. Many spatial optimization problems, including location-allocation and routing problems, are solved today by the use of sophisticated heuristics. Because of the complexity of road networks, for example, route finders will often use a variant of the A* algorithm, a heuristic, rather than exhaustively examine every possible route.

## 14.5 Hypothesis Testing

This last section reviews a major area of statistics—the testing of hypotheses and the drawing of inferences—and its relationship to GI science and spatial analysis. Much work in statistics is *inferential*; that is, it uses information obtained from samples to make general conclusions about a larger population, on the assumption that the sample came from that population. The concept of inference was introduced in Section 2.4 as a way of reasoning about the properties of a larger group from the properties of a sample. At that point several problems associated with inference from geographic data were raised. This section revisits and elaborates on that topic and discusses the particularly thorny issue of spatial hypothesis testing.

For example, suppose we were to take a random and independent sample of 1,000 people and ask them how they might vote in the next U.S. Presidential election. By *random and independent*, we mean that every person of voting age in the general population has an equal chance of being chosen and that the choice of one person does not make the choice of any others—parents, neighbors—more or less likely. Suppose that 45% of the sample said they would support Hillary Clinton. Statistical theory then allows us to give 45% as the best estimate of the proportion who would vote for Clinton among the *general* population, and it also allows us to state a *margin of error*, or an estimate of how much the true proportion among the population will differ from the proportion among the sample. A suitable expression of margin of error is given by the 95% confidence limits, or the range within which the true value is expected to lie 19 times out of 20. In other words, if we took 20 different samples, all of size 1000, there would be a scatter of outcomes, and 19 out of 20 of them would lie within these 95% confidence limits. In this case a simple analysis using the *binomial distribution* shows that the 95% confidence limits are 3%; in other words, 19 times out of 20 the true proportion lies between 42 and 48%.

This example illustrates the *confidence limits* approach to inference, in which the effects of sampling are expressed in the form of uncertainty about the properties of the population. An alternative that is commonly used in scientific reasoning is the *hypothesis-testing* approach. In this case our objective is to test some general statement about the population— for example, that 50% will support Clinton in the next election (and 50% will support the other candidate—in other words, there is no real preference in the electorate). We take a sample, and we then ask whether the evidence from the sample supports the general statement. Because there is uncertainty associated with any sample, unless it includes the entire population, the answer is never absolutely certain. In this example and using our confidence limits approach, we know that if 45% were found to support Clinton in the sample, and if the margin of error was 3%, it is highly unlikely that the true proportion in the population is as high as 50%. Alternatively, we could state the 50% proportion in the population as a *null hypothesis* (we use the term *null* to reflect the absence of something, in this case a clear choice) and determine how frequently a sample of 1,000 from such a population would yield a proportion as low as 45%. Again the answer is very small; in fact, the probability is 0.0008. But it is not zero, and its value represents the chance of making an error of inference—of rejecting the hypothesis when in fact it is true.

**Methods of inference reason from information about a sample to more general information about a larger population.**

These two concepts—confidence limits and inferential tests—are the basis for statistical testing and form the core of introductory statistics texts. There is no point in reproducing those introductions here, and the reader is simply referred to them for discussions of the standard tests—$F$, $t$, $\chi^2$, and so on. The focus here is on the problems associated with using these approaches with geographic data in a GI science context. The next section reviews the inferential tests associated with one popular descriptive statistic for spatial data, the Moran index of spatial dependence that was discussed in Section 2.7. The following section discusses the general issues and points to ways of resolving them.

## 14.5.1 Hypothesis Tests on Geographic Data

Although inferential tests are standard practice in much of science, they are very problematic for geographic data. The reasons have to do with fundamental properties of geographic data, many of which were introduced in Chapter 2, and others have been encountered at various stages in this book.

Many inferential tests propose the existence of a population, from which the sample has been obtained by some well-defined process. We saw in Section 2.4 how difficult it is to think of a geographic dataset as a sample of all of the datasets that might have been. It is equally difficult to think of a dataset as a sample of some larger area of the Earth's surface, for two major reasons.

First, the samples in standard statistical inference are obtained independently (Section 2.4). But a geographic dataset is often *all there is* in a given area—it *is* the population. Perhaps we could regard a dataset as a sample of a larger area. But in this case the sample would not have been obtained randomly; instead, it would have been obtained by systematically selecting all cases within the area of interest. Moreover, the samples would not have been independent. Because of spatial dependence, which we have understood to be a pervasive property of geographic

data (Chapter 2), it is very likely that there will be similarities between neighboring observations.

**A GIS project often analyzes all the data there is about a given area, rather than a sample.**

Figure 14.20 shows a typical instance of geographic sampling. In this case the objective is to explore the relationship between topographic elevation and vegetation cover class, based on a DEM and a map of vegetation, in the area north of Santa Barbara. We might suspect, for example, that certain types of vegetation are encountered only at higher altitudes. One way to do this would be by randomly sampling at a set of points, recording the elevation and vegetation cover class at each, and then examining the results in a routine statistical analysis. In Figure 14.20, roughly 50 sample points are displayed. But why 50? If we increased the number to 500, our statistical test would have more data, and consequently more power to detect and evaluate any relationship. But why stop at 500? Here as in Box 13.4 the geographic case appears to create the potential for endless proliferation of data. Tobler's First Law (Sections 2.2 and 13.3.4) tells us, however, that after a while additional data values will not be truly informative because they could have been predicted from previously sampled values. We cannot have it both ways—if we believe in spatial interpolation (Section 13.3.6), we cannot at the same time believe in independence of geographic samples, despite the fact that this is a basic assumption of statistical tests. In effect, a geographic area can only yield a limited number of truly independent samples, from points spaced sufficiently far apart. Beyond that number, the new samples are not truly independent and in reality add nothing to the power of a test.

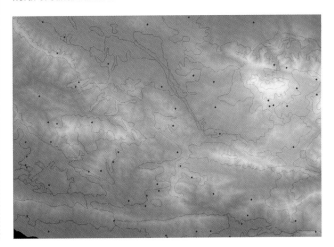

**Figure 14.20** A randomly placed sample of points used to examine the relationship between vegetation cover class (delimited by the boundaries shown) and elevation (whiter areas are higher), in an area north of Santa Barbara.

Finally, the issue of spatial *heterogeneity* (Section 2.2) also gets in the way of inferential testing. The Earth's surface is highly variable, and there is no such thing as an average place on it. The characteristics observed on one map sheet are likely to be substantially different from those on other map sheets, even when the map sheets are neighbors. So the census tracts of a city are certainly not acceptable as a random and independent sample of all census tracts, even the census tracts of an entire nation. They are not independent, and they are not random. Consequently, it is very risky to try to infer the properties of *all* census tracts from the properties of all the tracts in any *one* area. The concept of sampling, which is the basis for statistical inference, does not transfer easily to the spatial context.

**The Earth's surface is very heterogeneous, making it difficult to take samples that are truly representative of any large region.**

Before using inferential tests on geographic data, therefore, it is advisable to ask two fundamental questions:

- Can I conceive of a larger *population* that I want to make inferences about?
- Are my data acceptable as a *random* and an *independent* sample of that population?

If the answer to either of these questions is *no*, then inferential tests are not appropriate.

Given these arguments, what options are available? One strategy that is sometimes used is to discard data until the proposition of independence becomes acceptable—until the remaining data points are so far apart that they can be regarded as essentially independent. But no scientist is happy throwing away data.

Another approach is to abandon inference entirely. In this case the results obtained from the data are descriptive of the study area, and no attempt is made to generalize. This approach, which uses local statistics to observe the *differences* in the results of analysis over space, represents an interesting compromise between the nomothetic and idiographic positions outlined in Section 1.3. Generalization is very tempting, but the heterogeneous nature of the Earth's surface makes it very difficult. If generalization is required, then it can be accomplished by appropriate experimental design—by replicating the study in a sufficient number of distinct areas to warrant confidence in a generalization.

Another, more successful approach exploits the special nature of spatial analysis and its concern with detecting pattern. Consider the example in Figure 13.21, where the Moran index was computed at +0.4011, an indication that high values tend to be surrounded by high values and low values by low values—a positive spatial autocorrelation. It is reasonable to ask whether such a value of the Moran index could have arisen by chance because even a random arrangement of a limited number of values typically will not give the theoretical value of 0 corresponding to no spatial dependence (actually the theoretical value is very slightly negative). In this case 51 values are involved, arranged over the 51 features on the map (the 50 states plus the District of Columbia). If the values were arranged randomly, how far would the resulting values of the Moran index differ from 0? Would they differ by as much as 0.4011? Intuition is not good at providing answers.

In such cases a simple test can be run by simulating random arrangements. The software used to prepare this illustration includes the ability to make such simulations, and Figure 14.21 shows the results of simulating 999 random rearrangements. It is clear that the actual value is well outside the range of what is possible for random arrangements, leading to the conclusion that the apparent spatial dependence is real.

How does this *randomization* test fit within the normal framework of statistics? The null hypothesis being evaluated is that the distribution of values over the 51 features is random, each feature receiving a value that is independent of neighboring values. The population is the set of all possible arrangements, of which the data represent a sample of one. The test then involves comparing the test statistic—the Moran Index—for the actual pattern against the distribution of values produced by the null hypothesis. So the test is a perfect example of standard hypothesis testing, adapted to the special nature of spatial data and the common objective of discovering pattern.

**Randomization tests are uniquely adapted to testing hypotheses about spatial pattern.**

Finally, a large amount of research has been devoted to devising versions of inferential tests that cope effectively with spatial dependence and spatial heterogeneity. Software that implements these tests is now widely available, and interested readers are urged to consult the appropriate sources. GeoDa is an excellent comprehensive software environment for such tests (available via geodacenter.asu.edu), and many of these methods are available as extensions of standard GI systems.

**Figure 14.21** Randomization test of the Moran Index computed in Figure 13.20. The histogram shows the results of computing the index for 999 rearrangements of the 51 values on the map. The yellow line on the right shows the actual value, which is very unlikely to occur in a random arrangement, reinforcing the conclusion that there is positive spatial dependence in the data.

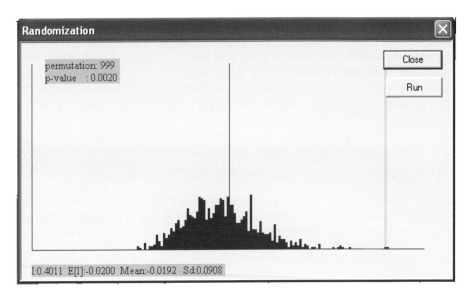

Randomization

permutation: 999
p-value    : 0.0020

Close

Run

I:0.4011  E[I]:-0.0200  Mean:-0.0192  Sd:0.0908

## 14.6 Conclusion

This chapter has covered the conceptual basis of many of the more sophisticated techniques of spatial analysis that are available in GI systems. Box 14.5 is devoted to Doug Richardson, Executive Director of the Association of American Geographers, who has done much to promote spatial analysis and spatial technologies at the U.S. National Institutes of Health and has himself developed and exploited many new

---

### Biographical Box 14.5

### Doug Richardson, A Leader in Academic Geography

Doug Richardson (Figure 14.22) is the Executive Director of the Association of American Geographers (AAG). During the past ten years, he has led a highly successful organizational renewal of the AAG and has built strong academic, research, publishing, and financial foundations for the organization's future.

Prior to joining the AAG, Doug founded and for 18 years was the president of GeoResearch, Inc., a private-sector scientific research company specializing in geographic science and technology, including GISS, spatial modeling, and GPS. GeoResearch developed and patented the world's first real-time interactive GPS/GIS technologies, leading to far-reaching changes in the ways in which geographic information is collected, mapped, integrated, and used within geography, as well as in society at large. The technologies and methods pioneered by GeoResearch are now at the heart of a wide array of real-time interactive mapping, navigation, location-based business, geographic research, mobile computing, military operations, and large-scale operations management applications of most major industries and governments. Doug sold his company and its core patents in 1998.

Doug continues to conduct research at the AAG and to publish across multiple dimensions of geography, ranging from GI science to the GeoHumanities, and from international health research to interactions between science, innovation, and human rights. As lead author of a 2013 *Science* article entitled "Spatial Turn in Health Research," for example, he explores how geographic science and technologies are opening profound new possibilities for understanding the prevalence, diffusion, and causes of disease, and its treatment. He holds a Bachelor's degree from the University of Michigan and a PhD in Geography from Michigan State University. He currently is a member of the U.S. National Geospatial Advisory Committee.

Courtesy: Doug Richardson

**Figure 14.22** Doug Richardson.

technological developments in the areas of GI systems and GPS.

The last section of the chapter raised some fundamental issues associated with applying methods and theories that were developed for nonspatial data to the spatial case. Spatial analysis is clearly not a simple and straightforward extension of nonspatial analysis, but instead raises many distinct problems, as well as some exciting opportunities. The two chapters on spatial analysis have only scratched the surface of this large and rapidly expanding field.

## Questions for Further Study

1. Parks and other conservation areas have geometric shapes that can be measured by comparing park perimeter length to park area, using the methods reviewed in this chapter. Discuss the implications of shape for park management, in the context of (a) wildlife ecology and (b) neighborhood security.

2. What exactly are *multicriteria* methods? Examine one or more of the methods in the Eastman chapter referenced in Further Reading, summarizing the issues associated with (a) measuring variables to support multiple criteria, (b) mixing variables that have been measured on different scales (e.g., dollars and distances), and (c) finding solutions to problems involving multiple criteria.

3. Besides being the basis for useful summary measures, fractals also provide interesting ways of simulating geographic phenomena and patterns. Browse the Web for sites that offer fractal simulation software, or investigate one of many commercially available packages. What other uses of fractals in GI science can you imagine?

4. Every point on the Earth's surface has an antipodal point—the point that would be reached by drilling an imaginary hole straight through the Earth's center. Britain, for example, is approximately antipodal to New Zealand. If one-third of the Earth's surface is land, you might expect that one-third of all of the land area would be antipodal to points that are also on land, but a quick look at an atlas will show that the proportion is actually far less than that. In fact, the only substantial areas of land that have antipodal land are in South America (and their antipodal points in China). How is spatial dependence relevant here, and why does it suggest that the Earth is not so surprising after all?

## Further Reading

De Smith, M. J., Goodchild, M. F., and Longley, P. A. 2009. *Geospatial Analysis: A Comprehensive Guide to Principles, Techniques and Software Tools* (3rd ed.). Winchelsea, UK: Winchelsea Press. www.spatialanalysisonline.com. See especially Section 6 on Surface and Field Analysis and Section 7 on Network and Location Analysis.

Eastman, J. R. 2005. Multicriteria methods. In Longley, P. A., Goodchild, M. F., Maguire, D. J., and Rhind, D. W. (eds.). *Geographical Information Systems: Principles, Techniques, Management and Applications* (abridged ed.). Hoboken, NJ: Wiley, pp. 225–234.

Ghosh, A. and Rushton, G. (ed). 1987. *Spatial Analysis and Location-Allocation Models*. New York: Van Nostrand Reinhold.

Mardia, K. V. and Jupp, P. E. 2000. *Directional Statistics*. New York: Wiley.

O'Sullivan, D. and Unwin, D. J. 2010. *Geographic Information Analysis* (2nd ed.). Hoboken, NJ: Wiley.

Steinitz, C. 2012. *A Framework for Geodesign: Changing Geography by Design*. Redlands, CA: Esri Press.

# Spatial Modeling with GI Systems

**15**

Models are used in many different ways, ranging from simulations of how the world works to evaluations of planning scenarios to the creation of indicators. In all these cases the geographic information (GI) system is used to carry out a series of transformations or analyses of geographic space, either at one point in time or at a number of intervals. This chapter begins with the necessary definitions and presents a taxonomy of models, with examples. It addresses the difference between analysis, the subject of Chapters 13 and 14, and this chapter's focus upon modeling. The alternative software environments for modeling are reviewed, along with capabilities for cataloging and sharing models, which are developing rapidly. The chapter ends with a look into the future of modeling and associated GI system developments.

## LEARNING OBJECTIVES

After studying this chapter you will:

- Know what modeling means in the context of GI systems.

- Be familiar with the important types of models and their applications.

- Be familiar with the software environments in which modeling takes place.

- Understand the needs of modeling and how these needs are being addressed by current trends in GI system software.

## 15.1 Introduction

This chapter identifies many of the distinct types of models supported by GI systems and gives examples of their applications. After *system* and *object*, *model* is probably one of the most overworked terms in the English language, especially in the language of GI systems, with many distinct meanings in that and even in this book. So first it is important to address the meaning of the term as it is used in this chapter.

**Model is one of the most overworked terms in the English language.**

A clear distinction needs to be made between *data models* discussed in Chapter 7 and the spatial models that are the subject of this chapter. A data model is a template for data, a framework into which specific details of relevant aspects of the Earth's surface can be fitted, and a set of assumptions about the nature of data. For example, the raster data model forces all

knowledge to be expressed as properties of the cells of a regular grid lain on the Earth. Data models are closely related to the concept of ontology, which is best understood as the study of the basic elements of description, or "what we talk about."

A data model is in essence a statement about form or about how the world looks, limiting the options open to the data model's user to those allowed by its template. Models in this chapter are very different and can be expressions of how the world is believed to work, or how a task is broken down into a sequence of operations; in other words they are expressions of process (see Section 1.3 on how these both relate to the science of problem solving). They may include dynamic simulation models of natural processes such as erosion, tectonic uplift, the migration of elephants, or the movement of ocean currents. They may include models of social processes, such as residential segregation or the movements of cars on a congested highway. They may include processes designed by humans to search for optimum alternatives, for example in finding locations for a new retail store. They may include standard workflows that are executed every day in organizations. Finally, they may include simple calculations of indicators or predictors, such as happens when layers of geographic information are combined into measures of groundwater vulnerability or social deprivation.

The common element in all these examples is the manipulation of geographic information in multiple stages, especially if these stages must be run repeatedly. In some cases these stages will perform a simple transformation or analysis of inputs to create an output, and in other cases the stages will loop to simulate the development of the modeled system through time, in a series of iterations. Only when all the loops or iterations are complete will there be a final output. There will be intermediate outputs along the way, and it is often desirable to save some or all of these, in case the model needs to be rerun or if parts of the model need to be changed.

All the models discussed in this chapter are digital or computational models, meaning that the operations occur in a computer and are expressed at the most fundamental level in a language of 0s and 1s. In Chapter 3 representation was seen as a matter of expressing geographic form in 0s and 1s; this chapter looks for ways of expressing geographic process in 0s and 1s. The term *geocomputation* is often used to describe the application of computational models to geographic problems.

All the models discussed in this chapter are also spatial models (and deal, of course, with geographic space). There are two key requirements of such a model:

1. There is variation across the space being manipulated by the model (an essential requirement of all GI system applications).
2. The results of modeling change when the locations of objects change—location matters (this is also a key requirement of spatial analysis as defined in Section 13.1).

Models do not have to be digital, and it is worth spending a few moments considering the other type, known as analog. The term was defined briefly in Section 3.7 as describing a representation, such as a paper map, that is a scaled physical replica of reality. Analog models can be very efficient, and they are widely used to test engineering construction projects and proposed airplanes. They have three major disadvantages relative to digital models, however: they can be expensive to construct and operate, their accuracy is limited by the effects of scaling the real world, and unlike digital models, they are virtually impossible to copy, store, or share. Nevertheless they can be

extremely efficient at modeling complex systems, such as airflow over proposed buildings.

**An analog model is a scaled physical representation of some aspect of reality.**

The level of detail of any analog model is measured by its *representative fraction* (Section 3.7, Box 2.2), the ratio of distance on the model to distance in the real world. Like digital data, computational models do not have well-defined representative fractions; instead, level of detail is measured as *spatial resolution*, defined as the shortest distance over which change is recorded. *Temporal resolution* is also important for models of processes, being defined as the shortest time over which change is recorded and corresponding in the case of many dynamic models to the time interval between iterations.

Spatial and temporal resolutions are critical factors in models. They define what is left out of the model, in the form of variation that occurs over distances or times that are less than the appropriate resolution. They also therefore define one major source of uncertainty in the model's outcomes. Uncertainty in this context can be best defined through a comparison between the model's outcomes and the outcomes of the real processes that the model seeks to emulate. Any model leaves its user uncertain to some degree about what the real world will do; a measure of uncertainty attempts to give that degree of uncertainty an explicit magnitude. Uncertainty has been discussed in the context of geographic data in Chapter 5; its meaning and treatment in the context of spatial modeling are discussed later in this chapter.

**Any model leaves its user uncertain to some degree about what the real world will do.**

Spatial and temporal resolution also determine the cost of acquiring the data because in general it is more costly to collect a fine-resolution representation than a coarse-resolution one. More observations have to be made, and more effort is consumed in making and compiling them. They also determine the cost of running the model because execution time expands as more data have to be processed and as more iterations have to be made, and fine spatial and temporal resolution may also raise ethical questions for many types of socioeconomic data. One benchmark is of critical importance for many dynamic models: They must run faster than the processes they seek to simulate if the results are to be useful for planning. For example, a model of the atmosphere is only useful if it runs faster than changes in the weather, and if the model can be rerun as changes occur in order to update predictions. One expects this to be true of computational models, but in practice the amount of computing can be so large that the model simulation slows to unacceptable speed. Supercomputers, and indeed the entire field of high-performance computing, must sometimes be used in order for models to run sufficiently fast, on sufficiently detailed data. This and other aspects of advanced computing, often termed *cyberGIS*, are addressed more extensively in Chapter 10.

**Spatial and temporal resolutions are major factors in the cost both of acquiring data for modeling, and of actually running the model.**

## 15.1.1 Why Model?

Models are built for a number of reasons. First, a model might be built to support a decision or design process in which the user wishes to find a solution to a spatial problem in support of a decision (Box 15.1 discusses some of the spatial problems that Budhendra Bhaduri

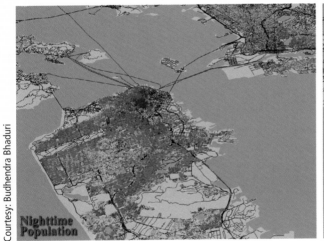 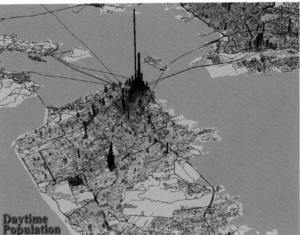

and his team at Oak Ridge National Laboratory are working to solve), perhaps a solution that optimizes some objective. This concept was discussed in Section 14.4. Often the decision or design process will involve multiple criteria, an issue discussed in Section 15.4. Second, a model might be built to allow the user to experiment on a replica of the world rather than on the real thing. This is a particularly useful approach when the costs of experimenting with the real thing are prohibitive, or when unacceptable impacts would result, or when results can be obtained much faster with a model, allowing impacts to be anticipated. Medical students now routinely learn anatomy and the basics of surgery by working with digital representations of the human body rather than with expensive and hard-to-get cadavers. Humanity is currently conducting an unprecedented experiment on the global atmosphere by pumping vast amounts of $CO_2$ into it. How much better it would have been if we could have run the experiment on a digital replica and understood the consequences of $CO_2$ emissions before the experiment began.

Experiments embody the notion of *what-if scenarios*, or policy alternatives that can be plugged into a model in order to evaluate their outcomes. The ability to examine such options quickly and effectively is one of the major reasons for modeling. Figure 15.3 shows an example of modeling to evaluate possible planning scenarios in Los Angeles.

**Models allow planners to experiment with what-if scenarios.**

Finally, models allow the user to examine dynamic outcomes by viewing the modeled system as it evolves and responds to inputs. As discussed in the next section, such dynamic visualizations are far more compelling and convincing than descriptions of outcomes or statistical summaries when shown to stakeholders and the public. Scenarios evaluated with dynamic models are thus a very effective way of motivating and supporting debates over policies and decisions and of communicating scientific findings to the general public.

## 15.1.2 To Analyze or to Model?

The problem of emergency evacuation provides a good example of GI system applications in the areas of transportation and logistics. Organizing the evacuation of neighborhoods illustrates a fundamental difference of approach that explains the relationship between Chapters 13 and 14 on the one hand, and Chapter 15 on the other. Using the methods of Chapters 13 and 14 we might analyze the pattern of streets and population, with the aim of mapping difficulty of evacuation, measured as the number of people evacuated per lane of road. This analysis is inherently static, capturing what planners might need to know in a single map display. On the other hand, we might simulate the process of evacuation using a program designed to replicate what would actually happen to individual vehicles and drivers in an emergency, using

Figure 15.3 The graphical user interface to the Virtual Co-Laboratory for Policy Analysis in Greater Los Angeles, a project funded by the University of California's Office of the President. In this illustration, traffic between nodes in the Los Angeles basin is being simulated under conditions controlled by the user, such as the price of gasoline in dollars per U.S. gallon.

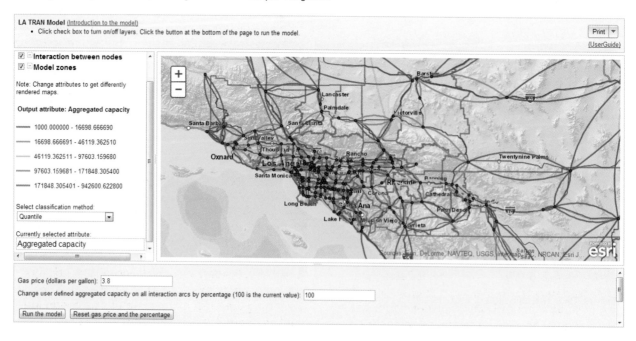

rules of human behavior that have been extracted from analysis of real-world traffic. In this way, a GI system would replicate a dynamic *process*. These simulations would allow the researcher to examine what-if scenarios by varying a range of conditions, including zoning controls and new highways. Simulations like these can galvanize a community into action far more effectively than static analysis.

**Models can be used for dynamic simulation, providing decision makers with dramatic visualizations of alternative futures.**

In summary, analysis as described in Chapters 13 and 14 is characterized by:

- A static approach, at one point in time or an average over time
- The search for patterns or anomalies, leading to new ideas and hypotheses, and perhaps predictions
- Manipulation of data to reveal what would otherwise be invisible

By contrast, modeling, as we define it in this chapter, is characterized by:

- Multiple stages, perhaps representing different points in time
- Implementing ideas and hypotheses about the behavior of the real world
- Experimenting with policy options and scenarios

## 15.2 Types of Models

### 15.2.1 Static Models and Indicators

A static model represents a single point in time, or a system that does not change through time, and typically combines multiple inputs into a single output. There are no time steps and no loops in a static model, but the results are often of great value as predictors or indicators. For example, the Universal Soil Loss Equation (USLE) falls into this category. It predicts soil loss at a point, based on five input variables, by evaluating the equation:

$$A = R \times K \times LS \times C \times P$$

where $A$ denotes the predicted erosion rate, $R$ is the Rainfall and Runoff Factor, $K$ is the Soil Erodibility Factor, $LS$ is the Slope Length Gradient Factor, $C$ is the Crop/Vegetation and Management Factor, and $P$ is the Support Practice Factor. Full definitions of each of these variables and their measurement or

estimation can be found in descriptions of the USLE (see, for example, www.danewaters.com/business/stormwater.aspx).

**A static model represents a system at a single point in time.**

The USLE passes the first test of a spatial model in that many if not all of its inputs will vary spatially when applied to a given area. But it does not pass the second test because moving the points at which $A$ is evaluated will not affect the results. Why, then, use a GI system to evaluate the USLE? There are four good reasons: (1) because some of the inputs, particularly $LS$, require a GI system for their calculation from readily available data, such as digital elevation models (see Section 14.3.1 for a discussion of the calculation of slope); (2) because the inputs and outputs are best expressed, visualized, and used in map form rather than as tables of point observations; (3) because the inputs and outputs of the USLE are often integrated with other types of data, for further analysis that may require a GI system; and (4) because the data management capabilities of a GI system, and its ability to interface with other systems, may be the best immediately available. Nevertheless, it is possible to evaluate the USLE in a simple spreadsheet application such as Excel; the Web site cited in the previous paragraph includes a downloadable Excel macro for this purpose.

Models that combine a variety of inputs to produce an output are widely used, particularly in environmental modeling. The DRASTIC model calculates an index of groundwater vulnerability from input layers, by applying appropriate weights to the inputs (Figure 15.4). Box 15.2 describes another application, the calculation of a groundwater protection model from several inputs in a karst environment (an area underlain by potentially soluble limestone, and therefore having substantial and rapid groundwater flow through cave passages). It uses Esri's ModelBuilder software, which is described later in Section 15.3.1.

### 15.2.2 Individual and Aggregate Models

The simulation models used by transportation planners to examine traffic patterns work at the individual level by attempting to forecast the behavior of each driver and vehicle in the study area. By contrast, it would clearly be impossible to model the behavior of every molecule in the Mammoth Cave watershed (Box 15.2). Instead, any modeling of groundwater movement must be done at an aggregate level by predicting the movement of water as a continuous fluid. In general, models of physical systems are

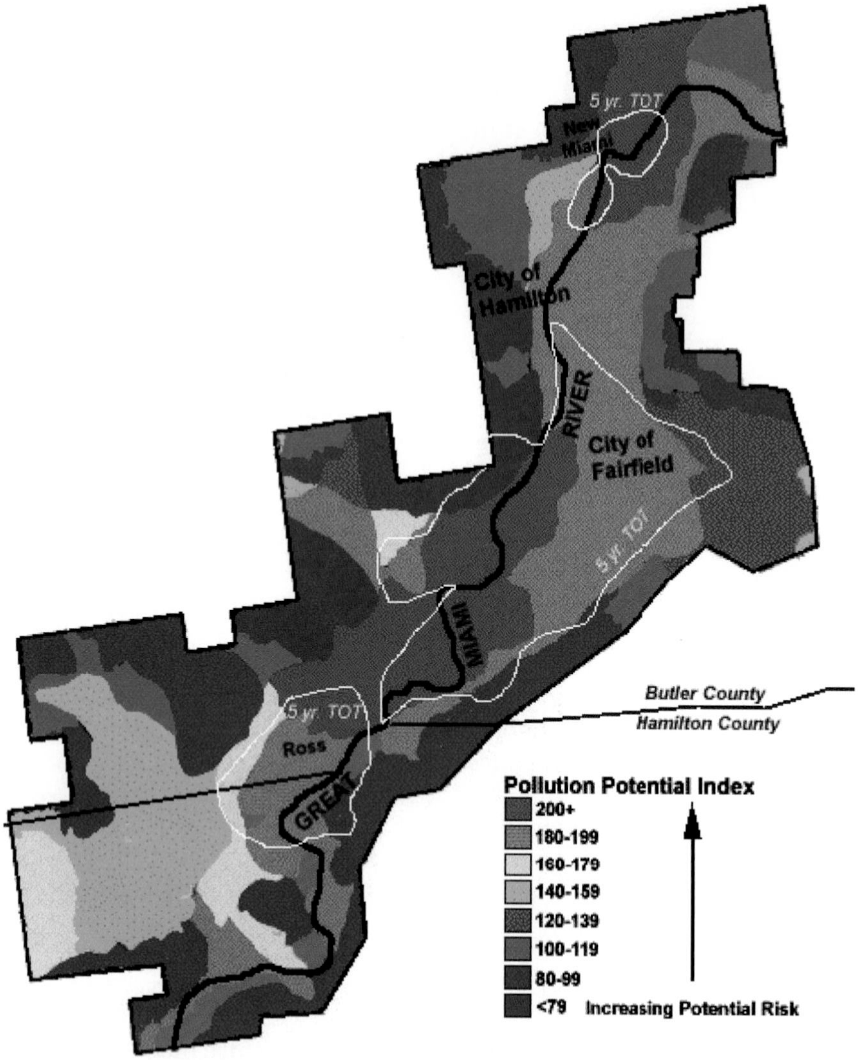

Source: Hamilton to New Baltimore Ground Water Consortium

**Figure 15.4** The results of using the DRASTIC groundwater vulnerability model in an area of Ohio. The model combines data layers representing factors important in determining groundwater vulnerability and displays the results as a map of vulnerability ratings.

**Pollution Potential Index**

- 200+
- 180-199
- 160-179
- 140-159
- 120-139
- 100-119
- 80-99
- <79  Increasing Potential Risk

## Application Box 15.2

### Building a Groundwater Protection Model in a Karst Environment

Rhonda Pfaff (Esri staff) and Alan Glennon (recently graduated PhD from the University of California, Santa Barbara) describe a simple but elegant application of modeling to the determination of groundwater vulnerability in Kentucky's Mammoth Cave watershed. Mammoth Cave is protected as a national park, containing extensive and unique environments, but is subject to potentially damaging runoff from areas in the watershed outside park boundaries and therefore not subject to the same levels of environmental protection. Figure 15.5 shows a graphic rendering of the model, in Esri's ModelBuilder. Each operation is shown as a rectangle and each dataset as an ellipse. Reading from top left, the model first clips the slope layer to the extent of the watershed, then selects slopes greater than or equal to 5 degrees. A land-use layer is

analyzed to select fields used for growing crops, and these are then combined with the steep-slopes layer to identify crop fields on steep slopes. A dataset of streams is buffered to 300 m, and finally this is combined to form a layer identifying all areas that are crop fields, on steep slopes, within 300 m of a stream. Such areas are particularly likely to experience soil erosion and to generate runoff contaminated with agricultural chemicals, which will then impact the downstream cave environment with its endangered populations of sightless fish. Figure 15.6 shows the resulting map.

A detailed description of this application is available at www.esri.com/news/arcuser/0704/files/modelbuilder.pdf, and the datasets are available at www.esri.com/news/arcuser/0704/summer2004.html.

▶

**Figure 15.5** Graphic representation of the groundwater protection model developed by Rhonda Pfaff and Alan Glennon for analysis of groundwater vulnerability in the Mammoth Cave watershed, Kentucky.

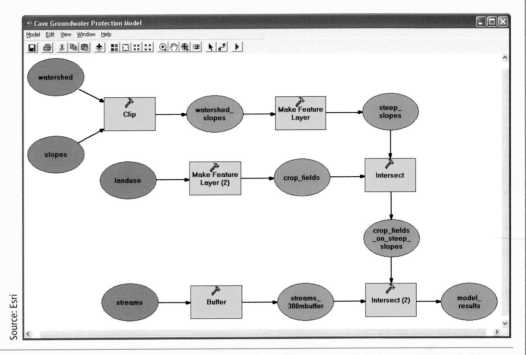

**Figure 15.6** Results of the groundwater protection model. Highlighted areas are farmed for crops, on relatively steep slopes and within 300 m of streams. Such areas are particularly likely to generate runoff contaminated by agricultural chemicals and soil erosion and to impact adversely the cave environment into which the area drains.

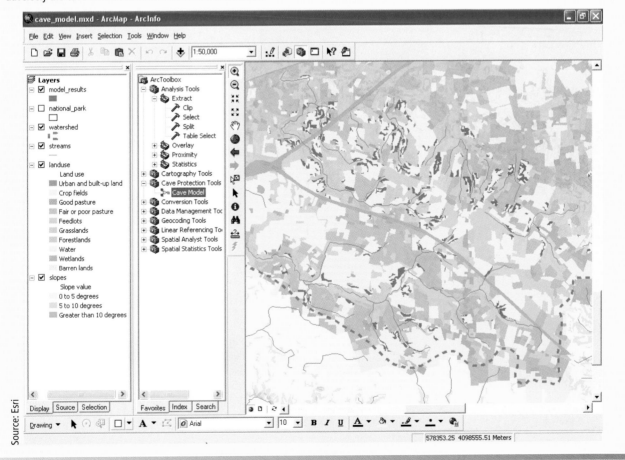

forced to adopt aggregate approaches because of the enormous number of individual objects involved, whereas it is much more feasible to model individuals in human systems or in studies of animal behavior. Even when modeling the movement of water as a continuous fluid, it is still necessary to break the continuum into discrete pieces, as it is in the representation of continuous fields (see Figure 3.7). Some models adopt a raster approach and are commonly called *cellular* models (see Section 15.2.3). Other models break the world into irregular pieces or polygons, as in the case of the groundwater protection model described in Box 15.2.

**Aggregate models are used when it is impossible to model the behavior of every individual element in a system.**

Models of individuals are often termed *agent-based* models (ABM) or *autonomous agent* models, implying the existence of discrete agents with defined decision-making behaviors (Box 15.3). Each agent might represent an individual, a vehicle and driver, a

---

## Technical Box (15.3)

### Agent-Based Models of Movement in Crowded Spaces

Working at the University College London's Centre for Advanced Spatial Analysis, geographer and planner Michael Batty uses GI systems to simulate the disasters and emergencies that can occur when large-scale events generate congestion and panic in crowds. The events involve large concentrations of people in small spaces that can arise because of accidents, terrorist attacks, or simply the buildup of congestion through the convergence of large numbers of people into spaces with too little capacity. He has investigated scenarios for a number of major events, such as the movement of very large numbers of people to Mecca to celebrate the Hajj (a major holy event in the Muslim calendar; Figure 15.7) and the Notting Hill Carnival (Europe's largest street festival, held annually in West Central London; Figure 15.8). His work uses agent-based models that take GI science to a finer level of granularity and incorporate temporal processes as well as spatial structure. Fundamental to agent-based modeling of such situations is the need to understand how individuals in crowds interact with each other and with the geometry of the local environment.

© Rabi Karim Photography/Getty Images, Inc.

**Figure 15.7** Massive crowds congregate in Mecca during the annual Hajj. On January 12, 2006, 346 pilgrims were trampled when panic stampeded the crowd. This was unfortunately not a unique occurrence: 244 pilgrims were killed in 2004, 50 in 2002, 35 in 2001, and 107 in 1998, and 1425 were killed in a pedestrian tunnel in 1990. ▶

Figure 15.8 shows one of Batty's simulations of the interactions between two crowds. Here a parade is moving around a street intersection—the central portion of the movement (walkers in white) is the parade and the walkers around this in gray/red are the watchers. This model can be used to simulate the buildup of pressure through random motion, which then generates a breakthrough of the watchers into the parade, an event that often leads to disasters of the kind experienced in festivals, rock concerts, and football (soccer) matches as well as ritual situations like the Hajj.

(A)                    (B)

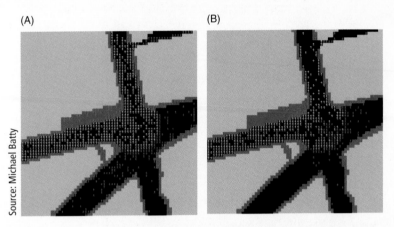

Source: Michael Batty

**Figure 15.8** Simulation of the movement of individuals during a parade. Parade walkers are in white, watchers in red. The watchers (A) build up pressure on restraining barriers and crowd control personnel, and (B) break through into the parade.

corporation, or a government agency and would have a defined location at any point in time. With the massive computing power now available on the desktop, together with techniques of object-oriented programming, it is comparatively easy to build and execute ABMs, even when the number of agents is itself massive. Such models have been used to analyze and simulate many types of animal behavior, as well as the behavior of pedestrians in streets, shoppers in stores, and drivers on congested roads. They have also been used to model the behavior of decision makers, not through their movements in space but through the decisions they make regarding spatial features. ABMs have been constructed to model the decisions made over land use in rural areas by formulating the rules that govern individual decisions by landowners in response to varying market conditions, policies, and regulations. Of key interest in such models is the impact of these factors on the fragmentation of the landscape, with its implications for the destruction of wildlife habitat. Figure 15.9 shows an example of the modeling of land-use transition in the Amazon Basin.

Models such as these find rich applications in the study of disease transmission and the evaluation of alternative policies and interventions aimed at reducing disease prevalence. Indy Hurt, for example, has constructed a model of tuberculosis transmission within a town in Kenya, based on the social behavior of individuals, their journeys to work and shop, and the stages of development and transmission of the disease. Figure 15.10 is an example of her work, visualizing the distribution of susceptibles, people with active disease, and people with latent forms of the disease. Although the accuracy of the predictions (Section 15.5) cannot be guaranteed, models such as these are nevertheless very useful in helping planners, decision makers, and citizens to examine the future impacts of alternative decisions and policy frameworks.

## 15.2.3 Cellular Models

Cellular models represent the surface of the Earth as a raster (note, however, the difficulties of doing this when the study area is a substantial fraction of the Earth's surface, as discussed in Chapter 4). Each cell in the fixed raster has a number of possible states, which change through time as a result of the application of transition rules. Typically, the rules are defined over each cell's neighborhood and determine the outcome of each stage in the simulation based on the cell's state, the states of its neighbors, and the values of cell attributes. The study of cellular models was first popularized by the work of John Conway, professor of mathematics at Princeton, who studied the properties of a model he

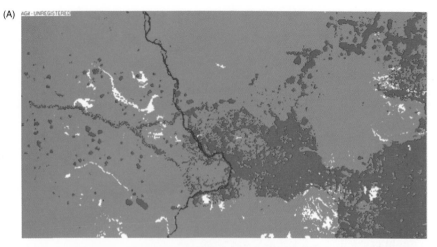

(A)

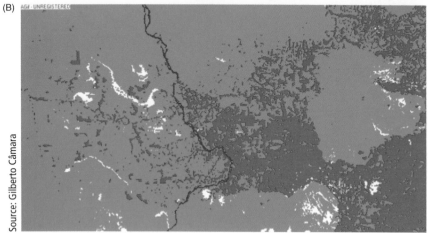

(B)

Source: Gilberto Câmara

**Figure 15.9** Simulation of land-cover transition in part of the Amazon Basin. (A) Predictions of a model based on eight years of transitions of individual cells starting with the observed pattern in 1997, using rules that include proximity to roads, changing agricultural conditions, and so on. (B) Observed pattern after eight years of transitions.

**Figure 15.10** An illustration from the work of Indy Hurt (Apple), using an agent-based model to simulate the spatial distribution of tuberculosis in a town in Kenya.

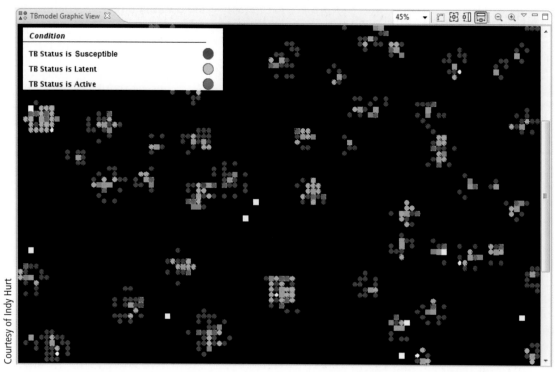

Courtesy of Indy Hurt

called the Game of Life (Box 15.4). This has been used by geographer Keith Clarke to model urban growth, using appropriate data and rules of behavior.

**Cellular models represent the surface of the Earth as a raster, each cell having a number of states that are changed at each iteration by the execution of rules.**

## 15.2.4 Cartographic Modeling and Map Algebra

In essence, modeling consists of combining many stages of transformation and manipulation into a single whole, for a single purpose. In the example in Box 15.2, the number of stages was quite small—only six—whereas in the Clarke urban growth model

---

### Technical Box (15.4)

## The Game of Life

The game is played on a raster. Each cell has two states, *live* and *dead*, and there are no additional cell attributes (no variables to differentiate the space), as there would be in GI system applications. Each cell has eight neighbors (the Queen's case, Figure 14.9). There are three transition rules at each time-step in the game:

1. A dead cell with exactly three live neighbors becomes a live cell.

2. A live cell with two or three live neighbors stays alive.

3. In all other cases a cell dies or remains dead.

   With these simple rules it is possible to produce an amazing array of patterns and movements, and some surprisingly simple and elegant patterns emerge from the chaos. (In the field of agent-based modeling these unexpected and simple patterns are known as

---

Figure 15.11 Three stages in an execution of the Game of Life: (A) the starting configuration, (B) the pattern after one time-step, and (C) the pattern after fourteen time-steps. At this point all features in the pattern remain stable.

(A)

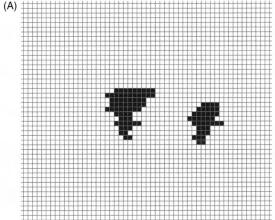

(B)

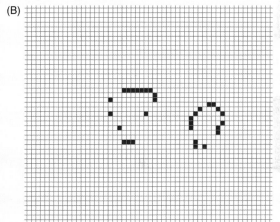

(C)

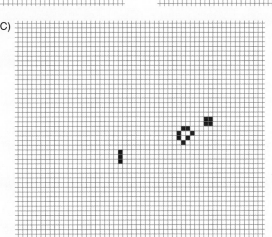

▶

*emergent properties*.) The page www.math.com/ students/wonders/life/life.html includes extensive details on the game, examples of particularly interesting patterns, and executable Java code. Figure 15.11 shows three stages in an execution of the Game of Life.

Several interesting applications of cellular methods have been identified, and particularly outstanding are the efforts to apply them to urban growth simulation. The likelihood of a parcel of land developing depends on many factors, including its slope, access to transportation routes, status in zoning or conservation plans, but above all its proximity to other development. These models express this last factor as a simple modification of the rules of the Game of Life—the more developed the state of neighboring cells, the more likely a cell is to make the transition from undeveloped to developed. Figure 15.12 shows an illustration from one such model, that developed by Keith Clarke and his coworkers at the University of California, Santa Barbara, to predict growth patterns in the Santa Barbara area through

2040. The model iterates on an annual basis, and the effects of neighboring states in promoting infilling are clearly evident. The inputs for the model—the transportation network, the topography, and other factors—are readily prepared in GI systems, and the GI system is used to output the results of the simulation.

One of the most important issues in such modeling is calibration and validation—how do we know the rules are right, and how do we confirm the accuracy of the results? Clarke's group has calibrated the model by *brute force*, that is, by running the model forward from some point in the past under different sets of rules, and comparing the results to the actual development history from then to the present. This method is extremely time consuming because the model has to be run under vast numbers of combinations of rules, but it provides at least some degree of confidence in the results. The issue of accuracy is addressed in more detail, and with reference to modeling in general, later in this chapter.

## Urban Growth Boundary Exclusion

## Only Parks are Excluded

Source: Keith Clarke

**Figure 15.12** Simulation of future urban growth patterns in Santa Barbara, California. (Upper) Growth limited by current urban growth boundary. (Lower) growth limited only by existing parks.

several stages are executed many times, as the model iterates through its annual steps.

The individual stages can consist of a vast number of options, encompassing all the basic transformations of which a GI system is capable. Various ways of organizing the options have been proposed,

including that used to structure Chapters 13 and 14. Perhaps the most successful is the one developed by Dana Tomlin and known as *cartographic modeling* or *map algebra*. It classifies all transformations of rasters into four basic classes, and it is used in several raster-centric GI systems as the basis for their analysis languages:

- *Local* operations examine rasters cell by cell, examining the value in a cell in one layer and perhaps comparing it with the values in the same cell in other layers.
- *Focal* operations compare the value in each cell with the values in its neighboring cells—most often eight neighbors (see the discussion of neighborhoods in Chapter 14).
- *Global* operations produce results that are true of the entire layer, such as its mean value.
- *Zonal* operations compute results for blocks of contiguous cells that share the same value, such as the calculation of shape for contiguous areas of the same land use, and attach their results to all the cells in each contiguous block.

With this simple schema, it is possible to express any model, such as Clarke's urban growth simulation, as a series of processing functions and to compile a sequence of such functions into a *script* in a well-defined language, allowing the sequence to be executed repeatedly with a simple command. The only constraint is that the model inputs and outputs be in raster form.

**Map algebra provides a simple language in which to express a model as a script.**

A more elaborate map algebra has been devised and implemented in the PCRaster package, developed for spatial modeling at the University of Utrecht (pcraster. geog.uu.nl). In this language, a symbol refers to an entire map layer, so the command $A = B + C$ takes the values in each cell of layers $B$ and $C$, adds them, and stores the result as layer $A$.

## 15.3 Technology for Modeling

### 15.3.1 Operationalizing Models in GI Systems

Many of the ideas needed to implement models at a practical level have already been introduced. In this section they are organized more coherently in a review of the technical basis for modeling.

Models can be defined as sequences of operations, and we have already seen how such sequences can be expressed either as graphic flowcharts or as scripts. The graphic flowchart has already been illustrated in Figure 15.5. In these interfaces datasets are typically represented as ellipses, operations as rectangles, and the sequence of the model as arrows. The user is able to modify and control the operation sequence by interacting directly with the graphic display.

Scripts allow a user to program combinations of GI system operations or to code new operations that are not possible within the standard GI system. Today Python has become a very popular scripting language for GI systems; many GI system tools are now coded in Python, and many users exploit Python to adapt and extend GI system functions to meet specific needs. Models such as the ones being discussed in this chapter can be expressed as sequences of Python commands, allowing the user to execute an entire sequence simply by invoking a script.

**Any model can be expressed as a script or visually as a flowchart.**

### 15.3.2 Model Coupling

The previous section described the implementation of models as direct extensions of an underlying GI system, through either graphic model-building or scripts. This approach makes two assumptions: first, that all the operations needed by the model are available in the GI system (or in another package that can be called by the model), and second, that the GI system provides sufficient performance to handle the execution of the model. In practice, a GI system will often fail to provide adequate performance, especially with very large datasets and large numbers of iterations, because it has been designed as a general-purpose software system, rather than specifically optimized for modeling. Instead, the user is forced to resort to specialized code, written in a lower-level language such as C, C++, C#, or Java. Clarke's model, for example, is programmed in C, and the GI system is used only for preparing input and visualizing output. Other models may be spatial only in the sense of having geographically differentiated inputs and outputs, as discussed in the case of the USLE in Section 15.2.1, making the use of a GI system optional rather than essential.

In reality, therefore, much spatial modeling is done by *coupling* a GI system with other software. A model is said to be *loosely coupled* to a GI system when it is run as a separate piece of software, and data are exchanged to and from the GI system in the form of files. Because many GI formats are proprietary, it is common for the exchange files to be in openly published interchange format and to rely on translators or direct-read technology at either end of the transfer. In extreme cases it may be necessary to write a special program to convert the files during transfer, when no common format is available. Clarke's is an example of a model that is loosely coupled to a GI system. A model is said to be *closely coupled* to a GI system when both the model and the GI system read and write the same files, obviating any need for translation. When the model is executed as a GI system script or through the graphic user interface of a GI system, it is said to be *embedded*. Section 6.3.3 describes the process of GI system customization in more detail.

## 15.3.3 Cataloging and Sharing Models

In the digital computer everything—the program, the data, the metadata—must be expressed eventually in bits. The value of each bit or byte (Box 3.1) depends on its meaning, and clearly some bits are more valuable than others. A bit in Microsoft's operating system Windows is clearly extremely valuable, even though there are hundreds of millions of bits in a single copy of the operating system, and hundreds of millions of copies in computers worldwide. On the other hand, a bit in a remote-sensing image that happens to have been taken on a cloudy day has almost no value at all, except perhaps to someone studying clouds. In general, one would expect bits of programs to be more valuable than bits of data, especially when those programs are usable by large numbers of people.

This line of argument suggests that scripts, models, and other representations of process are potentially valuable to many users and well worth sharing. But almost all the investment society has made in the sharing of digital information has been devoted to data and text. In Section 10.2 we saw how geolibraries, data warehouses, and metadata standards have been devised and implemented to facilitate sharing of geographic data, and parallel efforts by libraries, publishers, and builders of search engines have made it easy nowadays to share text via the Web. But comparatively little effort to date has gone into making it possible to share *process objects*, that is, digital representations of the process of GI system use.

**A process object, such as a script, captures the process of GI system use in digital form.**

Notable exceptions that facilitate sharing include Esri's ArcGIS Resource Center (resources.arcgis.com), a large collection of resources, including scripts, many of them contributed by users. The library is easily searched for scripts and other resources to perform specific types of analysis and modeling. For example, a search of the online center for the Storm Water Management Model (SWMM) led immediately to tools to implement the model in ArcGIS.

The term *GI service* is often used to describe a function being offered by a server, for use by any user connected to the Internet (see Chapters 6 and 10). In essence, GI services offer an alternative to locally installed GI systems, allowing the user to send a request to a remote site instead of to his or her GI system. GI services are particularly effective in the following circumstances:

- When it would be too complex, costly, or time consuming to install and run the service locally.

- When the service relies on up-to-date data that would be too difficult or costly for the user to constantly update.

- When the person offering the service wishes to maintain control over his or her intellectual property. This would be important, for example, when the function is based on a substantial amount of knowledge and expertise.

- When it is important for several users to collaborate simultaneously in the solution of a problem. By using online services it is possible for many users to view and interact with the problem at the same time.

To date, several standard functions are available as GI services, satisfying one or more of the preceding requirements, and the list is growing rapidly as online services expand. There are several instances on the Web of *gazetteer services*, which allow a user to send a place-name and to receive in return one or more coordinates of places that match the place-name. Several companies offer *geocoding services*, returning coordinates in response to a street address (try www.geocoder.us). Esri's ArcGIS Online (www.arcgis.com) provides many services, plus a searchable catalog to GI services. Such services are important components of the *service-oriented architecture* concept (Section 10.1), in which a user's needs are satisfied by chaining together sequences of such services.

## 15.4 Multicriteria Methods

The model developed by Pfaff and Glennon and depicted in Figure 15.5 rates vulnerability to runoff based on three factors—cropland, slope, and distance from stream—but treats all three as simple binary measures and produces a simple binary result. Land is vulnerable if slope is greater than 5%, land use is cropping, and distance from stream is less than 300 m. In reality, of course, 5% is not a precise cutoff between nonvulnerable and vulnerable slopes, and neither is 300 m a precise cutoff between vulnerable distances and nonvulnerable distances. Yet such clear cutoffs are essential if a plan is to be administratively workable. The issues surrounding rules such as these, and their fuzzy alternatives, were discussed at length in Section 5.2.

A more general and powerful conceptual framework for models like this would be constructed as follows. A number of factors influence vulnerability, denoted by $X_1$ through $X_n$. The impact of each factor on vulnerability is determined by a transformation of the factor $f(X)$. For example, the factor *distance* would be transformed so that its impact *decreases* with increasing distance (as in Section 2.6), whereas the impact of *slope* would be *increasing*. Then the combined impact of all of the factors is obtained by weighting and adding them, each factor $i$ having a weight $w_i$:

$$I = \sum_{i=1}^{n} w_i f(x_i)$$

In this framework both the functions *f* and the weights *w* need to be determined. In the example the *f* for slope was resolved by a simple step function, but it seems more likely that the function should be continuously decreasing, as shown in Figure 15.13. The U-shaped function also shown in the figure would be appropriate in cases where the impact of a factor declines in both directions from some peak value (for example, smoke from a tall smokestack may have its greatest impact at some distance downwind—less at shorter distances where the smoke is still descending from the top of the smokestack toward the ground, and less at longer distances where it has become diluted).

**Many decisions depend on identifying relevant factors and adding their appropriately weighted values.**

This approach provides a good conceptual framework both for the indicator models typified by Box 15.2 and for many models of design processes. In both cases it is possible that multiple views might exist about appropriate functions and weights, particularly when modeling a decision over an important development with impact on the environment. Different *stakeholders* in the design process can be anticipated to have different views about what is important, how that importance should be measured, and how the various important factors should be combined. Such processes are termed *multicriteria decision making*, or MCDM, and are commonly encountered whenever decisions are controversial. Some stakeholders may feel that environmental factors deserve high weight, others that cost factors are the most important, and still others that impact on the social life of communities is all-important.

An important maxim of MCDM is that it is better for stakeholders to argue in principle about the merits of different factors and how their impacts should be measured than to argue in practice about alternative decisions. For example, arguing about whether slope is a more important factor than distance, and about how each should be measured, is better than arguing about the relative merits of Solution A, which might color half of John Smith's field red, over Solution B, which might color all of David Taylor's field red. Ideally, all the controversy should be over once the factors, functions, and weights are decided, and the solution they produce should be acceptable to all because all accepted the inputs. Would it were so!

**Each stakeholder in a decision may have his or her own assessment of the importance of each relevant factor.**

Although many GI systems have implemented various approaches to MCDM, Clark University's Idrisi offers some of the most extensive functionality, as well as detailed tutorials and examples (www.clarklabs.org). Developed as nonprofit software for GI and image processing, Idrisi has many tens of thousands of users worldwide. One example of Idrisi's support for MCDM is the Analytical Hierarchy Process (AHP) devised by Thomas Saaty, which focuses on capturing each stakeholder's view of the appropriate weights to give to each impact factor. The impact of each factor is first expressed as a function, choosing from options such as those shown in Figure 15.13. Then each stakeholder is asked to compare each pair of factors (with $n$ factors there are $n(n-1)/2$ pairs) and to assess their relative importance in ratio form. A matrix is created for each stakeholder, as in the example shown in Table 15.1. The matrices are then combined and analyzed, and a single set of weights is extracted that represent a consensus view (Figure 15.14). These weights would then be inserted as parameters in the spatial model, to produce a final result. The mathematical details of the method can be found in books or tutorials on the AHP.

**Figure 15.13** Three possible impact functions: (red) the step function used to assess slope in Figure 15.5; (blue) a decreasing linear function; and (black) a function showing impact rising to a maximum and then decreasing.

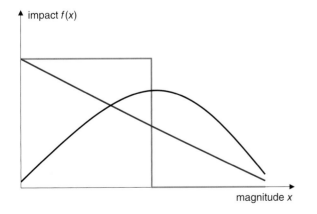

impact *f*(*x*)

magnitude *x*

**Table 15.1** An example of the weights assigned to three factors by one stakeholder. For example, the entry "7" in Row 1 Column 2 (and the 1/7 in Row 2 Column 1) indicates that the stakeholder felt that Factor 1 (slope) is seven times as important as Factor 2 (land use).

| | Slope | Land use | Distance from stream |
|---|---|---|---|
| Slope | | 7 | 2 |
| Land use | 1/7 | | 1/3 |
| Distance from stream | 1/2 | 3 | |

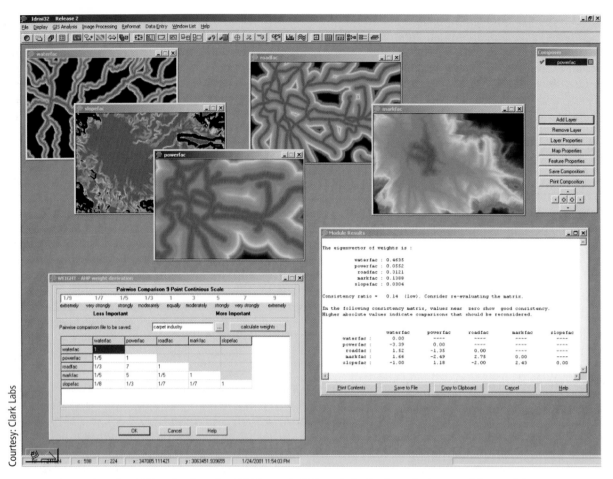

**Figure 15.14** Screenshot of an AHP application using Idrisi (www.clarklabs.org). The five layers in the upper left part of the screen represent five factors important to the decision. In the lower left the image shows the table of relative weights compiled by one stakeholder. All of the weights' matrices are combined and analyzed to obtain the consensus weights shown in the lower right, together with measures to evaluate consistency among the stakeholders.

## 15.5 Accuracy and Validity: Testing the Model

Models are complex structures, and their outputs are often forecasts of the future. How, then, can a model be tested, and how does one know whether its results can be trusted? Unfortunately, there is still an inherent tendency to trust the results of computer models because they appear in numerical form (and numbers carry innate authority) and because they come from computers (which also appear authoritative). Scientists normally test their results against reality, but in the case of forecasts reality is not yet available, and by the time it is there is likely little interest in testing. So modelers must resort to other methods to verify and validate their predictions.

**Results from computers tend to carry innate authority.**

Models can often be tested by comparison with past history by running the model not into the future, but forward in time from some previous point. But these are often the data used to *calibrate* the model, to determine its parameters and rules, so the same data are not available for testing. Instead, many modelers resort to *cross-validation*, a process in which a subset of data are used for calibration and the remainder for validating results. Cross-validation can be done by separating the data into two time periods or into two areas, using one for calibration and one for validation. Both are potentially dangerous if the process being modeled is also changing through time or across space (a statistician would call such a process *nonstationary*), but forecasting is dangerous in these circumstances as well.

Models of real-world processes can be validated by experiment by proving that each component in the model correctly reflects reality. For example, the Game of Life would be an accurate model if some real process actually behaved according to its rules, and Clarke's urban-growth model could be tested in the same way by examining the rules that account for each past land-use transition. In reality, it is unlikely that real-world processes will be found to behave as simply as model rules, and it is also unlikely that the model will capture every real-world process impacting the system.

If models are no better than approximations to reality, then are they of any value? Certainly, human society is so complex that no model will ever fit perfectly. As Ernest Rutherford, the experimental physicist and Nobel Laureate, is said to have once remarked, perhaps in frustration with social-scientist colleagues, "The only result that can possibly be obtained in the social sciences is, some do, and some don't." Neither will a model of a physical system ever perfectly replicate reality. Instead, the outputs of models must always be taken advisedly, bearing in mind several important arguments:

- A model may reflect behavior under ideal circumstances and therefore provide a norm against which to compare reality. For example, many economic models assume a perfectly informed, rational decision maker. Although humans rarely behave this way, it is still useful to know what would happen if they did, as a basis for comparison.

- A model should not be measured by how closely its results match reality, but by how much it reduces uncertainty about the future. If a model can narrow the options sufficiently, then it is useful. It follows that any forecast should also be accompanied by a realistic measure of uncertainty (see Chapter 5).

- A model is a mechanism for assembling knowledge from a range of sources and presenting conclusions based on that knowledge in readily used form. It is often not so much a way of discovering how the world works, as a way of presenting existing knowledge in a form helpful to decision makers.

- Just as Winston Churchill is said to have once remarked, that capitalism was the worst possible economic system "apart from all of the others," so modeling often offers the only robust, transparent analytical framework that is likely to garner any respect among decision makers with competing objectives and interests.

**Any model forecast should be accompanied by a realistic measure of uncertainty.**

Several forms of uncertainty are associated with models, and it is important to distinguish between them. First, models are subject to the uncertainty present in their inputs. Uncertainty *propagation* was discussed in Section 5.4.2. Here it refers to the impacts of uncer-

tainty in the inputs of a model on uncertainty in the outputs. In some cases propagation may be such that an error or uncertainty in an input produces a proportionate error in an output. In other cases a very small error in an input may produce a variable, sometimes small and sometimes massive change in output; and in other cases outputs can be relatively insensitive to errors in inputs. It is important to know which of these cases holds in a given instance, and the normal approach is through repeated numerical simulation (often called Monte Carlo simulation because it mimics the random processes of that city's famous gambling casino), adding random distortions to inputs and observing their impacts on outputs. In some limited instances it may even be possible to determine the effects of propagation mathematically.

Second, models are subject to uncertainty over their parameters. A model builder will do the best possible job of calibrating the model, but inevitably there will be uncertainty about the correct values, and no model will fit the data available for calibration perfectly. It is important, therefore, that model users conduct some form of *sensitivity analysis*, examining each parameter in turn to see how much influence it has on the results. This can be done by raising and lowering each parameter value, for example by +10% and −10%, rerunning the model, and comparing the results. If changing a parameter by 10% produces a less-than-10% change in results, the model can be said to be relatively insensitive to that parameter. This allows the modeler to focus attention on those parameters that produce the greatest impact on results, making every effort to ensure that their values are correct.

**Sensitivity analysis tests a model's response to changes in its parameters and assumptions.**

Third, uncertainty is introduced because a model is run over a limited geographic area or *extent*. In reality, there is always more of the world outside the extent, except in a few cases of truly global models. The outside world both impacts the area modeled and is impacted by it. Changes in land-use practices within the extent will influence areas downstream through water pollution and downwind through atmospheric pollution; in turn they will be impacted within the extent by what happens in other areas upstream or upwind. National strategies will be influenced by in-migration, whether legal or illegal. In all such cases the effect is to make the results of any modeling uncertain.

Fourth, and most important, models are subject to uncertainty because of the simplified communication or *labeling* of their results. Consider the model of Box 15.2, which computes an indicator of the need for groundwater protection. In truth, the areas identified by the model have three characteristics: slope greater than

5%, used for crops, and less than 300 m from a stream. Described this way, there is little uncertainty about the results, though there will be some uncertainty due to inaccuracies in the data. But once the areas selected are described as *vulnerable*, and in need of management to ensure that groundwater is *protected*, a substantial leap of logic has occurred. Whether or not that leap is valid will depend on the reputation of the modeler as an expert in groundwater hydrology and biological conservation; on the reputation of the organization sponsoring the work; and on the background science that led to the choice of parameters (5% and 300 m). In essence, this third type of uncertainty, which arises whenever labels are attached to results that may or may not correctly reflect their meaning, is related to how results are described and communicated, in other words to their metadata, rather than to any innate characteristic of the data themselves.

## 15.6 Conclusion

Modeling was defined at the outset of this chapter as a process involving multiple stages, often a repeated workflow, or perhaps an emulation of some real physical process. Modeling is often dynamic, and current interest in modeling is stretching the capabilities of GI system software, most of which was designed for the comparatively leisurely process of analysis, rather than the intensive and rapid iterations of a dynamic model. In many ways, then, modeling represents the cutting edge of GI systems, and the next few years are likely to see very rapid growth both in users' interest in modeling and in vendors' interest in software development. The results are certain to be interesting.

This chapter has provided a very small and limited sampling of the very rich potential of models, and of the many ways they are being used in a spatial (and temporal) context. Although much work with GI systems is concerned with how the geographic world looks, modeling of processes drives to the very heart of science, in asking how the social and environmental worlds work, and why they evolve as they do. The work of Denise Pumain provides a fitting, final illustration of the fundamental questions spatial modelers are asking (Box 15.5) using today's sophisticated tools and abundant data.

---

### Biographical Box (15.5)

#### Denise Pumain, Demographer, Urban Systems Modeler, and Quantitative Geographer

Denise Pumain (Figure 15.15) is Professor at Université Paris I Panthéon-Sorbonne, Director of the European Research Group "Spatial Simulations for Social Sciences" at CNRS (www.S4.parisgeo.cnrs.fr), and Science Editor of *Cybergeo, European Journal of Geography* (www.cybergeo.eu). She specializes in urban modeling and theoretical geography. Her main scientific contribution is an evolutionary theory of urban systems, transferring concepts and models from self-organizing complex systems to the social sciences. Her latest books *Hierarchy in Natural and Social Sciences* and *Complexity Perspectives in Innovation and Social Change* were published by Springer in 2006 and 2009. She has received several awards (e.g., International Prize for Geography Vautrin Lud 2010, Corresponding Member of the Austrian Academy of Science, Corresponding Fellow of the British Academy) and is the holder of an Advanced Grant (2010) from the European Research Council to research urban system dynamics in the world by analyzing and modeling the geographical diversity of cities and systems of cities. She writes:

Figure 15.15 Denise Pumain.

> Spatial analysis is essential for understanding and predicting how cities are expanding their buildings and material or their financial and informational networks at local as well as regional and global scales. Learning about this evolution not only means reconstructing the dynamics of this complex evolution but also looking at specific features of history and culture in several countries and regions whose effects are perceptible in their morphologies and destinies. This knowledge can be inferred both from calibration of analytic models and from experiments with simulation models. When teaching or doing research with students I enjoy trying to explain the surprising long-term existence and stability of cities all over the world, despite the dramatic variations in their appearance. I think that geographic diversity in our way of inhabiting the earth is a lesson about the creativity of humankind and a major resource for future generations!

## Questions for Further Study

1. Write down the set of rules that you would use to implement a simple version of the Clarke urban growth model, using the following layers: transportation (cell contains road or not), protected (cell is not available for development), already developed, and slope.

2. Review the steps you would take and the arguments you would use to justify the validity of a GI system-based model.

3. Select a domain of environmental or social science, such as species habitat prediction or residential segregation. Search the Web and your library for published models of processes in this domain, and summarize their conceptual structure, technical implementation, and application history.

4. Compare the relative advantages of the different types of dynamic models discussed in this chapter when applied to a selected area of environmental or social science.

## Further Reading

De Smith, M. J., Goodchild, M. F., and Longley, P. A. 2009. *Geospatial Analysis: A Comprehensive Guide to Principles, Techniques and Software Tools* (3rd ed.). Leicester: Troubador and www.spatialanalysisonline.com.

Heuvelink, G. B. M. 1998. *Error Propagation in Environmental Modeling with GIS*. London: Taylor and Francis.

Maguire, D., Batty, M., and Goodchild, M. F. (eds.). 2005. *GIS, Spatial Analysis, and Modeling*. Redlands, CA: Esri Press.

Saaty, T. L. 1980. *The Analytical Hierarchy Process: Planning, Priority Setting, Resource Allocation*. New York: McGraw-Hill.

Tomlin, C. D. 1990. *Geographic Information Systems and Cartographic Modeling*. Englewood Cliffs, NJ: Prentice Hall.

# 16 Managing GI systems

Much of the material in earlier parts of this book assumes that you have a geographic information (GI) system and can use it effectively to meet your organization's goals. Getting to that happy state is the focus of this chapter. Later chapters deal with how GI systems contribute to the business of an organization.

This chapter covers how to choose, implement, and manage operational GI systems successfully. We deliberately describe a tried and tested approach to procuring large systems because that is the most complex to get right, but parts of our approach can be used for smaller systems. Underpinning our whole approach are two interrelated key success criteria:

- Will risks be managed successfully and catastrophes avoided?
- Will you get a good return on the investment of money and other resources expended (like time of skilled staff)?

Procurement and operational management involves four key stages: the analysis of needs, the formal specification, the evaluation of alternatives, and the implementation of the chosen system. In particular, implementing GI systems requires consideration of issues such as return on investment, planning, support, communication, resource management, and funding. Successful ongoing management of a major operational GI system has five key dimensions: support for customers, operations, data management, application development, and project management.

## LEARNING OBJECTIVES

After studying this chapter you will understand:

- Return-on-investment (ROI) concepts.
- How to go about choosing a GI system to meet your needs.
- Key GI system implementation issues, including managing risk.
- How to manage an operational GI system effectively with limited resources and ambitious goals.
- Why GI projects fail—some pitfalls to avoid and some useful tips about how to succeed.
- The roles of staff members in a GI project.
- Where to go for more detailed advice.

## 16.1 Introduction

This chapter examines the practical aspects of obtaining and managing an operational GI system. It is deliberately embedded in the part of the book that focuses on high-level management concepts because success comes from combining strategy and implementation. It is the role of management in GI projects or programs to ensure that operations are carried out effectively and efficiently, and that a healthy, sustainable GI system can be maintained—one that meets the organization's strategic objectives.

Obtaining and running a GI system seems at first sight to be a routine and apparently mechanical process. It is certainly not rocket science, but neither is it simple or without tried and tested principles and best practices. The consequences of failure can be catastrophic, both for an organization and for careers. Success involves constant sharing of experience and knowledge with other people, keeping good records, monitoring the status of risks to the program, and making numerous judgments where the answer is not preordained.

## 16.2 Managing Risk

Every program has risks, and not all risks can be anticipated. But the identification of significant risks, how to mitigate them, and oversight of the process are critical to program success at every stage. This normally involves at least one risk register (a list of risks, their impacts and likelihood of occurring, and mitigation actions) which is reviewed frequently by those with oversight responsibilities (Figure 16.1).

Good governance processes—defining who is responsible for what and who answers to whom—are also crucial. Thus the program leader and his or

Figure 16.1 Excerpt from a hypothetical corporate risk register.

| Risk Number | Risk Description | Owner | Original Risk | | | Mitigation Plan | Contingency Plan | Progress Against Action | Residual Risk | | |
|---|---|---|---|---|---|---|---|---|---|---|---|
| | | | L | I | E | | | | L | I | E |
| 2 | New mobile mapping software may not be ready for publicly announced national launch. | Chief Operating Officer | 5 | 4 | 20 | Bring in extra resource, COO to monitor progress weekly and report to Chief Executive Officer, prioritize and action what must be present in first release. | Prepare draft press release explaining why – stress additional functionality, unique features, and define new launch date. Prepare private statement for key partners and licensees for use if needed before problem becomes public. Once emergency is over, review staff contributions to the problem and the solution. | 3 staff transferred from project 22. List of 'nice-to-have' but not essential functions identified. Quality Assurance and Quality Control mechanisms reviewed | 3 | 4 | 12 |
| 3 | Server scalability may not be sufficient for very large numbers of users. | Head of IT | 4 | 5 | 20 | Use soft launch of service to test likely uptake. Inform product management and release teams. | Work with cloud service provider to allow provision of additional capacity. Alert marketing and communications of need to manage client expectations. | Outline discussions held with cloud service provider, but business negotiations still ongoing. | 2 | 5 | 10 |

Key:
L: Likelihood
I: Impact
E: Exposure (=L times I)

**Residual risk matrix:** all red risks are reported to the Board and the CEO describes actions taken. The status of yellow risks is monitored by senior management, with appropriate actions taken. The Board reviews the entire risk register periodically. For residual risks below 10 tolerate but monitor direction of travel.

| | | | Impact | | | | |
|---|---|---|---|---|---|---|---|
| | | | 1 | 2 | 3 | 4 | 5 |
| Likelihood | | 5 | 5 | 10 | 15 | 20 | 25 |
| | | 4 | 4 | 8 | 12 | 16 | 20 |
| | | 3 | 3 | 6 | 9 | 12 | 15 |
| | | 2 | 2 | 4 | 6 | 8 | 10 |
| | | 1 | 1 | 2 | 3 | 4 | 5 |

Meaning of Likelihood and Impact terms

| | Likelihood | Impact |
|---|---|---|
| 5 | Very likely | Very severe/catastrophic |
| 4 | Likely | Major |
| 3 | Quite possible | Moderate |
| 2 | Unlikely | Minor |
| 1 | Very unlikely | Insignificant |

her team must ensure that tactical, technological, and operational risks are regularly quantified and mitigation strategies deployed. Risks remaining at a very high level after mitigation are reported to the Management Board (Section 16.5.1), which acts as the guardian of the program on behalf of the enterprise. That Board also oversees the strategic risks to the program. Figure 16.1 illustrates part of a simple hypothetical risk register owned by a GI system program leader and the Management Board. In each case risk has two components—the likelihood of coming to pass and the impact on the program if it did. The risk score is computed by multiplying these together. Mitigating actions are then designed for each risk above a level judged appropriate and the residual risk estimated once these have been implemented.

Risk registers are only useful if they are built into the way a whole program is run, treated seriously, updated regularly, and used to steer actions. They must be living documents and reflect careful judgments that are debated by the relevant team(s).

## 16.3 The Case for the GI System: ROI

The most fundamental question an organization can ask about a GI system is: Do we need one? This question can be posed in many ways, and it covers quite a number of issues (see Maguire, Kouyoumjian, and Smith in Further Reading). The key strategic questions that senior executives will have are likely to be:

- What value will an investment in a GI system have for our organization?
- When will the benefits of a GI system be delivered?
- Who will be the recipients of the benefits?
- What is the level of investment needed, both initially and on an ongoing operational basis?
- Who is going to deliver these benefits, and what resources are required—both internally and externally—to realize the expected benefits?
- What is the proven financial case—that is, does the investment in a GI system provide the financial or other value to make it worthwhile?

For many years GI projects have been initiated based largely on qualitative, value-added reasoning; for example, "if we implement this technology, then we will be able to perform these additional services." However, in recent years there has been greater pressure for financial accountability in both the public and private sectors. Combined with a better understanding of difficulties associated with implementing enterprise IT systems and processes, and a realization that proper project accountability is a key aspect of good management, these have all led to wider adoption of return-on-investment (ROI) methods in project planning and evaluation.

To be successful, GI system strategies must be aligned with business strategies, and GI system processes must reflect business processes. GI systems that exist in a vacuum and are disconnected from an organization's business processes may detract from business success more than having no GI system at all.

Although there are many ways to describe the success or failure of a GI project, the most widely used is return on investment. ROI studies use a combination of qualitative and quantitative measures to assess the utility that an organization will obtain from an investment. Much greater importance is almost always placed on fact-based, quantitative measures given their greater objectivity and persuasiveness as far as senior executives are concerned.

**ROI uses a combination of qualitative and quantitative measures to assess the utility that an organization will obtain from an investment.**

One ROI method suitable for use in GI projects (see Further Reading) comprises a sequence of ten interrelated steps designed to be performed by a GI professional supported by a small project team. Shown in outline form in Figure 16.2, the method begins with a series of planning and investigation activities that lay the groundwork for subsequent steps. Step 1, preparing for the ROI project, requires a review of an organization's mission statement(s) and an understanding of its past and present GI system landscape. Step 2 comprises a series of interviews with key stakeholders to elicit, with guidance, how a GI system can contribute to an organization's mission, collecting information concerning the high-level business issues and challenges that it faces. These insights are then organized into a series of business opportunities, which are prioritized in Step 3.

The next group of steps is concerned with GI program definition. The information gathered in earlier stages is used to define a program of GI projects in Step 4 and dictates how these projects will be governed and managed in Step 5. The following series of steps form the core of the method and are concerned with business analysis. In Step 6, the defined projects are broken down into constituent parts and

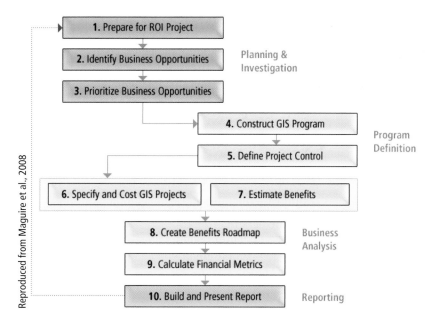

Figure 16.2 Overview of GI system ROI methodology.

the resource costs are determined, from which estimated benefits will be detailed in Step 7. In Step 8, a benefits road map is created that predicts when an organization will realize the benefits. In Step 9, ROI and other relevant financial metrics are calculated in order to demonstrate quantitatively the value of a GI system to an organization. In the final Step 10, a summary report is created by aggregating the information and research completed previously. If the GI system ROI project is to be seen as worth supporting, this report will concisely show how the GI system can contribute value to an organization, including its cost, benefits, time to implement, resources required, governance needed, the risks to be faced and the ROI an organization will realize.

In most IT projects—and GI projects are no exception—it is much easier to measure costs than benefits. Costs can simply be determined by making an inventory of all the goods and services required (hardware, software, data, consultants, etc.). The term *benefit* usually means any type of material value obtained from a GI project. It is useful to distinguish between tangible and intangible benefits. Tangible benefits, also sometimes referred to as hard or economic benefits, are those that can be precisely defined and to which we can assign a specific monetary value. Examples of tangible benefits include cost avoidance (such as increasing the number of valve inspections a water utility engineer can complete per route and therefore decreasing inspection time); increased revenue (e.g., from additional property taxes derived as a result of using a GI system to

manage land records); and the value of time saved (for instance by creating essential reports more quickly using a GI system).

Intangible benefits, also sometimes referred to as soft savings or institutional benefits, are those to which we cannot safely assign a monetary value. Examples of intangible GI system benefits include increased morale of employees due to improved information systems, improved citizen satisfaction with government as a result of readily available online access to maps and data, and better customer relationships through more efficient information management. Although it is important to include both types of benefits in a GI system business justification case, executive decision makers are usually much more easily persuaded by tangible, measurable benefits than by those that are intangible.

Table 16.1 describes some examples from the government and utility sectors of the main types of tangible and intangible benefits that have been used in the past to justify GI projects. Because the general principles are common across all these areas, it should easily be possible to extrapolate from the details of government and utility benefits to other sectors and industries. These lists are high level, are by no means exhaustive, and have a degree of overlap between several categories (for example, revenue growth, cost reduction, and cost avoidance). Nevertheless they can be a guide that will help focus the process of researching GI system program benefits. Table 16.1 begins with very tangible benefits and ends with very intangible benefits.

Table 16.1 Examples of tangible and intangible benefit types for public and private organizations. Both tangible and intangible benefits are ordered strongest to weakest.

| | Benefit Type | Government | Utility |
|---|---|---|---|
| **STRONG** ↓↓↓ Tangible ↓↓↓ **WEAK** | Revenue Growth—how can a GI system generate revenue (strictly speaking profit) for the organization? | Property taxes account for a substantial portion of the income for many local governments. A GI system is used to assess accurately the size of land parcels and keep an up-to-date record of property improvements. This typically results in additional tax revenue. The benefit is the total additional tax revenue that results from using a GI system. | There is great demand for accurate, detailed utility data for emergency management, construction coordination, and other purposes. Utilities sell network data products to other utilities and governments in the same geography for such purposes. The total benefit is the sum of all income. |
| | Revenue Protection and Assurance—how can a GI system save the organization money? | A GI system is used to optimize thenumber and location of fire stations and to comply with key government regulations about service response times. By using a GI system to analyze fire station locations, road networks, and the distribution of population, the location of existing or new stations can be optimized. Avoiding the cost of building a new fire station, or saving money by combining existing stations, saves governments millions of dollars a year. The benefit is the amount of money saved by optimizing the location of the fire stations. | Costly reactive and emergency repairs to sewer networks can be eliminated by creating a planned cycle of inspections and maintenance work orders. This enables network infrastructure to be properly maintained. The total benefit is the sum of the average costs of an emergency repair compared with those of planned repairs. |
| | Health and Safety—how can a GI system save the lives (or reduce injury) of employees or citizens? Although some might take the view that lives are invaluable, it is commonplace to ascribe a monetary value to loss of life in, for example, the insurance industry. | The most important role of governments is to protect the lives of citizens. Police forces are usually tasked with monitoring the security of major public events. A GI system is increasingly regarded as a key component of emergency operations centers where it is used to store, analyze, and visualize data about events. Data about, for example, suspicious packages can be used to help support decision making about the evacuation of surrounding areas. The benefit of a GI system is based on an estimate of the monetary value of lives saved as a result of using the GI system. | Integrated AM/FM/GI system electric network databases and work-order management systems are being used to create map products for use in Call-Before-You-Dig systems that reduce the likelihood of electrocution from hitting electrical conductors. Before-and-after comparison of lives saved multiplied by an agreed value of a life can lead to large benefits. |
| | Cost Reduction—this is different from Cost Avoidance because here we are assuming that this is an activity that an organization | Local government planning departments are reducing the cost of creating land-use plans and zoning maps by building | Multiple entry of "as-built" work orders by several departments is reduced by centralizing data entry in a |

Table 16.1 (*continued*)

| | Benefit Type | Government | Utility |
|---|---|---|---|
| **STRONG** | has to perform, and the objective becomes how to perform the activity with minimal net expenditure. | databases and map templates that can be reused many times. The benefit is the difference in the cost of manual and GI system-based plan and map creation. | single department. The total benefit is the cost of the as-built work-order entry multiplied by the reduction in the number of work orders entered. |
| ↓ ↓ ↓ ↓ | Cost Avoidance—rather than reducing costs, it is sometimes possible to avoid them altogether. | Local government departments issue permits for many things such as public events and roadside dumpsters (skips). By using a GI system, a government department can automate the process of finding the location, issuing the permit, and tracking its status. The benefit is the reduction in the cost of issuing and tracking each permit, multiplied by the number of permits issued. There are often additional benefits of reduced time to obtain a permit and improved tracking. | Forecasting the demand for gas by integrating geodemographic data with a model of the gas network will avoid costly overbuilding of distribution capacity. The benefit is the reduced cost of construction and operation of the unwanted gas facilities. |
| **Tangible** ↓ ↓ | Increase Efficiency and Productivity—how can the organization do more with less resource? | Fire is a major hazard to forests in many drier parts of the world. Firefighters need access in their fire trucks to maps of structures to better fight fires. Field-based GI systems are used to map structures more efficiently. These data are then transferred to a GI database, and map books are produced automatically, thus greatly improving the productivity of the surveyors and cartographers (not to mention the firefighters). The benefit of increased efficiency and productivity can be quantified by comparing manual with GI system-based operations. The hours saved can be converted into dollars based on the hourly rates of workers. | Automating map production improves the efficiency and productivity of staff, freeing them up to perform other tasks and in some cases avoiding hiring new staff. The benefit is equal to the number of person-hours that can be assigned to other tasks, multiplied by their hourly rate. |
| ↓ ↓ **WEAK** | Save Time—if a process is carried out using a GI system, how much faster can it be completed? | Searches are typically performed whenever a property is bought/sold to determine ownership rights and encumbrances (rights of way, mineral rights, etc.). This requires a search to be sent to many departments. Historically this has been a manual, paper-based process taking several weeks. By implementing this | The amount of time it takes to inspect outside plant will be reduced by providing better routing between inspection sites and field-based data-entry tools. The benefit is the additional number of inspections per inspector per day multiplied by the cost of an inspection. There are additional benefits of reduced vehicle use (less congestion, |

(*continued*)

Table 16.1 *(continued)*

| | Benefit Type | Government | Utility |
|---|---|---|---|
| **STRONG** ↓ ↓ ↓ ↓ **Tangible** ↓ ↓ ↓ ↓ ↓ **WEAK** | | in a GI system, the process can be managed electronically, performed in parallel, and responses automatically produced as a report. The time taken to complete the process can be reduced by several days or weeks. The benefit is the monetary value of the time saved multiplied by the number of searches performed. There are additional benefits of improved service to citizens. | pollution, and gasoline consumption). |
| | Increase Regulatory Compliance—if an organization has to comply with mandatory regulations, can it be done cheaper and faster with a GI system? | Various e-government initiatives around the world mandate that federal/central government agencies expand the use of the Internet for delivering services in a citizen-centric, results-oriented, market-based way. A GI system is a cost-effective way to achieve this goal because it is a commercial off-the-shelf system that can be relatively easily implemented across multiple departments. The objective of government is generally to minimize the cost of compliance. This is a cost (a negative benefit), and the benefit derives from minimizing the costs of compliance and avoiding fines for noncompliance. | Telecom utilities are bound by law to comply with several exacting operating regulations. A GI system can be used to maintain a database of the status of resources and to produce timely reports that summarize compliance levels. Given that this is a cost (a negative benefit) to the utility, the benefit derives from minimizing the costs of compliance and avoiding fines for noncompliance. |
| | Improve Effectiveness—if a process is carried out using a GI system, how much more effective will it be? | A GI system is used to fight crime using hotspot analysis and thematic mapping techniques to illustrate geographic patterns. The results from this work are being used to catch criminals and as the basis for crime prevention campaigns (e.g., increase police presence on the streets and advise members of the public to lock garages). The effectiveness of the GI system is quantified by assessing the reduction in crime and assigning a monetary value to it. | GI systems for outage management improve the speed and accuracy with which the location of outages (loss of service) can be identified and repair crews can be dispatched. The benefit is the value of total reduction in time for repairing outages, which in turn translates into electricity, water, or gas consumed. |

Table 16.1 (*continued*)

| | Benefit Type | Government | Utility |
|---|---|---|---|
| **STRONG** ↓ **Tangible** ↓ **WEAK** | Add New Capability—what new activities or services does a GI system make possible, and what value do they add to the organization? | Infection by West Nile virus has resulted in a number of deaths in the past few years. A GI system is a key component of the virus-spread monitoring process because it allows data to be analyzed and visualized geographically. This is a new capability that has allowed medical staff to understand and mitigate the spread of the virus. The benefit is the value of the GI system in terms of reduced medical costs for treating infected carriers. | A GI system allows utilities to create map products that show the status of construction, to facilitate coordination of participating contractors. The benefit is the value of the improved level of coordination. |
| **STRONG** ↓ ↓ **Intangible** ↓ ↓ **WEAK** | Improve Service and Excellence Image—in what ways will use of a GI system cast the organization in a better light (e.g., forward looking, better organized, or more responsive)? | A number of government departments have front-desk clerks that respond to requests for information from citizens (e.g., status of a request for planning consent, rezoning request, or water service connection). A GI system can be used to speed up responses to requests, and in some cases access can be provided in public facilities such as libraries or on the Web. The benefit is the value of improved organizational image. | By creating a publicly accessible Web site that shows planned electrical network maintenance and outages, an electrical utility can improve its image. The benefit is determined by the value of a page view, multiplied by the total number of page views. |
| | Enhanced Citizen/Customer Satisfaction—can a GI system be used to enhance the satisfaction levels of citizens/customers? | Providing access to accurate and timely information is not only a requirement for governments, but it is also desirable because more information generally leads to higher levels of satisfaction. Changes in the satisfaction of citizens can be measured by questionnaires. | Utilities can enhance their customer care capability by providing access to accurate and up-to-date outside-plant network information. This enables customers to find out about themselves and their services. |
| | Improve Staff Well-Being—can a GI system be used to enhance the well-being and morale of employees? | Staff are generally happier if they can see that an organization is progressive and is improving effectiveness and efficiency, as well as spending public money wisely. A GI system is a tool that can contribute to all these and therefore to worker satisfaction and well-being. | Automating boring and repetitive data-entry tasks can increase staff retention. There is a tangible benefit in reducing recruitment costs and an intangible value in having happy and contented staff. |

### Ross Smith, GI System Consultant

Ross Smith's (Figure 16.3) interest in GI systems started during his undergraduate studies at Wilfrid Laurier University in Ontario, Canada, while working in the Computer Science Department as an internal IT consultant. Upon graduating, Ross went to work for the Ministry of Natural Resources, where he developed GI system applications and digitized massive amounts of data—in those days on VAX/VMS and Tektronix machines.

Recognizing the potential of the technology, Ross pursued his M.Sc. in GI systems at the University of Edinburgh, Scotland, and then embarked on a UK-based consulting career. After several years in the UK, Ross moved to New Zealand, where he worked independently for major utilities, telecommunications providers, and GI system vendors. He then went to Poland, where he helped launch and lead an offshore network broadband design center that was heavily reliant on GI technology. While based in Poland, Ross also became involved in contracts for the U.S. Air Force in Europe, working in Italy and Germany on the Air Force bases and running multiple GI projects that ranged from antiterror/force protection to land-use management. Ross returned to North America in 2003 and joined PA Consulting Group, a global management consulting firm, where he led the Denver-based IT consulting team, as well as the global geospatial consulting team.

Ross is known for structuring and delivering large complex technology programs, and he is recognized as an industry leader in geospatial strategy development and return-on-investment models. His primary focus is

**Figure 16.3** Ross Smith, GI System Consultant.

on the telecommunication, utilities, and defense sectors and on IT-enabled business transformation. Ross is a coauthor of a GI system business case and strategy book (*The Business Benefits of GIS: An ROI Approach*), and he is also the author of numerous articles on strategic planning, ROI modeling, and organizational design. Ross comments:

It has become obvious to me that GI projects are not just about technology, although this is clearly important; it is also critical that business and organizational strategies are given due consideration if a project is to be successful and sustainable in the long term.

---

The best benefits are specific, measurable, and relevant to an organization (so-called SMART, for specific, measurable, attainable, relevant, and time-bound). Examining the generic benefits in Table 16.1 shows that these textual descriptions are imprecise rather than being expressed in monetary terms, time saved, or processes improved. In real-world cases it is important to avoid generalized statements such as "a GI system will reduce costs" or "with a GI system we can improve operational performance." More strongly tangible benefits should be defined through further analysis. The Further Reading section gives more details on how to achieve that end.

Finally, the implementation of a GI system may lead to costs rising in some parts of the organization but substantial benefits accruing in other parts. Thus it is important that the final analysis takes a holistic, enterprise-wide view of the balance between all costs and all benefits of a major GI system.

## 16.4 The Process of Developing a Sustainable GI System

GI projects are similar to many other large IT projects in that they can be broken down into four major life-cycle phases (Figure 16.4). For our simplified purposes, these are

- Business planning (strategic analysis and requirements gathering)
- System acquisition (choosing and purchasing a system)

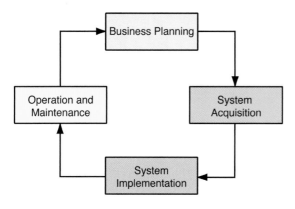

Figure 16.4 GI project life-cycle stages.

- System implementation (assembling all the various components and creating a functional solution)
- Operation and maintenance (keeping a system running)

These phases are iterative. Over a decade or more, several iterations may occur, often using different generations of GI system technology and methodologies. It follows that risks will change and need to be reevaluated accordingly. Variations on this model include prototyping and rapid application development but space does not permit much discussion of them here.

**GI projects comprise four major life-cycle phases: business planning; system acquisition; system implementation; and operation and maintenance.**

Roger Tomlinson, often regarded as the father of GI systems (see Table 1.4), developed a methodology for obtaining a GI system that is likely to fulfill user needs (see Further Reading). This high-level approach is very practical, easy to understand, and designed to ensure that a resulting GI system will match user expectations. Here we use a variant of that approach.

Table 16.2 The Tomlinson methodology for getting a GI system that meets user needs. (Source: adapted from R. Tomlinson 2007)

| Stage | Action | Commentary |
|---|---|---|
| 1 | Consider the strategic purpose | This is the guiding light. The system that gets implemented must be aligned with the purpose of the organization as a whole. |
| 2 | Plan for the planning | Because the GI system planning process will take time and resources, you will need to get approval and commitment at the front end from senior managers in the organization. |
| 3 | Conduct a technology seminar | Think of the technology seminar as a sort of town-hall meeting between the GI system planning team, the various staff, and other stakeholders in the organization. |
| 4 | Describe the information products | Know what you want to get out of it. |
| 5 | Define the system scope | Scoping the system means defining the actual data, hardware, software, and timing. |
| 6 | Create a data design | The data landscape has changed dramatically with the advent of the Internet and the proliferation of commercial datasets. Developing a systematic procedure for safely navigating this landscape is critical. |
| 7 | Choose a logical data model | The new generation of object-oriented data models is ushering in a host of new GI system capabilities and should be considered for all new implementations. Yet the relational model is still prevalent, and the savvy GI system user will be conversant in both. |
| 8 | Determine system requirements | Getting the system requirements right at the outset (and providing the capacity for their evolution) is critical to successful GI system planning. |
| 9 | Return on investment, migration, and risk analysis | The most critical aspect of doing an ROI analysis is the commitment to include all the costs that will be involved. Too often managers gloss over the real costs, only to regret it later. |
| 10 | Make an implementation plan | The implementation plan should illuminate the road to GI system success. |

## 16.4.1 Choosing a GI System: The Classical Acquisition Model

A general model of how to specify, evaluate, and choose a GI system is shown in Figure 16.5. Variations of it have been used by many organizations over the past 25 years so we use this "classical model." It is based on fourteen steps grouped into four stages: analysis of requirements, specification of requirements, evaluation of alternatives, and implementation of the system. Such a process is both time consuming and expensive. It is only appropriate for large GI system implementations (contracts over $250,000 in initial value) where it is particularly important to have investment and risk appraisals. We describe the model here so that those involved with smaller systems can judiciously select those elements relevant to them. On the basis of painful experience, however, we urge the use of formalized approaches to evaluate the need for any subsequent acquisition of a system. It is amazing how small projects, carried out quickly because they are small, evolve into big and costly ones!

**Choosing a GI system involves four stages: analysis of requirements, specification of requirements, evaluation of alternatives, and implementation of system.**

For organizations undertaking acquisition for the first time, huge benefits can be accrued through partnering with other organizations that are more advanced, especially if they are in the same field (see Chapter 18). This is often possible in the public sector, for example, where local governments have similar tasks to meet. But a surprising number of private-sector organizations are also prepared to share their experiences and documents.

***Stage 1: Analysis of Requirements***   The first stage in choosing a GI system is an iterative process for identifying and refining user requirements and for determining the business case for acquisition. The deliverable for each step is a report that should be discussed with users and the management team. It is important to keep records of the discussions and share them with those involved so that there can be no argument at a later stage about what was agreed! The results of each report help determine successive stages.

Step 1:   **Definition of objectives**
This is often a major decision for any organization. The rational process of choosing a GI system begins with and spins out of the development of the organization's strategic plan—and an initial supposition that a GI system might play a valuable role in the implementation of this plan. Strategic and tactical objectives must be stated in a form that is understandable to managers. The outcome from Step 1 is a document that managers

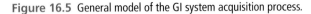

**Figure 16.5** General model of the GI system acquisition process.

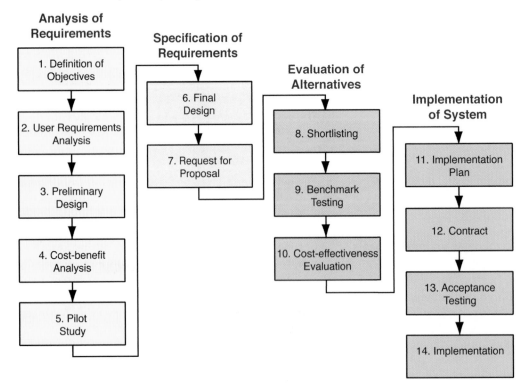

and users can endorse as a plan to proceed to the next stage; that is, the relevant managers believe there is sufficient promise to commit the initial funding required.

Step 2: **User requirements analysis**

The analysis will determine how the GI system is designed and evaluated. Analysis should focus on what information is presently being used, who is using it, and how the source is being collected, stored, and maintained. This is a map of existing processes (which may possibly be improved before being replicated by the GI system) or of processes newly designed after a business reengineering exercise. The necessary information can be obtained through interviews, documentation, reviews, and workshops. The report for this phase should be in the form of workflows, lists of information sources, and current operation costs. The clear definition of likely or possible change (e.g., future applications—see Figure 16.6), new information products (e.g., maps and reports—Figure 16.7), or different utilization of functions and new data requirements is essential to subsequent successful GI system implementation.

Step 3: **Preliminary design**

This stage of the design is based on results from Step 2. The results will be used for subsequent cost-benefit analysis (Step 4)

---

APPLICATION

Display zoning map information for a user-defined area.

FUNCTIONS USED IN THE APPLICATION:

Review and prepare zoning changes.

DESCRIPTION OF APPLICATION:

This application uses zoning and related parcel-based data from the database to display existing information related to zoning for a specific area that is defined by the user. The application must be available interactively at a workstation when the user invokes a request and identifies the subject land parcel. The application will define a search area based upon the search distance defined and input by the user, and will display all required data for the area within the specified distance from the outer boundary of the subject parcel.

DATA INPUTS:

User defined: Parcel identifier
              Search distance

Database:     Zoning boundaries
              Zoning dimensions
              Zoning codes
              Parcel boundaries
              Parcel dimensions
              Parcel numbers
              Street names
              Addresses

PRODUCTS OUTPUT:

1. Zoning map screen display with subject parcel highlighted, search area boundaries, search distance, all zoning data, parcel data, street names and addresses.

2. Hard copy map of the above.

**Figure 16.6** Sample application definition form.

and will enable specification of the pilot study. The four key tasks are: develop preliminary database specifications, create preliminary functional specifications, design preliminary system models, and

---

**Figure 16.7** A report produced by a local government GI system.

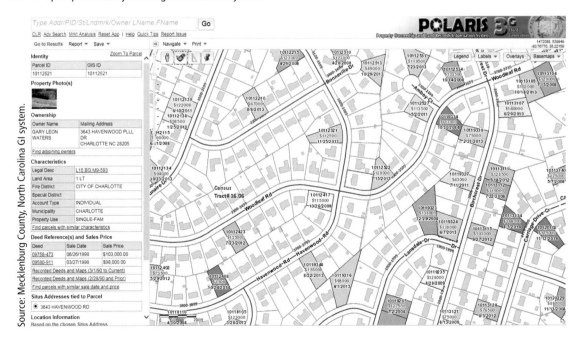

Source: Mecklenburg County, North Carolina GI system.

survey the market for potential systems. Database specifications involve estimating the amount and type of data. Many consultants maintain checklists, and vendors frequently publish descriptions of their systems on their Web sites. The choice of system model involves decisions about raster and vector data models and a survey should be undertaken to assess the capabilities of commercial off-the-shelf (COTS) systems. This might involve a formal request for information (RFI) to a wide range of vendors. A balance needs to be struck between creating a document so open that the vendor has problems identifying what needs are paramount and a document that is so prescriptive and closed that no flexibility or innovation is possible.

Whether to buy or to build a GI system used to be a major decision (Section 6.4). This occurred especially at "green field" sites—where no GI system technology had hitherto been used—and at sites where a GI system had already been implemented but was in need of modernization. But the situation is now quite different: use of general-purpose COTS solutions is the norm. COTS GI systems have ongoing programs of enhancement and maintenance and can normally be used for multiple projects. Typically, they are better documented, and more people in the job market have experience of them. As a consequence, risks—such as those arising from loss of key personnel—are reduced.

**There has been a major move in GI systems away from building proprietary products toward buying COTS solutions.**

Step 4:   **Cost-benefit analysis**
Purchase and implementation of a GI system is a nontrivial exercise because of the expense involved in both money and staff resources (typically management time). It is very common for organizations to undertake a cost-benefit analysis to justify the effort and expense and to compare it against the alternative of continuing with the current data, processes, and products—the status quo. Cost-benefit cases are normally presented as a spreadsheet, along with a report that summarizes the main findings and suggests whether the project should be continued or halted. Senior managers then need to assess the merits of this project in comparison with any others that are competing for their resources. Cost-benefit analysis can be considered a simplified form of ROI calculation (see Section 16.3), and we urge the use of that particular approach.

Step 5:   **Pilot study**
A pilot study is a miniature version of the full GI system implementation that aims to test several facets of the project. The primary objective is to test a possible or likely system design before finalizing the system specification and committing significant resources. Secondary objectives are to develop the understanding and confidence of users and sponsors, to test samples of data if a data capture project is part of the implementation, and to provide a test bed for application development.

**A pilot is a miniature version of a full GI system implementation designed to test as many aspects of the final system as possible.**

It is normal to use existing hardware or to lease hardware similar to that which is expected to be used in the full implementation. A reasonable cross section of all the main types of data, applications, and product deliverables should be used during the pilot. But the temptation to try to build the whole system at this stage must be resisted—regardless of how easy the "techies" may claim it to be! Users should be prepared to discard *everything* after the pilot if the selected technology or application style does not live up to expectations.

The outcome of a pilot study is a document containing an evaluation of the technology and approach adopted, an assessment of the cost-benefit case, and details of the project risks and impacts. As stressed earlier, risk analysis is an important activity even at this early stage. Assessing what can go wrong might help avoid potentially expensive disasters in the future. The risk analysis should focus on the actual acquisition processes as well as on implementation and operation.

***Stage 2: Specification of Requirements***   The second stage is concerned with developing a formal specification that can be used in the structured process of soliciting and evaluating proposals for the production system.

**Step 6: Final design**

This creates the final design specifications for inclusion in a Request for Proposals (RFP: also called an invitation to tender, or ITT) to vendors. Key activities include finalizing the database design, defining the functional and performance specifications, and creating a list of possible constraints. From these, requirements are classified as mandatory, desirable, or optional. The deliverable is the final design document. This document should provide a clear description of essential and desirable requirements—without being so prescriptive that innovation is stifled, costs escalate, or insufficient vendors feel able to respond.

**Step 7: Request for proposals**

The RFP document combines the final design document with the contractual requirements of the organization. These documents will vary from organization to organization but are likely to include legal details of copyright of the design and documentation, intellectual property ownership, payment schedules, procurement timetable, and other draft terms and conditions. Once the RFP is released to vendors by official advertisement or personal letter, a minimum period of several weeks is required for vendors to evaluate and respond. For complex systems, it is usual to hold an open meeting to discuss technical and business issues.

*Stage 3: Evaluation of Alternatives*

**Step 8: Short-listing**

In situations where several vendors are expected to reply, it is customary to have a short-listing process. Submitted proposals must first be evaluated, usually using a weighted scoring system, and the list of potential suppliers needs to be narrowed down to between two and four. Good practice is to have the scoring done by several individuals acting independently and to compare the results. This whole process allows both the prospective purchaser and supplier organizations to allocate their resources in a focused way. Short-listed vendors are then invited to attend a benchmark-setting meeting.

**Step 9: Benchmarking**

The primary purpose of a benchmark is to evaluate the proposal, people, and technology of each selected vendor. Each one is expected to create a prototype of the final system that will be used to perform representative tests. The prospective purchaser scores the results of these tests. Scores are also assigned for the original vendor proposal and the vendor presentations about their company. Together, these scores form the basis of the final system selection. Unfortunately, benchmarks are often conducted in a rather secretive and confrontational way, with vendors expected to guess the relative priorities (and the weighting of the scores) of the prospective purchaser. It is essential to follow a fair and transparent process, maintain a good audit trail, and remain completely impartial. However a more open cooperative approach usually produces a better evaluation of vendors and their proposals. If vendors know which functions have the greatest value to customers, they can tune their systems appropriately.

**Step 10: Cost-effectiveness evaluation**

Next, surviving proposals are evaluated for their cost-effectiveness. This is again more complex than it might seem. For example, GI software systems still vary quite widely in the type of architectural options and pricing models available; some need additional database management system (DBMS) licenses, some run in a Cloud computing environment with a pay-as-you go business model, customization costs will vary, and maintenance will often be calculated in different ways. The goal of this stage is to normalize all the proposals to a common format for comparative purposes. The weighting used for different parts must be chosen carefully because this can have a significant impact on the final selection. Good practice involves debate within the user community—for they should have a strong say—on the weighting to be used and some sensitivity testing to check whether very different answers would have been obtained if the weights were slightly different. The deliverable from this stage is a ranking of vendors' offerings.

*Stage 4: Implementation of System* The final stage is planning the implementation, contracting with the selected vendor, testing the delivered system, and actually using the GI system "in anger."

## Step 11: Implementation plan

A structured, appropriately paced implementation plan is an essential ingredient of a successful GI system implementation. The plan commences with identification of priorities, definition of an implementation schedule, and creation of a resource budget and management plan. Typical activities that need to be included in the schedule are installation and acceptance testing, staff training, data collection/purchase, and customization. Implementation should be coordinated with both users and suppliers.

## Step 12: Contract

An award is subject to final contractual negotiation to agree on general and specific terms and conditions, what elements of the vendor proposal will be delivered, when they will be delivered, and at what price. General conditions include contract period, payment schedule, responsibilities of the parties, insurance, warranty, indemnity, arbitration, and provision of penalties and contract termination arrangements.

## Step 13: Acceptance testing

This step is to ensure that the delivered GI system matches the specification agreed in the contract. Part of the payment should be withheld until this step is successfully completed. Activities include installation plus tests of functionality, performance, and reliability. A system seldom passes all tests the first time so provision should be made to repeat aspects of the testing.

## Step 14: Implementation

This is the final step at the end of what can be a long road. The entire acquisition period for a major GI system can stretch over many months or even longer. Activities include training users and support staff, data collection, system maintenance, and performance monitoring. Customers may also need to be "educated" as well! Once the system is successfully in operation, it may be appropriate to publicize its success to give recognition to the staff involved and enhance the brand image or political position.

### 16.4.1.1 Discussion of the Classical Acquisition Model

The general model outlined earlier has been widely employed as the primary mechanism for large GI system procurements in public organizations. It is rare,

however, that one size fits all, and although the model has many advantages, it also has some significant shortcomings:

- The process is expensive and time consuming for both suppliers and vendors. A supplier can spend as much as 20% of the contract value on winning the business, and a purchaser can spend a similar amount in staff time, external consultancy fees, and equipment rental. This ultimately makes systems more expensive—though competition does drive down cost.

- Because it takes a long time and because GI technology is a very fast-developing field, proposals can become technologically obsolete within several months.

- The short-listing process requires multiple vendors, which can end up lowering the minimum technical selection threshold in order to ensure enough bidders are available.

- In practice, the evaluation process often focuses undue attention on price rather than on the long-term organizational and technical merits—some at least of which are intangible—of the different solutions.

- This type of procurement can be highly adversarial. As a result, it can lay the foundations for an uncomfortable implementation partnership (see Chapter 18), and it often does not lead to full development of the best solution. Every implementation is a trade-off between functionality, time, price, and risk. A full and frank discussion between purchaser and vendor on this subject can generate major long-term benefits.

- Many organizations—especially those with little previous experience of a GI system—have little idea about what they *really* need. Furthermore, it is very difficult to specify precisely in any contract what a system must perform. As users learn more, their aspirations also rise, resulting in "feature creep" (the addition of more capabilities) often without any acceptance of an increase in budget. On the other hand, some vendors take a minimalist view of the capabilities of the system featured in their proposal and make all modifications during implementation and maintenance through chargeable change orders. All this makes the entire system acquisition cost far higher than was originally anticipated; the personal consequences for the budget holders concerned can be unfortunate.

- Increasingly, most organizations already have some type of GI system; the classical model works best in a "green field" situation.

As a result of these problems, this type of acquisition model is not used in small or even some larger procurements, especially where existing facilities can be augmented rather than totally replaced. A less complex and formal selection method is *prototyping*. Here a vendor or pair of vendors is selected early on using a smaller version of the evaluation process outlined in the preceding part of this chapter. The vendors are then funded to build prototypes in collaboration with the user organization. This fosters a close partnership to exploit the technical capabilities of systems and developers, and it helps to maintain system flexibility in the light of changing requirements and technology. This approach works best for those procurements—sometimes even some large ones—where there is some uncertainty about the most appropriate technical solution and where the organizations involved are mature, able to control the process, and not subject to draconian procurement rules.

**Prototyping is a useful alternative to classical, linear system acquisition exercises. It is especially useful for smaller procurements where the best approach and outcome is more uncertain.**

## 16.4.2 Implementing a GI System

This section provides a checklist of important management issues to consider when implementing a GI system.

### 16.4.2.1 Plan Effectively

Good planning is essential through the full life cycle of all GI projects. Both strategic planning and operational planning are important to the success of a project. Strategic planning involves reviewing overall organizational goals and setting specific GI project objectives. Operational planning is more concerned with the day-to-day management of resources. Several general project-management productivity tools are available that can be used in GI projects. Figure 16.8 shows one diagramming tool called a Gantt chart. Several other implementation techniques and tools are summarized in Table 16.3.

### 16.4.2.2 Obtain Support

If a GI project is to prosper, it is essential to garner support from *all* key stakeholders. This can entail several activities, including establishing executive (director-level) leadership support; developing a

**Figure 16.8** Gantt chart of a basic GI system project. This chart shows task resource requirements over time, with task dependencies. This example presents a straightforward chart, with a small number of tasks.

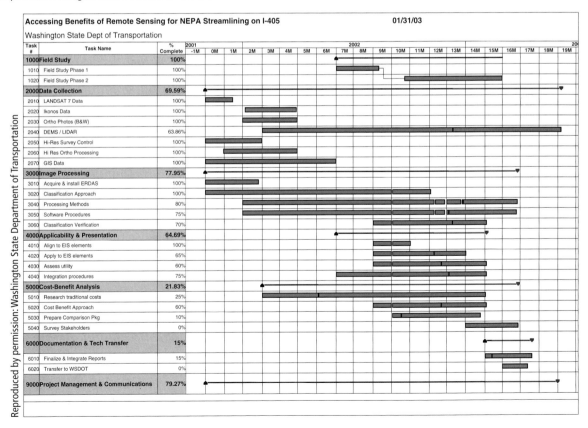

**Table 16.3** GI system implementation tools and techniques (After Heywood et al., 2012, with additions).

| Technique | Purpose |
|---|---|
| SWOT analysis | This is a management technique used to establish strengths, weaknesses, opportunities, and threats (hence SWOT) in a GI system implementation. The output is a list and narrative. |
| ROI analysis | This methodology is used to assess the value added by comparing outputs with inputs (see Section 16.3). |
| Rich picture analysis | Major participants are asked to create a schematic/picture showing their understanding of a problem using agreed conventions. These are then discussed as part of a consensus-forming process. |
| Demonstration systems | Many vendors and GI project teams create prototype demonstrations to stimulate interest and educate users/funding agencies. |
| Interviews and data audits | These aim to define problems and determine current data holdings. The output is a report and recommendations. |
| Organization charts, system flowcharts, and decision trees | These are all examples of flowcharts that show the movement of information, the systems used, and how decisions are currently reached. |
| Data flow diagrams and dictionaries | These are charts that track the flow of information and computerized lists of data in an organization. |
| Project management tools | Gantt charts (see Figure 16.8) and PERT (program evaluation and review technique) are tools for managing time and resources. |
| Object-model diagrams | These show objects to be modeled in a GI database and the relationships between them (see, for example, Figure 7.18). |

public relations strategy by, for example, exhibiting key information products or distributing free maps; holding an open house to explain the work of the GI system team; and participating in GI system seminars and workshops, locally and—depending on the organization—sometimes nationally.

### 16.4.2.3 Communicate with Users
Involving users from the very earliest stages of a project will lead to a better system design, reduce some risks, and help with user acceptance. Seminars, newsletters, and frequent updates about the status of a project are good ways to educate and involve users. Setting expectations about capabilities, throughput, and turnaround at reasonable levels is crucial to avoid any later misunderstandings with users and managers.

### 16.4.2.4 Anticipate and Avoid Obstacles
These obstacles may involve staffing, hardware, software, databases, organization/procedures, time frame, and funding. Be prepared: make your risk register(s) the vehicle for coping with adversity!

### 16.4.2.5 Avoid False Economies
Money saved by not paying staff a reasonable (market value) wage or by insufficient training is often manifested in reduced staff efficiencies. Furthermore, poorly paid or poorly trained staff often leave through frustration. This situation cannot be prevented by contractual means and must be tackled by paying market-rate salaries and building a team culture where staff enjoy working for the organization.

Cutting back on hardware and software costs by, for example, obtaining less powerful systems or canceling maintenance contracts may save money in the short term but will likely cause serious problems in the future when workloads increase and the systems get older. Failing to account for depreciation and replacement costs, that is, by failing to amortize the GI system investment, will store up trouble ahead. The amortization period will vary greatly—hardware may be depreciated to zero value after, say, four years while buildings may be amortized over 30 years.

### 16.4.2.6 Ensure Database Quality and Security
Investing in database quality is essential at all stages from design onward. Catastrophic results may ensue if any of the updates or (especially) the database itself is lost in a system crash or corrupted by hacking, and the like. This requires not only good precautions but also contingency (disaster recovery and business continuity) plans and periodic serious trials of them.

There are now many options for securing data, such as Cloud storage. Losing data through lack of a good backup strategy is a very serious matter.

### 16.4.2.7 Accommodate the GI System within the Organization

Building a system to replicate old and inefficient ones is not a good idea; nor is it wise to go to the other extreme and expect the whole organization's ways of working to be changed to fit better with what the GI system can do! Too much change at any one time can destroy organizations just as much as too little change can ossify them. In general, the GI system must be managed in a way that fits with the organizational aspirations and culture if it is to be a success. All this is especially a problem because GI projects often blaze the trail in terms of introducing new technology, interdepartmental resource and risk sharing, and generating new sources of income.

### 16.4.2.8 Avoid Unreasonable Time Frames and Expectations

Inexperienced managers often underestimate the time it takes to implement a GI system. Good tools, risk analysis (see Section 16.2), and time allocated for contingencies are important methods of mitigating potential problems. The best guide to how long a project will take is experience in other similar projects—though the differences between the organizations, staffing, tasks, and so on need to be taken into account.

### 16.4.2.9 Funding

Securing ongoing, stable funding is a major task of a GI system manager. Substantial GI projects will require core funding from one or more of the stakeholders. None of these will commit to the project without a sound business case, risk analysis, and consideration of how much uncertainty they can tolerate (Section 16.5.3). Additional funding for special projects, and from information and service sales, is likely to be less certain. In many GI projects the operational budget will often change significantly over time as the system matures. The three main components are staff, goods and services, and capital investments. A commonly experienced distribution of costs between these three elements is shown in Table 16.4.

### 16.4.2.10 Prevent Meltdown

Avoiding the cessation of GI system activities valued within an organization is the ultimate responsibility of the GI system manager and his or her superior. According to Tomlinson (see Further Reading), some of the main reasons for the failure of GI projects are as follows:

- Lack of executive-level commitment
- Inadequate oversight of key participants

**Table 16.4** Percentage distribution of GI system operational budget elements over three time periods (after Sugarbaker, 2005).

| Budget item | Year 1–2 | Year 3–6 | Year 6–12 |
|---|---|---|---|
| Staff and benefits | 30 | 46 | 51 |
| Goods and services | 26 | 30 | 27 |
| Equipment and software | 44 | 24 | 22 |
| Total | 100 | 100 | 100 |

- Inexperienced managers
- Unsupportive organizational structure
- Political pressures, especially where these change rapidly
- Inability to demonstrate benefits
- Unrealistic deadlines
- Poor planning
- Lack of core funding

You have been warned!

## 16.4.3 Managing a Sustainable, Operational GI System

Sugarbaker (see Further Reading) has characterized the many operational management issues throughout the life cycle of a GI project as customer support, effective operations, data management, and application development and support. Success in any one—or even all—of these areas does not guarantee project success, but they certainly help to produce a healthy project. Each is now considered in turn.

> **Success in operational management of a GI system requires customer support, effective operations, data management, and support for applications development.**

Applications Box 16.2 describes a very successful and well-managed GI system in South Korea.

### 16.4.3.1 Customer Support

In progressive organizations *all* users of a system and its products are referred to as customers. A critical function of an operational GI system is a customer support service. This could be a physical desk with support staff, or, increasingly, it is a Web, email, and telephone service. Because this is likely to be the main interaction with GI system support staff, it is essential that the support service creates a good impression and delivers the type of service users need. The unit will typically perform key tasks, including technical support and problem logging plus supplying requests

## Managing Land Information in Korea through a GI System

In South Korea, a complex and rapidly changing society, local government authorities administer the public land through assessment of land prices, management of land transactions, land-use planning and management, and civil services. In many cases, more than one department of a local government authority produces and manages the same or similar land and property information; this has led to discrepancies in the information held across local government. With the large number of public land administration responsibilities and the control of each given to the local authorities, many problems arose in the past. This led to the decision to develop a GI system-based method for sharing the information produced or required for administering land in the public and private sectors (Figure 16.9B).

The purpose of this Korean Land Management Information System (LMIS) is to provide land information, increase productivity in public land administration, and support the operation of the land planning policies of the Korean Ministry of Construction and Transportation (MOCT). The LMIS database contains many spatial data layers including topographic, cadastral, and land-use districts.

Hyunrai Kim, vice director of the Land Management Division of Seoul Metropolitan City, summarizes the advantages of this system thus: "By means of the Internet-based Land Information Service System,

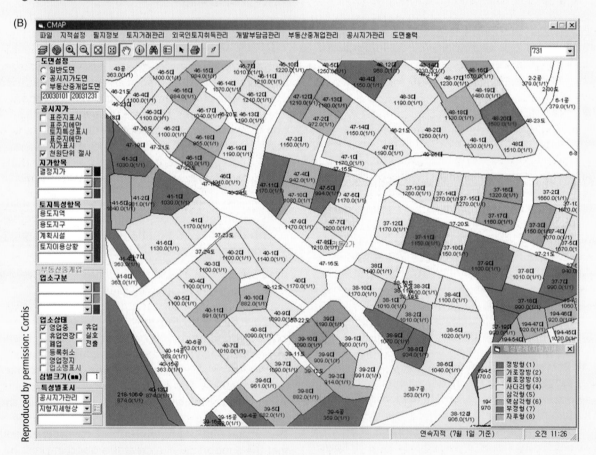

(A) © Sean Pavone/Shutterstock

(B) Reproduced by permission: Corbis

Figure 16.9 (A) Seoul by night. (B) Land information map for part of Seoul.

citizens can get land information easily at home. They don't have to visit the office, which may be located far from their homes." The system has also resulted in time and cost savings. With the development of the Korean Land Price Management System, it is also possible to compute land prices directly and produce maps of variations in land price. Initially, the focus was mainly on the administrative aspects of data management and system development; however, attention then turned to the expansion and development of a decision support system using various data analyses. It is intended that the Land Legal Information Service System will also be able to inform land users of regulations on land use. In essence, LMIS is becoming a crucial element of e-government. This case study highlights the role that GI systems can play beyond the obvious one of information management, analysis, and dissemination. It highlights the value of a GI system in enabling organizational integration and the reality of generating benefits through improved staff productivity.

for data, maps, training, and other products. Performing these tasks will require both GI system analyst and administrative skills. It is imperative that *all* customer interaction is logged and that procedures are put into place to handle requests and complaints in an organized and structured fashion. This is both to provide an effective service and to correct systemic problems.

Customer support is not always seen as the most glamorous of GI system activities. However, a GI system manager who recognizes the importance of this function and delivers an efficient and effective service will be rewarded with happy customers. Happy customers remain customers. Effective staff management includes finding staff with the right interests and aspirations, rotating GI system analysts through posts, and setting the right (high) level of expectation in the performance of all staff. Managers can learn much by taking a turn in the hot seat of a customer support role!

### 16.4.3.2 Operations Support

Operations support includes system administration, maintenance, security, backups, technology acquisitions, and many other support functions. In small projects, everyone is charged with some aspects of system administration and operations support. But as projects grow beyond five or more staff, it is worthwhile designating someone specifically to fulfill what becomes a core, even crucial, role. As projects become larger, this grows into a full-time function. System administration is a highly technical and mission-critical task requiring a dedicated, properly trained, and paid person.

Perhaps more than in any other role, clear written descriptions are required for this function to ensure that a high level of service is maintained. For example, large, expensive databases will require a well-organized security and backup plan—perhaps exploiting the Cloud (Section 6.2)—to ensure that they are *never* lost or corrupted. Part of this plan should be a disaster recovery strategy. What would happen, for example, if there were a fire in the building housing the database server or some other major problem?

### 16.4.3.3 Data Management Support

The concept that geographic data are an important part of an organization's critical infrastructure is becoming widely accepted. Large, multiuser geographic databases use DBMS software to allocate resources, control access, and ensure long-term usability (see Section 9.2). DBMS can be sophisticated and complicated, requiring skilled system administrators for this critical function.

A database administrator (DBA) is a person responsible for ensuring that all data meet all the standards of accuracy, integrity, and compatibility required by the organization. A DBA will also typically be tasked with planning future data resource requirements—derived from continuing interaction with current and potential customers—and the technology necessary to store and manage them. Similar comments to those outlined earlier for system administrators also apply to this position.

### 16.4.3.4 Application Development and Support

Although a considerable amount of application development is usual at the onset of a project, it is also likely that there will be an ongoing requirement for this type of work. Sources of application development work include improvements/enhancements to existing applications, as well as new users and new project areas starting to adopt GI systems.

Software development tools and methodologies are constantly in a state of flux, and GI system managers must invest appropriately in training and new software tools. The choice of which language to use for GI system application development is often a difficult one. Consistent with the general movement away from proprietary GI system languages, wherever possible GI system managers should try to use mainstream, open languages that are likely to have a long lifetime. Ideally, application developers should be assigned full-time to a project and should become permanent members of the GI system group to ensure continuity (but often this does not occur).

## 16.5 Sustaining a GI System— The People and Their Competences

Throughout this chapter we have sometimes highlighted and sometimes hinted at the key role of staff as assets in all organizations. If they do not function well—both individually and as a team—nothing of merit will be achieved.

### 16.5.1 GI System Staff and the Teams Involved

Several different staff will carry out the operational functions of a GI system. The exact number of staff and their precise roles will vary from project to project. The same staff member may carry out several roles (e.g., it is quite common for administration and application development to be performed by a GI system technical person), and several staff members may be required for the same task (e.g., there may be many data compilation technicians and application developers). Figure 16.10 shows a generalized view of the main staff roles in medium to large GI projects.

All significant GI projects will be overseen by a management board composed of a senior sponsor (usually a director or vice-president), members of the user community, and the GI system manager. It is also useful to have one or more independent members to offer disinterested advice. Although this group may seem intimidating and restrictive to some, used in the right way it can be a superb source of funding, advice, support, and encouragement.

Typically, day-to-day GI system work involves three key groups of people: the GI system team itself, the GI system users, and external consultants. The GI system team comprises the dedicated GI system staff at the heart of the project, with the GI system manager designated as the team leader. This individual needs to be skilled in project and staff management and have sufficient understanding of GI technology and the organization's business to handle the liaisons involved. Larger projects will have specialist staff experienced in project management, system administration, and application development.

GI system users are the customers of the system. There are two main types of user (other than the leaders of organizations who may rely on the GI system indirectly to provide information on which they base key decisions). These are professional users and clerical staff/technicians. Professional users include engineers, planners, scientists, conservationists, social workers, and technologists who utilize output from the GI system for their professional work. Such users are typically well educated in their specific field, but may lack advanced computer skills and knowledge of the GI system. They are usually able to learn how to use the system themselves and can tolerate changes to the service.

**Figure 16.10** The GI system staff roles in a medium to large size project.

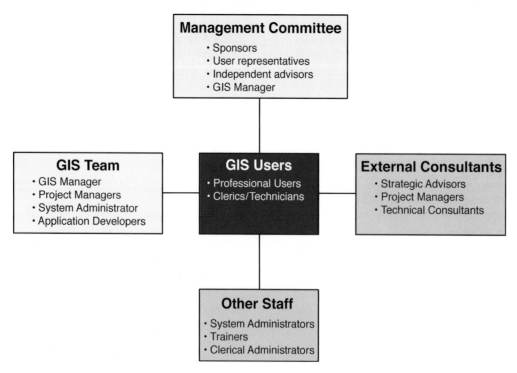

Clerical and technical users are frequently employed as part of the wider GI project initiative to perform tasks such as data collection, map creation, routing, and service-call response. Typically, the members of this group have limited training and skills for solving ad hoc problems. They need robust, reliable support. They may also include staff and stakeholders in other departments or projects that assist the GI project on either a full- or part-time basis—for example, system administrators, clerical assistants, or software engineers provided from a common resource pool, or managers of other databases or systems with which the GI system must interface.

Finally, many GI projects utilize the services of external consultants. They could be strategic advisors, project managers, or technical consultants able to supplement the available staffing. Although these consultants may appear expensive at first sight, they are often well trained and highly focused. They can be a valuable addition to a project, especially if internal knowledge or resources are limited and for benchmarking against approaches elsewhere. But the in-house team must not rely too heavily on consultants lest, when they go, all key knowledge and high-level experience goes with them.

**The key groups involved in GI projects are the management board, the GI system team (headed by a GI system manager), the users, external consultants, and various customers.**

## 16.5.2 Project Managers

A GI project will almost certainly have several subprojects or project stages and hence require a structured approach to project management. The GI system manager may take on this role personally, although in large projects it is customary to have one or more specialist project managers. The role of the project manager is to establish user requirements, to participate in system design, and to ensure that projects are completed on time, within budget, and according to an agreed-upon quality plan. Good project managers are rare creatures and must be nurtured for the good of the organization. One of their characteristics is that, once one project is completed, they like to move on to another, so retaining them is only possible in an enterprising environment. Transferring their expertise and knowledge into the heads and files of others is a priority before they leave a project (see also Section 1.2).

## 16.5.3 Coping with Uncertainty

As we have seen, GI varies hugely in its characteristics, and rarely is the available information ideal for the task in hand. Staff need a clear understanding of the concepts and implications of uncertainty (see Chapter 5) and the related concepts of accuracy, error, and sensitivity. An understanding of business risk arising from GI system use and of how a GI system can help reduce organizational risk is also essential; this presupposes a prior definition of the organization's risk "appetite" or tolerance. This section focuses on the practical aspects relevant to managers of operational GI systems—and hence on the skills, attitudes, and other competences they need to bring to their work.

Organizations must determine how much uncertainty they can tolerate before information is deemed useless. This can be difficult because it is application-specific. An error of 10 m in the location of a building is irrelevant for business geodemographic analysis, but it could be critical for a water utility maintenance application that requires digging holes to locate underground pipes. Some errors in a GI system can be reduced but sometimes at a considerable cost. It is common experience that trying to remove the last 10% of error typically costs 90% of the overall sum. As we concluded in Section 5.5, uncertainty in GI representations is almost always something that we have to live with to a greater or lesser extent. The key issue here is identifying the amount of uncertainty (and risk) that can be tolerated for a given application and what can be done at least partially to eliminate it or ameliorate its consequences. Some of this, at least, can only be done by judgment informed by past experience.

A conceptual framework for considering uncertainty was developed in Section 5.1. This discussion also introduced the notion of measurement error. Some practical examples of errors in operational GI systems are as follows:

- Referential errors in the identity of objects (e.g., a street address could be wrong, resulting in incorrect property identification during an electric network trace).

- Topological errors (e.g., a highway network could have missing segments or unconnected links, resulting in erroneous routing of service or delivery vehicles).

- Relative positioning errors (e.g., a gas station incorrectly located on the wrong side of a divided highway or dual-carriageway road could have major implications for transportation models).

- Absolute errors in the real location of objects in the real world (e.g., tests for whether factories are within a smoke control zone or floodplain could provide erroneous results if the locations are incorrect). This could lead to litigation.

- Attribute errors (e.g., incorrectly entering land-use codes would give errors in agricultural production returns to government agencies).

Managing errors requires use of quality assurance techniques to identify them and assess their magnitude. A key task is determining the error tolerance that is acceptable for each data layer, information product, and application. It follows that both data creators and users must make analyses of possible errors and their likely effects, based on a form of ROI analysis (Section 16.3). And sitting in the midst of all this is the GI system manager, who must know enough about uncertainty to ask the right questions; if the system provides what subsequently turns out to be nonsense, the manager is likely to be the first to be blamed! In no sense is the good and long-lasting GI system manager simply someone who ensures that the "wheels go round."

## 16.6 Conclusions

Any management function in GI system (and indeed elsewhere) is mostly about motivating, organizing, or steering project teams; enhancing skills; and monitoring the work of other people.

Managing a GI project *is* different from using a GI system in decision making. Normally, managing a GI project requires good GI system expertise and first-class project management skills. In contrast, those involved at different levels of the organization's management chain need some awareness of a GI system, its capabilities, and its limitations—scientific and practical—alongside their substantial leadership skills. But our experience is that the division is not clear-cut. GI project managers cannot succeed unless they understand the objectives of the organization, the business drivers, and the culture in which they operate, as well as something of how to value, exploit, and protect their assets (Section 17.1.2). Moreover, decision makers can make good decisions only if they understand more of the scientific and technological background than they may wish to do: running or relying on a good GI service involves much more than the networking of a few PCs running one piece of software.

So, good management of a GI project requires excellent people, technical and business skills, and the capacity to ensure mutual respect and team working between the users and the experts. And it is underpinned by following management processes proved in other GI projects!

## Questions for Further Study

1. How can we assess the potential value of a proposed GI system?

2. Prepare a new sample GI system application definition form using Figure 16.6 as a guide.

3. List ten tasks critical to a GI project that a GI system manager must perform and the roles of the main members of a GI project team.

4. Why might a GI project fail? Draw on information from various chapters in this book.

## Further Reading

Douglas, B. 2008. *Achieving Management Success with GIS.* Hoboken, NJ: Wiley.

Heywood, I., Cornelius, S., and Carver, S. 2012. *An Introduction to Geographical Information Systems* (4th ed.). Harlow: Pearson Education Ltd.

Maguire, D. J., Kouyoumjian, V., and Smith, R. 2008. *The Business Benefits of GIS: An ROI Approach.* Redlands, CA: Esri Press.

Sugarbaker, L. J. 2005. Managing an Operational GIS. In Longley, P. A., Goodchild, M. F., Maguire, D. J., and Rhind, D. W. (eds.) *Geographical Information Systems: Principles, Techniques, Management and Applications* (abridged ed.). Hoboken, NJ: Wiley.

Tomlinson, R. 2013. *Thinking about GIS: Geographic Information System Planning for Managers* (5th ed.). Redlands, CA: Esri Press.

# Information and Decision Making

This chapter sets out the ways in which geographic information (GI) can make it possible to use the techniques described in earlier chapters in order to make sound decisions. GI is considered both as a good and as a value-adding service. We review the economic and other characteristics of GI and consider how different data types and business drivers have implications for GI system use. We highlight the numerous trade-offs and uncertainties inherent in using GI for decision making and the ways in which value is added through data linkage. We describe the implications of Open Data concepts and practice for users and developers of GI technology functionality in many countries. Finally, we use a military example to illustrate how all of these can be brought together in an information infrastructure created to aid decision making.

The discussion complements the "hard science" perspective of previous chapters by explaining the relevance of economics, human behavior, public policy, and sometimes politics in GI management. The objective of efficient and effective management is to arrive at and implement decisions without losing public trust, incurring excessive costs, or being challenged successfully by lawyers.

## LEARNING OBJECTIVES

After studying this chapter, you will understand:

- The role of information in decision making.
- Trade-offs, uncertainty, and risk in decision making.
- Characteristics of information and GI.
- Added value through GI linkage.
- Different types of GI.
- Open Data, Big Data, and Open Government.
- An example of a major information infrastructure.
- Where to go for more detailed information and advice.

## 17.1 Why We Need Information

Most readers of this book live in capitalist societies, with all their advantages and disadvantages (see Chapter 19). In such societies, manufacturing has diminished as a source of employment, and information-based services have become dominant. According to the UN Economic Commission for Europe, in 2011 the median proportions of the workforce employed in some 20 developed countries was approximately 3% in agriculture, 20% in industry, and 75% in services. Even in countries where agriculture and manufacturing industry are still strategically important employers, there is movement up the value chain associated with increasing employment in information-based services. Much of that information is geographic in nature. It tells us where things are happening, where there is high attraction to live and work, where a critical mass of specialists operates or natural resources exist, which is a generally safe area, how to get from point A to point B, and so on. GI systems as services (Section 1.5.2.3) already have great value and are growing rapidly. Figure 17.1

summarizes one estimate of these areas carried out for Google by a respected economic consultancy. This estimates that the global geospatial services sector generates $150 to $270 billion annually. The Boston Consulting Group estimated that the U.S. geospatial

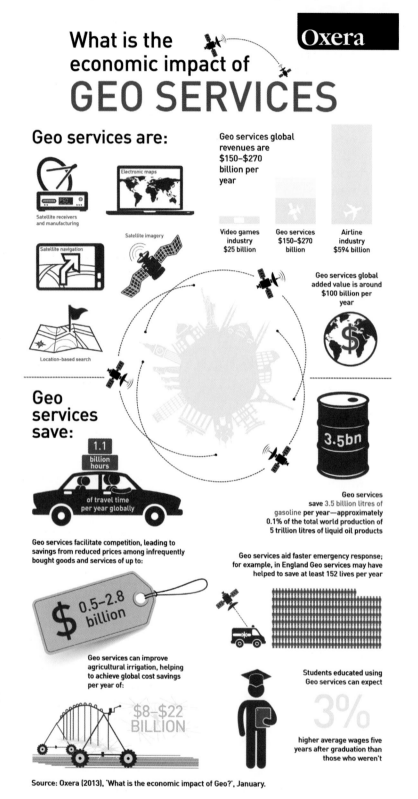

Source: Oxera (2013), 'What is the economic impact of Geo?', January.

**Figure 17.1** The economic impact of geo-services. (Source: Oxera, 2013)

## Some uses of geographic information

These include

- Describing the current status of some phenomenon (e.g., the geographic distribution of population), the historical changes in that distribution, or projections of what will happen to it in the future. The identification of geographic patterns can suggest possible explanations.

- Seeking reasons why the distribution is as it is:
  - By establishing correlations between it and other variables.
  - By seeking to isolate causality.
- Acting on the results by making decisions or facilitating decision making by others, based on an understanding of the processes that led to the distribution or that will create a different pattern in future.

---

industry generated some $73bn in 2011 and was composed of at least 500,000 jobs. Box 17.1 sets out some of the specific objectives we have in using geographic information.

> **Management without relevant information is like driving in the dark without headlights.**

But even if the "hidden hand" of capitalism shapes how societies evolve and prosper, most of us also live in some form of a managed economy—whether this is in China, Russia, France, or many other countries. Managers exist to make decisions and implement them successfully. To do this requires analytical tools such as multicriteria decision-making techniques (MCDM: Section 15.4), but these tools need to be fed with information. Management without relevant information is like driving in the dark without headlights. Increasingly, there is a demand to ensure that decision making in managed economies relies on demonstrably good evidence. In some countries there is also pressure to ensure that the evidence is available for widespread scrutiny in order to hold governments to account (Sections 1.5.3 and 17.4). Every organization—businesses, governments, voluntary organizations, and even universities—has managers, and everyone at some stage is going to be a manager in some function. Given that much of the information we use is geographic, it follows that GI and GI systems are central to many management functions and much decision making.

> **Geography shapes decisions and their consequences.**

## 17.1.1 Trade-Offs, Uncertainty, and Risk

Good decision making can be difficult and entails taking account of many environmental or contextual factors as well as simpler operational ones (Chapter 18). All decisions involve trade-offs—for example, the relatively simple personal decision of whether to purchase a more expensive house if that reduces the commute to work. More generally, a trade-off almost always exists between the benefits and disbenefits of any decision and its subsequent implementation. Sometimes these decisions bring near-universal benefits (like the decision to build GPS; see Box 17.7). But in many cases we are operating in a zero-sum game or something close to it. Benefits accrue to one group and disbenefits fall on another (at least in the short term). Sometimes these benefits or disbenefits provide advantages only to a small number of people such as the owners selling a GI services enterprise. Sometimes the benefits accrue to the whole of society, though at some cost to something else that society holds dear. In some cases the benefits take a long period to appear, whereas costs are experienced immediately. And finally, the balance between private gain and public benefit that is acceptable differs in different societies; there is a geography of trust, privacy, openness, tolerance, and commercial exploitation founded on different cultures and laws.

> **Everything is interconnected; usually, some gain and some lose when a decision is made.**

In the real world, rarely is geography the *only* important issue with which managers have to grapple. And rarely do managers have the luxury of being solely in charge of anything or have all the information needed to make truly excellent decisions. The reality is about operating in a situation where knowledge of the options and the likely outcomes of different decisions is incomplete. Uncertainty of many types is inevitable, and hence risk exists and needs to be mitigated (see Chapters 5, 16, and 19).

## 17.1.2 Organizational Drivers

Organizations and individuals operating in different sectors have some shared and some different drivers for their actions. Table 17.1 summarizes these in a simplified way for the business and government sectors.

Over the past 20 years we have seen some breakdown of previously discrete sectors. There is some convergence between the activities and operations of commerce and industry, government, the not-for-profit sector, and academia. Some governments have outsourced some of their functions (e.g., the operations of utilities). Increasingly,

**Table 17.1** Some business drivers and typical responses. Note that some of the responses are common to different drivers, and some drivers are common to different types of organization.

| Sector | Selected Business Drivers | Possible Response | GI-Related Example |
|--------|---------------------------|-------------------|--------------------|
| Private | Create bottom-line profit and return part of it to shareholders. Build tangible and intangible assets of firm. Build brand awareness. | Get first-mover advantage; create or buy best possible products, hire best (and most aggressive?) staff; take over competitors or promising start-ups to obtain new assets; invest as much as needed in good time; ensure effective marketing and "awareness raising" by any means possible; reduce internal cost base. | Purchase and exploitation of MapGuide software by Autodesk and subsequent development—one of the earliest Web mapping tools. |
| | | | Engagement of Esri in collaboration with educational sector since about 1980, leading to 80%+ penetration of that market and most students then becoming Esri software-literate. |
| | | | Purchase of GDT by Tele-Atlas to obtain comprehensive, consolidated U.S. database and to remove major competitor. |
| | Control risk. | Set up risk management procedures, arrange partnerships of different skills with other firms, establish secret cartel with competitors (illegal)/gain de facto monopoly. Establish tracking of technology and of the legal and political environment. Avoid damage to the organization's "brand image"—a key business asset. | Typically, GI service firms will partner with other information and communication technology organizations (often as the junior partner) to build, install, or operate major information technology (IT) systems. |
| | | | Many GI system software suppliers have partners who develop core software, build value-added software to sit "on top," act as system integrators, resellers, or consultants. |
| | | | Data creation and service companies often establish partnerships with like bodies in different countries to create pan-national seamless coverage. |
| | | | Know what is coming via network of industry, government, and academic contacts. |
| | Get more from existing assets. | "Sweat assets," e.g., find new markets, which can be met from existing data resources, reorganized if necessary. | Target marketing: use data on existing customers to identify like-minded consumers and then target them using geodemographic information systems. |
| | Create new business. | Anticipate future trends and developments and secure them. | Go to relevant conferences and monitor developments, for example, via competitors' staff advertisements. Buy start-ups with good ideas (e.g., Esri and the City Engine technology). Network with others in GI industry and adjacent ones and in academia to anticipate new opportunities. |

Table 17.1 (*continued*)

| Sector | Selected Business Drivers | Possible Response | GI-Related Example |
|---|---|---|---|
| Government | Seek to meet the policies and promises of elected representatives or justify actions to politicians to get funding from taxes. | Identify why and to what extent proposed actions will have an impact on policy priorities of government and lobby as necessary for tax appropriations. Obtain political champions for proposed actions: ensure they become heroes if these succeed. | Attempts to create national GI-related strategies and capacities, e.g., via NSDIs (see Section 18.6.1). Force the pace of progress on interoperability, e.g., to meet the needs of homeland security, minimize environmental hazards such as flooding. Bring data together from different government departments to demonstrate how funding is distributed geographically in relation to need or agreed political imperative (e.g., additional UK funding per capita in Northern Ireland). |
|  | Protect citizens from threats, e.g., war, crime. | Ensure equipment, military and other skills and manpower, and information infrastructure is adequate to warn off aggressors or triumph in conflict. | Obtain surveillance capabilities such as UAVs (Section 17.5.1), build integrated information infrastructure using geographic locators to link information derived from SIGINT, HUMINT, OPENINT, etc. (Section 17.5.2). Ensure that human capital is adequate to keep up with fast-moving developments in GI-related areas. |
|  | Provide good value for money (VfM) to taxpayers. | Demonstrate effectiveness (meeting specification, on time, and within budget) and efficiency (via benchmarking against other organizations). | Constant reviews of VfM of British, Norwegian, and Swedish government bodies (including British Ordnance Survey, Meteorological Office) over 15+ years, including comparison with private-sector providers of services. National performance reviews of government (including GI system users) in the United States and the impact of e-government and other initiatives. |
|  | Respond to citizens' needs for information or for enhanced services via the Web. | Identify these; set up/ encourage delivery infrastructure; set laws that make some information availability mandatory. | Setting up the U.S. National Geospatial Clearing House (1992), Geospatial One-Stop (2001), Geospatial Line of Business (2006), Geospatial Platform (2010), and Geospatial Shared Services (2013) + equivalent developments elsewhere (Chapter 18). Seek to create international level playing field for information trading, for example, through European Union INSPIRE initiative (see Section 18.6.2). Hold competitions to obtain new ideas from innovators (typically over half the entries use GI). |
|  | Control risk. | Avoid Congressional/ Parliamentary exposure for projects going wrong and media pillorying. | Get buy-in from superiors at every stage. Adhere to risk-management strategies and processes. Do numerous pilot studies in several areas. |
|  | Act equitably and with propriety at all times. | Ensure that all citizens and organizations (clients, customers, suppliers) are treated identically and that government processes are transparent, publicized, and followed strictly. | Treat all (including freedom of information) requests for information equally. Put all suitable material on Web but also ensure material is available in other forms for citizens without access to the Internet. |

everyone is concerned with explicit goals (many of which share similar traits) and with knowledge management. Moreover, in the GI world at least, we also see significant overlap in functions. For instance, both government and the private sector are important producers of geographic information. The not-for-profit sector increasingly acts as an agent for or supplement to government and often operates in a very business-like fashion. The result is that previous distinctions are becoming blurred, and movement of staff from one sector to another is becoming more common.

In practice then, virtually every organization now has to listen and respond to customers, clients, patients, or other stakeholders in their area of operations and find new ones. Every organization also has to pay attention to citizens whose power sometimes can be mobilized successfully against even the largest corporations: In the Internet era, it is easier than ever for citizens to provide feedback, such as volunteered geographic information (VGI; Sections 1.5.6 and 17.3.1.3). Every organization has to plan strategically and deliver more for less input, meeting (sometimes public) targets (e.g., profitability or service quality). Everyone is expected to be innovative and to deliver successful new products or services much more frequently, cheaply, and rapidly than in the past. All managers have to act and be seen to be acting within the law, regulatory frameworks, and some conventions. Finally, everyone has to be concerned with risk minimization, knowledge management, and protection of the organization's reputation and assets. Failure to do so can have disastrous organizational and individual consequences.

The emphasis placed on these objectives, however, differs significantly between organizations in different sectors. This translates into the parameters defining the decision-making space within which an individual operates (see Table 17.1). Thus commonality between the different sectors should not be exaggerated. For instance, mergers and acquisitions are more common in the commercial than the government sector. One example of this is the purchase of Navteq, the car guidance data supplier, by Nokia, the Finnish mobile telecommunications business, in July 2008 for $8.1 billion. Another is the $1bn purchase by Google in June 2013 of Waze, an Israeli creator of a traffic and navigation app for smartphones. Yet another is the merger of GeoEye and DigitalGlobe, two major providers of fine-resolution satellite imagery, in January 2013; this was driven by cutbacks in the U.S. federal government's budgets. But, irrespective of such acquisitions being primarily a feature of the commercial world, it is still realistic to regard all good organizations as operating in a business-like fashion.

For these reasons, we use *business* as a single term to describe the corporate activities in all four sectors identified earlier—commerce and industry, government, not-for-profit, and academia. Accordingly, we think of the GI world as being driven by organizational and individual objectives, using scientific understanding and raw material (data, information, evidence, knowledge, or wisdom; see Section 1.2), tools (GI system software and hardware), and human capital (skills, insight, attitudes, and experience) to achieve them.

## 17.2 Information as Infrastructure

The importance of maintaining a sound physical infrastructure—roads, railways, utilities, and so on—is well recognized. Thus the interstate highways of the United States transformed commerce, employment, and recreational travel from the 1950s on, and China's sustained growth since the 1990s has been driven by investment in physical infrastructure. In times of recession, many governments across the world see suitable "shovel-ready" infrastructure projects as a means of stimulating the economy and generating long-term efficiencies. Despite this, relatively few governments have seen the parallel between physical and information (content) infrastructures. We believe this to be a close (but not exact) parallel.

Yet a coherent interoperable national information infrastructure (NII)—of which GI is a key part—brings substantial economic benefits and competitive advantage. Such NII must utilize strategically important datasets to meet current and near-term future needs in government and commerce, consistent with relevant policies, procedures, standards, directories, metadata, tools, user guidance, and skill sets. Increasingly it will include information derived from the entire public sector—not just central or federal government—and the private sector.

The case for ensuring the existence, coherence, and quality of a national information infrastructure is made more, rather than less, important by the data deluge (see Table 1.2). Changing technology has enabled us to collect ever more data. It was claimed by reputable scientists in 2012 that 90% of all information ever created had been collected in the previous two years. Not all of this is of high quality. Some is of unknown provenance and has little metadata (Section 10.2). Examples of large-volume GI include point clouds from LiDAR, satellite imagery, and cell phone and credit card records geotagged by use of the Global Positioning System (GPS; Section 4.9 and Box 17.7).

Few of us live in some socialist planned economy where the state decides everything. The private sector is a crucial player in creating information and exploiting it. Nevertheless the state has a role to ensure that certain key information sets exist—core reference data—and that need to be of good quality and widely (arguably freely) accessible. These act as frameworks to which other data are attached (see Section 17.2.1.4).

There is a precedent for such action: The GI community and many national governments pioneered the concept of a certain subset of NII through the development of what have been termed national spatial data infrastructures (NSDI). These are discussed in more detail in Section 18.6.

But the information community, of which the GI community is one part, is not just a random collection of sets of information and tools to handle them. Because there are many interactions between what is available and other factors, we are dealing with an ecosystem. It has biotic components (humans and their skills) and abiotic ones (data, interdependencies, network of communications, institutional arrangements). It is strongly influenced by internal and external factors such as leadership, access to data (analogous to nutrients), and financing. And it is growing in size and mutating in other characteristics. Like other ecosystems, its future is difficult to predict in any detail even though it is reproducing rapidly at present.

## 17.2.1 Information for Management

Information, as seen from a management perspective, has a number of unusual characteristics as a commodity. In particular, it does not wear out through use, though it may well diminish in value as time passes. On occasions its value may rise again somewhat, where it is used for historic comparisons. The U.S. government decision in 2008 to make the entire Landsat archive freely available created an opportunity for researchers across the world to analyze temporal changes in the environment over a forty-year period (see, for example, Section 19.6.7).

**Information does not wear out but may become outdated.**

### 17.2.1.1 Information as a Public Good
Information is in general a public good. A pure public good has very specific characteristics:

- Even though the initial cost of collecting, quality assuring, and documenting information may be very high, the marginal cost of providing an additional digital unit is close to zero. Thus, in effect, copying a small amount of GI adds nothing to the total cost of production (see Section 1.2); however, where large datasets and high response rates are involved, the costs (and rewards) of computing, storage, and power for data analysis, dissemination, and exploitation may still be significant. One

example is high-frequency trading in the foreign exchange markets where close geographic proximity of the source of release of market-sensitive data (e.g., government statistical institutes) and the location of analysis tools can provide an opportunity for algorithms to trade many millions of dollars in very short time windows (milliseconds or less). Such activity would ensure that there is not a level playing field in a market where over $4 trillion is traded every day.

**The marginal cost of providing an additional digital copy is close to zero.**

- Use by one individual does not reduce availability to others (termed nonrivalry). This characteristic is summarized in the famous Thomas Jefferson quotation: "He who receives an idea from me, receives instruction himself without lessening mine; as he who lights his taper at mine, receives light without darkening me."

**The use of information by one individual does not reduce availability to others.**

- Individuals cannot be excluded from using the good or service (termed nonexcludability). Ways of achieving such a ban include designating information as restricted (e.g., the habitats of endangered species) or through pricing mechanisms.

In practice, information is an optional public good, in that—unlike national defense—it is possible to opt to take it or not; not everyone chooses to use freely available U.S. Geological Survey data, for example! To be pedantic, it may also be best to define information as a quasi-public good because it may be nonrival (see second bullet point listed earlier), but its consumption can in certain circumstances be excluded and controlled. The business cases and vast investments of a number of major commercial GI purveyors such as GeoEye/DigitalGlobe are based on this proposition. If everything they produced could be copied for free and redistributed at will by anyone, their business would be at risk (but different business models are used by other commercial players such as Google, as discussed in Section 18.2).

Thus the monetary value of information may depend on restricting its availability, whereas its social value may be enhanced by precisely the opposite approach—another trade-off. To complicate matters, a particular set of information is also often an "experience good" that consumers find hard to value unless they have used it before.

### 17.2.1.2 Externalities
An externality is a cost or benefit resulting from an activity or transaction that affects an individual or

community without their direct involvement. A pure public good is a special form of externality. Negative externalities arise where production or consumption of a good (in this case, information) by one agent imposes unavoidable costs on other producers or consumers. Pollution of water or air and pollution by ambient noise are classic examples of externalities arising from external costs—that is, disbenefits. Figure 17.2 is of four video-clips showing the dispersion of atmospheric release of caesium-137 in

**Figure 17.2** A to D The progress of the radioactive plume after the Chernobyl nuclear accident—a negative (geographic) externality. (Source and copyright: Institut de Radioprotection et Sûreté Nucléaire (IRSN), France) (*continued*)

(A)

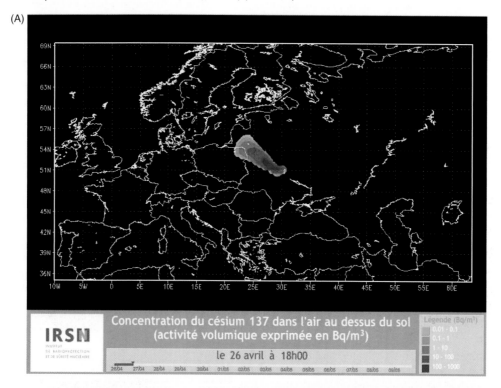

(B)

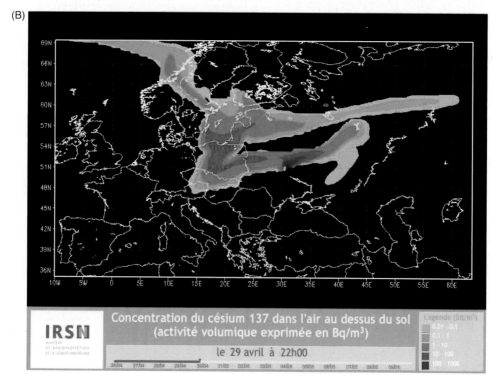

(C)

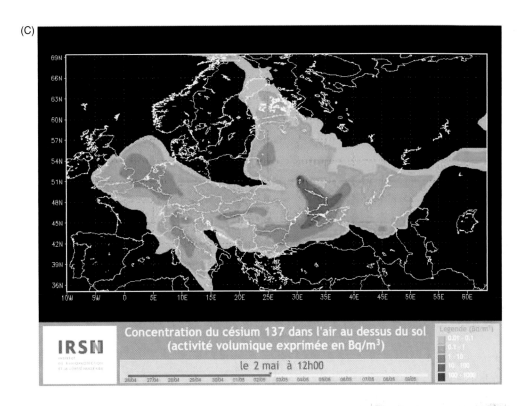

(D)

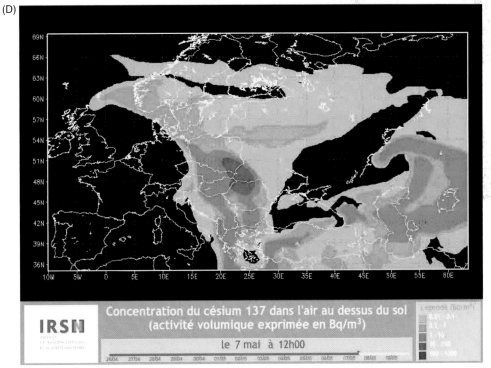

**Figure 17.2** (*continued*)

becquerels/m³ from Chernobyl. The model calculated the distribution of air contamination at ground level on a European scale at 15-minute intervals, from 26 April to 10 May 1986. There was good agreement between calculations and ground measurement data.

Classic examples of positive externalities (benefits) include refuse collection, education, and public health. Although individuals who benefit from positive externalities without paying are considered to be free-riders, it may be in the interests of society to encourage free-riders to consume goods that also generate substantial external benefits for others.

There are different types of externalities in the information world:

- Ensuring consistency in the collection of information creates *producer externalities* by reducing the costs of creating and using data. This in turn broadens the range of potential applications.
- Providing users access to the same information produces *network externalities*. It is, for example, desirable that all the emergency services use the same geographic framework data.
- Promoting the efficiency of decision making generates *consumer externalities*. For example, access to consistent information allows pressure groups to be more effective in influencing government policy or monitoring activities in regard to pollution.

### 17.2.1.3 Price Elasticity and Commoditization

If information is not free to the user, price elasticity becomes an important characteristic. The demand for a good is said to be *inelastic* when changes in price have a relatively small effect on the quantity sought. Hence for some users it is critical that they must have the most up-to-date or highly detailed GI, such as emergency services. The demand for a good is said to be *elastic* when changes in price have a relatively large effect on the quantity sought. Individuals might buy many different hiking maps or datasets if the unit price declined the more they bought. Substitution effects come into play if the price is too high, when users will seek alternative sources of information.

A *commodity* is a good—such as coarse-resolution GI—that is inexpensive and extremely widely used. This has impacts on the business model of the seller (see Section 18.2). The user's choice of what to purchase is then influenced by perceptions of reliability and quality, the nature of the marketing, and convenience.

### 17.2.1.4 Distinctive Characteristics of GI

GI has the following characteristics, which differ in degree from other data:

- Some GI acts as frameworks to which other GI are fitted; many call these core reference data. The most basic frameworks are geodetic data and coordinate systems (Section 4.8). For many organizations and users, however, the everyday framework consists of geocoded post- or zip codes, official topographic maps and imagery,

plus other GI like the distribution of population derived from censuses, etc. All such data have been sampled (Section 2.4), generalized (Section 3.8), and projected (Section 4.8) in ways that ensure that they may differ from other apparently similar GI. Where a high-quality, maintained, and widely available data framework exists, there are hugely positive network externalities (see Section 17.2.1.2) in fitting other data to it. The more people do this, the better datasets will fit each other. There is, however, a complication. In the past most of these frameworks were provided by governments. The advent of such products as Google Maps and Google Earth has changed the situation; their global availability and free noncommercial use have made them a de facto framework used by many organizations on which to "pin" their own data (see Section 4.12).

**The more people who use a framework dataset, the better datasets will fit each other and the more one source of uncertainty is reduced.**

- Linking multiple datasets provides added value, at almost no cost. The number of possible overlays of two datasets rises very rapidly as the number of input datasets rises (Figure 17.3). Assuming order is not important and repetition is not allowed, one combination exists of two variables, and 1,048,555 exist with 20 variables. Many of the latter are of course likely to be of little value (i.e., they include all groups of 2, 3, 4 . . . 20 variables). Such data linkage necessitates a common linkage key. In many cases the location of the data objects provides that key. Thus many more applications can be tackled and new products and services created using linked GI than when they are held separately as multiple files.
- It is particularly difficult to quantify the quality of some types of GI, for example, area classification data like soil type. This has ramifications when combining data by overlay (Section 5.4.4) and in modeling. Quite often we have to use proxy measures of quality, such as the established reputation of the data creator in deciding which data to use. As in almost all GI system operations, we are wise to vary the inputs systematically in sensitivity analyses to see whether the results are stable (see Section 15.5).
- There is huge geographic variation in the need for and use of GI. Thus GI for urban areas is much more heavily exploited than that for rural areas—in the main because more than half of the world's

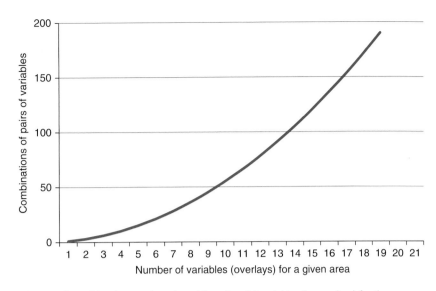

**Figure 17.3** The numbers of pairs of variables that can be selected from 2 to 20 variables (= overlays) for the same area.

population now lives in cities but the urban extent of cities is probably less than 2.7% of the land area (excluding Antarctica). Nevertheless, even where there is no commercial case for provision of data in some rural areas, there is sometimes a strong social case. This may therefore require government support. Figure 17.4 makes this point: when Pan American flight 103 was brought down over Scotland by a bomb in 1988, wreckage was scattered over a 40-km-wide area. The emergency services immediately needed detailed and up-to-date maps to ensure they missed no areas in their hunt for bodies and wreckage. Such rural mapping is difficult to cost-justify because of the small number of users in normal circumstances. Here it provided a form of insurance—vital in that case.

**Added value—almost for free—can be created by data linkage using location as the linkage key.**

### 17.3 Different Forms of GI

The characteristics of available information and GI in particular shape the opportunities and pitfalls that GI system users face. We have already described some classifications of GI (see Sections 3.4 and 3.5). But all classifications are approximations to reality and are best tailored to particular purposes: none is universally useful for all purposes. Thus far we have tended to think of GI as highly structured data, typically

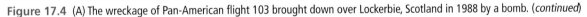

**Figure 17.4** (A) The wreckage of Pan-American flight 103 brought down over Lockerbie, Scotland in 1988 by a bomb. (*continued*)

(A)

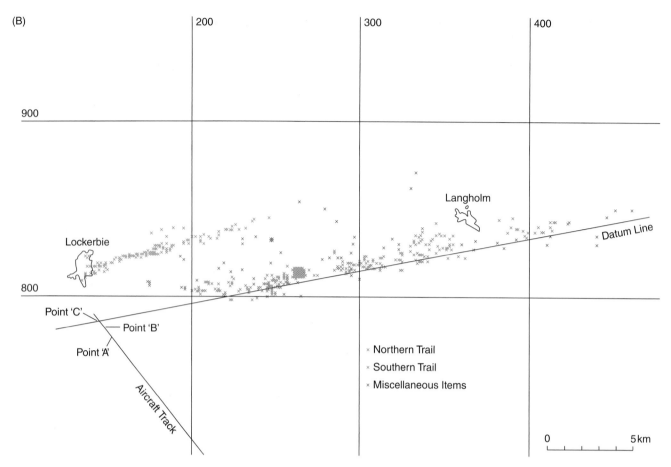

**Figure 17.4** (*continued*) (B) Map of the debris trail; grid squares shown are 10 km in size. Detailed and up-to-date rural mapping facilitated collection of evidence from which the cause of the event was deduced. (Source: UK Air Accidents Board. Crown Copyright. Open Government License v2.0)

arranged in rows and columns (where each row relates to a discrete physical or human-defined object and the columns contain attributes of that entity) or as "spot samples" of a continuous field (Section 3.5.2). This is in reality a gross simplification; landforms are not always continuous, sometimes containing discontinuities or cliffs. The extent and naming of area objects is often inherently ambiguous (Chapter 4) and sometimes highly contentious (for example, the boundaries of disputed places such as Kashmir; see also Box 12.1); wars have often been triggered by such disputes.

The reality is that GI now exists in a growing number of different forms, each with different characteristics. Set out in Table 17.2 is a classification that we use in considering how GI is used in various domains. The first two rows are "traditional geographic information" dealt with in more detail earlier in this book; the others are more novel but increasingly of real significance. "Big Data" (Box 17.2) is often seen as encapsulating such diverse data types.

We obviously need to ensure our tools can deal with different forms of GI, but we also need to understand the implications for their use. As will become obvious, the trade-offs between the variables that may be collected, the accuracy, the spatial granularity, and the currency of GI change when administrative data and VGI replace traditional data collection mechanisms like surveys.

**Trade-offs between variables collected, accuracy, spatial granularity, and currency change when administrative data and VGI replace surveys.**

It is important to understand that transformations are often applied to a number of the data types in Table 17.2 to convert them into another type, as discussed in Section 12.3. This facilitates their visualization or comparison with other data. Thus it is relatively easy to convert suitably georeferenced lists of the locations of the homes of human individuals into aggregate data in field

Table 17.2 Classification of GI for the purposes of Chapter 17. Note that all categories below may be produced by official sources or commercial enterprises or provided by volunteers. Note also Section 2.4 on sampling, and Chapter 3 on representation.

| Type | Subtype | Example(s) | Comments |
|---|---|---|---|
| Aggregate (1) | Continuous fields | Landscape/digital elevation model, remote sensing imagery | Usually sampled as array of heights or reflectance values or patchwise formulae (implicit $x$ and $y$) |
| Aggregate (2) | Collection of individuals in area aggregate, each area considered as a discrete object | People in a given area | May be converted into a field, e.g., a density function |
| Lists of individuals | Individuals (people, fauna, etc.) considered as discrete objects | Each in form of $(x, y, z_1$ to $z_n)$, e.g., characteristics and locations of human individuals, dwellings, or fauna at defined times | Can sometimes be highly sensitive because of privacy or national security considerations (Chapter 18) |
| Ambient and remote sensor data | | Sensor information integrated to enable safe use of driverless cars | Key sensors involve measuring proximity to other vehicles, position on the road, etc. |
| Photographs of places: geopics | | Technically similar to aerial or satellite imagery, but geometry often more complex and geography may be implicit rather than explicitly defined, | IARPA research project to geotag predigital imagery (Section 17.3.2) |
| Geovideos | | e.g., Street View | Mostly just used for visual inspection? Limited use if these are out of date. Locally sourced street imagery is widely used in real-estate marketing and for monitoring the safety and insurance risk of closed premises |
| Verbal descriptors, including geography | | Free text, e.g., certain historical novels, diaries | Extracting geographic location and descriptions of places into more conventionally structured form handled by GI systems can be tricky |
| Aural geographies | | Spoken descriptors of place or people in the place; may well be in local language | Cultural-specific meanings may well be embedded. Can be very important to military activities. |

## Technical Box (17.2)

### Big Data and Science

The exponential growth in data collected has already been described (see Chapter 1). This has been going on for some time. Librarians and others have argued since at least 1944 that data volumes would outstrip capacity to store them—based on the observation that U.S. university libraries were then doubling in size every 16 years. The shift from analog storage—99% of all storage capacity in 1986—to digital (94% in 2007) has transformed the situation, at least temporarily.

There are many different definitions of Big Data, but the most widely used one is summarized by the 3 Vs—data volume, data velocity, and data variety. The first is simply defined as being too large to handle by standard contemporary analytical tools; the second is about how fast data are being collected; and the third is a way of describing the many different forms of data which are used—structured and unstructured (the majority), which are held in different types of databases as text documents, emails, imagery, videos, and much else.

▶

It is clear that the definition of "Big Data" is subjective and relational, that is, what are Big Data will differ from one organization and perhaps even one sector such as multinational science to another such as local government.

Problem solving is all about separating the signal from the noise in data, usually statistically (see Further Reading). The essence of the case for Big Data is that data are now widely available, computer power is cheap, it is readily possible to hunt for associations between variables, and these give clues to the signal. Indeed Mayor-Schönberger and Cukier have claimed that:

- We can usually dispense with the need for new sample surveys (the cornerstone of most official statistics) given that we now often have access to huge existing volumes of data.

- Causality is less important than easily computed correlations based on large data volumes (from these we can predict the future provided we recompute frequently to cope with any change in the underlying relationships between variables; see also Chapter 1).

- We must get used to accepting "messiness" rather than expecting or searching for "privileged exactitude" from our data.

Much of this is uncomfortable for classical analysts brought up on the importance of sampling and identifying and understanding possible causal links. Fortunately it is also exaggerated and even misleading. Many Big Data sets do not cover entire populations (e.g., in the UK health retailers issue loyalty cards mostly to female customers) and hence the data are not representative of the whole population. It is also nonsense that significant imprecision can always be tolerated. In applications such as allocating resources to poor families, precisely matching what the law states is essential. Nevertheless, proponents of Big Data raise an important question in asking, "How good is good enough?" Answering the question is rarely easy—especially with GI—and ultimately involves judgment founded on a very clear idea of the users' needs, intimate understanding of the characteristics of the data, professionalism, and a code of ethics (see Section 18.5.2).

Finally, Big Data and Open Data only partially overlap. The first is predicated upon data volume, complexity, and massive analyses. The second is characterized by ease of access and reuse of data without significant penalties or difficulty. Open Data (see Section 17.4) can exist and be valuable without being voluminous. Many Big Data are held by private-sector firms like Google, Amazon, and satellite imagery companies, but the public sector—especially big science experiments—also hold and analyze colossal data volumes including GI pertaining to the 510 million km$^2$ of the Earth's surface and the 7 billion people living upon it in 2013.

form (i.e., a variable with a value everywhere in two dimensions). Overlaying multiple aggregate area datasets, each with different sets of boundaries, or combining one with a field variable is another matter (Section 13.2.4). The mechanics of the first process are embedded in most GI systems but this requires a set of assumptions about the spatial characteristics of the measurements within each of the areas. In some population-related cases this is facilitated by knowledge of the location of urban and rural areas (Section 12.3.3). In all such cases, however, uncertainty is being introduced into the newly created data in addition to what existed in the original dataset.

The characteristics of "traditional GI types" have already been described earlier in this book (Chapter 2). Now we consider those data types where changes in technology, use, and policy make them attractive for particular uses and where new problems arise.

### 17.3.1 GI About Individuals

Area-based aggregate data are relatively easy to handle but suffer two related methodological problems: the Modifiable Areal Unit Problem (MAUP; see Section 5.4.3) and the Uncertain Geographic Context Problem (UGCoP). The first is where any analyses of data (for example, correlation structures between geographic variables) are affected by the particular zoning system used. The second is where the analytical results are affected by the extent of any mismatch between the reporting zones used and the neighborhoods in which the data subjects are operating and whose properties are influencing their actions. Census tracts and city blocks are, for instance, convenient data-reporting geographies but do not necessarily reflect functioning neighborhoods. Dasymetric mapping (Section 12.3.3) seeks to reduce the latter mismatch but is normally based on simplistic constructs such as outlines of built areas.

For these and other reasons, where data about individuals is suitably geocoded and is available (e.g., not constrained by privacy considerations), it is much more valuable than area aggregate data (to which it may be readily converted). It can, for instance, be aggregated and mapped in multiple different zone systems to check the stability of the results. It is particularly valuable where it also records changes through time (such as changes of location of the individual and hence exposure to different environmental hazards; see Section 17.3.3.1).

There are three broadly different ways of collecting information about human individuals: through surveys to collect new information, by reuse of existing administrative data, or by summarizing volunteered information. We now review each of the subcategories.

### 17.3.1.1 Survey Approaches

National statistical institutes (NSIs) exist in almost all countries, and many have long collected much of their information by way of formal, carefully organized sample surveys. All such surveys produce estimates of the (unknown) true value and of their accuracy. Unfortunately some NSIs do not have strong independence from political pressure, and this can lead to biased results favorable to the government. The International Monetary Fund, for instance, has severely criticized the quality of the Argentinean inflation data. Just as serious, a widely observed characteristic of statistical surveys across the world is that response rates are falling as citizens become ever more loath to fill in government questionnaires. In the United States response rates on average have fallen by around 20% over two decades. Finally, high-quality surveys are expensive because they necessitate a large field force. For some purposes, such as polling, Internet-based surveys have proved valuable. But given all the above, it is no surprise that official bodies are taking increased interest in information already held for other purposes by government.

### 17.3.1.2 Administrative Data

Much information in contemporary society about individual people, and their actions and transactions, wealth, and relationships and about individual businesses is now collected through administrative systems. Detailed personal data exist in almost all governments (and in many commercial organizations). In many cases the individuals have an incentive to report any changes (e.g., if they have another child and this entitles them to additional benefits). Such detailed information is now increasingly aggregated to produce information used in GI systems.

As usual, there is a trade-off here. The actual or potential loss of privacy (see Section 18.4) through misuse or loss of personal data has to be traded against the considerable benefits to be gained from use of individual data. Potential benefits include the following:

- Reducing the burden on people or businesses to fill in multiple questionnaires, with data collected for one purpose being spun off from databases originally collected for another.

- Knowing where everyone is based and hence the location of potential victims in the event of natural hazards occurring.

- Ensuring that wherever an individual travels, his or her detailed health records could be available to any doctor in the event of a sudden illness or accident.

- Allocating resources (e.g., social security payments) on a fair basis related to the individual's characteristics.

- Reducing the incidence of fraud by comparing living standards and so on of each individual with the norm for people or businesses of their type, often by merging multiple administrative datasets together (e.g., tax records and social security benefits).

- Tracking the life histories and geographies of individuals to study correlations between, for example, exposure to environmental hazards and subsequent illnesses.

- Profiling people on the basis of their background, personal characteristics, or contacts as being predisposed toward acts of crime or terrorism or for marketing or credit-rating purposes.

- Studying inequality in society through analysis of life experiences by people in small, specifically defined groups (e.g., communities or ethnic groups) from which new policies and actions might flow.

A geographic (or other) code of some kind attached to each record is required to produce the aggregate information (when needed) from the individual's details.

Given all the preceding points, the combination of GI systems and personal data brings many benefits for decision makers. Yet it is obvious that these benefits are bought at a price, at least for some people. The downside is potentially that:

- An individual's deeply valued privacy may be compromised.

- Fear of misuse of the data may undermine trust in the organization collecting it.

- Errors in data linkage could lead to incorrect judgments and policies.

- Administrative data are normally collected to meet specific purposes and hence are classified accordingly. Unsurprisingly, therefore, apparently similar datasets show different things: crime information is typically collected both from police recording systems and also via a sample survey of the total population. Box 17.3 illustrates this situation of multiple conflicting sources of information and what needs to be done to harmonize information from different administrative sources.

- The body collecting administrative data may decide to change what is collected to suit governmental policy purposes. This may well cause discontinuities in time series valued by other users.

### 17.3.1.3 Volunteered GI

We have already discussed VGI in various sections. Such information can assume many different forms, may be highly variable in quality and coverage, and may pertain to the physical or man-made environment or people themselves. It ranges from the often (but not invariably; see Box 12.1) astonishingly good quality of much Open Street Map (OSM) information to the partial and inconsistent. It can often be difficult to ascertain the quality of VGI even if published guidelines exist for those compiling it. Such data

---

**Technical Box (17.3)**

### Beyond a Traditional Census

Censuses of population have been held in the U.S. since 1790, in Britain since 1801, and in many other countries over periods of 100 years or more. The results of the 2011 UK Census were used to allocate the equivalent of many billions of dollars annually between local governments, health authorities, and other arms of the state—as well helping us to understand societal change.

The cost of that census was about some $650m, and the first results appeared over a year after the census date. Unsurprisingly, cheaper and faster means of producing the required data were demanded by politicians. To explore what was possible, the UK Office for National Statistics set up a major program to study the combined use of administrative datasets held elsewhere in government, individual data held by private-sector bodies (where those were available), and novel forms of survey. The challenges faced were numerous—highly technical, legal constraints on data sharing and political antipathy plus coping with public concerns about privacy and feared misuse of their data for detecting fraud, and so on.

The results are very specific to the national context. In countries with existing public registers (see Section 18.4), many of these challenges do not exist. For example, neither the UK nor the U.S.—unlike many mainland European countries—have a registration system whereby individuals must reregister with the state on moving to a different house.

Using the detailed 2011 Census data as a benchmark, different approaches were trialed. The results of using snapshots for the same date as the census from two major administrative databases—the number of people registered with a physician and those on the social security register—in raw form are compared in Figures 17.5A and 17.5B. The differences of the matches to the 2011 Census in different areas reflect the presence of military personnel in some large clusters, the low propensity of young males to register with physicians until they are ill, and other factors. The team subsequently devised methods of coping with these different counts, demonstrating a good ability to match census counts of population, though collecting certain attributes of the population would still require some form of survey. Based on this and other research the UK government has agreed that the 2021 Census will be a hybrid of Internet-based survey, some field survey, and some administrative data. The aim is to reduce the survey components over the years afterwards.

This approach differs from many of the commonly cited Big Data stories in recognizing conceptual and measurement differences between different administrative datasets and seeking to harmonize them to enhance the quality of the final result.

▶

---

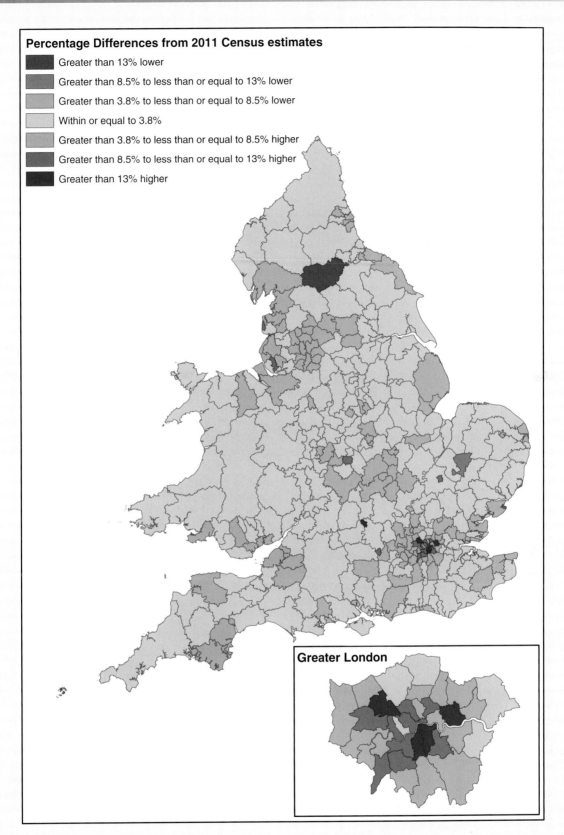

**Percentage Differences from 2011 Census estimates**

- Greater than 13% lower
- Greater than 8.5% to less than or equal to 13% lower
- Greater than 3.8% to less than or equal to 8.5% lower
- Within or equal to 3.8%
- Greater than 3.8% to less than or equal to 8.5% higher
- Greater than 8.5% to less than or equal to 13% higher
- Greater than 13% higher

**Greater London**

**Figure 17.5A** The geographic distribution of percentage differences between the 2011 Census population in local government areas in England and Wales and the totality of people registered with local physicians under the National Health Service on the same date.

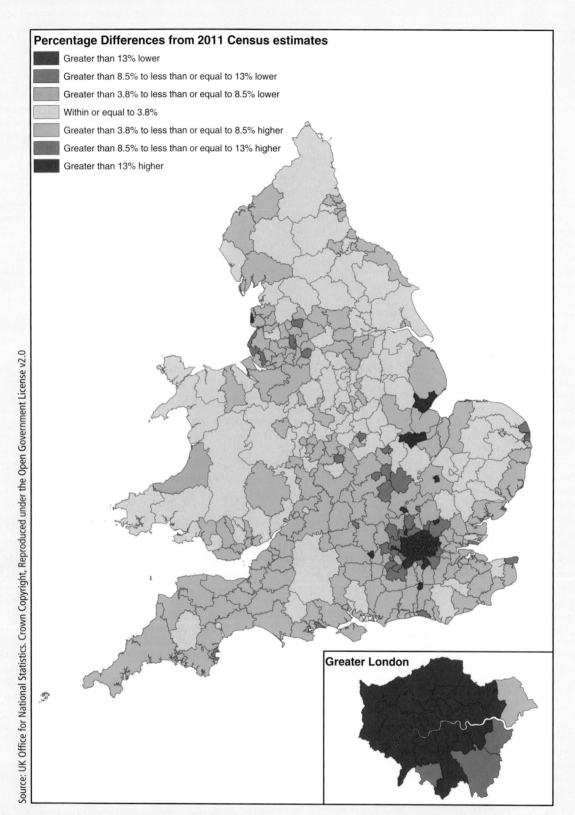

**Percentage Differences from 2011 Census estimates**

- Greater than 13% lower
- Greater than 8.5% to less than or equal to 13% lower
- Greater than 3.8% to less than or equal to 8.5% lower
- Within or equal to 3.8%
- Greater than 3.8% to less than or equal to 8.5% higher
- Greater than 8.5% to less than or equal to 13% higher
- Greater than 13% higher

**Greater London**

**Figure 17.5B** is an equivalent based on the numbers on the UK government's social security database at the same date. The gray areas are where the matches are within a tolerance of 3.8%. These are raw counts before modeling resulted in much closer matching.

**Figure 17.6** Open Street Map for part of Pyongyang, the capital of North Korea.

are often produced despite officialdom, rather than with its support or approval. Figure 17.6 illustrates a remarkable example of an OSM mapping of North Korea's capital city despite the regime's draconian security policies.

Here, however, we are mainly concerned with information about individuals, especially that volunteered about themselves. Perhaps the most active exploiter of high-frequency individual data is the retail industry (see Box 17.4). Individuals willingly provide personal data—some provided directly by them and much collected electronically from their purchasing habits—because of marketing incentives. Thus much use is made of store loyalty cards to analyze what many millions of individuals purchase, and where and when this is done. From this special offers are made, tailored to the individuals' interests, shopping habits, and activity patterns. The technology already exists to link mobile phone and credit card records of consenting individuals in order to monitor their movements and purchasing behavior—enormously valuable information to retailers. This technology and the information base

held by Google and others enables geographically contingent marketing to be tailored to individuals, based on past purchasing preferences, similar profiles to other consumers, the location of friends, or other factors.

What makes this "volunteered information" is ultimately the law. Individuals agree to disclose their information and allow it to be aggregated for commercial purposes by the act of signing a contract. Usually this is achieved simply by ticking a box to accept the (often unread) many pages of terms and conditions. For this, the individuals gain access to services and incentives but trade off details about themselves that they would often resent giving to the government. This contractual commitment permits the service provider to exploit the information so long as it does not breach national privacy laws.

Increasing amounts of VGI are also derived from aggregations of highly disaggregated data obtained incidentally from individuals sending messages via Twitter, uploading photographs to Flickr, or simply having their cellphones on.

### Exploiting Customer Data

Created by its eponymous founders in 1989, dunnhumby (www.dunnhumby.com) is now owned by Tesco, the world's third-largest retailer. The firm also provides similar services to other major retailers in 28 countries (e.g., Macy's in the U.S.). Globalization of retailing is ensuring that many cutting-edge practices are spreading widely, and dunnhumby's business is expanding accordingly.

The main service dunnhumby provides is based on the data mining of information from sales, loyalty card use, and other data to enable retailers to understand customers and their needs, wants, and preferences. The scale of the mining, analysis, and exploitation is huge: Tesco, for example, has 15 million regular customers, and the geographical and other patterns of their purchasing are created from their shopping records and responsiveness to tactical marketing initiatives. More than 30,000 Tesco products are categorized individually to help build up a "Lifestyle DNA Profile" of each customer. On

the basis of such analyses, each individual shopper is assigned to a group, and vouchers are printed for discounts on goods purchased in the past or for goods that other shoppers with similar characteristics have also purchased.

This results in a mailing to all 15 million Tesco Clubcard customers at least four times a year, with a summary of their rewards and vouchers tailored to encourage them to return and try new goods. Some seven million different variations of product offerings are made in each mailing. Customer take-up is between 20% and 50%, in contrast to the norm of about 2% in most direct marketing. The ranges of goods in store are also adjusted geographically in response to the habits of those who shop there. And the characteristics of new stores are planned on the basis of knowledge of people living nearby (including use of Census of Population and other externally provided data).

---

Figures 17.7A and 17.7B show fascinating spatial patterns at different scales. Such disaggregations can offer the potential for valuable insights into crowd behavior, the languages used by local residents, and even population migration (or at least local populations and those of visitors). The patterns, however, can be hard to interpret given the many interacting factors. Moreover for many purposes the data are only a numerator. Such raw data are dangerously seductive

because they are often wholly unrepresentative of the underlying population: women, for example, are much more likely to have health store cards than males, and the great bulk of tweets are created by people between 20 and 40 years of age. To assess the real significance we often need at least to standardize such data with a denominator such as the total local population with cell phones or those using Twitter and geotagging their Tweets. Correcting for this bias may

Courtesy: Eric Fischer

**Figure 17.7A** The European distribution of geotagged Twitter and Flickr locations. Red dots are Flickr picture locations; blue dots are locations of Tweets, whereas white areas posted to both. The distribution reflects differing penetrations of Twitter and Flickr use, varied base populations, holiday travel, and other factors.

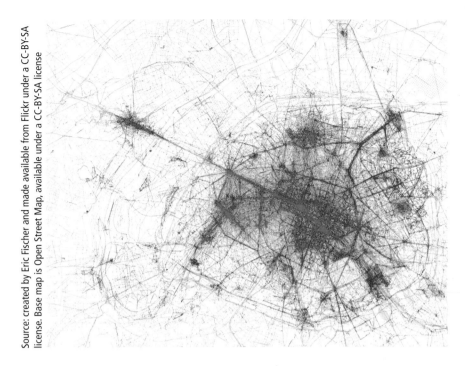

**Figure 17.7B** Visitors and locals in Paris. The latter (blue) are defined as not having taken any photographs in this area in the previous month, whereas visitors (red) congregate in tourist areas taking many pictures. Mixed areas are shown in yellow.

also require other information without the biases (e.g., official survey or administratively based information).

All this differs somewhat from crowdsourcing, where individuals select the cause to which they wish to contribute. This is now commonplace. One such GI example is the georeferencing of a variety of early historical maps, some dating back 400 years. Following the first public appeal by the British Library, some 700 maps from around the ancient world were geo-referenced in less than a week using current mapping and the local knowledge of volunteers. Results are displayed using Google Earth to overlay the trans-formed map on contemporary mapping or imagery.

In some cases, however, involuntary information is collected without knowledge of the owner or data subjects. Mostly this is achieved by scraping Web sites to extract information (Section 17.3.3.2). Though in principle copyright restrictions exist (see Section 18.3.2.1.2), the reality is that once material—text, numbers, or images—has been posted on the Internet, subsequent control is limited.

### 17.3.1.4 Aggregate Data from Synthetic Individuals

The global coverage, quality, and currency of some crucial datasets—especially those pertaining to human beings at fine spatial granularity—are highly variable. Population censuses provide perhaps the best overall source but are held infrequently, and the variables summarized are often inconsistent between nations.

As a result, some enterprising organizations have sought to create information about the numbers of people distributed across the Earth by synthesizing it from a number of variables, some acting as prox-

ies. The Landscan data product produced by the U.S. Oak Ridge National Laboratory GI system facility is a good example. This uses a mixture of data from many sources to model the global picture of ambient population (a 24-hour average) at 30 arc seconds resolution (c $1 \text{ km}^2$ at the equator). The input data include national midyear estimates of population published by the U.S. Bureau of Census, land cover, topographic mapping, administrative boundaries, and various forms of imagery data. The data are merged via a form of dasymetric mapping (Section 12.3.3). The Landscan team acknowledges that the multiple data sources vary considerably in characteristics between and within countries. So this may be the most consistent global information and the best population data that exists for some areas, but it is inevitably of highly variable accuracy.

A highly innovative example of synthetic individualized data relates to public health. In 2009, a new strain of flu H1N1 appeared in Asia. Health authorities were concerned that this would become a pandemic (see Section 19.6.3.2). At the time there was no effective known vaccine. A critical piece of information needed to take whatever action was possible was where the flu had spread; the sooner this was available the better. Typically quality-assured official statistics derived from reports by doctors take weeks to be available. In the United States, Google took the 50 million most common search terms in the three billion search queries they received each day. They then correlated the changing incidence of these search terms with the Centers for Disease Control's data files on the spread of seasonal flu between 2003 and 2008. They found that a small number of these search terms correlated with the spread of flu as recorded in official figures in 2007 and 2008. Using this "the past is the guide to the present" approach, they were able to describe the spread of flu in near real time. This begs the question of how good was the proxy? Was it good enough for the purpose? And will the past always be a good guide to the present? Part of the answer was provided in a paper in *Science* in March 2014, which showed that the Google Flu Trends tool was overestimating flu cases by 30% or more when calibrated against flu reports provided to the U.S. Centers for Disease Control and Prevention (CDC) by doctors (which had a 2-week delay). The cause seems to have been a combination of human factors—people searching for information when they only had flu or because of a "snowballing effect" on hearing others had found apparently good guidance—and methodological shortcomings of the algorithm. The overall conclusion seems to be that such approaches are best regarded as exploratory and that there is value in comingling Google trends and official data.

### 17.3.2 More Novel Forms of GI

The BBC Domesday Project of 1986 (Box 17.5) was a conscious attempt to capture details of the whole UK 900 years after the original Domesday Survey. It involved bringing together government data, crowd-sourced text and photographs, sounds, maps and imagery, and much else, and linked these on various criteria, especially location. It was the precursor for many contemporary Web-based systems.

Ever since Domesday, the forms of GI we seek to exploit have been multiplying. Table 17.2 highlights some of these. But the handling of some of these for certain tasks is often still difficult. For example, manual extraction of geographic characteristics from artistic images and written or (especially) spoken text is quite common, especially in regard to historical information. But automated extraction of such information for use in standard GI systems is difficult. Indeed, we have no agreed-upon tools and procedures for such automated extraction of "facts" from qualitative information, not least because this may be strongly influenced by cultural factors and language constructs (e.g., the huge range of words in the Inuit language for forms of snow). Potential applications of such GI are, however, widespread, including archaeology, historical comparisons, and in military intelligence. For instance the U.S. Intelligence Advanced Research Products Activity (IARPA) has advertised funding for research to find better ways to geotag predigital images.

### 17.3.3 The Changing World of GI

Here we highlight two major changes in recent years, over and beyond the growth of data about individuals.

#### 17.3.3.1 The Rise of Geotemporal Data in GI

Some datasets are collected much more frequently than others; for example, meteorological data are generally recorded much more often than people are counted. Such data introduce temporal as well as spatial sampling issues (see Section 15.1). Thus Figure 17.2 shows a rapidly mutating radioactive plume emanating from Chernobyl.

---

## Application Box (17.5)

### The Domesday Project

In November 1986 the BBC marked the 900th anniversary of the original Domesday Survey by William the Conqueror by carrying out a multisource survey of Britain. The result was arguably the first personal GI system. The BBC and its partners brought together 54,000 images (maps, photographs, satellite images), 300 mB of compressed official statistical data, and millions of words of text about 4 km by 3 km areas provided by one million members of the public—an early example of crowdsourcing. Included in the data were 21,000 files of spatial data showing national coverage down, in some cases, to 1 km$^2$ granularity. This data included geology, soils, geochemistry, population, employment and unemployment, agriculture, and land use/land cover. The different types of data were cross referenced by location or theme. Access to the data was by pointing on maps, inserting place-names or coordi-nates, or through use of a thesaurus. The file structures and location of the data were invisible to the user.

The platform used was the BBC microcomputer, which was then almost ubiquitous in British schools. The data were stored on a Philips Laser-Vision ROM. At the time the government's pricing policy was to charge for much of official data (Section 18.3.2.3.1). To have purchased all the official data held on Domesday would have cost about $375,000; the charge for the whole system to schools was $4500.

Domesday was essentially an early information infrastructure encapsulated in a box. Developments of the Internet and Web plus the huge increase in information available in digital form distributed across a multiplicity of sources rendered it obsolete. But it pioneered many of the concepts that we now take for granted.

In times past we have typically been content with—or at least had no option but to use—GI collected infrequently even if sometimes at fine spatial granularity. Thus most countries have traditionally updated their topographic mapping infrequently and collected detailed information about their population most commonly every 10 years. In some Western countries some of the population results are made available for hundreds of thousands or more small areas (such as city blocks). But this trade-off between space and time is now changing as new technology, changing user needs, and the public's growing unwillingness to fill in government forms evolves. Both official administrative data and volunteered GI provide the opportunity for increased frequency of information and greater analysis of geotemporal trends.

The most striking of these changes relates to transport data. Thanks to the ubiquity of smartphone apps and GPS, it is now routine in many jurisdictions to have access not only to public transport timetables but also to the actual location of the next bus, train, or plane in real time. In the U.S. and UK drivers may be charged for their insurance based upon where, when, and how they drive, computed using GPS in the car. This provides an incentive for careful driving. Nor is transport data only useful in individualized form. Computing speed of car movement from individual phone locations over time and aggregating these figures to provide advice to others approaching the same road gives a proxy for traffic congestion.

More directly individual is the use of GPS-enabled tags, often on the ankles, of those recently released

conditionally from jail. This enables officials to ensure these potential recidivists do not go to areas from which they are excluded and to correlate their location and time with the occurrences of newly committed crimes. For different reasons, tags may also be fitted to those who are suffering from mental illness or some other debilitating condition such as dementia. Here the aim is to protect them from becoming lost.

### 17.3.3.2 Private or Public Sector?

Until recently, most of the benefits of GI arose from government collecting, holding, and analyzing raw data. But the situation has changed dramatically. For example, using Web scraping technologies to monitor online prices every day, PriceStats—a spin-off from MIT research—collects consumer price information from about 900 retailers across more than 70 countries. It publishes daily inflation series for over 20 countries, including both developed and developing economies. The organization offers a commercial service and claims that the trends in its data are very similar to those of more conventional data series. In effect it is seeking to provide a substitute for traditional sources of data.

One area where the private sector currently leads is in "nowcasting." This involves defining the current situation some weeks or months before official data are produced. The first papers to suggest that Web search data might be useful in forecasting economic and other statistics were published in 2005. The technique, mostly employing Google Trends, has since been used in epidemiology (Section 17.3.1.4 ), consumer sentiment, retail sales, and housing applications. Figure 17.8 shows a comparison between two

**Figure 17.8** Comparison of the numbers of Polish people migrating to England between 2004 and 2011 derived from Google Trends and from the UK Labour Force Survey.

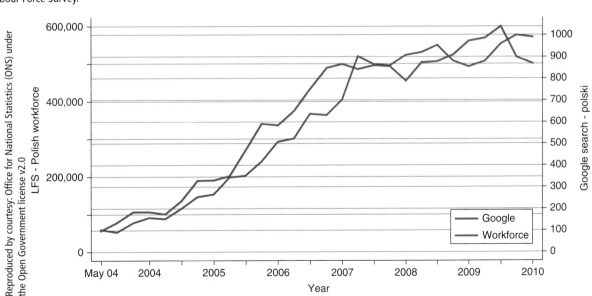

measures of migration—one based on conventional surveys, the other based on Google Trends.

All this raises a question of whether private-sector information is better able to describe many of the characteristics of human individuals and societies than traditional official statistics or other public-sector information. The answer is that it all depends—and is mutating. Certainly the social media tools provided by the private sector are now being used by billions worldwide. The information assembled using these tools is proving hugely valuable to the commercial sector. It is also of considerable interest to security organizations seeking to identify possible terrorists and to armed forces operating in foreign terrain.

But still greater benefits could certainly be realized if public- and private-sector information could be merged or comingled. The ability to merge customer preferences (e.g., for food, alcohol, or cigarettes) or spending patterns with health information could provide many benefits to the treatment of disease. Such cross-sector exchanges of data are fraught with difficulties given national legislation, commercial confidentiality, the nonrepresentative sample coverage of much commercial data, and the possible reactions of citizens and customers.

## 17.4 Open Data and Open Government

In Chapter 1 we introduced the concept of Open Data (OD); here we examine it as well as the drivers that have led to it becoming important in GI and far beyond in more detail. As ever, definitions vary between different organizations and authors. A simple description of Open Data, however, is "Data that can be freely accessed, used, reused, and redistributed without hindrance by anyone—subject only at most to the requirement for attribution and share-alike." Some bodies say that there may be some charge for OD, usually no more than the cost of reproduction.

Predicted benefits of Open Data—and prime drivers for political action—have included enhanced transparency and government accountability to the electorate, improving public services, better decision making based on sound evidence, and enhancing the country's competitiveness in the information business. The common finding worldwide is that GI is the cornerstone of success in making government information widely useful.

The U.S. federal government has always taken the view that the great bulk of information that it holds—other than that restricted for security or environmental protection purposes—should be made available at marginal cost or less and free of copyright restrictions

(Section 18.3.2.1.2). In fact, that country pioneered the concept of a national spatial data infrastructure (Section 18.6.1), in many respects a partial precursor of national information infrastructures now under discussion. President Obama launched an Executive Order authorizing Open Data on his first day in office in 2009; the number of OD sets has risen from 47 in March 2009 to over 156,000 in late 2014.

The global policy picture has always been much more complex and variegated but has changed very rapidly in some countries since about 2007. The most evident manifestation of this is the growth of the Open Data and Open Government movements. In Britain a series of bodies to represent the user needs in relation to additional datasets and to carry out research and entrepreneurial activities with start-ups was set up. By late 2014 nearly 20,000 datasets had been made available in the UK as Open Data under a standard, simple Open Government License. These included many datasets useful to GI system or service practitioners such as numerous mapping files produced by the national mapping organization and real-time (e.g., transport) information.

Related and contemporary developments have also taken place in many other countries. Perhaps the most obvious demonstration of the internationalization of the Open Data movement is the charter signed by the leaders of Canada, France, Germany, Italy, Japan, Russia, the United Kingdom, and the United States at the G8 meeting in June 2013. This commits those nations to a set of principles ("Open by default," "Usable by all," etc.), to a set of best practices, including metadata provision, and national action plans with progress to be reported publicly and annually. Other global bodies—notably the World Bank and the UN Economic Statistics Directorate plus some U.S. states and cities around the world—have also committed to the principles of Open Data and enhanced access to their data stores.

These developments have forced debates about the extent to which government departments can be compelled to make available information at a time of austerity, the extent to which a national information infrastructure and strategy is required, and the trade-offs between making available very "raw data" quickly with little quality assurance (described by some as "fly tipping," i.e., dumping trash illegally) or slower delivery and better QA and metadata. It is now widely accepted that there is also a shortage of data scientists who are able to manipulate such Open Data but also understand its domain-specific characteristics.

A related development is the Open Government Partnership (OGP). This is an international organization with some 55 countries founded in September 2011. The objective is to promote transparency, increase civic participation, fight corruption, and harness new technologies to make government more open, effective, and accountable. OGP is overseen by a multinational steering committee of governments and civil society organizations. To become a member of OGP, participating countries must embrace a high-level Open Government Declaration, deliver a country action plan developed with public consultation, and commit to independent reporting on their progress going forward. It is evident that progress can only be made through the use of widely accessible data—hence the link to Open Data and GI in particular.

### 17.4.1 The Metadata Issue

Throughout this chapter we have referred frequently to the quality and other characteristics of GI. Yet getting sound, consistent and user-focused metadata describing such characteristics has long been a problem. Previous schemes produced by government bodies have had limited success. Typically costs fell on the producers, whereas any benefits accrued to the users. One more recent Web-based scheme is the data certification mechanism proposed by the Open Data Institute.

One element of metadata concerns OD interoperability and avoidance of use of proprietary tools. Sir Tim Berners-Lee, founder of the World Wide Web, has argued strongly that all OD should be migrated toward more sophisticated formats to facilitate more effective use (Chapter 1). He devised a star-based rating system for summarizing the utility of Open Data, which is now in use in many public-sector organizations.

★ One star means the data are accessible on the Web. They are readable by the human eye, but not by a software agent, because they are in a "closed" document format such as PDF.

★★ Two stars mean that the data are accessible on the Web in a structured, machine-readable format. Thus, the reuser can process, export, and publish the data easily, still depending, however, on proprietary software like Word or Excel.

★★★ Three stars mean that reusers will no longer need to rely on proprietary software (e.g., use CSV instead of Excel data). Accordingly, reusers can manipulate the data without being confined to a particular software producer.

★★★★ Four stars mean that the data are now *in* the Web as opposed to *on* the Web through the use of a URI, a Uniform Resource Identifier. As a URI is completely unique, it gives a fine-granular control over the data, allowing for things like bookmarking and linking. An example would be an RDFa file containing URIs.

★★★★★ Five stars mean that the data are not only in the Web but are also linked to other data, fully exploiting the Web's network effects. Through this interlinking, data get interconnected whereby the value increases exponentially because they become discoverable from other sources and are given a context (e.g., through links to Wikipedia). An example would be an RDFa file containing URIs and semantic properties (allowing for linked data reuse).

It is relatively easy to imagine a corresponding ordinal scale describing the geographic characteristics of spatial data, such as the form of geocoding used.

---

## Biographical Box (17.6)

### Sir Tim Berners-Lee—Web founder and Open Data champion

A physics graduate of Oxford University, Tim Berners-Lee (Figure 17.9) invented the World Wide Web, an Internet-based hypermedia initiative for global information sharing while at CERN, the European Organization for Nuclear Research, in 1989. He wrote the first Web client and server in 1990. His specifications of URIs, HTTP, and HTML were refined as Web technology spread.

He is the 3Com Founders Professor of Engineering in the School of Engineering with a joint appointment in the Department of Electrical Engineering and Computer Science at the Laboratory for Computer Science and Artificial Intelligence (CSAIL) at the Massachusetts Institute of Technology (MIT). He also heads the MIT

Source: World Wide Web Consortium

**Figure 17.9** Sir Tim Berners-Lee.

Decentralized Information Group (DIG). He is also a Professor in the Electronics and Computer Science Department at the University of Southampton, UK.

Tim is the Director of the World Wide Web Consortium (W3C), a Web standards organization founded in 1994 that develops interoperable technologies (specifications, guidelines, software, and tools) for the Web. He is also a Director of the World Wide Web Foundation, launched in 2009 to coordinate efforts to further the potential of the Web to benefit humanity.

He has promoted Open (government) Data globally, persuading two successive British Prime Ministers (and many other senior players worldwide) to become its advocates on the basis of enhancing transparency, good governance, and the creation of new technology-based industries by entrepreneurs. Tim is president of London's Open Data Institute. Among the many awards that he has received was the newly founded Queen Elizabeth II $1.5m prize in 2013 in Engineering for "ground-breaking innovation in engineering that has been of global benefit to humanity." He shared this with Vinton Cerf, Robert Kahn, Louis Pouzin, and Marc Andreesen, fellow pioneers of the Internet or Web.

Source: Berners-Lee biography at www.w3.org/People/Berners-Lee/

## 17.5 Example of an Information Infrastructure: The Military

Thus far we have talked in the abstract or given short examples of how an information infrastructure is crucial to organizations or governments. We now give a more concrete example.

We can trace the beginnings of much national mapping to the needs of the military. This is not surprising: the first duty of the state is to protect its citizens and their interests. Sometimes such defense of national interests also involves operations far from home. Unsurprisingly, therefore, many countries maintain archives of current and historical maps of areas outside their own terrain. The Soviet Union had an extensive mapping program outside its own borders until its collapse. Effective defense necessitates the state having the wherewithal to make such protection effective, including a decision-making apparatus and an information infrastructure. Decisions, often based on the best available geographic information, are made from top to bottom of every military organization every single day.

This section is based on information in the public domain and relates only to developments in Western countries. It excludes any consideration of the substantial GI and GI technology activities of the security services, such as the U.S. National Geospatial-Intelligence Agency (NGA), which has overall responsibility for GEOINT (see Section 17.5.2) in the U.S. intelligence community. We can also be certain that the military information infrastructures in many countries are, together with those of the security services, by far the most extensive, sophisticated, multinational, and integrated of any in the world.

### 17.5.1 Technological Change and the Military

The most familiar role of the military is in warfare. There have been around 200 wars since 1945. Technology has long been influential in reshaping intelligence gathering and the way warfare is executed. It has radically changed mechanization and communications; the evolution of air power since 1918 has been a particularly important factor. In spatial terms these developments have collapsed distance strategically, operationally, and tactically. Now the entire world is under surveillance from satellites and some of it from unmanned autonomous vehicles (or UAVs; see Figure 17.10); armed drones are guided from command centers thousands of miles away. Elite infantry and armored units can be transported quickly by air and deployed in a matter of a few days yet be in constant touch with headquarters.

The changes have been so profound that some historians have argued that geographic space has ceased to be an encumbrance, let alone a friction. This extreme view ignores the reality of operating in hostile terrain, with equipment that does not always work as intended. Moreover the increasing need to engage verbally with local communities has been a feature of recent conflicts since tank warfare across the European plains has given way to urban and guerrilla warfare. Social science and mapping of community characteristics have become increasingly used to help military personnel understand, differentiate, and engage successfully with local populations.

Finally, underpinning all the obvious military tasks is a variety of other roles that require information of different types and latency. For example, estate

Source: (A) estt/iStockphoto; (B) Richard Watt/UK Ministry of Defence under the Open Government License 2.0; (C) U.S. Geological Survey

**Figure 17.10** A to C Examples of Unmanned (flying) Autonomous Vehicles used for imaging or other surveillance purposes. Some of these are now increasingly being used for a variety of civilian purposes as well as military ones.

management and logistics are of huge consequence. Running many military bases requires management of a complex array of resources, assets, facilities, lands, and services similar to those of a medium-sized city—and using GI systems in the same way as in the civilian environment. Getting stores (including ammunition) to the right area when needed is similar in principle to the role of transport logistics companies such as FedEx but complicated by the hostile environment in which the military sometimes has to operate.

## 17.5.2 The Military Information Infrastructure

Success in operationalizing the concept of command, control, communication, and coordination in military operations is substantially dependent on the availability of accurate GI, typically integrated from multiple sources, to arrive at quick operational decisions. Though interoperability is crucial, it is also essential that it does not extend to any enemy forces! Table 17.3 sets out a very simple summary

**Table 17.3** Levels of decision making in the military and information required.

| Level of decision making | Information required |
| --- | --- |
| Grand strategic | Overviews suitable for discussions with politicians (e.g., Common Operating Picture graphics) |
| Strategic | Synoptic views of assets and challenges to them plus plans for action (e.g., battle theatre plans) |
| Operational | Information needed by one-star general leading the activities of a regiment or brigade (battle scenario maps and graphics) |
| Tactical | Detailed information on the local area needed by front-line troops (e.g., tactical maps) |

of the types of geographic information needed by Western forces for different purposes.

As in other application areas, GI systems allow the user to capture, manage, analyze, visualize, and exploit geographically referenced information about physical features and much else. For instance in planning campaigns military field commanders would like to know terrain conditions, especially elevations, for maneuvering armored carriers and tanks and for use of various weapons. In addition, they need information on vegetation cover, road networks, and communication lines to deploy resources effectively. All these details must be available to field commanders in a coordinate system that matches with the equipment they use for position fixing. Any discrepancy in these inputs may endanger the operation and the lives of troops.

In military parlance, GI is often called geospatial intelligence, or GEOINT. This is normally defined as "information about any object—natural or man-made—that can be observed or referenced to the Earth and that has national security implications." It is derived from the full range of current and historical information sources described in earlier chapters.

In addition to GEOINT, signals intelligence (SIGINT), human intelligence (HUMINT), and open-source intelligence (OSINT) contribute to the military information infrastructure. These are most useful when geocoded and integrated with each other. SIGINT, for example, played a significant role in identifying the location of the leader of Al-Qaeda. HUMINT is intelligence derived from information collected and provided by human sources. Finally, OSINT is collected from publicly available sources, most typically now from analysis of material on the Internet, including social media. The integration of such a variety of information of very different characteristics and uncertainty is challenging but essential.

#### 17.5.2.1 MGCP: Example of Multinational Collaboration

Military organizations are typically parts of a major alliance; NATO is the most well-known example. This means that additional resources may be brought to bear in times of stress, including access to mapping and GI.

A striking example of such collaboration is the Multinational Geospatial Co-Production Program (MGCP). This program had 29 nations as members in 2013. It involves the creation and maintenance of fine-resolution vector data at accuracies equivalent to 1:50,000-scale mapping. The basic data unit or "exchange unit" is a 1-degree cell. Marketing-style incentives are used to maximize inputs of units by the partners to the database; the numbers of units that can be extracted by a member country rises on a sliding scale as the number they input increases. After inputting above 200 units, the donor country may access all the units available provided they also take part in quality assurance. Such an arrangement reduces duplication of effort and has been demonstrated to work effectively in emergencies. For example, the UK's Defence Geographic Centre now has the capability to produce standard map sheets from the data to an internationally agreed specification and distribute them to its military customers in less than 20 hours.

Whereas all systems are now digital, different sections of the armed forces require different products including paper-based ones. Those in armored vehicles or aircraft have the capacity to use heads-up and other digital displays. But for the infantry, any extra weight to be carried, the danger of being located by light emission, and the possibility of digital jamming all ensure that paper maps are often favored, especially with topographic maps and imagery arranged back to back.

The range, magnitude, complexity, and interrelationships of information that is within scope for military personnel and the frequent need for speedy analyses is such that "Big Data" (see Box 17.2) is inevitably of great interest to the armed services. Yet large-scale data mining and delivery capabilities have some challenging consequences. For instance, the wider the access to information the greater the likelihood of leaks of confidential information (such as appeared on Wikileaks).

### 17.5.3 Civilian Spin-Offs

There are at least three areas where spin-offs of military capability benefit the civilian sector. These are the development of new technologies, availability of GI for civilian purposes, and the use of military personnel in civil emergencies.

The most influential military technology by far used in the civilian sector is, of course, the Global Positioning System, or GPS (see Table 1.4, Section 4.9, and Box 17.7). This has totally transformed the entire GI industry through changing the way humans—and machines—can describe and interact with geography.

Substantial taxpayer investment in military R&D led to the development of ARPANET, the precursor to the Internet. The success of the U.S. dominance in

### The History of GPS

In the early 1960s the U.S. Department of Defense (DoD) began pursuing the idea of developing a global, all-weather, continuously available, highly accurate positioning and navigation system to meet the needs of a broad spectrum of military users. After much experimentation with different approaches, approval was given for full-scale development in August 1979. This was punctuated with problems: budget cuts led to the program being halted for two years.

The crucial demonstration of GPS's military value came in the 1990–1991 war in the Persian Gulf. It enabled coalition forces to navigate, maneuver, and fire with high accuracy in the extensive desert terrain almost 24 hours a day despite frequent sandstorms, few paved roads, no vegetative cover, and few natural landmarks. So effective was it that the DoD had to hastily procure more than 10,000 commercial units to meet field demands.

The first U.S. pronouncement regarding civil use of GPS came in 1983 following the downing of Korean Airlines Flight 007 after it strayed over territory of the Soviet Union. President Reagan announced that the Global Positioning System would be made available for international civil use once the system became fully operational. In 1991 the U.S. Federal Aviation Administration promised that GPS would be available free of charge to the international community beginning in 1993 on a continuous, worldwide basis for at least 10 years.

All this led to the expansion of the surveying market by the mid-1980s even while GPS was still in development and only a small number of operating GPS satellites were in orbit. Surveyors also pioneered some of the more advanced differential GPS techniques, such as kinematic surveying. This development generated commercial revenue for U.S. equipment manufacturers to invest in new developments—a positive feed-back mechanism. The later development into a ubiquitous and free service has transformed the lives of many of the world's people.

Source: *GPS history, chronology, and budgets* at www.cs.cmu.edu/~sensing-sensors/.../GPS_History-MR614.appb.pdf

---

commercial fine-resolution satellite imagery industry is due in part to the military being the guaranteed "anchor tenants." UAVs (Section 17.5.1 and Figure 17.10) are becoming very widely used in the civilian domain. The U.S. Geological Survey has used them for wildlife monitoring, checking fencing, and tracking fires; others have used them for collecting samples from volcanic eruptions, surveying, and much else.

Inherent in military operations is the need to create GI if that does not already exist. An early example of this was making publicly available the Digital Chart of the World in 1993. This was the first globally complete 1:1 million standard map series in computer form. Another example is the detailed 1:5000-scale mapping of all Helmand Province and of cities across Afghanistan. Such information (as well as physical) infrastructures created for war are potentially of great value in subsequent peacetime.

The military also provide support of and intimate interoperability with civilian agencies. Providing search-and-rescue missions after natural disasters is a common and vital role worldwide for military forces.

 ## 17.6 Conclusions

We have shown how understanding the characteristics of information, especially that of GI, is crucial to many everyday tasks. The situation is not stable: nontraditional forms of GI—such as detailed administrative or personal data—are evolving as technology and user needs mutate. Information from difference sources and with different specifications is being comingled, with implications for its safe use. More positively, many governments have come to realize the potential of Open Data and begun to commit to making it more of a reality: GI, especially core reference data, is central to its success. But success also depends upon the existence of enough suitably trained data scientists who combine analytical skills with domain knowledge. In particular, GI expertise is essential to make our visions a reality.

## Questions for Further Study

1. Suppose you are in charge of a business providing GI services or you are an army general. What kind of information and software tools would you need to make effective decisions?

2. What are the main characteristics of information, and how do these differ from those of physical goods? Are there ways in which GI characteristics differ from those of other information?

3. What is a national information infrastructure? Does it already exist? And why is it important?

4. Search the Web for national policies and activities on Open Data and summarize their differences. Why are there differences in progress between countries?

## FURTHER READING

Institut de Radioprotection et Sûreté Nucléaire (IRSN). 2011. *Chernobyl's accident.* See www.irsn.fr/EN/publications/thematic/chernobyl/Pages/overview.aspx and simulation.

Kolvoord, R. and Keranen, K. 2011. *Making Spatial Decisions Using GIS: A Workbook.* Redlands, CA: Esri Press.

Mayor-Schönberger, V. and Cukier, K. 2013. *Big Data: A Revolution That Will Transform How We Live, Work and Think,* London: John Murray.

Shapiro, C. and Varian, H. R. 1999. *Information Rules: A Strategic Guide to the Network Economy.* Cambridge, MA: Harvard University Press.

Silver, N. 2012. *The Signal and the Noise: The Art and Science of Prediction,* London: Penguin Press.

Issues of *Trajectory: The Official Magazine of the U.S. Geospatial Intelligence Foundation.* Herndon, VA: U.S. Geospatial Intelligence Foundation.

# 18 Navigating the Risks

This chapter assumes knowledge of what is contained in Chapter 17. Here we describe the pitfalls to be avoided in being a geographic information (GI) system practitioner. We start by summarizing the business models that have been adopted in GI-related enterprises and their respective advantages and risks. Then we describe briefly major constraints arising from regulation of GI-based businesses in different countries, including laws on competition, human rights, information access, intellectual property rights, and liability.

We review the role of the nation state in GI-related matters. Examples are given where the state competes with the private sector in the supply of GI and related services and the challenges this can pose to those involved. The past and possible future roles of the state in creating national spatial data infrastructures are discussed, especially in a world of Open Data and national information infrastructures (Chapter 17).

The different concepts of privacy in different countries and hence the focus of its preservation are outlined—especially where data are being comingled. Success in all this is achieved by awareness and conscious mitigation of the risks and by engaging in partnerships whenever appropriate to acquire complementary expertise. An important contribution is to act ethically, actively manage relationships with the media, and foster public trust. Finally, we illustrate the importance of human frailty, showing how some users have ignored risk and managed to misuse GI technology and GI, with unfortunate consequences.

## LEARNING OBJECTIVES

After studying this chapter, you will understand:

- How personal or organizational disasters may befall GI system practitioners or scientists if they behave unprofessionally or communicate badly.
- The various business models used in GI services and their risks.
- Legal and regulatory constraints on actions.
- Privacy issues involving geographic information.
- The importance of public trust, ethics, and the media.
- Spatial data infrastructures.

## 18.1 Clashes Between Scientists and the Judiciary

The Italian Major Risks Committee is an expert panel that advises local governments on risks of natural disasters. The UK *Guardian* newspaper reported that on 31 March 2009, the committee met in the Italian town of L'Aquila, a medieval settlement built on an ancient lake nestled in the Apennines. This was at the request of the Italian Civil Protection Agency to discuss whether a major earthquake was imminent. It has been reported that the minutes of the meeting show that at no point did any of the scientists say that there was "no danger" of a big quake. After the meeting, Bernardo De Bernardinis from the Civil Protection Agency reportedly walked out and told the media representatives: "The scientific community tells me there is no danger because there is an ongoing discharge of energy."

On April 6, an earthquake struck L'Aquila. More than 300 people died, and 20,000 buildings were destroyed (Figure 18.1). A year later, De Bernardinis and six scientists from the committee were indicted for manslaughter. They were found guilty and sentenced to six years in prison (though the six were acquitted on appeal in late 2014). The judge explained his decision on the basis that the committee's members had analyzed the risk of a major quake in a "superficial, approximate and generic" way and that they were willing participants in a "media operation" to reassure the public. This extreme and extraordinary event has profound lessons for any professional involved in analyzing data and providing advice to the public—especially through a third party.

Such clashes between scientists or data analysts and politicians and the judiciary are not uncommon. For example, the International Statistical Institute has reported that statisticians have been prosecuted for some years in two countries. In Argentina, private statisticians and economists working for universities or consulting firms are producing alternative consumer price indices (CPI) to provide users with more reliable data than the official CPI that has allegedly been manipulated by the government. These statisticians have faced fines and threats of imprisonment following prosecution. In Greece, after years of underestimation of the public deficit and debts by the Greek Statistical Office, the EL.STAT agency established since 2010 is now producing figures that are compliant with the high professional standards practiced elsewhere in the European Union. However, the Greek Chief Statistician and two of his colleagues—who were hired to bring Greece's debt statistics in line with European norms—have faced prosecution for producing what are alleged to be inflated budget deficit numbers.

In this chapter we seek to identify and minimize these and other risks and show how best to maximize opportunities.

## 18.2 Business Models for GI-Related Enterprises

In the beginning (see Section 1.5.1) the GI system world was dominated by big institutional players. In many parts of the world government investment and initiatives played a major role through some private companies, such as utilities that pioneered GI system applications for record keeping and operational efficiency reasons. Since then, the commercial sector has grown to dominance, though the scale of government support via research and military funding remains considerable. Thus, the scale and composition of the private sector in GI systems has changed dramatically over the years. Many new businesses have sprung up, some then bought out by bigger players. Some serious players have all but disappeared, such as Siemens' SICAD system. One common factor has been the continuing success of Esri Inc.; on the basis of estimates published by various market research firms, this company has some 40% of the global market for GI systems, and its professional users top one million.

The market has changed qualitatively as well as quantitatively—many new and global players like Google, Oracle, and Microsoft have changed the nature of what we understand by GI systems and services (see Section 1.5.2.3). As a result, much GI system functionality has been commoditized. Over the period since ca.1995, the delivery vehicle has mutated from large mainframe computers to desktop machines to mobile devices, with Cloud-based Web services playing an increasingly central role (see Chapter 6). And the user base has evolved from a small number

**Figure 18.1** The Governor's offices after the destruction of the city of L'Aquila, central Italy, Monday, April 6, 2009, by a powerful earthquake.

Source: Alessandro Bianchi/Reuters/Landov Media

of well-resourced governments or commercial enterprises to a diverse one including tens of millions of individuals, each one of whom selects just the particular functionality most useful to them and frequently accesses these functions via a tablet or a cell phone.

As described earlier, many interrelated factors have been responsible for these changes—the introduction of disruptive technologies, cost reductions of computing power, the development of new applications, changing user needs, and the role of education in enhancing geographic information science and systems (GISS) human capital among them. But, however clever and hardworking are the staff of a business, the business model(s) to which they are working will play a major role in determining their market success. Few firms prosper by adhering slavishly to one business model over many years. For example, if a firm is successful and grows, the model must be scalable to be economic. It follows that, to survive and prosper, leaders and managers of companies must seek to anticipate changes in the marketplace and user needs and decide whether to mutate their business model before their competitors have won first mover or other advantage.

Yet any significant change to the business model can be extremely expensive and painful—even fatal. Good judgment, bravery, the right corporate culture, and some luck are also essential for success. Furthermore, in some cases "betting the shop" on the latest development may not be the best solution. This applies especially for organizations with an already large installed base of users. The approach of organizations such as Apple and Esri with mass-market products that ship in large volumes and that have built loyal communities of users is not to seek first mover advantage. Rather they seek to build robust solutions, often improving on the innovation of others. This is not a manifestation of poor management—rather that the business model is continuously reviewed and changes only made when they have long-term benefits.

**Being clear about the business you are in—and having the right business model—is the first step to success.**

Crudely put, we can distinguish several different families of business model (Table 18.1). None is totally

**Table 18.1** Business models used by GI-related enterprises

| Nature of business model | Characteristics | Comment |
| --- | --- | --- |
| Selling or licensing software (and also sometimes hardware) | Typically involves up-front charge and annual maintenance for updates. Selling content works best where data are updated frequently, e.g., financial, traffic, or meteorological data | Difficult to scale in early days: sales often involved much human interaction. Advent of broadband for delivery of updates helped. The model prospered originally when GI systems were small scale, involved experts, and everyone necessarily had the capacity to collect their own data. Sometimes this was implemented or run by consultants to a business. Now, however, this model is followed by major firms such as Esri, SAS, Adobe, and Oracle. It also serves well when it involves simple systems that are best sold using a one-off business model, i.e., buy software or data once and walk away. |
| Selling or licensing content (GI) and elements of GI system functionality | Increasingly important through provision as a Cloud service, e.g., routing, store finder, providing embedded mapping in corporate sites. | Often works best where commercial sector adds value to hard-to-understand/poorly packaged datasets provided free by governments. Integration of software tools and data often needed significant expertise in early days. |
| Advertising-based | Often at no direct charge to end user. Mapping has become a mandatory consumer application even if the end user can't be directly charged—hence use of advertising. | Can work well for large organizations, notably Google, who can tailor advertising and location-based services to personal characteristics of users derived from social media, cell phone locations, and volunteered data. Revenue generation is often indirect, e.g., through number of visitors to Web sites of advertised retailers. |

*(continued)*

**Table 18.1** (*continued*)

| Nature of business model | Characteristics | Comment |
|---|---|---|
| Subscription-based | Evolving to a situation where all services are based in the Cloud and a monthly fee enables user to access integrated hardware, software apps, centrally and user-provided data (sometimes shared with others), and sophisticated analytical tools with support services | Developing rapidly. Users can also become service companies on the back of this platform. Mostly hosted on megainfrastructures supplied by Apple, Google, Microsoft, or other major firms so has high fixed costs. Fostering volunteered contributions to content and ideas for new functionality may strongly influence success. |
| Hybrids | Freemium services form one example: base services are provided free (e.g., open-source software and online map services) whilst "professional versions" (such as Google Maps APIs) are charged, especially if incorporated seamlessly in retailers' Web sites. Some firms offer a mix of software and subscription, e.g., through a desktop and Cloud service | Many smaller developer-based firms offer free tools and map services to consumers, as do major vendors such as Esri and Autodesk. |

distinct, some can be run concurrently, and elements of one may be transplanted into another. No one model exclusively characterizes any one size or type of business. All are associated with risks, but the magnitude of these risks often depends on local circumstances. What, for example, happens to your service business if the Global Positioning System (GPS) signal is jammed (as is easy to do locally?). Understanding the options and why successful enterprises have chosen one rather than another is an important management task.

 ## 18.3 Legal and Regulatory Constraints

The use of GI can support implementation of the law. It underpins the efforts of many police forces. Examples include planning effectively for emergency responses, creating and testing risk mitigation strategies, analyzing current spatial patterns (e.g., crime), and projecting future ones. But the law also strongly influences the use of GI systems.

During a career in GI services or technology, we may have to deal with many manifestations of the law. These include copyright and other intellectual property rights (IPR), competition law, data protection laws, public access issues enabled, for example, through Freedom of Information Acts (FOIA) and human rights laws, and legal liability issues. As an example of a GI-related—if outdated—law, many countries have long enacted constraints on collecting aerial photography for state security reasons. The Dutch government formally lifted

their ban on this in June 2013 as a consequence of the ready availability of maps and imagery from Google, Microsoft, and other sources. The U.S. government has also long restricted the sale of very high resolution satellite imagery. In June 2014 Reuters reported that the government—following support from the intelligence community—had licensed the sale of data at 40 cm resolution rather than the previous 50 cm limit. In addition, it had licensed the sale of DigitalGlobe's WorldView-3 satellite's 31 cm resolution data from early 2015.

**Beware! The law touches everything and has direct and indirect effects.**

Our contacts with the law may occur through regulators or through the courts. But because laws of various sorts have several roles—to regulate and incentivize the behavior of citizens and organizations and to help resolve disputes and protect the individual citizen—almost all aspects of the operations of organizations and individuals are steered or constrained by them.

### 18.3.1 Geography and the Law

Laws vary from country to country, and this directly affects us all. For example, the Swedish tradition of open records on land ownership and many personal records dates back to the Seventeenth Century, but much less open frameworks exist even in other countries of Europe. In particular, the creation, maintenance, and dissemination of "official" (government-produced) geographic information are strongly influenced by national laws and practice, though this may be diminishing.

In addition, however, geography can be used to minimize taxes and legal challenges. For example, transfer pricing is alleged to be used within many multinational businesses such as Apple, Starbucks, and Google. This involves the subsidiary being charged for use of IPR or similar held by the parent company. The effect is that revenues are transferred from where they are earned to the enterprise headquarters registered in a lower tax jurisdiction. The obvious result is tax losses to many nation–states of billions of dollars and gains to low-tax havens such as Luxembourg and the Cayman Islands. Such transfer pricing is presently legal in many countries, but this may change.

A second example is that some international organizations argue that the only jurisdiction in which they have to obey the law is that where their headquarters are domiciled. A 2013 example involved Google being sued in Britain for bypassing security settings on the Safari browser on users' iPhones and Mac computers without consent. Google argued that because their software operates from California, the alleged breaches of privacy law should not be heard in a British courtroom. Here we largely ignore such geographic effects of the law, but they are often important.

## 18.3.2 Three Aspects of the Law and GI

Accordingly, in this chapter we introduce three aspects of the law particularly relevant to GI-related activities. These are protecting innovation, coping with liability, and ensuring access to information. More formal considerations of law and GI are given by Cho and by Obermeyer and Pinto (see Further Reading), focusing especially on the situations in Australia and the United States. *Readers are strongly advised, however, not to rely on textbooks and to take up-to-date legal advice in their own jurisdiction when relevant issues arise.*

**Always take legal advice in the relevant jurisdiction if faced with legal challenge.**

An important complication in areas such as GI services and technology is that the law is always doomed to trail behind the development of new technology. Laws are only enacted (sometimes long) after a technology appears.

### 18.3.2.1 Protecting Innovation and Exploitation

Innovation is the key to progress. GI examples include Harrison's chronometer and the Global Positioning System (GPS). Much innovation currently arises from commercial enterprises, but underpinning this is substantial basic research funded by governments. Many of the results from that research are published openly. There is, however, tension in universities and elsewhere between the open scientific approach and the merits of generating revenue and national jobs from the expenditure of government research funds.

**Protecting innovation underpins much commerce—and governments.**

Central to most commercial activity is protecting the fruits of innovation (at least for a time) from any competition which simply seeks to clone a product (such as a new algorithm, software package, or database), a service, or a process invented by others. Protection from unauthorized cloning is through intellectual property rights agreed on by most countries of the world. There are various categories under which IPR is protected (see Box 18.1), and

## Application Box 18.1

### Intellectual Property Rights

Intellectual property rights (IPR) are the rights given to persons over their intellectual creations, and they usually give the creator an exclusive right to the use of his or her creation for a certain period of time. The length of time establishes a balance between the individual's rights to benefit from the inventions and the benefits to wider society. There are two main types of IPR:

#### Copyright and Rights Related to Copyright
The rights of authors of literary and artistic works (such as books and other writings, musical compositions, paintings, sculpture, computer programs, and films) are protected by copyright, for a minimum period of 50 years after the death of the author. The main social purpose of protecting copyright and related rights is to encourage and reward creative work.

#### Industrial Property
Industrial intellectual property can be divided into two main areas:

● The protection of distinctive signs, in particular trademarks (which distinguish the goods or services of one undertaking from those of other undertakings) and geographic indicators (see following). This protection aims to stimulate and ensure fair competition ▶

and to protect consumers by enabling them to make informed choices between various goods and services. It may last indefinitely, provided the sign in question continues to be distinctive.

- Other types of industrial property. These rights are protected primarily to stimulate innovation, design, and creation of technology. Into this category fall inventions (protected by patents), industrial designs, and trade secrets. The social purpose of these rights is to provide protection for the results of investment in developing new technology, thus giving an incentive

and means to finance research and development activities. The protection is usually given for a finite term (typically 20 years in the case of patents).

Geographic place-name labels applied to a product may also be protected under World Trade Organization regulations. These indicators denote that the product was—and can only be—created in a particular (usually limited) geographic area. Examples include Champagne, Tequila, Parma Ham, and Roquefort or Feta cheese.

Source: www.wto.org

the type of protection available varies between countries. IPR is in constant flux as a result of changes in the law and challenges facilitated by new technologies.

In some industries, however—notably in the services sector, where innovation often occurs in close association with a particular customer—IPR protection is not sought, and the "first to market" advantage is exploited in products with a short shelf life. Alternatively, developers may choose a nontraditional business model and seek the widest take-up of their products, with few revenue benefits obtained directly from users (e.g., open-source software).

### 18.3.2.1.1 Patents

Patent wars—notably over how software has been implemented in a user interface on smartphones—have occurred for some time between major technology developers such as Apple, Google, Microsoft, and Samsung. Big changes to the structure of industry sometimes come about because of the desire to secure the patents of others (e.g., Google's $12.5bn purchase of key elements of Motorola's business, bringing them 17,000 patents, some of which involved location capabilities).

A patent is a government-granted monopoly to an invention. An example of patent wars specifically in the GI world featured Facet Technology Corporation and the Dutch navigation device maker TomTom. The businesses reached a settlement after a 16-month legal battle. The case was settled through a patent agreement under which TomTom acquired a limited license to Facet's patents on tools that facilitate capturing road sign and other data needed to underpin advanced driver assistance systems.

### 18.3.2.1.2 Copyright and Database Protection

How does IPR fit into the world of GI systems and geographic information? The answer is that it mostly

involves copyright law, and it is complex, especially when seen from a global rather than a national perspective. There are at least four reasons:

- National or regional IPR laws vary in some elements of substance and in many details.
- The formal relationship between the private-sector and public-sector data suppliers also varies between countries (Section 18.3.2.3).
- Global databases have been built in some domains (e.g., for satellite navigation purposes), whereas many other geographical databases are national or subnational in scope, content, ontology, and reference systems; different legal systems impact on each.
- Many commercial databases incorporate data from multiple sources, sometimes global.

As a result, GI system users must be careful to observe IPR when downloading data or software from the Internet.

**Ignore IPR when downloading data or software from the Internet at your peril!**

The simplest way to summarize the complex situation is to answer some frequently asked questions:

***Can geographic data and information be regarded as property?*** Normally yes, at least under certain legal systems and under either copyright or database legal protection. Who owns the data is sometimes difficult to define unequivocally, notably in aggregated personal data such as geocoded health data. Some place-name labels may also be property (see Box 18.1).

***Can all geographic information be legally protected?*** There is little argument about copyright where it is manifestly based on great originality and creativity, for example, in relation to a painting or mapping carried out personally using GPS and that is distinctive in content and form. Rarely, however, is the situation as clear-cut, especially in the GI world, where some information is widely held to be in the public domain, some is regarded as facts (see later discussion), and some underpins either the commercial viability or the public rationale of government organizations.

In the United States, the Supreme Court's ruling in the famous *Feist* case of 1991 has been widely taken to mean that factual information gathered by "the sweat of one's brow"—as opposed to original, creative activities—is not protected by copyright law. Names and addresses are regarded widely as geographic facts. Nonetheless, several jurisdictions in the United States and elsewhere have found ways to protect compilations of facts, provided that these compilations demonstrate creativity and originality. The real argument is about which compilations are sufficiently original to merit copyright protection. A number of post-*Feist* instances have occurred in which the courts have recognized maps not as facts but as creative works, involving originality in selection and arrangement of information and the reconciliation of conflicting alternatives. Thus some uncertainty still exists in U.S. law about what GI can be protected by copyright law.

In Europe, different arrangements pertain—both copyright and database protection exists. For copyright protection to apply, the data must have originality in the selection or arrangement of the contents, and for a database right to apply, the database must be the result of substantial investment. It is entirely possible that a database will satisfy both of these requirements so that both copyright and database rights may apply. A database right is like copyright in that there is no need for registration; in both cases it is an automatic right, and it begins as soon as the material that can be protected exists in a recorded form. Such a database right can apply to both paper and electronic databases. It lasts 15 years from making, but if published during this time, then the term is 15 years from publication. Protection may be extended if the database is refreshed substantially within the period. As stressed earlier, geographic differences in the law have potential geographic consequences for commercial GI activity. Protection equivalent to the database right will not necessarily exist in the rest of the world, although all member countries of the World Trade Organization (WTO) do have an obligation to provide copyright protection for some databases.

**In some jurisdictions, some GI databases are seen merely as facts and are not protected by law; elsewhere they are protected because of the investment made to create or update them.**

***Can information collected directly by machine, such as a satellite sensor, be legally protected?*** Given the need for originality, it might be assumed that automated sensing should not be protectable by copyright because it contains no originality or creativity. Clearly, this interpretation of copyright law is not the view taken by the major content providers, such as satellite imagery companies like DigitalGlobe, because they include copyright claims for their material and demand contractual acceptance of this proposition before selling products to firms or individuals.

From the point of view of enforcing copyright, it helps that there are at present only a modest number of original suppliers of satellite imagery—especially that of high spatial resolution. Thus any republication or resale of an image could probably be tracked back to the source with relative ease. This situation is changing as small operators such as Skybox Imaging and Planet Labs launch constellations of small satellites trading off ultra-high spatial resolution for frequent imaging. If the data source can be identified definitively, imagery up to 15 years old would be protected under the EC's Database Directive, at least within the European Union area itself.

***How can tacit geographic and process knowledge—such as that held in the heads of employees and gained by experience—be legally protected?*** "Know-how" can migrate very rapidly. This situation has happened in GI systems/GI organizations—and many others—from the earliest times. The simplest way to protect tacit knowledge is to write some appropriate obligations into the contracts of all members of staff—and be prepared to sue the individual if he or she abrogates that agreement. The most popular "weapon" is the noncompete agreement, which bans an employee from working for a rival for a fixed period, often two years. It has been estimated that 90% of management and technical staff in the United States have signed them. In some countries human rights legislation may complicate winning such action. Keeping staff busy and happy is a better solution!

***Who owns information derived by adding new material to source information produced by another party?*** Suppose a dataset has been obtained that is useful but, by the addition of an extra element or by overlaying data, becomes immensely more valuable or useful. So who owns and can exploit the results in derivative works? The general answer is both the originator of the first dataset and the person who has added value. Without the originator's input, the latter would not have been possible. This is true even if extracted data are only implicit—if, for instance, road center lines are computed from the digitized road casings on the source maps. Moreover, combining datasets may result in deep trouble if permission is not sought from the originator. Where moral rights exist (e.g., in European countries), the author has the right to integrity (i.e., to object to any distortion or modification of his or her work that might be prejudicial to his or her honor). He or she also has the right to attribution as the author, to decide when a work will be published, and to withdraw a work from publication under defined circumstances.

**To protect your GI under law, it is wise to watermark and fingerprint it.**

*How can you prove theft of your data or information?* It is now fairly common to take other parties to court for alleged theft of information. To prove theft it is crucial to have good evidence, especially as digital technology makes it easy to disguise the look and feel of GI and to reproduce products to very different specifications, perhaps generalized. The solution is to be both proactive and reactive. The data should be "watermarked" in both obvious and nonvisible ways. It is wise to have a good audit trail to demonstrate that the watermarking was established by management action for that purpose. In addition to watermarking, "fingerprinting" can also be used. At least one major commercial mapping organization has added occasional fictitious roads to its road maps for this purpose (though comparison with satellite imagery normally discloses such practices). In areas subject to temporal change (such as due to tides or through vegetation change), nature's fingerprinting ensures it is very unlikely that any two surveys or uses of different aerial photographs could ever have produced the same result in detail.

*How do I seek redress if I have been badly treated?* If theft of intellectual property has occurred, go to the police. For other forms of unfair treatment—such as price gouging, unfair competition, illegal restrictions of access, or manifestly inaccurate product descriptions—approach the relevant government regulator.

*Do I need a license to use GI or software provided by another organization?* In most cases, users of GI explicitly or implicitly have to accept the terms of a license for the information they are accessing and reusing. Even where it is not explicit and use of the information is free, licensing still exists (e.g., under terms and conditions that prevent

editing of the data and then misrepresenting the work of the originator or in regard to "creative commons"; see creativecommons.org). Thus it is always wise to read the terms of any licensing agreement with great care before simply clicking on an "I agree" box. Google Maps, for instance, has different licensing arrangements for use of its APIs, depending on whether this is for nonbusiness or business purposes. The criteria for the first purpose require that the service based on your information must be freely and publicly available; strict limits are also placed on the resolution of the maps produced and the number of Web requests made per day. Its use is free, but the professional use version requires payment.

In some cases, licensing is an onerous process and can necessitate signing up to hundreds of pages of contractual details limiting the extent and nature of reuse and even the charging mechanism. In some other cases, however, licensing is now extremely straightforward, at least for Open Data (see Section 17.4). The UK Open Government License (OGL) is a good example of simplicity and is being cloned in some other countries.

**Almost all GI is licensed somehow. Read the license before using the GI!**

### 18.3.2.2 Coping with Liability from Use of GI and GI Systems

Liability is a creation of the law to support a range of important social goals, such as avoiding injurious behavior, encouraging the fulfillment of obligations established by contracts, and distributing losses to those responsible for them. It too is a huge and complicated area and covers negligence, fraud, and product liability. Though few jurisdictions seem to have legislated specifically for liability in regard to GI systems and GI use, a number of cases have been reported in regard to air and nautical charts and in boundary disputes (see following discussion and Section 19.6.4).

Nevertheless, the potential for legal action is very real, and two examples may be cited. The first is described by Cho as follows. Four Australian National Parks and Wildlife Service (NPWS) officers were killed from smoke suffocation in a burn-off operation. The officers were given maps that showed two possible escape routes. The maps showed a cleared hilltop, which potentially could have provided shelter from the fire. Unfortunately, the map did not show a 30-m cliff that stood between anyone trying to escape the fire and the cleared area. The map also showed a path known as Wallaby Track running directly towards a local motorway. In reality this path twists into impenetrable bush. It was found by a court that the original botanical map had not been "ground-truthed" to include specific details and did not mark areas with safe refuges should they be required.

A second instance is more general and often relates to historic GI. Disputes over boundaries in land or at sea are common and can have catastrophic consequences. The ongoing dispute between China, Japan, and other South East Asia nations over the sovereignty of the islands known as Senkaku in Japanese and Diaoyu in Chinese is particularly dangerous. Many of these disputes are inflamed because of the understandable but inadequate quality of historical maps and incomplete records. In such cases there is obviously no contractual basis for action—only legal arbitration if all parties agree to it.

In many cases liability in data, products, and services related to GI will be determined by resort to contract law and warranty issues. This assumes that gross negligence has not occurred. If it has occurred, the situation can be far worse, and individuals as well as corporations can be liable for damages. Other liability burdens may also arise under legislation relating to specific substantive topics such as intellectual property rights, privacy rights, antitrust laws (or noncompetition principles in a European context), and open records laws.

Clearly, liability is a serious issue for all GI practitioners. Minimizing exposure to risk is achieved primarily through performing competent work and keeping all parties informed of their obligations. Any contract to provide software, data, services, or consultancy should make clear the limits of your responsibility, though disclaimers rarely count for much. And you should adhere to well-recognized professional codes of conduct (sometimes called Codes of Practice or Ethics; see Section 18.5.2).

### 18.3.2.3 Access to GI and the Role of the State

Access to GI provided by commercial suppliers is relatively straightforward and usually involves a contract and some payment or exposure to advertising (Section 18.2). Data available from the private sector generally reflects the sector's ability to monetize it. Hence car guidance systems based on digital mapping are now almost ubiquitous: Major commercial players such as Apple, Google, and Microsoft have all bought their way into accessing or owning the relevant GI as an essential part of their normal business. In some cases roads have been newly surveyed to avoid copyright infringements and other restrictions associated with using data from governments.

It might be thought from Box 17.2 and Table 1.2 that because we are now "drowning in data" and the Open (government) Data movement is growing worldwide, access to GI held by the public sector would be routinely easy, cheap or free, unencumbered by bureaucratic hurdles, and defined by law (as in U.S. states' public record laws). Notwithstanding many impressive national visions for the use of GI systems—such as that for India, the world's second most populous country, set out in moes.gov.in/national_gis.pdf—nothing could be further from the truth.

### National security concerns limit the availability of some GI in a surprising number of countries.

In India there are numerous examples of impressive GI system applications: The tsunami warning system has been built around GI systems and is capable of providing information within minutes about travel time and run-up heights at 1600 locations along the coast of the Indian Ocean. Yet, despite this sophistication, there has been a long history in India of preventing access to topographic mapping of coastal and border areas on security grounds. This was relaxed somewhat in the 2005 Indian National Map Policy, but the terrorist attacks on Mumbai in 2008 made the situation difficult again. For example, the Survey of India lodged a complaint with the Delhi police against Google India's plans for a Mapathon in 2013. This was a competition open to all to map their neighborhoods, with the data then being uploaded to the Google server in the United States. This contravened the 2005 policy, which said that maps intended for publication must be sent to the Ministries of Defense and of Housing and to intelligence agencies for security checks. To those in Europe and North America, this may seem strange. But such security concerns still exist in various areas of the world.

In summary, the reasons why access is not universally straightforward include the following:

- National policies or statutes like India's National Map Policy contain restrictions, often legitimated on security grounds.
- The lack of appropriate technology in government bodies, which are the traditional custodians of the GI.
- The costs of preparing data to be made available, including creation of metadata.
- A lack of incentives to make such GI available, especially in the face of other government priorities.
- A wish in some countries to ensure that the taxpayer manifestly benefits from revenues generated by any GI made available to the business sector—often international, rather than national, and in some cases accused of not paying reasonable levels of tax (see Section 18.3.1). One approach to this is for the government to become an information trader.

#### 18.3.2.3.1 Governments as Information Traders

Governments have been and remain major collectors (and sometimes providers) of many types of information, some of it collected under a statutory monopoly. We should recognize that the normal remit of government departments is to implement government policies, and few of them are specifically funded as data providers. What data they hold may generally be characterized as "exhaust data," created as a by-product of the department's core functions. There are, however, some special-purpose government bodies, such as mapping and meteorological agencies. Their rationale is to collect and, increasingly, to exploit data for the nation. A number of them, incentivized by government policy or reductions in funding, have become information traders.

Despite many research studies, there is no unambiguous economic demonstration of which of the following options is in the best interests of a country:

- To pay for the creation and essential processing of data needed for a government's purposes, then make it freely available in the hope that this will stimulate the economy and provide greater transparency (see Section 17.4), or
- To pay for the first stage, then for the government to seek to generate revenues by leasing the data to end-users or value-added resellers.

Option 1 places immediate costs on the taxpayer with the possibility of large but unpredictable future economic and social benefits. There is much evidence that the level of use of some government data—notably GI—expands substantially when made readily available and at marginal cost or less. However the use/accrued revenue relationship is complex. Option 2 internalizes costs and benefits. Its success depends

on the organs of the nation–state being efficient and entrepreneurial, having possibly competitive advantage through access to superior data and skills funded by sunk costs, and being able to deny access to those who cannot or will not pay for the data. All that has to be achievable under national laws.

But this is more than a matter of economics: It involves a matter of principle about the role of the nation–state. What should the state do, and what should be left to private enterprise? In some cases in some countries the answer is uncontroversial, such as the state providing and guaranteeing definitive information about parcels of land and houses upon them. Elsewhere these issues are controversial. There are practical concerns as well: If a government body starts to trade in information, how is mission creep averted; for example, if it then begins to provide added-value services (education and training, data linkage, and analysis of data from their own and other sources), thereby competing with national and international businesses?

All this also ignores important external factors. An obvious one is the international agreements to which a nation–state is committed by its government. These include commitments made under international treaties (such as the agreements on data sharing that emerged from the UN 1992 Rio treaty). The 28 European Union member states have obligations to convert EU directives, such as that on reuse of public-sector information and the INSPIRE Directive (see Section 18.6.2), into national law. Perhaps the best example of how this complex of time-varying factors has played out in one country is that of the UK. Box 18.2 summarizes the oscillations of government

## Application Box 18.2

### Mutations in UK Information Policy

Following the policies of the British government, the Ordnance Survey (OS)—the national mapping agency—first took action as early as 1817 to prevent "free riding" or illegal use of the government's maps. The OS expressly warned anyone infringing their copyright that this would lead to legal action. In the 1930s world of mapping, public policy was on a "charge only for ink and paper" basis; in the 1960s it changed to charging users a fraction of the total costs of operations. In 1999, the move to a Trading Fund status effectively led to the OS having to become profitable and thus meet their full costs and interest on their capital employed.

In 2008, the UK government was seeking to nurture the use of the Web for social, economic, and educational purposes (Section 17.4). Over the next four years rapid developments occurred under

two different administrations, stimulated by Sir Tim Berners-Lee (see Box 17.6) and others. The Open Data concept was elaborated and around 20,000 datasets have been made available via data.gov.uk. These included many sets of GI; notably all bar the most detailed mapping of the UK produced by Ordnance Survey. They also included large numbers of statistical datasets and administrative ones designed to support "armchair accountants" keeping a watch on how national and local governments were spending taxpayers' money. All these were made available for free under a new and simple Open Government License.

In some areas the changes were rapid—notably in apps being created for a multiplicity of location-based services, especially those making available real-time travel data on air, rail, and bus services and

▶

congestion data. The new technologies helped make possible a variety of new services, such as a national road works register, created and constantly updated by a commercial body.

All of this was powered by a direct interest taken by two successive Prime Ministers. This and President Clinton's endorsement of the U.S. National Spatial Data Infrastructure (NSDI) illustrate the importance of having support from the highest political level in any public sector initiatives.

Austerity associated with the financial crisis from 2007 onward, however, has ensured that four government bodies (Ordnance Survey, the Meteorological Office, Land Registry, and Companies House) have all been required as government Trading Funds to continue to generate revenue to cover their costs though income from Parliament is now counted in that figure. Continuing debates revolve around the regulation of these Trading Funds and their ability to compete with the private sector and how strategic a view government should take on which datasets held by departments should be prioritized for "freedom." In early 2014 the UK government consulted publicly on privatizing the Land Registry but drew back. In summary, policy oscillations evident over 200 years continue.

policy over a 200-year period, with particular reference to the official mapping of Britain.

Some core reference GI form perhaps the most valuable data in any national information infrastructure. These typically form the framework to which other data are linked. All the datasets using this framework will then be spatially congruent. This has both theoretical and practical advantages. But the longer governments delay in making such data freely available, the more users turn to alternatives, and the opportunity for congruence may be lost. The growth of use of Google Maps, thanks to its sophistication, low or zero cost, ease of use, and global coverage, has been remarkable and is outflanking the use of government official data for many applications.

Finally, it has been argued that access to certain information—largely about oneself—is a basic human right and therefore subject to human rights law in jurisdictions where these are enacted. It can, by extension, be argued that access to many other government data (such as how it is performing against its own targets and what its branches spend our money on) is a basic democratic right. We touch on this in part in Section 17.4.

 ## 18.4 Privacy and GI Systems

Privacy in regard to personal information is a matter of law in many countries. Here we discuss it separately because of the breadth of factors involved.

Our personal privacy has in practice become sharply undermined since the bombing of the Twin Towers in New York in 2001. This led to a global increase in surveillance and the development of new technologies for capturing the movements of people.

Privacy considerations are germane to both the private and public sectors: Loss or theft and subsequent publication of private information can cause great reputational damage. Personal information collected by government bodies is often mandated by law: Citizens have no right to refuse but benefit from government using the data in planning facilities or supplying social services (as well as collecting income-based taxes). In contrast, individuals voluntarily—and often happily—provide personal information to many private-sector bodies (e.g., retailers: see Section 17.3.1.3 and Box 17.5) for gain of some form. The explosion of use of social media has multiplied the personal information being used semipublicly; the more geotagging becomes common, the wider is the possible GI use, but the greater are some threats to privacy. New technologies that permit locating individuals within buildings may well exacerbate the threats.

Some commentators are resolutely opposed to the tracking of position of individuals via cell phones or other mechanisms. Jerry Dobson and Peter Fisher have coined the term *geoslavery* for it. Another term is *geopiracy*, used to describe when personal GI is allegedly being seized without consent.

Attitudes to privacy may be time dependent. Some people abhor any broadcasting of their location. Others welcome having locations shared for social reasons. Parents concerned about the safety of their children might well contemplate giving them a smartphone with a location-reporting app. An injured hill walker or a wounded soldier would be happy for the rescue services to know his or her location. In these latter examples, an element of privacy is being traded to ensure enhanced safety at some moment in time. Thus privacy is a complex concept, varying between individuals and at different moments in time.

**Privacy is a complex concept, varying between individuals and at different moments in time.**

What information is considered appropriate for the state to hold varies greatly between different countries. Nordic countries operate a series of linked registers to provide large amounts of administrative data about individuals for multiple government purposes. In general, three base registers exist containing details of each "data unit." These are a population register; a business register (enterprises and establishments); and a register of addresses, buildings, and dwellings. A unique identifier for each unit in the base registers, and links between them, is used. The unique identifier is also used in many other administrative registers for these units (see Figure 18.2), such as educational registers and taxation registers. An extract from the population register serves as a basic document that is needed when applying for a passport, when getting married or divorced, or when a funeral is held or an estate is distributed; this is provided free of charge. It is thus in the interest of the individual to make sure that all the data within administrative register systems are accurate. Most important, although the purpose of the registers is primarily for government bodies, the public can access some registered information via a terminal. From these registers a wide variety of aggregate-level data are spun off and analyzed and mapped using GI systems. Such an intricate and largely public infrastructure of linked personal data would not be acceptable

in some other countries with lower levels of trust in government than in Scandinavia.

Google's introduction of car- and tricycle-based imaging to produce Street View in 2008 led to many concerns worldwide about its impact on privacy. The Greek data protection agency, for example, banned Street View's expansion in the country until it obtained acceptable commitments on how long the images would be kept on Google's database and what measures the firm would take to make people aware of privacy rights. The initial furor about deployment of Street View has largely subsided, as it has become part of the taken-for-granted information infrastructure and as its currency has declined.

**Most socioeconomic data are collected for individuals, and so privacy matters.**

## 18.4.1 Preserving Privacy without Losing the Use of Personal Information

In many cases, the resolution of GI pertaining to an area of land or the altitude of a point on the earth's surface is not a matter of contention over privacy. Central to privacy concerns are information about identifiable human individuals or small firms. We have

Figure 18.2. A visualization of how many Swedish registers are linked using unique identifiers for people, businesses, and buildings; those without lines are not linked at present.

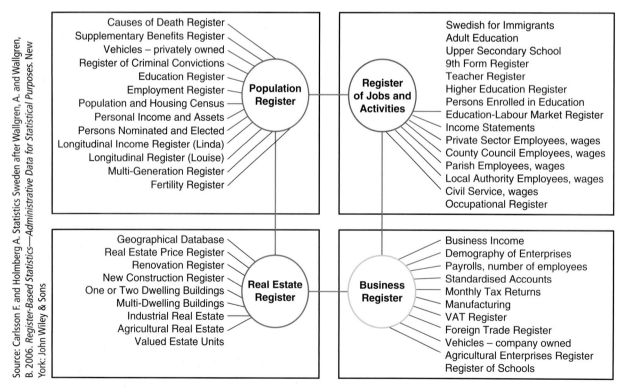

Source: Carlsson F. and Holmberg A. Statistics Sweden after Wallgren, A. and Wallgren, B. 2006. *Register-Based Statistics—Administrative Data for Statistical Purposes.* New York: John Wiley & Sons

**Figure 18.3** The "market footprint" of Portsmouth Hospitals based upon analysis of annual Hospital Event Statistics for one class of morbidity in the UK. The shaded areas include 90% of all outpatients treated in the year; a percent volume contour associated with a kernel function has bounded these 90% to ensure no individuals with a particularly sensitive disease are identified in rural areas or very small settlements.

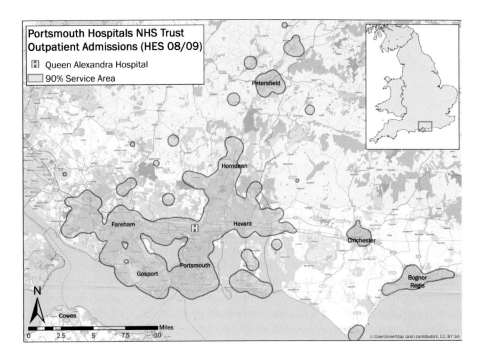

already highlighted the numerous benefits from being able to analyze personal data, notably for research into causes of poor health (Section 17.3.1.2). But five issues need to be addressed:

- How to avoid compromising privacy when mapping any individual data. This is particularly sensitive when displaying the incidence of some medical conditions. Fortunately there are well-established GI system tools for doing this (Figure 18.3) and tuning the level of privacy desired—albeit at the price of losing some valuable information.

- How to avoid disclosure of information generated by the "mosaic effect" when overlaying GI (Section 5.4.4). In creating added value, small differences in the boundaries of different datasets may result in details about individuals being identifiable (Figure 18.4).

- How to persuade members of the public that their data are safely held and their privacy is not compromised. In Britain the tax authorities lost a CD containing 25 million records of individual taxpayers in 2007, which strongly influenced public opinion against the holding or sharing of personal information.

- How to cope with any legislation—such as that under discussion in the European Union at the time of writing—that would require every individual data subject to give explicit permission for their records in databases to be used for any purpose other than the original public one, for example, for subsequent research purposes (i.e., the "opt-in" requirement). Demonstrating that very robust data anonymization procedures are effective and that there are large public benefits from carefully organized research may be the best solutions in the longer term to build public acceptance.

- Making routine and easy the process by which those who wish to remove their existing records from a database run by a commercial entity (e.g., Facebook or Google) can do so. It is not always easy to retreat into anonymity and reestablish privacy, especially if that involves wiping criminal records.

**Figure 18.4** Diagrammatic representation of the problem of disclosure of information about individuals or small groups through the "mosaic effect" (overlay, in GI system terms). If population details are known for areas A to E and V to Z, then overlay will compute the counts of the sliver areas in pink.

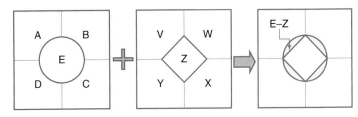

# Public Trust, Ethics, and Coping with the Media

## 18.5.1 Public Trust

Public trust is essential to the effective running of civil society, to maintaining the success of a business, and to preserving reputations. If the public distrusts what is being done, big problems arise. These could be expensive in money or new business terms. High levels of trust minimize time-wasting debates and facilitate the implementation of policies.

Normal levels of public trust vary greatly for different institutions and groups of people—between governments of different levels, physicians, politicians, scientists (e.g., those engaged in nuclear power or stem cell research), in banking and other businesses, journalists, and those in professional bodies (Figure 18.5). In many Western countries there have been significant falloffs in the past three decades to the public's acceptance of some expert scientific advice. This more generally is the "death of deference." But too high a level of trust is also bad: Mindless acceptance of what governments, scientists, or businesses propose is bad for democracy. The question is: how much trust is appropriate?

This has significant implications for those working in the GI domain. Insofar as the GI system-based analyses are illustrated by maps, we have an advantage over cost/benefit studies that simply produce tables of numbers; maps are often regarded (incorrectly) as easy to understand and convincing.

Widespread acceptance of embedded GI system functionality in telephones, satellite navigation, and Web sites providing "where is" factual-type information have also produced a positive spillover effect on other GI system functionality. The common GI system practice of drawing together data from multiple sources can also be seen as an advantage in suggesting we do not rely solely on one view of the world. And the stunning capabilities of GI technology driven by skilled experts dazzle the onlooker into acceptance and trust.

**The GI community seems to enjoy high public trust—but it could evaporate.**

Yet we are not totally in control of our reputation. For instance, distrust spills over from what people think of their government (or business) into products (like statistics) produced by it. This is particularly true where the statistics are used to buttress government claims of progress; economic and health statistics are typically trusted less than are transport ones, presumably because the public thinks governments have little to gain by manipulating transport statistics!

Such lack of trust does not in general occur with producers of reference mapping (such as government national mapping agencies) or in regard to some major commercial providers of software or data. Yet we know that the quality of such data varies considerably across the globe and that many of the data series are as difficult to collect as good-quality statistics. Soil maps, for instance, have typically been based upon field sampling principles and estimation techniques as much as are population statistics (see Section 2.4). Greater

**Figure 18.5** Changes in public confidence in US institutions 1973–2008

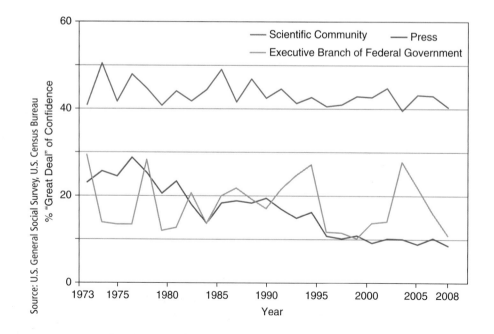

awareness of this complexity could lead to a diminution of public trust; that would be extremely dangerous for the continuing success of GI-related activities.

We minimize the risk of loss of trust by following a published Code of Practice (or Ethics: see following section); by being transparent about the inputs, processes, and algorithms underlying any decisions we make; and by running sensitivity analyses as part of any "what-if" analysis (see Section 15.5), describing as best we can what levels of uncertainty remain in the results and being prepared to answer questions from those affected or interested (e.g., the media).

## 18.5.2 Ethics

Professional competence and acceptance of contractual and other responsibilities are readily understood as being essential for any GI system, service, or technology professional. The Association of American Geographers posits that there are at least four reasons why ethics should also be essential to any GI professional:

- Geographic technologies are surveillance technologies. The data they produce may be used to invade the privacy, and even the autonomy, of individuals and groups.

- Data gathered using geographic technologies are used to make policy decisions. Erroneous, inadequately documented, or inappropriate data can have grave consequences for individuals and the environment.

- Geographic technologies have the potential to exacerbate inequities in society, insofar as large organizations enjoy greater access to technology, data, and technological expertise than smaller organizations and individuals.

- Georeferenced photos, tweets, and volunteered (and unvolunteered) geographic information can reveal private information. Those data that are increasingly publicly available and used to study societal phenomena raise significant privacy concerns.

More prosaically, technology in general, and GI system technology in particular, is neither good nor evil, and it certainly cannot be held responsible for the sins of society. But technology can empower those who choose to engage in either good or bad behavior.

**Ethics matter in operating GI systems safely.**

Ethics refers to principles of human conduct, or morals, as well as to the systematic study of such human values. An act is considered to be ethical if it agrees with approved moral behavior or norms in a specific society. GI technology provides ample scope for unethical behavior, notably to produce results that knowingly benefit one individual or group rather than another or society as a whole. As described earlier, this behavior is facilitated by the apparent simplicity and believability of map outputs—despite the high complexity of the assumptions made in combining data, the various different algorithms that might be employed, and well-known ways of misleading the eye by use of map scale, color, projection, and symbolism (see Chapter 11).

How do we know who to trust? How do we know, for instance, that the person who has offered to forecast environmental risk or provide disaster management at a microlevel is appropriate for the job? The simplest way to find out is to review candidates' educational attainment and track record and to use only registered professionals who have gone through an extensive formal qualifications process and who can be disbarred for unethical conduct.

GI systems are a field where until recently few directly relevant professional certification programs existed, and many GI system practitioners operated on an ad hoc basis. Fortunately various helpful Codes of Ethics have now been established. Valuable early work was done by the Urban and Regional Information Systems Association. Contemporary Codes of Ethics include those published by the GIS Certification Institute and the Association of American Geographers. The former, an independent not-for-profit organization, has been particularly active. A short summary of their code is given in Box 18.3. Much greater detail on what it includes is given by Obermeyer and Pinto and on various Web sites (see Further Reading). Such codes can never be perfect because they cannot legislate for every possible eventuality. The essence of a true professional, however, is that he or she abides by the spirit of the Code, even when it does not proscribe a questionable action.

There is now scarcely a major university around the world that does not have some form of GI systems education program, and many include a module on ethics. It is also becoming commonplace for GI system professionals to attend certification courses within their own domains. In these, individuals who demonstrate possession of specified competencies through some combination of education, experience, examination, and study of ethics gain certification. For example, the U.S. intelligence community has been mandated to provide certification processes for the entirety of their professional employees, and the Open Data Institute in London is providing certification in Open Data exploitation.

## An overview of the GIS Certification Institute Code of Ethics

This is a high-level summary of the document at www.gisci.org/code_of_ethics.aspx:

### I. Obligations to Society

The GIS professional recognizes the impact of his or her work on society as a whole, on subgroups of society including geographic or demographic minorities, on future generations, and inclusive of social, economic, environmental, or technical fields of endeavor. Obligations to society shall be paramount when there is conflict with other obligations. Therefore, the GIS professional will:

1. Do the Best Work Possible
2. Contribute to the Community to the Extent Possible, Feasible, and Advisable
3. Speak Out About Issues

### II. Obligations to Employers and Funders

The GIS professional recognizes that he or she has been hired to deliver needed products and services. The employer (or funder) expects quality work and professional conduct. Therefore, the GIS professional will:

1. Deliver Quality Work
2. Have a Professional Relationship
3. Be Honest in Representations

### III. Obligations to Colleagues and the Profession

The GIS professional recognizes the value of being part of a community of other professionals. Together, we support each other and add to the stature of the field. Therefore, the GIS professional will:

1. Respect the Work of Others.
2. Contribute to the Discipline to the Extent Possible

### IV. Obligations to Individuals in Society

The GIS professional recognizes the impact of his or her work on individual people and will strive to avoid harm to them. Therefore, the GIS professional will:

1. Respect Privacy
2. Respect Individuals

## 18.5.3 Coping with the Media

The print and electronic media are a crucial influence on public trust. Very different legalities and practices characterize the media in different countries: some are under state control and thus invariably supportive of the government. Others—such as the print media in the UK—can be extraordinarily aggressive and combative, determined to seek out what they see as wrongdoing and humiliating those whom they believe should be blamed. Redress against false accusations is often difficult and expensive. The story told in Section 18.1 illustrates just how important the media can be.

In an ideal world, only one person in any organization—with the right skills and experience—should be allowed to talk to the media. Any briefings must be available in written form as well as spoken; otherwise some invented or misconstrued report is not plausibly denied. Yet minimization of contact with the media and other key influencers is also a mistake. Building relationships through keeping in touch regularly with journalists and politicians and keeping the latter informed about how better services are being provided, new jobs are created, or the quality of their constituents' lives is being improved is part of a manager's role. It is a mistake to become too cozy with the media, but their drivers (e.g., to sell newspapers) must be understood and supported where possible and ethical.

## 18.6 Partnerships, Up-Scaling Activities, and Risk Mitigation

We live in a competitive world. But there are many occasions when the GI system user finds it best to collaborate or partner with other institutions or people. In principle, partnerships can bring missing skills, know-how, technology, finance, and better branding to the enterprise. Because the need for partnerships applies at the personal, institutional, local area, national, and global levels, there is great scope for different approaches. The form of the partnerships can range from the enforced (e.g., through legislation) and the highly formal based on contract to the informal where participation is entirely voluntary. Some partnerships result from short-term events that span national boundaries (such as the Chernobyl nuclear reactor accident in 1986 or the Asian tsunami of 2004). Others arise from long-term relationships like membership of NATO—which has had a profound effect on the standardization of GI within the military (and beyond) in those nations that participate.

Technological change makes it possible to achieve partnerships between geographically dispersed organizations. But technology does not resolve all difficulties. Rarely are partnerships trouble free and totally successful. It simplifies matters if someone is in overall charge (such as the Department of Homeland Security, in charge of counterterrorism in the United States). This facilitates setting clear project milestones, monitoring of risk and progress, and providing enforcement action when things go wrong. In contrast, ad hoc partnerships tend to be created on the fly in the case of an unexpected disaster.

We begin with an extraordinary success story: the Open Street Map (OSM) project. The resulting global database of mapping information can be used for many different purposes with very few restrictions (www.openstreetmap.org). It is licensed under the Open Database License. In essence, OSM mapping and data may be used for any purpose without charge subject to two conditions: that proper attribution is given and that users agree to share any corrections or other changes they make with the project so all other users can benefit. The entire database may be downloaded and held on local servers, but for many purposes the OSM servers provide good facilities.

Launched in 2004 by Steve Coast, it operates through thousands of individuals collecting and volunteering copyright-free data (see Sections 1.5.6

and 17.3.1.3 on volunteered geographic information [VGI] and Figures 17.7 and 18.6A). The tools used are also open source. The growth in use of OSM has been spectacular (Figure 18.6B), including that by businesses, governments, and nonprofits as well as individuals. Underpinning the OSM project is the OpenStreetMap Foundation, a not-for-profit organization. The OSM Foundation owns and maintains the infrastructure of the OpenStreetMap project.

We now draw lessons from more official contemporary partnerships focusing on national and supranational ones. Broadly, many of these large-scale collaborations are encompassed within the collective title of spatial data infrastructures (SDI).

## 18.6.1 Spatial Data Infrastructures: The U.S. Experience

The U.S. NSDI is far from being the only one worldwide; important SDI developments occurred even earlier in Australia and some other countries. There has even been an organization set up to take forward the concept of a global spatial data infrastructure (GSDI). Moreover, many other developments have occurred at

**Figure 18.6A** Number of nodes added to the OpenStreetMap database each day.

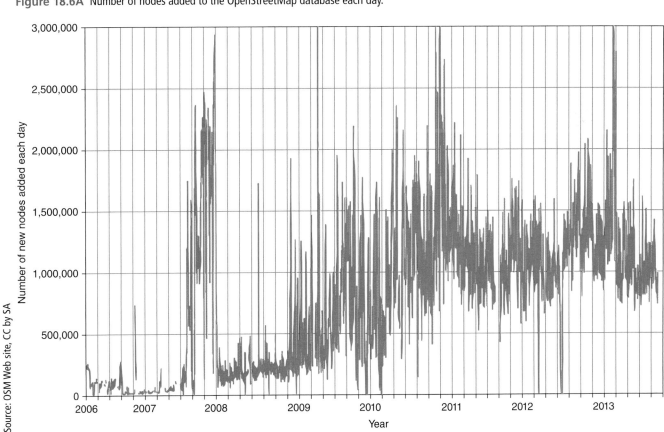

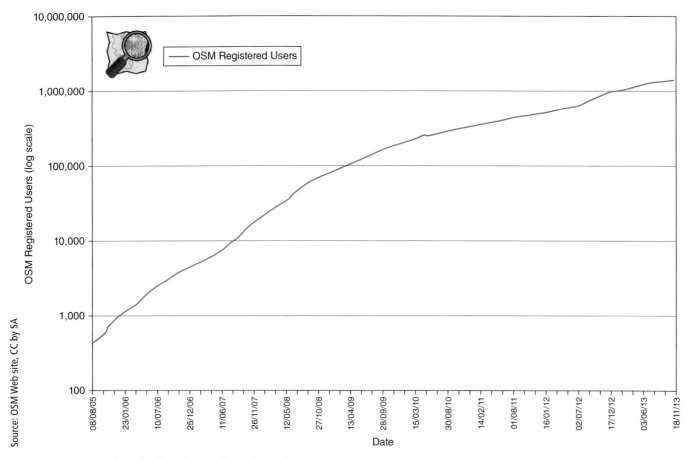

**Figure 18.6B** Number of registered users of OSM (log scale).

state and local levels in the United States and at various levels in other countries (Figure 18.7). For practical purposes, however, the NSDI concept and reality originated in the United States in 1994 with President Clinton's *Executive Order 12906—Co-ordinating Geographic Data Acquisition and Access: The National Spatial Data Infrastructure*. This order directed that federal agencies, coordinated by the Federal Geographic Data Committee (FGDC), carry out specified tasks to implement the NSDI; it created an environment within which new partnerships were not only encouraged but also required of federal bodies. This was all embedded further through the contents of the Office of Management and Budget's (OMB) Circular A-16 and the U.S. E-Government Act of 2002.

The original concept of an NSDI was as a national endeavor: it was viewed as a comprehensive and coordinated environment for the production, management, dissemination, and use of geospatial data, involving the totality of the relevant policies, technology, institutions, data, and individuals.

**The potential value of an NSDI in the United States was demonstrated by disasters like 9/11 and Hurricane Katrina.**

A substantial impetus for the NSDI and local SDIs was provided by a number of emergencies in the United States such as 9/11 in 2001 and the destruction wrought by Hurricane Katrina in New Orleans in 2005. The 9/11 experience in which the New York City Emergency Operations Center was located in a destroyed building, and data and mapping support had to be provided initially by staff from Hunter College, was particularly telling. Among the lessons learned was the need to duplicate and geographically distribute data and metadata, to have wide-area access to them, and to have better arrangements in place to circumvent bureaucracy in emergencies. Subsequent changes in technology have facilitated meeting the first two of these needs.

At the outset, the role of the private sector was scarcely factored into the NSDI concept—this was to be a federal government–led endeavor. Yet private-sector developments—many involving commercial products and partnerships—such as Google Maps and Google Earth, the availability of fine-resolution satellite imagery and SatNav systems based on road data—plus increasing activity by the states and local government and much else made the original concept seem very out of date.

Source: Web sites of responsible agencies

**Figure 18.7** Some examples of SDIs and portals to them.

Nevertheless, governments still have a role to play because of their data holdings (Section 17.4) and because of their control of relevant laws and regulations (Section 18.3).

**The biggest changes in GISS since the NSDI launch are due to commercial organizations—and yet these played minimal roles in the original vision.**

In late 2012 the U.S. Government Accountability Office (GAO) issued a strongly critical report on the progress of NSDI. The report concluded that "While the President and the Office of Management and Budget (OMB) have established policies and procedures for coordinating investments in geospatial data, government-wide committees and Federal departments and agencies have not effectively implemented them . . . because these efforts have not been a priority. FGDC's strategic plan is missing key elements, such as performance measures for many of its defined objectives." Six months later the FGDC issued a new draft strategic NSDI plan for public consultation.

The new FGDC strategic plan in response proposed three goals—to "develop national shared GI services . . . ensure accountability and effective development and management of Federal geospatial resources . . . and . . . convene leadership of the national geospatial community." These are backed

up by nine objectives and underpinned by 27 specific actions. Responses to the public consultation on this plan included fierce comment by a national association of private-sector geospatial firms, which complained this was not a strategy and did not address performance management. The commentary further argued that major elements were missing such as a critical part of the framework data (Section 17.2.1.4)—a national land parcel system.

The NSDI implementation difficulties highlighted by the GAO report are scarcely surprising for an activity where no one is really in charge, the activity was not adequately prioritized, and inadequate finances were ring-fenced for the activity. In a country with over 80,000 different public-sector governing bodies (e.g., counties, school boards)—perhaps the most complex geographical structure anywhere in the world, with some different jurisdictions overlapping in space—and a vigorous but neglected private sector, the task was huge. That said, the pioneering U.S. NSDI made some significant impacts; these are discussed in Section 18.6.4.

## 18.6.2 INSPIRE

Probably the most ambitious SDI-like scheme in the world is the INSPIRE (INfrastructure for SPatial InfoRmation in Europe) project (inspire-geoportal. ec.europa.eu/). This is based on a legal act set up by the European Council and the Parliament and covering the 28 member countries and some 508 million people of the European Union (in 2013). There is no European Union (EU) equivalent to the data collecting federal bodies like the Bureau of Census in the United States; all EU-wide data is sourced nationally and coordinated by agreement within the legal structures of treaties.

The purpose of INSPIRE is to support environmental policy, and overcome major barriers still affecting the availability and accessibility of relevant data. These barriers include

- Inconsistencies in spatial data collection
- Lack or incomplete documentation of available spatial data
- Lack of compatibility among spatial datasets that cannot therefore be combined with others
- Incompatible SDI initiatives in the Member States that often function only in isolation
- Cultural, institutional, financial, and legal barriers preventing or delaying the sharing of existing spatial data

The solutions to these problems involved creating metadata; harmonizing key data themes; building agreements on network services to provide discovery, viewing, and downloading of information; and policy agreements on sharing and data access. The scale of the enterprise is awesome: The U.S. NSDI covers seven data themes but INSPIRE covers 30; the internal heterogeneity in the EU is greater than the United States—Germany, for example has 17 SDIs, one for each of the lander (states) and one federal one, each requiring a separate legal act. All this has to be accomplished through working in 23 languages.

It was not possible to ask the Member States and their national and local organizations to reengineer all their existing databases. This ensured it was necessary to develop not only translation tools to help overcome the natural language barriers, but also agreed reference frameworks, classification systems and ontologies, data models, and schemas for each of the data themes. It took seven years of work to develop an agreed methodology (the Generic Conceptual Model) and tools, mobilize hundreds of experts in different domains, and deliver and test the specifications for the datasets falling in each of the themes. Thousands of pages of specifications and tens of thousands of comments had to be addressed individually during the stakeholder consultations. No less than 491 Spatial Data Interest Communities and 295 Legally Mandated Organizations were created, 259 experts were involved in drafting teams and working groups, and over 3000 user organizations have registered on the INSPIRE website. The scale of the task

and work has been described by Italians as Herculean and by Russians as Stakhanovite!

**INSPIRE may well be the most complex GI program yet enacted.**

INSPIRE is coming to the conclusion of all this work at the time of writing, so its success cannot yet be judged. It has, however, certainly provided valuable insights, which we describe in Section 18.6.4.

## 18.6.3 UN Initiative on Global Geospatial Information Management

GGIM is a very different form of partnership. It comprises a group of experts drawn from the national mapping and charting agencies (NMCAs) of over 60 UN member countries. They seek to gather case study evidence from around the world of GI system– and GI-related activities and share ideas about future developments and experiences. Case studies on Egypt's use of GI to help grow its economy and deliver better public services, on Brazil's GI use to reduce crime rates, and Korea's use of GI systems to enhance its cadastral system and manage land systems are examples of what the group has produced.

Perhaps unsurprisingly, this group is particularly fascinated by the future needs of governments and the future roles of mapping and charting bodies. A 2013 report produced for the group recognizes the changing relationship with private-sector GI providers and users and the growth of volunteered GI. It even speculates that the future role of NMCAs might well be in setting standards and operating more in a policy, advisory, and procurement role, rather than a production one.

## 18.6.4 Have SDIs Worked?

The SDI movement was *the* big development of GI in the 1990s. Two decades after its launch it is now possible to make an assessment of whether the many SDI schemes have worked effectively.

Not surprisingly, implementation of many NSDIs (and SDIs more generally) has not gone smoothly. Anything so broadly defined and forming such an intangible asset, whose creation involves huge numbers of different organizations and individuals—each with their own views, values, and objectives—and having no one in overall charge was always going to have difficulty achieving universal approval. In the United States, the post-Clinton administration had different priorities. There was also internal dissension within the federal government about who led on what; some shortcomings in agreements between federal, state, and local government and with the private sector on how to progress; and budget constraints, plus a host of other, competing new initiatives demanding staff and resources.

At a high level, few would dispute the merits of achieving data sharing, reduction of duplication, and risk minimization through the better use of good-quality GI. But how to make it happen for real, how to avoid excessive bureaucracy while being accountable to stakeholders, and how to measure success are different matters. A number of external-to-GI researchers have, for instance, criticized the way evaluations of SDIs have been carried out for not focusing on governance matters and for using only qualitative measures of success.

Craglia (see Further Reading and Box 18.4) has provided an example where the traditional NSDI approach did not work well initially but was then fixed. The Global Earth Observation System of Systems (GEOSS) was set up by the G8 countries to provide a framework for integrating the Earth observation efforts of the Group on Earth Observation, comprising 60 countries. GEOSS was a voluntary, "best efforts," multidisciplinary, and cash-limited enterprise. Three years after its launch only a few hundred datasets and services had been made available. However a step change was achieved by changing the operational model to one in which brokering was used to build bridges between different disciplines. It introduced a new middleware layer of service offerings in the SDI that achieve the necessary mediation, adaptation, distribution, semantic mapping, and even the quality checks required to address the complexity of an interconnecting multidisciplinary infrastructure. The approach is claimed to have resulted in discovering 28 million datasets in a few months!

Drawing on numerous reviews worldwide and particularly that of Craglia, we can conclude that:

- SDIs are more complex than many first thought. They are more about building social networks and reaching agreements on the semantics of data than technologies and standards. Applied social science is a crucial element in making SDIs work.

- The context has changed: SDIs have remained a relatively niche affair for public administrators and policy analysts, whereas a few large private-sector companies have come to dominate the global GI systems and services activities. This has enabled the dissemination and use of spatial information to an extent that could not be foreseen in the 1990s.

- There will always be different standards and practices across different communities, and it is better to accept diversity than trying to impose uniformity that is unlikely to work.

- There are multiple models for an SDI. In many countries such as Italy, Spain, Belgium, and Germany, the regional level of SDIs often forms the key building blocks of the national SDIs, with the national level providing a thin layer on the regional infrastructures. Others are rather more centralized, reflecting the institutions and history of the state.

- Adequate and continuing funding is crucial. In many SDIs this support has been lacking, in part because so many organizations are involved, and all see themselves as contributors rather than taking lead responsibility and/or providing significant funding. A widely agreed business plan with measurable targets and a clear statement of which organizations will benefit is vital.

- Inter- and intraorganizational conflict is inevitable in implementing an SDI. Thus successful SDIs build and maintain networks of people and organizations, in which technology only plays a supporting but nevertheless important role.

Curiously another contextual change may actually nurture SDIs in future—the growth of the Open Data movement and the recognition by G20 leaders that this has potentially major benefits for governments, businesses, and citizens (Section 17.4). Thus we may come to see SDIs, focused on spatial data or GI, as precursors to the wider national information infrastructures now being conceived (in which GI is playing a key role).

**Something as diffuse as NSDI will never be seen as a success by everyone, but it has been a catalyst for many positive developments and was a precursor of national information infrastructures.**

Perhaps the greatest success of SDIs has been as a catalyst, acting as a policy focus, publicizing the importance of geographic information, and focusing attention on the benefits of collaboration through partnerships.

## 18.7 Coping with Spatial Stupidity

We live in a world where we expect technology to work, and it usually does: Making satellite calls from the summit of Mount Everest is now routine. However, in some trying circumstances, technology may fail or be disabled by enemy forces (Section 17.5), hackers, or failures of organizational processes. Disasters may also occur through misuse by individual users.

**Beware the user's ability to create disasters using GI and GI technology.**

Technology alone rarely causes disasters. The disastrous replacement by Apple of Google maps with their own service in September 2012, with some towns being erroneously positioned by as much as 50 km (31 miles), demonstrated that even the most technologically competent organizations make serious management mistakes. Beyond that, Figure 18.9 shows that users can also misuse GI and GI technology to cause disasters. There is much speculation and some research on what causes an individual to ignore or misinterpret what the technology says. Studies of cognitive processes in relation to driving have stressed the importance of attention control, but most of the conclusions are theoretical at present. What we do know is that humans are good at producing post hoc justifications, as the captions to Figures 18.9A and 18.9B testify. Where consistent bad judgments occur, however, it may well be that there is something wrong with the satellite navigation system, the GI therein, or the design of the human interface. But spatial stupidity exists everywhere and is manifested even without help from technology. In May 2010 a novice sailor intended to sail his new motor boat—bought online—from Gillingham to Southampton round the English coast. To guide him, he only had a road atlas, and he navigated by keeping the land on his right. Sadly, he went around an island, the Isle of Sheppey, several times and ran out of fuel before being rescued. We must strive to make our systems safe for use, but there are limits to our ability to anticipate such stupidity on the part of *homo geographicus*!

**Figure 18.9A** Three Japanese tourists followed their GPS's instructions to drive directly at low tide through Moreton Bay to Stradbroke Island, Queensland, Australia. One student said, "The GPS told us we could drive there."

**Figure 18.9B** A bus driver responsible for a high school girls softball team drove into a bridge in the Washington Park Arboretum, Seattle—despite flashing lights and yellow warning sign. The president of the charter bus company blamed the GPS: "We just thought it would be a safe route, because why else would they have a [route] selection for a bus?"

**Figure 18.9C** After numerous vehicles got stuck in narrow lanes near Winchester, England, authorities erected a warning sign to disregard GPS instructions.

## 18.8 Conclusions

In this chapter we have highlighted the importance of operating as transparently as possible—about the tools used, the experience of staff, the appropriate business model and methods used, and the commitment to a professional code of ethics. Knowledge of all these needs to be in the public domain. Any organization also needs to have its own data and metadata in good shape, and its staff must understand the data's foibles, provenance, and those applications for which it is fit for purpose.

All the processes used must be documented and a good audit trail preserved of what was done and when. As hardly needs to be said, good working relationships with clients and good customer care are essential. Before anything goes wrong, however, the managers and leader need to know where to go for legal advice in any area of their activity touched by the law (i.e., all of it).

To win business the GI professional will need to be articulate, persuasive, and honest in his or her presentations as well as understanding the science, social science, and technology of GI. He or she will need to ensure that the organization's reputation is burnished by keeping it positively in the public eye through the media.

Enthusiasm and commitment are characteristics of the GI world. But to survive and prosper, a degree of skepticism is helpful, as is a knowledge of others in the business and the ability to make a judgment of what is good enough to meet or surpass the client's requirements without bankrupting the organization.

As pointed out earlier, it is sometimes wise to join with others to achieve desired ends. The lessons from GI partnering are clear: Small and focused partnerships are much more controllable and successful than large and diffuse ones. Nevertheless, there is no escape from some national-level thinking, planning, and action if resources are not to be wasted and benefit is to be maximized. Do not, however, expect to be popular or progress to be rapid in such schemes!

All this sounds formidable, but much of it is not. It becomes part of the normal world once anyone has been appropriately trained and has accrued management responsibilities. Good luck!

## QUESTIONS FOR FURTHER STUDY

1. What business models have been used in commercial GI systems enterprises? What might be a good business model for a government GI systems unit?

2. Describe, using an example culled from the Web, one example where intellectual property rights have figured in disputes involving geographic information.

3. Is privacy a doomed concept? If not, how can it best be protected in the GI world?

4. What is meant by a spatial data infrastructure, and what have GI system practitioners learned from SDI experience over the period since President Clinton's edict for the United States to create one?

5. What forms of GI-related partnership can you identify in your local area? What have they achieved?

## FURTHER READING

Cho, G. 2005. *Geographical Information Systems: Mastering the Legal Issues* (2nd ed.). Chichester: WileyBlackwell.

Craglia, M. 2015. Spatial Data Infrastructures Revisited. In Wright, J. (ed.). *International Encyclopedia of the Social and Behavioral Sciences* (2nd ed.). Amsterdam: Elsevier.

GISCI. 2008. *A GIS Code of Ethics*. www.gisci.org/code_of_ethics.aspx

GGIM. 2013. *Future Trends in Geospatial Information Management: The Five to Ten Year Vision*. United Nations Global Geospatial Information Management Initiative, ggim.un.org

INSPIRE web site. inspire.jrc.ec.europa.eu/

L'Aquila Earthquake and Jailing of Scientists. See www.guardian.co.uk/commentisfree/2012/oct/23/italian-scientists-charged-laquila-earthquake; news.sciencemag.org/scienceinsider/2013/01/judge-in-laquila-earthquake-tria.html; www.nature.com/news/l-aquila-verdict-row-grows-1.11683 and www.economist.com/news/science-and-technology/21632459-six-seven-scientists-convicted-earthquake-advice-have-been-cleared-laws

Obermeyer, N. J. and Pinto, J. K. 2008. *Managing Geographic Information Systems* (2nd ed.). New York: Guilford.

# 19 Epilog: GISS in the Service of Humanity

LEARNING OBJECTIVES

The world is a very differentiated place: for example, 1% of the world's population owns 50% of the world's ownable assets. Although many people live in relative comfort, many millions of others live in great poverty, at risk of natural or human-made disasters, and succumb to avoidable diseases that kill or handicap; children often suffer the most. Warfare and strife are commonplace. The causes of this differentiation are many and varied. Moreover, we are increasingly interconnected: A disaster suffered by one community often extends to others directly or indirectly.

Most of us would wish the world functioned more equitably and with greater resilience. This chapter describes how we can use our geographic information science and systems (GISS) skills, tools, experience, and the knowledge set out in this book to make this happen. We start with a process for how we should operate then go on to suggest the Grand Challenges to whose solution GISS and GISS professionals can contribute.

**LEARNING OBJECTIVES**

After studying this chapter, you will understand:

- Just how differentiated is world geography, notably in the incidence of poverty and health.
- The biggest challenges facing humanity and how these are pervasive and dynamic.
- The interdependency between many factors causing these problems.
- Why geography is not merely the context for such challenges but also part of the problem and the solution.
- How GISS help us tackle these challenges.

## 19.1 GISS, the Active Citizen, and Citizen Scientists

In earlier chapters we have identified the science underlying the use of geographic information (GI) systems, the technologies that are transforming what we can do, and some of the constraints that need to be circumvented. We have given details of individual projects where the application of GI technology has made a real difference, and we have celebrated the work of people who have advanced the cause.

In this final chapter, however, we ask what we who work in GISS can, should, or will be able to do for humanity, which is faced with profound global challenges—a world that is divided, unequal, and troubled. To set the scene and ensure that everyone starts from a common base, we sketch out a process for maximizing beneficial use of GISS. Then we provide a thumbnail sketch of what we (and others) see as the Grand Challenges. Linked to this we draw on past experience of GI system deployment, the GI literature, foresight, and intuition to show how GISS practitioners can help improve matters.

### 19.1.1 Who Can Help?

We stress that contributions can be made by people in many walks of life: the GI systems designer or technician, their manager(s), presidents of firms and their Board members, and public servants and politicians. Commercial enterprises have a vital role to play. What is necessary is a commitment to using sound evidence to inform decisions and an understanding of how central is geography to the life experience of all of us. The biographical boxes in this book demonstrate that this is realistic, not utopian. Nor are we a tiny community: Because many GI technology functions are embedded in services, notably those in Web-based mapping and vehicle guidance systems, all users of these systems are therefore users of "our" functionality—even if they do not appreciate it (Section 10.1). In total, such users already may well number over a billion people worldwide. Beyond this ubiquity are the professional users of GI systems. Our guesstimate is that there are over a million of those around the world.

Some citizens are also amateur scientists. In the words of *Scientific American* "Research often involves teams of scientists collaborating across continents. Now, using the power of the Internet, nonspecialists are participating, too. Citizen Science falls into many categories. A pioneering project was SETI@Home (Section 10.1), which has harnessed the idle computing time of millions of participants in the search for extraterrestrial life. Citizen scientists also act as volunteer classifiers of heavenly objects [and] they make observations of the natural world."

We have already identified (Section 18.6) the role of active citizens in building Open Street Map. A good example of the citizen scientist in the GI domain is given in Box 19.1.

---

## Application Box (19.1)

### GI System–Enabled Field Science on a Budget

One example of what can be achieved with very modest means is the science work carried out in all its expeditions by British Exploring (BE). Formed in 1937, the society runs about five expeditions annually to very tough environments mainly for a total of around 200 people aged 17 to 21. Fund-raising is done by those on each expedition and by corporate means; the total staff of the Society is about 6 full-time equivalents but with contributions from many more volunteers. The overall aim is personal development and nurturing of leadership abilities of young people but carrying out good science (under expert guidance) and publicizing it afterward through scientific papers is a key measure of success. Figures 19.1A and 19.1B were

(A)

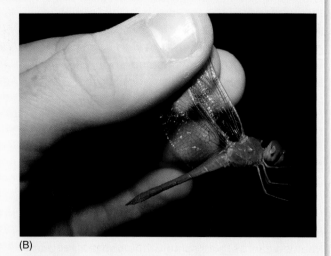

(B)

**Figure 19.1** Work by citizen scientists. (A) An infrared photograph of the Arabian leopard, taken by the noninvasive means of a camera trap. *Panthera pardus nimr* is a critically endangered species—under 250 are believed to exist in the 3.2 m km² of Arabia (one-third of the area of the entire United States). (B) shows a male *Tholymis tillarga* dragonfly—never previously recorded in Arabia, found at a shallow pool in Wadi Sayq, Dhofar.

▶

◀

taken on a 2012 expedition to the Dhofar desert areas of Oman, with the active support of the Oman Office for Conservation of the Environment and use of GI systems to map and analyze research results. These show fauna in that area previously unknown to the "outside world."

**Increasing Distance Inland from Wadi Mouth** →

| | A | B | C | D | E | F | G | H | I |
|---|---|---|---|---|---|---|---|---|---|
| *Anax ephippiger* | | 1.5% | | 2.6% | | 2.2% | | | |
| *Anax imperator* | | 0.8% | | | 1.3% | 1.1% | | | |
| *Azuragrion nigridorsum* | | | | | | 5.5% | | | |
| *Ceriagrion glabrum* | 16.7% | 21.5% | 18.0% | 20.5% | 17.5% | | | | 7.2% |
| *Crocothemis erythraea* | 16.7% | 13.1% | 2.0% | 17.9% | 1.3% | 2.2% | 6.3% | 1.0% | 9.3% |
| *Diplacodes lefebvrii* | | 7.7% | | 7.7% | | | | | |
| *Ischnura evansi* | | 0.8% | | | | | | | |
| *Ischnura senegalensis* | | 2.3% | | 2.6% | 10.0% | | | | |
| *Nesciothemis farinosa* | | 1.5% | | | | | | | |
| *Orthetrum chrysostigma* | 16.7% | 18.5% | 64.0% | 41.0% | 7.5% | | 6.3% | 17.6% | 18.6% |
| *Orthetrum ransonnetii* | | | | | | | 6.3% | 17.6% | 2.1% |
| *Orthetrum sabina* | 8.3% | 4.6% | 8.0% | 2.6% | 8.8% | 2.2% | | | |
| *Pantala flavescens* | 16.7% | 14.6% | | | 3.8% | 2.2% | 12.5% | 4.9% | |
| *Rhyothemis semihyalina* | | 0.8% | | | | | | | |
| *Tholymis tillarga* | | | | | 2.5% | | | | |
| *Trithemis annulata* | | 0.8% | 8.0% | | 10.0% | | | | |
| *Trithemis arteriosa* | 25.0% | 11.5% | | 5.1% | 37.5% | 84.6% | 68.8% | 58.8% | 62.9% |

0%
<2.5%
2.5-5%
5-10%
10-20%
20-50%
>50%

**Survey Sites**
A - brackish lagoon
B - vegetated freshwater lagoon
C - estuarine marshland
D - estuarine marshland
E - large temporary pool
F - small temporary pool
G - stagnant temporary pool
H - rocky pools
I - small pool at spring

(C)

**Figure 19.1C** shows the species composition of nine dragonfly communities in Wadi Sayq. All three figures are courtesy of the British Exploring Society and the Oman Office for Conservation of the Environment.

## 19.1.2 Areas Where GISS Contributes

Put simply, this approach is built on four big contributions that geographic information science and systems can make:

- Supporting our ability to help discover and share new understandings in the physical, environmental, and social sciences
- From these new understandings, providing the means to help us devise new products, processes, and services that improve the quality of life, especially for the disadvantaged of the world
- Using these to enhance the efficiency of public and private tasks to release resources for other valued things
- Achieving all this in as sustainable a way as possible

## 19.2 Context: Our Differentiated World

As suggested earlier, the world is highly differentiated and heterogeneous (Section 2.2). If this is true for the physical environment (Figures 1.9 and 19.2), it is at least as true for human societies that are very unequal in wealth, education, and much else. The best single guide to this is probably the World Bank's Development Indicators for the years published from 2008 to 2013. Given the difficulty of collecting reliable, consistent, and up-to-date statistics from 214 countries, some of these will be less good than is desirable. Nevertheless, the picture they paint is a bleak one, if improving in certain respects. And it is clear that, from topography through natural hazards to political systems, economic well-being, and adequacy of food,

(A)

Source: Wikipedia

(B)

**Figure 19.2** Examples of the world's differentiated physical landscapes—with very different spatial autocorrelation properties of their topography. (A) Wheatlands showing very little variation in elevation. (B) Mount Everest North Face.

our life experiences and opportunities are shaped—and in some cases determined—by geography. For example, our expectation of *how long* we will live differs greatly depending on *where* we live (Tables 19.1 and 19.2); life expectancy is often used as one indicator of life quality. Figure 19.3 shows the highly differentiated local geodemography of Glasgow.

Wealth and consumption are very unequally distributed globally (Figures 19.4 and 19.5). In 2011 Qatar had Gross National Income per capita of US$86,000 in Purchasing Power Parity (PPP) terms, 340 times that of the Democratic Republic of Congo. In 2010 there were 1.2 billion people living on less than $1.25 per day—the definition of extreme poverty. The rich consume massively more than the poor: for instance, electricity consumption per capita is over 13 times greater in high-income countries than in low-income ones. Figure 19.5 shows the consumption

**Table 19.1** Life expectancy at birth in 8 selected countries. (Source: www.cia.gov/library/publications/the-world-factbook/)

| Country | Estimated life expectancy at birth in 2013 |
|---|---|
| Japan | 84.19 |
| Singapore | 84.07 |
| Switzerland | 82.28 |
| Australia | 81.98 |
| Afghanistan | 50.11 |
| Guinea-Bissau | 49.50 |
| South Africa | 49.48 |
| Chad | 49.07 |

disparity by income level. Figure 19.6 illustrates the startling difference in power consumption between North Korea and its neighbors. The poorer you are and the poorer is the country in which you live, the lower tend to be the levels of education attained; this in turn affects much else, including statistics on mortality at birth.

Two important qualifications need to be made: The first is that some dire situations are becoming less so, notably through multinational efforts to reduce extreme poverty (Section 19.6.2) and the growth of some economies, notably the Chinese. The second point is that, despite these overall improvements in certain countries, variations within countries are often widening (notably in wealth and income). This is manifested geographically; for example, most rich people are geographically concentrated within their own countries.

> **The world is highly differentiated on almost every criterion and at many levels; if it were not, geography would not exist!**

**Table 19.2** Life expectancy at birth in small administrative areas. (Source: World Health Organization, 2008. *Closing the gap in a generation.*)

| Small area | Estimated life expectancy at birth |
|---|---|
| Calton, Glasgow, UK | 54 |
| Lenzie North, Glasgow, UK | 82 |
| Washington, DC, USA (black only) | 63 |
| Montgomery County, Washington Metropolitan Area (white only) | 80 |

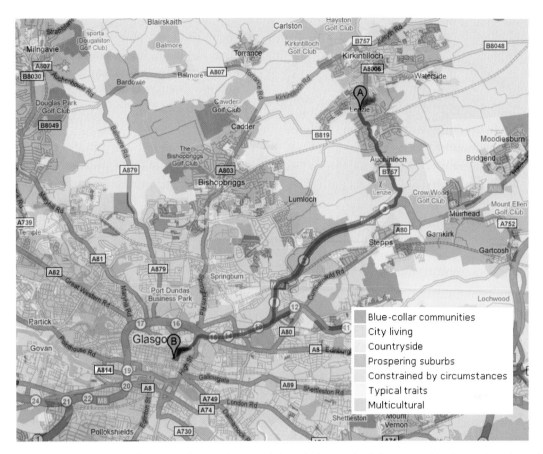

**Figure 19.3** The local geodemography of Glasgow, showing the 7.8-mile (12.6-km) route that links communities (A: Lenzie and B: Calton) with life expectancies at birth that vary by 28 years (see Table 19.2). (Background data: Google Maps)

**Figure 19.4** The global wealth pyramid: wealth held by adult individuals of different net worth in 2013. Individual net worth is defined as the marketable value of financial assets plus nonfinancial assets, mainly housing and land, less debt.

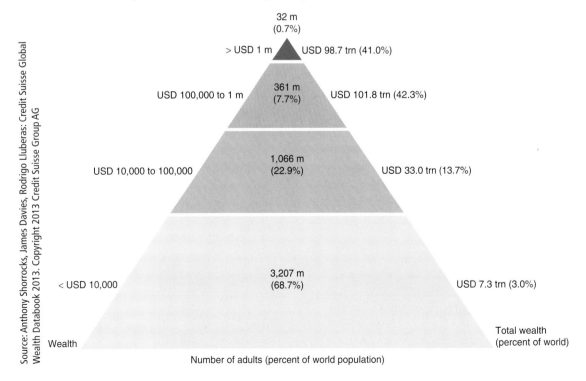

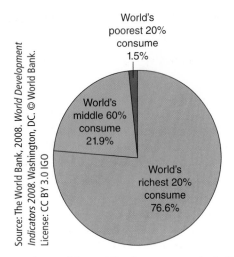

**Figure 19.5** Shares of the world's private consumption in 2005 by wealth.

# Context: Our Interdependent World

Interdependency operates at many different levels. For instance, there is clearly much interdependency at the local level between shortage of food, access to facilities like fresh water and sewage, and disease. It operates at all geographic levels: At the regional level the 2004 Indonesian tsunami killed around 225,000 inhabitants of several countries around the Indian Ocean.

Just as striking are the interdependencies that arise from human systems, notably financial and trade ones. The globalization of business has had major effects, some beneficial, and has led to big changes in the structure of employment and hence to human geography. Much employment in the developed world has moved out of manufacturing into service provision; low-margin manufacturing and many commoditized services have become the preserve of low-wage economies. Global firms or partnerships have been able to outsource or to move work offshore through the advent of the Web; the creation of workflow software has enabled collaborative working across the globe including integrated supply chains. The consequences of all this have included growing urbanization, international migration, and increasing global competition for talent and innovation.

Part of the taken-for-granted world that undergirds all this is the global trade and financial system. Over a 50-year period, trade between nations has multiplied hundreds of times. Seasonal availability was all but eradicated as we imported exotic flowers (and much else) by air from countries on the other side of the world. Measures of the world's wealth based on GDP grew dramatically between 1945 and 2006. Underlying all that was an increasingly globalized financial system in which stocks, commodities,

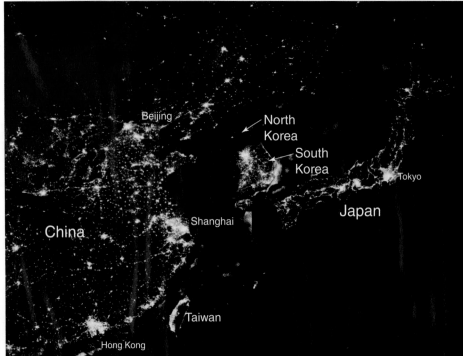

**Figure 19.6** Differences in consumption of power in different countries: North Korea is a black area compared to its neighbors in this nighttime imagery collected by the Suomi National Polar-orbiting Partnership (Suomi NPP) satellite. The image has been processed to remove auroras, fires, and other stray light to emphasize the city lights.

foreign exchange, and much else were traded constantly, especially through the three major financial centers in New York, London, and Tokyo (their success owing much to their locations equally distributed around the world and permitting 24-hour trading). The entire system provided credit. Without credit, societies cannot change, business grow, or individuals acquire property. Technology played a major role: The ability to move funds near-instantaneously across the world in response to opportunities large and small enabled speculators to profit from minute differentials in pricing.

All of this, however, has led to bouts of massive instability. The financial boom, followed by a crisis from 2007 on, had dire consequences. In the United States, and especially Europe, economies were severely hurt and the lives of millions of individuals disrupted: in Spain and Greece for example, unemployment among those aged under 25 reached over 50% in 2013. If nothing else, all this demonstrates that global interdependency on most fronts—economic, environmental, and much else—is a fact of life, that virtually everyone on Earth is affected by it, and that consistent stability is an illusory concept. But the impacts are not the same everywhere: geography matters. To be successful we need a sound process that takes into account our knowledge, analytical skills, tools, and information.

**Today's world is tightly interlinked: A disaster in one geographic area can trigger global impacts.**

## 19.4  The Process

The problems on which we have focused (Section 19.5) are bewildering in the range and linkage of their causal factors and their interdependencies. The feedback loops involved are only partly understood. Some of the factors are not even easily quantifiable. And the situation is often dynamic. All this indicates we need some sort of process that enables us to prioritize our actions and a recognition of the uncertainty involved.

What follows is aimed at active GISS professionals. The essence of working in GISS is that we are committed to the use of evidence to underpin decisions. Evidence must be assembled from the best available sources; tested to ascertain its veracity; used in defined ways with tools whose internal mechanics are well understood; and shared so that our findings can be tested and replicated by others. In short, we must operate as scientists. The nature of the Grand Challenges outlined later enables those

working with GISS to make particular contributions because:

- Many problems are manifested initially through geographical variations.

- Studying the geographical manifestation can help us to propose and test causal factors and hence identify possible solutions to the problems. Even if causal factors cannot be isolated, we can compute correlations between GI variables as a second-best outcome (see Box 17.2 on Big Data).

- The human systems through which we have to tackle problems are normally geographically structured (e.g., administrations that control access or provide resources or that need persuading).

- There has been a significant change, at least in Western democracies, toward the requirement for quantifiable and published evidence to support and justify policy making. Audit trails of analyses are a crucial part of such a way of working, and integrated GI systems can provide such audit trails.

As shown in earlier chapters, GISS has developed greatly in the last 30 years and provides demonstrable benefits:

- The integrative capacity of GI technology, enabling us to link multiple datasets and generate added value, analyze the results spatially, then redo the whole operation at short notice whenever the situation changes or we wish to vary our assumptions.

- Use of the same GI system tools at local, regional, national, and global levels simplifies linking models operating at different levels of abstraction.

- GI systems' growing modeling capability now forms part of common tools for data mining, mathematical modeling, and simulation. These tools are central to work in sectors from retailing through financial services to health care, environmental modeling, and transport planning and design.

- Superb visualization and user interaction capabilities, especially in regard to a widely appreciated communication medium—mapping.

- Our ability and willingness to share data, ideas, and concepts, increasingly on a global basis via a combination of workflow software, common standards, interoperability, and a GI community where mutual support is commonplace.

- A growing awareness and acceptance of the value of GI systems among policy makers/government leaders, public servants, and the business and education communities.

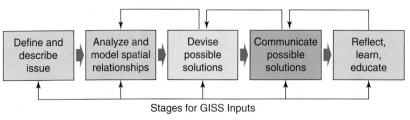

**Figure 19.7** Simplistic illustration of the process by which inputs by GISS professionals can contribute to tackling the Grand Challenges. Note that this ignores many feedback loops and other complications; GI systems have a more limited area of contributions.

We can best structure our potential contributions to the Grand Challenges under the simple model of different stages shown in Figure 19.7. Rarely, of course, is such a process so linear in practice: this ignores many feedback loops and the "soft factors" involved in innovative resolution of problems. But it will serve for present purposes.

### 19.4.1 Stage 1: Defining and Describing the Issue

This seems basic, even prosaic, but defining the objective and describing a problem are fundamental to success. In some cases we simply seek to know what (e.g., the type of geology or what is the most common disease) at a particular place or where specified incidences exist (e.g., rocks of type X or a high prevalence of certain diseases are to be found). In other cases we will wish to compute the trafficability of terrain or the distribution of wealth by combining multiple datasets, some of them proxies for what we would really use if only they were available. In a Big Data approach we may however simply let the computer find high correlations among many datasets and then seek to isolate ones of interest for further analysis.

To do this involves assembling and reusing any possibly relevant data from whatever sources are available—internal to an organization, from national or international statistical bodies, from commercial sources, or from the Web. To maximize the reliability of our answers this necessitates assembling and integrating the data in as consistent, documented, and professional a way as is possible in the circumstances. In addition, making them as widely available as possible to others with an interest is potentially invaluable because it provides for independent scrutiny and critique: Where data are widely used they become of higher quality as shortcomings become identified and better investigations ensue. We recognize that the confidentiality of work in the commercial sector and the need to be first to market sometimes precludes such widespread sharing. Yet in today's service-driven economy, the speed of innovation is such that it is often churlish, time consuming, and ultimately pointless to become totally consumed by issues of

intellectual property rights. This is acknowledged, for example, in commercially funded work of the Willis Research Network. Willis is a global commercial insurance organization that has successfully embraced an open publication approach involving more than 50 universities worldwide.

### 19.4.2 Stage 2: Analyzing and Modeling Spatial Interrelationships

Many examples of spatial analysis have been presented in this book. Beyond the simple queries such as "where is . . ." we normally wish to find the root causes of problems. But the reality is that detailed analysis of correlations and apparent relationships, of leads and lags between changes in different variables, and of changes in these relationships through time and over space are required if we are to go beyond superficial judgments.

As a result, some of the team members at least must be mathematically and statistically literate and expert in managing and exploiting data. Modeling drought frequency on a global basis or even demonstrating the relationship between the incidence of diarrhea and the availability of piped water in Africa is not a trivial endeavor. It is all too easy to create false positives and false negatives in circumstances where data at different spatial and temporal resolutions and different accuracies are involved.

### 19.4.3 Stage 3: Devising Possible Solutions

To devise possible solutions requires not only technical GI skills but also much domain knowledge. It is rare that one individual has both, unless someone has been trained in a relevant substantive discipline and then been immersed in GI systems. So normally this stage requires effective teamwork and continuing challenge. The criteria for deciding what is a possible solution involve not only sound analysis steered by the set objectives but also whether the solution is deliverable, the risks (e.g., the possible unintended consequences of adopting it), and its cost and ease of implementation.

## 19.4.4 Communicating Results and Possible Solutions to Decision Makers

The communication of the results of analyses and of possible solutions (including options) is critical to success as Section 18.1 illustrates. The skills of the decision makers may be very different to those of the GISS specialist. It is essential that those communicating results to decision makers are skilled in verbal and presentational skills as well as technical ones. The communication must be made in terms that the decision maker clearly understands and also made available in written form. The need to ensure that the decision maker knows enough safely to brief the media (where required) is essential.

Although this is generic to dealing with the challenges we have described earlier and will develop in Section 19.6, this counsel of perfection is difficult when dealing with emergencies. The role of small, technologically accomplished, and savvy organizations (such as MapAction; Section 19.6.5) in natural disasters has proven hugely important. Fast in action and less encumbered by bureaucracy than many state organizations, they typically exploit whatever information is available. They succeed only insofar as they do not alienate the local authorities who determine access to the area (compare this with the Bowman expedition: Section 1.5.5). This means that they too have to be able to communicate the nature of the problem to those authorities in a way that minimizes misunderstandings, facilitates discussion, and enables effective decision making—but in a truncated timetable.

## 19.4.5 Stage 5: Reflect, Learn, and Educate

After any project we must review what happened, the things that went well, and those that did not. This is an essential part of learning and continuous improvement. But there is a much wider issue about education for, in our view, the world of education needs to evolve substantially because of the rapid changes in technology, the diverse challenges we face, and globalization of enterprises. Ideally, we want technologically alert individuals who have specialist skills and also broad understanding of many other facets which determine success—the so-called T-shaped people.

For the past several decades, however, it has been standard practice to think of GI systems education as a process of training professionals to acquire substantial skills in manipulating the user interfaces of GI systems and (sometimes) in understanding the science that lies behind them. But today many of these GI system capabilities are familiar even to young children, who have become accustomed to the user interfaces of Google Maps and similar services. So we need to shape our education systems to a very different world full of digital natives. What skills, if any, are needed by a general public that is now increasingly able both to produce and to consume geographic information? Is there any point in teaching navigation when our handheld GPS receivers do it all for us (Section 10.3)? Do we need to carry out fieldwork if or when augmented reality can deliver an adequate understanding of parts of the real world with minimum cost and reduced risk? Our conclusion—on which this book is based—is that much greater levels of understanding of science principles and also wider societal issues are required than in the past. In particular, one of the fundamental forms of human reasoning—spatial thinking—is given little attention in the school or college curriculum. Yet there is abundant research to show that early attention to these concepts can lead to improved performance in a range of subjects. The delivery mechanism for all this may involve a different educational paradigm to that in place until now. We note the dramatic innovations in how education can be delivered and the form it takes through such developments as the Khan Academy and MOOCs (massive open online courses).

**Traditional GI system education needs to be replaced by more science-based curricula.**

In the longer term such education can only be achieved through schools and through inspiring teachers. But official recognition helps: in England, for instance, a revision of the national curriculum for Geography in 2014 made clear that all pupils must be "... competent in the geographical skills needed to ... interpret a range of sources of geographical information, including maps, diagrams, globes, aerial photographs and Geographical Information Systems (GIS)." This was mandated for all 11- to 16-year-old pupils in all state-funded schools.

Finally, we need all those who are involved in GISS to become ambassadors for what we can contribute. This is easiest where we converse with those commissioning work. Beyond those, however, we have a need to reach the general public. To do so requires recognizing that our technical lexicon is a foreign language to others and requires us to use their own language to inform or persuade them.

## 19.5 The Grand Challenges

There are many different views on the most important challenges facing humanity: some might point to the growth of urbanization and its consequences. Whereas only 13% of the world's population lived in cities in 1900, over 50% now do so, creating particular hazards (and opportunities). Indeed almost all the Grand Challenges we describe are encountered in the world's major cities.

There is no shortage of problems that GISS experts can help to ameliorate! An obvious set was formed at the global level through the adoption of the Millennium Development Goals (MDGs) following the Millennium Declaration in 2000 by all United Nations Member States. Progress in meeting the Goals is tracked annually by the UN against 21 targets and 60 indicators addressing extreme poverty and hunger, education, women's empowerment and gender equality, health, environmental sustainability, and global partnership (see, for example, mdgs. un.org/unsd/mdg/Resources/Static/Products/ Progress2014/English2014.pdf). A new set of goals—the Sustainable Development Goals—is likely to succeed the MDGs when the latter expire in 2015. The 2013 World Bank Development Indicators Report (see Further Reading) summarizes progress since 1990. This has been good on certain indicators—for example, reducing numbers in extreme poverty by 100 million since 2008—but 1.2 billion are still in such poverty.

Another assessment of global risks—and hence Grand Challenges—is given by the World Economic Forum, which seeks to look up to 10 years ahead. Quantifying most risks more than 5 years ahead is extraordinarily difficult in any meaningful way. Given the periodicity of many natural hazards, this often makes anticipation beyond that period informed speculation at best, but is still vital because lead times for remedial action often take many years.

Some Grand Challenges are quintessentially national or subnational in scope. An example of a (rarely public) National Risk Register is set out in Table 19.3. Even where such risks of national and international importance—such as climate change—are identified, anticipation of the detailed threat, adaptation, and immediate action needed to remedy a disaster are often local. In the United States, for instance, land use, zoning, much construction, and transport are typically under state or very local control. This provides many opportunities for GISS professionals to achieve benefits in planning, communicating threats, and carrying out operations efficiently.

**Even where challenges, for example, from climate change, are national or international, anticipation, adaptation to, and remedy of disasters are often largely local—and GISS expertise can help.**

**Table 19.3** UK high-level National Civilian Risk Register as of March 2013. (Source: Beddington, J. B., *Threats and Opportunities—the scientific challenges of the 21st Century.* www.foundation.org.uk/events/pdf/20130206_Beddington.pdf)

| Relative impact | Relative likelihood of occurring in the next 5 years | | | | |
|---|---|---|---|---|---|
| | Low Between 1 in 20,000 and 1 in 2,000 | Low-Medium Between 1 in 2,000 and 1 in 200 | Medium Between 1 in 200 and 1 in 20 | Medium-High Between 1 in 20 and 1 in 2 | High Greater than 1 in 2 |
| 5 Catastrophic | | Catastrophic terrorist attacks | | Pandemic influenza | |
| 4 Significant | | | Coastal flooding Effusive volcanic eruption | | |
| 3 Moderate | Major industrial accidents | Major transport accidents | Other infectious diseases Inland flooding Smaller scale CBR attacks | Severe space weather Low temperatures and heavy snow. Heatwaves | |
| 2 Minor | | | Zoonotic animal diseases Drought | Explosive volcanic eruption Storms and gales Public disorder | |
| 1 Limited | | | Non-zoonotic animal diseases | Disruptive industrial action | |

Our own selection of challenges, described in the following section, overlaps with but is not identical to governmental views. Our aim is to describe the challenges in such a way as to highlight interdependencies and to identify where, with our skills, knowledge, and technologies and determination, we GISS practitioners might make a contribution to ameliorating them.

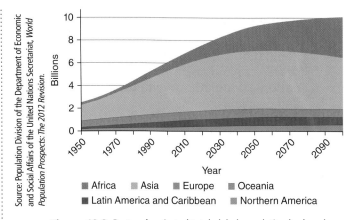

Source: Population Division of the Department of Economic and Social Affairs of the United Nations Secretariat, *World Population Prospects: The 2012 Revision.*

**Figure 19.9** Past and projected total global population broken down by major regions.

## 19.6 Grand Challenges Whose Effects We Can Help to Ameliorate

### 19.6.1 Population Growth

Reaching the first billion population took almost all human history; the second billion took just over a century up until 1930; whereas the third and fourth billions took 30 years and 15 years, respectively. The global population as of 8 October 2014 was estimated by the U.S. Bureau of Census to be 7.197 billion.

Eighteenth-Century views that population growth would outrun food supply have been averted or delayed by crop breeding, technological improvements, and other factors. But the problem of feeding the world's population remains. The size of global, national, and local populations, and economic or political migrations between them, manifestly drives consumption of food and many natural resources, some of which are finite though in practice may be substitutable. Furthermore, positive feedbacks occur: If any of the inequalities described in this section can be rectified, consumption will increase still further, placing even greater stresses on the environment and its sustainability.

The good news is that the *rate of increase* of global population has been falling (Figure 19.8). Until the publication of a new UN report in September 2014 many demographers expected the total to level

off to somewhere under 10 billion after the middle of the Twenty-First Century, that is, with about 35% more people on the planet than now. Much of the net increase will occur in the poorest areas of Africa and Asia (Figure 19.9). The new report, however, uses a Bayesian probabilistic methodology and suggests world population is unlikely to stop growing this century. There is claimed to be an 80% probability that world population, now 7.2 billion, will increase to between 9.6 and 12.3 billion in 2100. Notwithstanding this new development, the factors seemingly influencing a deceleration of growth as compared to previous decades are highly relevant for this epilog: they include the consequences of better sanitation and health care, increased education and employment opportunities for women, and the availability of family planning and contraception.

The geography of population change, as well as population density, is far from uniform. In many developed countries such as Germany and Russia, population growth is zero or negative. This has serious implications for sustaining the tax base to provide public services, providing enough labor for commerce, or supporting the increasing numbers of aged people.

In a narrow sense, international migration provides a simple solution, and it more than doubled between 1970 and 2000, with the largest proportion of migrants moving to countries in the developed world. In Europe, for example, almost all population growth is due to in-migration and the fecundity of the new arrivals. However, political tensions frequently arise from such migrations and from the changes they make in local culture and values. It is also the case that migrants are often disproportionately the most able and skilled in their country of origin, particularly given that developed countries place entry restrictions on in-migration of low-skilled workers. This means that migration ensures that the origin countries themselves usually become less well equipped to

**Figure 19.8** World population annual growth rates 1950–present and projected to 2050.

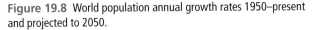

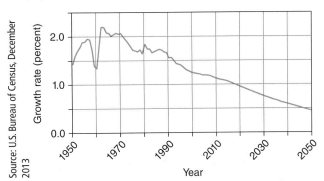

develop, thereby further accelerating the differentiation between rich and poor countries.

**Population migration is a major contributor to the size and average age of many national populations, with both benefits and problems.**

In principle, GI systems provide an excellent instrument for measuring the movement flows and the stock of population. However, measuring people numbers is becoming more difficult as response rates to censuses and surveys decline in most countries (Section 17.3.1.1) and proxy methods (e.g., remote sensing) remain less reliable (see Section 17.3.1.4). What works in some societies (e.g., the population registers of Scandinavian countries) will not be acceptable or effective in others. Thus there is a major challenge for scientists and GISS and other computer specialists to produce better methodologies—almost certainly based on linkage of many different kinds of data together—for measuring populations and changes to them over time (see Box 1.2).

Once again, however, the use of GI systems to measure and represent what is need not be the end of the story. Understanding *why* people move, *what* entices and constrains them, and *how* land-use patterns and social structures change as a consequence of population change is a major challenge for social scientists and GI experts. Together, GISS leverages the benefits of better data and modeling tools that are necessary to predict the impacts of changed laws, barriers, or incentives. Without this, deep understanding of migration in particular and population change in general is impossible, and policies to cope with them are likely to be ill-founded.

## 19.6.2 Poverty and Hunger

Poverty is a difficult concept to measure in a meaningful way, but the UN and other bodies have devised widely used thresholds based on measures of average personal income adjusted for purchasing power parity. Figure 19.10 shows how the numbers in poverty in 2005 varied depending on the income threshold used. Thanks to much effort, extreme poverty—experienced by those living on $1.25 per day or less—has more than halved to 22% compared with the situation in 1990. Yet almost half the world—nearly 3 billion people—still live on less than $2.50 a day, whereas about 5.2 billion people live on incomes of less than $10 per day. In addition, over 80% of the world's population lives in countries where income differentials are widening. Numerical comparisons are valuable: Box 19.2 presents some further, selected, grim figures concerning the state of global poverty and inequality. Mapping using GI systems adds greater richness and diversity to statistics such as these. GI science can further enrich the representation of geographic variation. For example, Figure 19.11 compares

**Figure 19.10** The global proportion and numbers of people in poverty in 2005 when that is defined at different income thresholds.

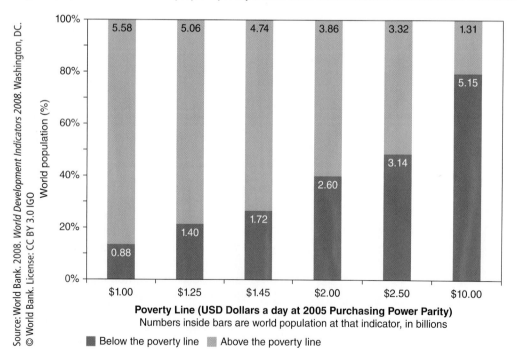

### Poverty and Inequality

- Almost 3 billion people live on less than $2.50 a day.

- The poorest half of the world's adults, 3.5 billion people, own a total of about $1.7tn worth of assets (about $400 per adult). That is similar to the wealth owned by the world's richest 85 people. Income disparities between rich and poor in most countries are widening.

- Nearly a billion people entered the Twenty-First Century unable to read a book or sign their names.

- Less than 1% of what the world spent every year on weapons was needed to put every child into school.

- One billion children live in poverty (1 in 2 children in the world), 640 million live without adequate shelter, 400 million have no access to safe water, 270 million have no access to health services, and 10.6 million died in 2003 before they reached the age of 5 (or roughly 29,000 children per day).

- According to UNICEF, 22,000 children die each day due to poverty.

Source: www.globalissues.org/article/26/poverty-facts-and-stats and policy-practice.oxfam.org.uk/blog/2014/01/working-for-the-few.

**Figure 19.11** The geography of human poverty. (A) The size of each territory shows the relative proportion of the world's population living there. (B) Territory size shows the proportion of all people living on less than or equal to US$1 in purchasing power parity a day in 2000. (Reproduced under a Creative Commons License. © Copyright Sasi Group (University of Sheffield) and Mark Newman (University of Michigan).)

(A)
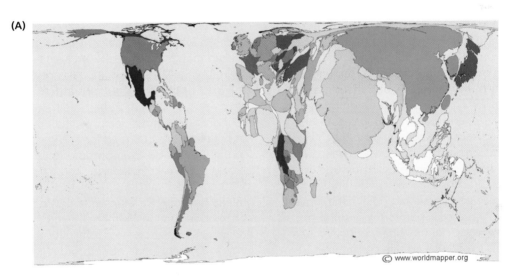

(B)
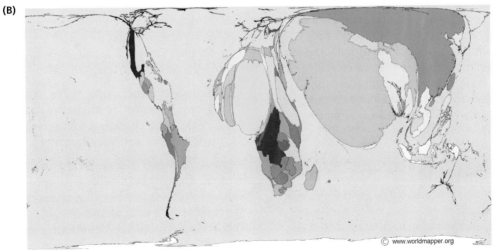

a cartogram (Box 11.5) showing each country's proportion of the world's population (A) by the area allocated to it with another (B) where the country's area is scaled in proportion to its fraction of the global population living there which is in extreme poverty (i.e., living on $1 or less a day).

**The combined wealth of the world's seven richest people exceeds the gross domestic product of 567 million people in 41 indebted countries.**

Poverty also generally means poor levels of nutrition or, in rural areas at least, dependence on one crop. Failure of that crop can mean starvation and death or enforced migration. Food security is therefore a key risk faced by many people worldwide. Research published in mid-2013 by the Institute on the Environment at the University of Minnesota highlighted the threat of growing global hunger. It is widely accepted that to feed the projected global population growth by 2050, global food production will need to increase between 60 and 100%. The researchers' results suggest that, based on current trends, production of maize, rice, wheat, and soybeans will increase only by 38 to 67%. This does not take into account possible consequences of climate change.

GISS professionals can contribute to our understanding of poverty and inequality and the scale of these (see Figure 19.3). We do this through our ability to integrate, analyze, and portray multiple datasets and to support logistics and communications through the use of our tools. Thus the Minnesota researchers argued that the approaches most likely to cope with a looming agricultural crisis are more efficient use of current arable lands and spreading best management practices—areas where precision farming utilizing GPS and GI technology can be hugely effective. Our abilities to do this in practice remain constrained by data that are of highly variable quality and inadequate modeling tools. Nevertheless new civilian developments like the use of low-cost unmanned aerial vehicles (UAVs) to collect timely local data (Section 17.5.3) and experience of the use of GI systems in developed countries suggest that we can develop better evidence-based approaches. Implementing them on a widespread scale would inevitably require us to engage closely with political and bureaucratic processes.

## 19.6.3 Human Health

The spread of disease, access to care, and the prevention and treatment of illness are all unevenly distributed across the globe, partly related to poverty. The substantial progress in overall global health improvement over recent decades has, however, been deeply unequal, with convergence toward improved health in a large part of the world but with a considerable number of other countries falling further behind. Every year there are 350–500 million cases of malaria in the world, with 1 million fatalities: Africa accounts for 90% of malarial deaths, and African children account for over 80% of malaria victims worldwide. Furthermore, there is now ample documentation, which was not available 30 years ago, of considerable and often growing health inequalities within countries. Even within a relatively prosperous country such as the United States, for example, cancers of various types are much more prevalent in some places than in others.

There are three areas in which GISS interacts closely with human health. These are

- Public health policies and campaign activity to prevent illnesses becoming rampant and to encourage action by individuals to stay healthy.

- Analysis, modeling, and understanding of patterns of disease.

- Operating the health system. Because health systems worldwide cost between 2 and 19% of total GDP (Table 19.4), it is crucial that they are operated effectively and efficiently.

### 19.6.3.1 Supporting Public Health Planning

Improving the level of public health has benefits for the individual (better personal health) and for the nation (reducing the need for and cost of acute treatment in hospitals). Examples of highly effective campaigns of this sort include the reduction in smoking and associated cancers, the reduction in incidence of malaria through use of insecticide-treated mosquito nets, and reductions in mortality at birth through enhanced education—especially for women. By and large such campaigns work through increasing the awareness of risks, changing attitudes, beliefs, motivations, and social norms. But to make them effective, such campaigns must be based on an understanding of current morbidity patterns and be targeted based on a detailed socioeconomic understanding of the geography of different populations.

**Public health planning saves lives and money. It works by increasing awareness of risks, and changing attitudes, beliefs, motivations, and social norms.**

A simple example is given by the great variations in the impact of certain cancers (see, for example, Figure 13.6)—largely because some groups present themselves for treatment much earlier than others. This occurs even where treatment is free. Another example relates to the growth of obesity in many countries. GI systems have been used to analyze how different environments present people with different opportunities to be healthy and hence can lead to different health behaviors and health outcomes. The "obesogenic environment" has been defined as one involving proximity to fast food. Increasingly studies are seeking to measure the impact of a wide range of environment factors. These include alcohol provision, food, green space, and places to do physical activity, right through to ways of characterizing urban form as complicit in population health. Bringing all this together in GI systems is an obvious benefit.

### 19.6.3.2 Analysis and Modeling of Disease Patterns

Many studies of the geographical variations in morbidity and the treatment they receive have been made. Perhaps the best-known work is that by the Dartmouth Institute of Health Policy and Clinical Practice and published in the Dartmouth Atlas of Health Care. The geographical variations apparent in Figure 19.12 strongly suggest the need for more detailed analysis of multiple datasets to understand the factors causing the variations.

Big Data approaches and GI technology have been argued to be effective in mapping the spread of certain diseases even before they can be detected by official mechanisms (Section 17.3.1.4). Typically, however, research to identify causes—and hence identify effective treatments—requires access to individual data (Section 17.3.1) and, ideally, geocoded data from large longitudinal studies.

Central to understanding of disease patterns is also the need to examine trends through time as well as geographic patterns. As a coarse spatial resolution example, studies have shown that populations in Spain, France, and Italy all suffered increased rates of prostate cancer from the 1980s; Spain showed the lowest increases. Incidence started to rise around 1985 in France and after 1990 in Italy and Spain. Age-adjusted mortality increased until the late 1980s in all these countries, then declined in France and Italy, but not in Spain. Younger people showed a much higher rise in incidence than the elderly. Why such temporal differences occurred in adjacent countries is a serious research question.

### Modeling the likely course of future pandemics is an exercise in applied geography.

The most pressing clinical health challenge of our times is perhaps the study of past pandemics and preparation for forthcoming ones. The world has suffered six pandemics in the last 120 years. For three years from 1918, between 20 and 100 million individuals worldwide were killed by the so-called Spanish Flu—many of them healthy young adults. More died in this pandemic than in World War I. It has been estimated that there is a 65% chance of a pandemic within the next 20 years. Because of the much greater extent of international travel and expansion of high-density urban environments, the spread of a new virus strain for which there is as yet no protection could be devastating—hence its appearance in the

**Figure 19.12** The percentage of cancer patients who died in hospitals in the period 2003 to 2007 by Hospital Referral Region. All rates are adjusted with the following patient level characteristics: age, race (black/nonblack), sex, cancer type, and noncancer chronic conditions.

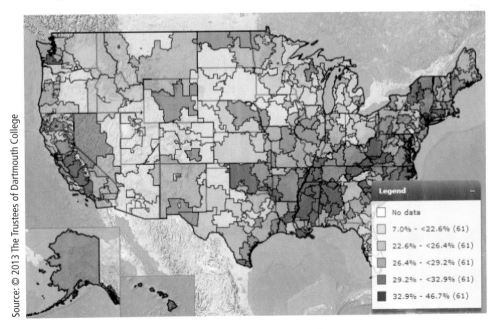

Source: © 2013 The Trustees of Dartmouth College

Legend
- No data
- 7.0% - <22.6% (61)
- 22.6% - <26.4% (61)
- 26.4% - <29.2% (61)
- 29.2% - <32.9% (61)
- 32.9% - 46.7% (61)

## Eyam's Self-Sacrifice and the Black Death

Eyam is a small village in Derbyshire in England. Bubonic plague was discovered there in August 1665. The source of the infection was fleas in a bundle of cloth delivered from London to the tailor. On the advice of the local rector, villagers chose to isolate themselves, staying within the confines of the village in order to minimize the spread of the disease to nearby areas. About three-quarters of the village population perished, but deaths in the surrounding area were less frequent.

UK national civilian risk register in Table 19.3. Unsurprisingly, therefore, great effort is being put into rapid identification of the location of outbreaks of infectious diseases—crucial to containment or remedial action. This response requires good information networks and the ability to aggregate information geographically and report different aggregations routinely and speedily, to different audiences. Much of this is essentially an exercise in applied geography in which GI systems play an important part. It also, however, requires suspension of belief in economists' "rational expectations" about human behavior. These are not always a sound basis for modeling: Not all individuals or groups behave selfishly (see Box 19.3).

### 19.6.3.3 Operating a Health System

As we said earlier, health systems are expensive and becoming more so. The World Bank estimated that the median cost of such national health systems in 2011 was 6.4% of national GDP. Table 19.4 shows some of the extremes. Such costs have major impacts on national taxation and influence the level of poverty when the services have to be paid for by poor people. Effective and efficient treatment of

Table 19.4 The costs of national health systems as a percentage of national GDP—some selected examples from a World Bank analysis (Source: data.worldbank.org/indicator/SH.XPD.TOTL. ZS?order=wbapi_data_value_2011+wbapi_data_value+ wbapi_data_value-last&sort=desc)

| Country | % of GDP |
|---|---|
| USA | 17.9 |
| Netherlands | 12.0 |
| France | 11.6 |
| Cuba | 10.0 |
| UK | 9.3 |
| South Africa | 8.5 |
| China | 5.2 |
| India | 3.9 |
| Myanmar | 2.0 |

patients is therefore a management task as well as a clinical one.

**In some countries, health costs are greater than total education and defense costs. GISS professionals can support more effective health management.**

Recent studies have begun to show that huge geographical variations exist in costs of treatment even within a single country. They suggest means by which these can be reduced, while ensuring quality of treatment is maintained. This applies everywhere, not least in the United States, which spends some $2.1 trillion annually on health care. According to the McKinsey consultants, this is $650bn more than would be expected in taking into account the whole U.S. economy. In *Health Services Research* in 2009 Mays and Smith argued that "On balance, very little empirical evidence exists about the extent and nature of geographic variation in public health spending [in the USA]." They analyzed community level variation and change in per capita public health agency spending between 1993 and 2005 in the 2900 health agencies in the United States. Using multiple regression models with panel data, they estimated associations between spending, institutional characteristics, health resources, and population characteristics. They showed that the top 20% of communities had public health agency spending levels over 13 times higher than communities in the bottom quintile. Most of this variation persisted even after adjusting for differences in demographics and service mix.

Various other studies, notably by the Dartmouth Institute for Health Policy and Clinical Practice, have also shown that costs of hospital health care vary greatly by area in the United States. Figure 19.13 shows how the average costs of surgery in the last two years of life have varied by a factor of four. Moreover, researchers found that high costs are not always associated with higher quality of treatment. The Dartmouth approach was to ask how much might be saved if all regions could safely reduce care to the level observed in low spending regions with equal quality. They made estimates ranging from 20–30%,

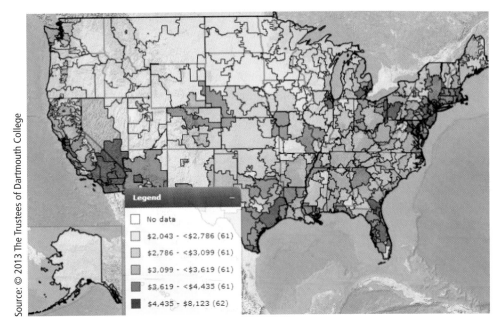

**Legend**
- ☐ No data
- ☐ $2,043 - <$2,786 (61)
- ☐ $2,786 - <$3,099 (61)
- ☐ $3,099 - <$3,619 (61)
- ☐ $3,619 - <$4,435 (61)
- ■ $4,435 - $8,123 (62)

**Figure 19.13** Total payments for physician visits per decedent during the last two years of life for the period 2003–2007; by Hospital Referral Region.

but viewed these as an underestimate given the potential savings even in low-cost regions. Such colossal geographical variations are certainly not restricted to the United States.

Where competition is permitted in national health systems, the use of GI systems to map referrals from physicians to different hospitals, often broken down by groups of types of illness and the flows of patients (and hence money), is now commonplace. These support marketing functions and strategic decisions on the hospital's business model.

Central to all health service activity is the need to maintain confidentiality of patient details. The publication of mapping of individual-level data can disclose such confidential data (e.g., a patient suffering from sexually transmitted disease). Figure 18.2 shows how a percent volume contour associated with a kernel function (Section 13.3.5) can be used to represent the density of features within a neighborhood, avoiding disclosure of the locations of any individual with a specific morbidity—hence preserving privacy.

A common feature in the developed world is the rapid increase in use being made of emergency facilities in hospitals. This overloads them and is very expensive, especially if the patient is being admitted on a "just in case" basis. Sophisticated geographically based "spatial hurdles" modeling of factors contributing to the overcrowding is becoming more common. This models the different effects of proximity, ethnic differences, private health insurance, and other factors on access to emergency departments and suggests solutions based on 24-hour community health facilities.

The World Health Organization (WHO) has argued that there are several main causes of ineffective health treatment: Poor people have the least access to health care and leave treatment to the last moment because they cannot easily pay for it. Many health services for the poor are highly fragmented and thus provide poor service, often where hospital-acquired infections flourish. The WHO is calling for the redirection of available resources toward the poor and toward primary rather than expensive tertiary services, that is, toward the local everywhere. This presupposes that we can identify the highest-priority areas for such support and create systems that are efficient and effective. Use of GI systems to describe the geographies of need and identify good access points could be a good contribution to a more equitable health system.

### Many of the Grand Challenges—poverty, disease, hunger—are closely connected.

Given their impact on the experienced quality of life and the costs of treatment, health issues form one of the most urgent and compelling areas in which GISS can contribute to society. Whether the focus is malaria, HIV-AIDS, cholera, avian influenza, children's obesity, access to health care, cancer, or the impact of climate change on health, developing a more thorough understanding of and response to critical health issues cannot be achieved without the approaches and concepts of different branches of science. GISS provides the organizing and analytical framework that accommodates and focuses our different approaches to science in ways amenable to practical problem solving. Examples of past successes include work to identify disease clusters; computation of correlations between disease outcomes and their possible causes; the formulation and testing of spatially explicit models of disease spread that incorporate concepts such as distance and connectivity directly into the model (often in the form of cellular automata or as agent-based models: see Sections 15.2.2 and 15.2.3); and research on the significance of proximity in determining successful treatments (e.g., the distance between the home of a stroke victim and the nearest emergency center). GISS has played a significant role in almost all contemporary work of this kind.

## 19.6.4 Access to Food, Potable Water, and Boundary Disputes

Water problems affect half of humanity: Some 1.1 billion people in developing countries have inadequate access to water, and 2.6 billion lack basic sanitation. About 1.8 million child deaths occur each year as a result of diarrhea. Close to half of all people in developing countries are suffering at any given time from a health problem caused by water or poor sanitation. GISS professionals can help identify problem areas and possible causes of ill health just as Snow did in 1854 (Box 13.1).

Water shortages are already increasing in many parts of the world as a result of competing uses, growing populations, and international disputes over riverine water flowing across national boundaries. Moreover, if the Intergovernmental Panel on Climate Change (the IPCC) is correct, the situation can only get worse. The Panel's 2013 and earlier reports projected that most dry regions will get drier, most wet regions will get wetter, drought-affected areas will become larger, heavy precipitation events are likely to become more common and will increase flood risk, and water supplies stored in glaciers and snow cover (Figure 19.14) will be further reduced.

**Wars have often been caused by competition for water or because of other boundary disputes.**

Rivers have long been a popular choice for boundary makers because of their defensive nature, their clarity on the ground, and some belief in their permanence. However, as dynamic natural features, the movement of rivers has generated frequent disputes over the position of boundaries that are often understood to require rigidity. In these circumstances, disputes over who has the rights to water resources can only grow still further.

One example of attempts to minimize such water-related disputes is the agreement between China and India announced in August 2009 to monitor jointly the state of glaciers in the Himalayas. Seven of the world's greatest rivers, including the Ganges and the Yangtse, are fed by glaciers. They supply water to about 40% of the world's population. In 1962 the two countries went to war over disputed territory in this sensitive area.

In 2013 Egypt and Ethiopia agreed to hold further talks to quell tensions over the building by Ethiopia of a new hydroelectric dam on the Blue Nile, following threats of military action. The Nile waters have been crucial to the survival of Egyptian agriculture for several millennia (see Figure 19.15). The strife began when Ethiopia's parliament ratified a controversial treaty to replace colonial-era agreements that gave

**Figure 19.14** (A) A photograph of Muir Glacier taken on August 13, 1941, by glaciologist William O. Field; (B) is a photograph taken from the same vantage point on August 31, 2004, by geologist Bruce F. Molnia of the U.S. Geological Survey (USGS). According to Molnia, between 1941 and 2004 the glacier front retreated more than 7 miles (12 km) and thinned by more than 800 m. Ocean water has filled the valley, replacing the ice of Muir Glacier; the end of the glacier has retreated out of the field of view.

Source: U.S. National Snow and Ice Data Center

(A)

(B)

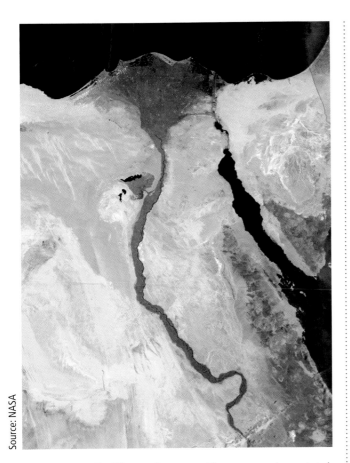

Source: NASA

Figure 19.15 The Nile seen from space. The green area is vegetated or urban areas (including Cairo and the delta). Eighty six million people live there. The remaining areas are desert and very sparsely populated.

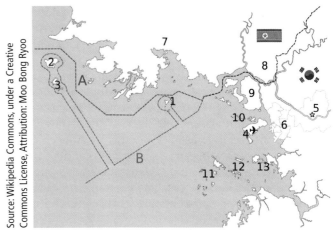

Source: Wikipedia Commons, under a Creative Commons License, Attribution: Moo Bong Ryoo

Figure 19.16 Unresolved maritime boundaries of North and South Korea. The area was the site of a North Korea torpedo attack in 2010 that killed 46 South Korean sailors and an artillery attack on South Korea territory eight months later. Line A (blue) is the Northern Limit Line, created by the United Nations in 1953. Line B (red) is the "Inter-Korean MDL in the Yellow Sea," declared by North Korea in 1999. The locations of specific islands are reflected in the configuration of each maritime boundary, including 1–Yeonpyeong Island, 2–Baengnyeong Island, and 3–Daecheong Island.

Egypt and Sudan the biggest share of the Nile's water. The original 1929 treaty written by Britain awarded Egypt veto power over any project involving the Nile by upstream countries. The new treaty had earlier been signed by five other Nile-basin countries— Rwanda, Tanzania, Uganda, Kenya, and Burundi. This is a classic example where the skills and tools of GISS practitioners could be invaluable.

Another example of a potentially catastrophic boundary dispute—but not involving potable water— is that between North and South Korea. The differences between the UN's and North Korea's versions of the boundary between the two Koreas is shown in Figure 19.16. In this highly politicized case, however, it is not clear how GISS professionals can contribute much beyond recording the different versions of the boundary.

GI system practitioners have already made many significant contributions to the definition of many boundaries. For instance, the International Boundaries Research Unit (IBRU) at Durham University has built up internationally recognized expertise over several decades in supporting boundary demarcations (many defined in relation to river courses), training those involved in such demarcations anywhere in the world, and maintaining in GI system form a geographical record of (often disputed) boundaries such as of Israel–Palestine and Cyprus–Northern Cyprus.

## 19.6.5 Coping with Natural Disasters

Television brings us frequent updates on natural disasters from many parts of the world, though reporting is usually massively biased toward reporting low-frequency but high-impact events. The most extreme natural disaster in recent years was probably the December 2004 tsunami originating from a submarine earthquake, whose epicenter was off Banda Ache in Indonesia; approximately 225,000 people were killed. Subsequently a real-time tsunami monitoring system was built spanning the Indian Ocean (Section 18.3.2.3).

In some cases the impact of natural hazards is exacerbated by human factors. An example was the aftermath of a 9.0 magnitude earthquake that struck the Fukushima prefecture north of Tokyo on March 11, 2011 (Box 1.1). Mismanagement exacerbated the problem. A second—and different—example followed an earthquake that devastated the city of Port-au-Prince, Haiti, on 12 January 2010, causing the death of up to 300,000 people. This is a city with a booming population, cramped and informal housing, poor sanitation, and extremes of wealth and poverty. The earthquake was followed by a severe outbreak of cholera with over a quarter of a million cases and more than 7,000 reported deaths. Many Haitians blamed the outbreak on a sewage spill from a UN peacekeeping base; a group of human rights lawyers sued the UN on behalf of those Haitians who contracted cholera. Based in part on geographical detective work by the U.S. Centers for Disease Control, the suit alleges that the cholera pathogen was introduced to Haiti inadvertently by a group of UN peacekeepers who traveled to the island from Nepal to provide aid after the earthquake.

If earthquakes and resulting tsunamis are the most destructive natural hazards, most natural disasters are caused by hydrological and/or meteorological factors. These usually result in flooding on a massive scale, such as caused by Hurricane Katrina and Hurricane Sandy in the United States. As Figure 19.17 shows, natural disasters have had huge economic, as well as human, impact; on average about three-quarters of these losses were due to weather-related events. The global reinsurer Munich Re summarized insurance losses in 2011 as around $125bn. The potential losses from flooding in global cities like New York are particularly massive: The Rockefeller Foundation has funded a large study on how to protect that city (see Further Reading). But even such dire figures disguise the real extent of natural disasters because many of the losses are uninsured or otherwise unquantifiable in areas with poor populations. Munich Re estimated that the worldwide uninsured losses in 2011 were over twice as large as the insured ones and amounted to over $300bn.

## Natural disasters have had huge economic, as well as human, impact.

Given the human misery and economic costs of such disasters, it is no surprise that GI systems are becoming widely used in attempts to predict areas where disasters may occur—taking into account natural circumstances plus deforestation, construction, and other human activities likely to contribute. The vastly experienced Belgian coordinating center for details of natural disasters recognized as long ago as 2008 that "GIS technology has imposed itself as an essential tool for all actors involved in the different sectors of the disaster and conflict management cycle." Unfortunately while real-time monitoring of hurricane tracks has proved relatively valuable—both for enhancing evacuations and predicting insurance losses—forecasting the occurrence of natural disasters remains mostly unreliable. Even where probabilistic statements of the likelihood of a disaster have been made, these have often been inadequately persuasive for governments to take action. The prime example of this is the warnings by geologists of the high risk of an earthquake off the shoreline of Banda Ache years before the catastrophe. Thus there is much for GISS experts, in partnership with other earth and social scientists, to do, though the lessons of the L'Aquila affair must be heeded (Section 18.1).

A relatively new development is the use of crowd-sourcing (Section 1.5.6). Online disaster response communities have grown in support of the traditional aspects of disaster preparedness, response, recovery,

**Figure 19.17:** Natural catastrophes worldwide 1980–2012: Overall and insured losses and their trend lines. Losses adjusted to inflation based on country CPI.

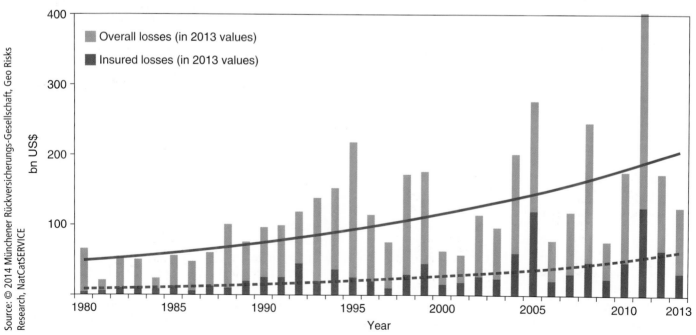

mitigation, and policy as facilitated by governmental agencies and relief response organizations. But these communities now use the Internet to donate money quickly and efficiently, create blogs, upload pictures, and disseminate information—often much faster than government agencies—and use message boards to seek family members and identify shelters. In essence they create a living and easily accessible geography.

### GISS experts have made major contributions to improving the effectiveness of post-disaster recovery work.

Where GI system experts have made the greatest contributions in recent years is through improving the effectiveness of postdisaster recovery work. The charity MapAction is a prime example of good practice, now being copied by many others. A standing organization founded in 2002, MapAction has made much use of highly trained volunteers for disaster relief. It argues that aid that ends up in the wrong place is of no use in relieving human suffering and that, in a humanitarian crisis, relief agencies need rapid answers

to questions such as "Where are the greatest needs?" and "Where are the gaps that need to be filled?" To provide such information, the volunteer teams are skilled in use of GI systems and GPS, produce frequently updated situation maps for all other agencies active in the area, and publish them on the Web (Figure 19.18); disaster relief agencies can also access free-of-charge maps from all current and previous deployments. The MapAction teams have operated in countries such as Angola, Bolivia, the Central African Republic, Cote d'Ivoire, Guatemala, Indonesia, Jamaica, Kenya, Libya, Madagascar, Myanmar, Japan, Pakistan, Paraguay, Philippines, Sri Lanka, Suriname, Syria, and Tajikistan. That this has major benefits is shown by an independent report produced after a deployment in Haiti following the passage of a hurricane. The review concluded that such GI system products made decision making by aid agencies much quicker, simpler, and more effective.

All the preceding relates to contemporary disasters. GI systems, however, can be used to model and depict long-term scenarios. The National

**Figure 19.18** Example of one type of map produced by MapAction to facilitate and report on rescue work after typhoon Haiyan (Yolanda) in the Philippines: Who-what-where snapshot of cash transfer relief assistance payments as of 30 November 2013.

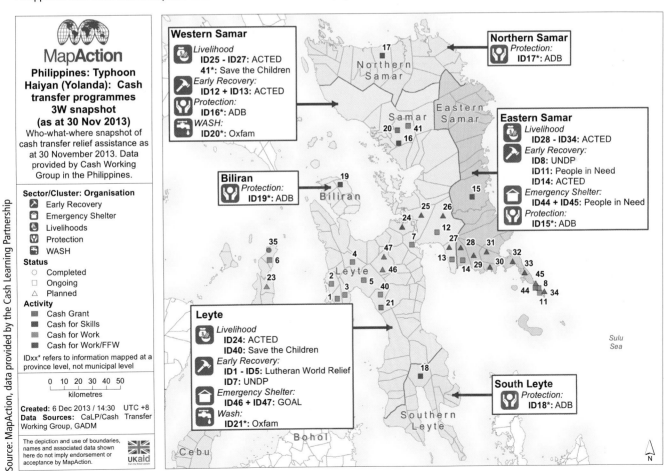

Geographic Society has published an interactive map (ngm.nationalgeographic.com/2013/09/rising-seas/if-ice-melted-map) that shows the impact in Asia of the extreme event of the melting of all land ice. The Society's calculations suggest about a billion people would be displaced in Asia alone. Similarly disastrous consequences would occur elsewhere, as New York, London, and many other major cities would be submerged.

## 19.6.6 Coping with Terrorism, Crime, and Warfare

Coping with terrorism, crime, and warfare all ideally require combinations of intelligence on the likely threats, frequently updated risk registers and plans to mitigate risk, operational plans for deployment of forces to control the threats, excellent communications, and descriptions of the affected geography occupied by the terrorists and the populace alike. As described in more detail earlier (Section 17.5), the result of growing security threats has been increasing use of surveillance technologies and linkage of all the sensors used together through GI systems to portray the movements of suspects or monitor suspicious activities around sensitive sites. The sensor arrays range from those on satellites and unmanned airborne vehicles to closed-circuit TVs, telephone interceptions, and much else. Even at the less serious level of crime against individuals or their properties, monitoring of activity in high-risk areas is important, as is post hoc analysis of the characteristics of the crime scene.

Increasingly, all security-related activities need socioeconomic and other social science information as well as that describing the physical landscape. Geodemographic profiling of the characteristics of local populations to quantify risk is commonplace, and matching of DNA samples from suspects or databases with that found at crime scenes is permitted in some jurisdictions. The integration and analysis of such admixtures of data with different characteristics, often to very short time scales, can be very demanding. Chicago's police pioneered such predictive policing in 2008, based on historical crime data and street intelligence then targeting resources where crime was predicted; there was a rapid reduction in homicide rates—for a time at least. While encouraging, we should, however, note unintended consequences: There is contention, for instance, that crime mapping and prediction has simply displaced crimes such as burglary from one location to another rather than stopping them.

**Surveillance of individuals can reduce risk of harm but endangers privacy.**

The descriptions of military GI infrastructure in Section 17.5 demonstrate that such information and contemporary GI system tools play a key role in efforts to safeguard life and well-being. All this, however, raises difficult trade-offs between potentially enhanced security and loss of privacy (see Section 18.4.1).

## 19.6.7 Environmental Sustainability

The book *Silent Spring*, published by Rachel Carson in 1962, is widely credited with helping launch the environmental movement. Within a few years of the publication, satellite remote sensing and GI technology began to play a significant role in monitoring the state of the environment globally and more locally. But the perceived seriousness of environmental sustainability and, in particular, the threats caused by climatic change—which seem to have been at least exacerbated by human action—are now widely perceived as major challenges. A further complicating factor is that some of the worst-affected areas are likely to be those in which some of the poorest people on Earth live. The most vulnerable regions are Africa, the Asian mega-deltas, small islands, and the Arctic. The biggest challenges are held to be the availability of water (especially in the dry tropics); agriculture (especially in low latitudes); human health in countries with low adaptive capacity; and some ecosystems, notably coral, sea-ice biomes, coastal mangrove and salt marshes, and those in tundra/boreal/mountain areas. The debate and controversies are now wide-ranging and encompass coping strategies as well as avoidance ones. As GI scientists, however, we have no doubt that climate change is underway, that some ameliorative and adaptive action is essential, if difficult, and that we in GISS have a role to play.

Even aside from climate change, human-induced environmental stress is now much more widespread and severe, whether it is manifested in the deforestation of the Amazon Basin or the progressive disappearance of wild areas and the extinction of species. Since 1970, over 600,000 square kilometers of Amazon rainforest have been destroyed. More recently, between May 2000 and August 2006 Brazil is said to have lost nearly 150,000 km$^2$ of forest—an area larger than Greece. Most definitively, in 2013 a multiorganizational team skilled in GI systems and led from the University of Maryland used 650 thousand satellite images to map (Figure 19.19) and publish a globally consistent portrayal of forest loss (2.3 million km$^2$) and gain (0.8 million km$^2$) from 2000 to 2012 at a spatial resolution of 30 meters. The tropics were the only climate domain to exhibit a trend, with forest loss increasing by some 2100 km$^2$ per year. Brazil's well-documented reduction in the rate of deforestation from around 2004 was offset by increasing forest loss in Indonesia, Malaysia, Paraguay, Bolivia, Zambia, Angola, and elsewhere. Intensive forestry practiced within subtropical forests resulted in the highest rates of forest

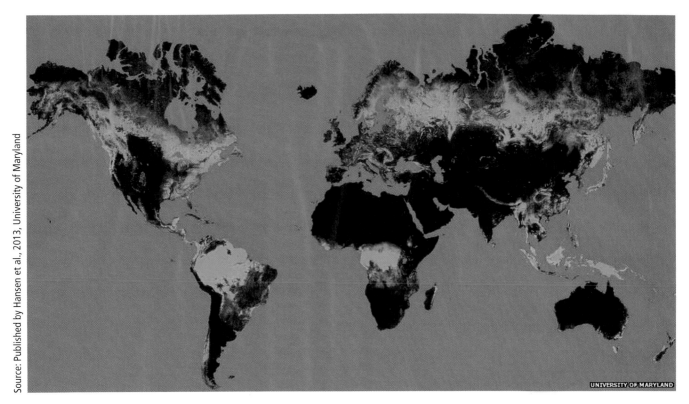

**Figure 19.19** Forest cover of the world and changes to it between 2000 and 2012. Red: reduction in forested area; green: forested areas; blue: gain in forest area; pink: areas of both loss and gain.

Source: Published by Hansen et al., 2013, University of Maryland

change globally. Boreal forest loss due largely to fire and forestry was second to that in the tropics in absolute and proportional terms. Clearly, then, we only know the true scale of the problem through use of GI systems. It seems self-evident to us that ways of reducing this stress and at least adapting to climate change (if amelioration proves impossible to enact) are essential and, again, that GISS practitioners will play a role at every scale from the local to the global.

Although mapping and measurement of environmental change is valuable, still greater value can be added through explanatory and predictive modeling. One area where GISS has made and will continue to make a contribution is in regard to preserving biological diversity and protecting endangered ecosystems. Nearly 850 known species have become extinct or disappeared in the wild over the past 500 years, apparently because of human activity. About two-thirds of endangered mammal and bird species owe their threatened status to habitat destruction and fragmentation. This matters because ecosystem functioning and the services provided by ecosystems are critical to human welfare. These functions include sequestration of carbon, production of oxygen, primary production of chemical energy from sunlight, soil formation, nutrient cycling, food production, wood and fiber production, fuel production, and regulation of water flow and transfer to the atmosphere.

Scientists from many backgrounds working in GISS are playing a significant role in documenting and explaining the current state of biodiversity and the rate and magnitude of ecosystem loss and in developing possible coping strategies. Astonishingly, we still know surprisingly little about such matters. Successful preservation of biodiversity and ecosystem functioning also requires anticipation of how and where land use and climate change will produce increased extinction rates. Built on that, it also requires the development of models that can estimate potential biodiversity and the selection of conservation areas and conservation management strategies to mitigate extinctions and declining ecosystem functioning. Typically, the models have to be at different scales and levels of generalization—global, national, and local—and linked. The same geographical scientists have already contributed significantly in recent years to advances in methods and applications linking land cover changes not only with biophysical processes and consequences, but also with human circumstances.

The U.S. National Research Council (see Further Reading) has argued that we need reliable and relatively fine-resolution global atlases of the world's threatened species and habitats/ecosystems in order to develop a series of geographically explicit, biophysically sound conservation strategies for the most endangered areas. Even that is not enough. We also

Chapter 19 Epilog: GISS in the Service of Humanity **457**

need socially acceptable strategies that take into account locally differing cultural approaches to nature valuation, conservation practices, and compliance. In short, GI systems provide a framework for integration of knowledge and information about the Earth or parts of it and have a crucial role to play in so doing. We again stress that in this area, as in all others, the GISS specialist cannot resolve all the problems on his or her own. Typically, close partnerships (Section 18.6) are needed with physical, biological, and social scientists to describe and understand what is going on and to seek solutions. Expert leadership of the institutions charged with fostering environmental sustainability is also essential (Box 19.4).

**To succeed, GISS specialists must work with other natural, environmental, and social scientists, and with decision-makers.**

### 19.6.7.1 Geoengineering

Elsewhere in this book we have tended to regard at least the physical environment as a given, though it can be modified (and often degraded) locally by humans. In preceding pages we have described how GI systems have been used to monitor changes through a range of applications. But in principle we now have the capacity to change certain aspects of the physical environment far beyond such local activities as quarrying.

One area where this *may* be desirable is in relation to climate change. Such change is happening, and part of it at least has been influenced by human action. On the basis of much scientific work—though disputed by some—the impacts and costs of climate change will be large, serious, and unevenly spread. We may already be seeing consequences on human health from extreme weather events of the past decade,

---

## Biographical Box 19.4

### Jacqueline McGlade, Earth Scientist

After serving two terms as Executive Director of the European Environment Agency (EEA), Jacqueline (Figure 19.20) retired in May 2013. The EEA is an agency of the European Commission serving the interests of 32 EU member countries and others. Her tenure of that post was widely judged to have been a considerable success.

In her 10 years there, she presided over a fundamental shift in the thinking behind environmental reporting and indicator development. Manifestations of this were a shared European spatial environmental information system and the production of integrated environmental assessments, analyzing the state of Europe's environment as a whole and providing soundly based projections.

Professor McGlade oversaw a significant increase in the coverage and extent of data and information processed and analyzed by the EEA and a doubling of the resources for the agency to support this work. She has been a strong advocate of using new technologies to improve information gathering and make it accessible to an increasing number of users.

As Executive Director, she worked to strengthen the link between science and policy. She urged policy makers to start developing and applying adaptation measures to climate change. To complement this, Professor McGlade encouraged dialogue with citizens, including indigenous peoples in remote regions such as the Arctic. In 2008, the EEA launched Eye on Earth, which brought together environmental data in the

Figure 19.20 Jacqueline McGlade, earth scientist.

form of dynamic maps of air and water quality. Eye on Earth was recognized in the 2012 Rio+20 summit declaration as a key public information platform on the environment.

Prior to her leadership role in the EEA, Jacqueline McGlade was educated in the UK and Canada in marine biology and aquatic zoology. Her PhD research was on the importance of spatial dynamics in determining evolutionary divergence and ecological sustainability in freshwater and marine fish populations. Later her work helped establish research in spatial modeling and artificial intelligence and applying space-based observations to biological oceanography at the Bedford Institute of Oceanography.

including droughts, floods, and superstorms. The impacts may be reduced by adaptation and moderated by mitigation, especially by reducing emissions of greenhouse gases. However, global efforts to reduce emissions have not yet been sufficiently successful to provide confidence that the reductions needed to avoid dangerous climate change will be achieved. This has led to growing interest in geoengineering—the deliberate large-scale manipulation of the planetary environment to counteract anthropogenic climate change.

Geoengineering methods fall into two basic categories:

- Carbon dioxide removal (CDR) techniques, which remove $CO_2$ from the atmosphere. As they address a root cause of climate change—rising $CO_2$ concentrations—they have relatively low uncertainties and risks. However, these techniques work slowly to reduce global temperatures.

- Solar radiation management (SRM) techniques, which reflect a small percentage of the sun's light and heat back into space. These methods act quickly and so may represent the only way to lower global temperatures quickly in the event of a climate crisis. However, they only reduce some, but not all, effects of climate change, while possibly creating other problems. They also do not affect $CO_2$ levels and therefore fail to address the wider effects of rising $CO_2$, including ocean acidification.

Given that attempts to reduce atmospheric $CO_2$ by international treaty have proved disappointing, geoengineering is seemingly attractive. We should recognize, however, that under present knowledge it remains unproven and potentially dangerous. That said, any such trials or larger scale application of such techniques will inevitably necessitate use of GI technology and GISS expertise to model and project consequential changes at all geographic levels from the global to the local.

## 19.7 Conclusions

This epilog could simply have set out to anticipate the future developments of science and technology in our field and their implications. These will impact on us all. In particular, the very way we do much science is mutating from largely a hypothesis-verification model to a data "dredging" Big Data one (Box 17.2). In the latter, relationships are computed between a mass of variables and interesting relationships isolated and studied further. As it happens, this is how much of GI science has always operated, so we are mainstream in new science. This is not new: We are traditionally

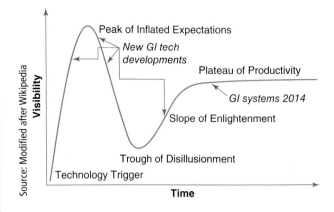

**Figure 19.21** The Gartner Hype Cycle and GI technology. In general GI systems are where shown but some new developments are at various earlier stages.

Source: Modified after Wikipedia

ahead of the mainstream in data science, having developed methods of data sharing, with interoperability and metadata, decades ago.

We have chosen in this final chapter to focus on Grand Challenges to humanity. After 50 years of evolution, GI systems in general and GI technology are now well established: We seem to be at a plateau of productivity in Gartner's Hype Cycle (Figure 19.21). Yet we know that further technological change is coming, but we do not know what that is or what its effects will be other than that employment will change. Over the last 30 years bank tellers, typists, and many other jobs have disappeared—but new ones and new capabilities have replaced them. We are equally confident that the evolution already underway from Terascale to Yottascale computing and beyond will certainly have major impacts on our ability to model complex realities like energy flows in the atmosphere and in ocean currents at finer spatial resolutions—and hence contribute to understanding and remedial actions.

The most casual reading of this chapter should tell us several things. The first is that, whatever their local manifestations, certain problems face us all: We cannot isolate ourselves from the impact of global pandemics, climate change, some natural hazards, or the consequences of financial collapses. Moreover these propagate increasingly rapidly. The second conclusion is that, although the realization of some Grand Challenges seems improbable, sudden changes of a catastrophic nature will occur. In an ideal world, therefore, precautions should be planned beforehand. Although accepting that it is difficult to anticipate the next crisis, business continuity planning is important for every organization. GI systems can often help with "what-if" scenario planning.

The third conclusion is that many of the challenges—and solutions—are interconnected, notably poverty, food and water supply, disease, and population numbers.

These challenges are truly formidable and complicated. Analyzing and anticipating threats involves use of both quantitative and qualitative information, which varies widely in reliability. No one set of skills—of geographers, earth or computer scientists, economists, or others—can resolve these problems on their own. Indeed, the collaborations involved must be multinational, multidisciplinary, and geographically extensive in scale. There is already a vast and partly connected set of organizational silos charged with responsibility for tackling these problems. That responsibility is discharged mainly through national and local governments, multinational bodies like the United Nations and its agencies, commercial enterprises, and nongovernmental organizations (NGOs). We need to be adept at working with organizations as well as being technically competent.

**All physical and human systems on the Earth are interconnected. Geography is the study of the interactions, and GISS helps us to understand and manage them.**

A fourth conclusion follows from everything that has been said earlier: Good communications of all sorts are a necessary (but not sufficient) condition for success, especially in emergencies. The impact of the Internet and Web have already been astonishing in providing new tools to describe, communicate, and tackle problems, some of it via volunteered GI or that collected without consent.

Another conclusion is that, although there has been a huge global increase in use of GI system functionality, this has not been paralleled by widespread development of a fundamental form of human reasoning—spatial thinking. We argue the need for encapsulating this in school and college courses in many disciplines but also advocate a rethinking of many courses in GI systems and a move to incorporate more GI science.

Finally, based on the earlier part of this chapter and indeed on the biographical examples throughout the rest of the book, we contend that as GISS practitioners we can manifestly each make a (mostly) small but important contribution to meeting these challenges big and small. Not all of us can be major players in regulating financial institutions or leading the World Health Organization or global environmental bodies to play key roles in the diminution of risk to humanity. But our skills, technology, knowledge of what works, plus our commitment and working with others enable us individually and collectively (Box 19.1) to make a real difference. We are convinced that such an activist, science-based approach is the right one. We wish you well in that noble endeavor.

**Multinational problems require multinational collaborations between scientists and governments.**

## QUESTIONS FOR FURTHER STUDY

1. Why is the world so differentiated?

2. What do *you* think are the biggest challenges to humanity where GISS can make a contribution?

3. You are invited to a local radio or TV station to talk about what GI systems can do to solve local problems. What would you say, and how would you say it?

4. How can GISS practitioners help to enhance the health of the general populace?

## FURTHER READING

Coleman, N. 2013. *The NYS 2100 Commission Report: Building Resilience in New York*. Rockefeller Foundation. www.rockefellerfoundation.org/blog/nys-2100-commission-report-building

NRC. 2010. *Understanding the Changing Planet: Strategic Directions for the Geographical Sciences*. Washington, DC: National Research Council.

World Bank. 2013. *World Development Indicators 2013*, Washington DC, World Bank. License: Creative Commons Attribution CC BY 3.0:

United Nations. 2013. *Millennium Development Goals*. www.un.org/millenniumgoals/bkgd.shtml

View Hans Rosling video on how to communicate complex technical material to lay audiences: www.youtube.com/watch?v5YpKbO6O3O3M

# INDEX

**Note:** Page numbers in *italics* refer to figures; page numbers in **bold** refer to tables; ***bold, italicized*** page numbers refer to boxes.